SYMPOSIA OF THE SOCIETY
FOR EXPERIMENTAL BIOLOGY

NUMBER XXIX

SYMPOSIA OF THE SOCIETY
FOR EXPERIMENTAL BIOLOGY

The Journal of Experimental Botany
is published by the Oxford University Press
for the Society for Experimental Biology

SYMPOSIA OF THE
SOCIETY FOR EXPERIMENTAL BIOLOGY

NUMBER XXIX

SYMBIOSIS

Published for the Society for Experimental Biology

CAMBRIDGE UNIVERSITY PRESS

CAMBRIDGE

LONDON · NEW YORK · MELBOURNE

Published by the Syndics of the Cambridge University Press

The Pitt Building, Trumpington Street, Cambridge CB2 IRP

Bentley House, 200 Euston Road, London NWI 2DB

32 East 57th Street, New York, NY 10022, USA

296 Beaconsfield Parade, Middle Park, Melbourne 3206, Australia

© Society for Experimental Biology 1975

Library of Congress Catalogue Card Number: 75–9093

ISBN: 0 521 20819 X

First published 1975

Printed in Great Britain
by R. & R. Clark Ltd, Edinburgh

CONTENTS

PREFACE

This volume contains the contributions to the first symposium on symbiosis to be held in this country since the one in 1963 organised by the Society for General Microbiology. The Committee responsible for organising this recent meeting were mindful of the considerable advances which had been made in the field but were aware that the construction of the programme could be hindered by semantics. Though the Committee never quite fully resolved the problem, it proved wise to use the de Bary definition of symbiosis, 'the living together of dissimilarly named organisms', as a working basis. This definition has always been more acceptable to zoologists than it has been to botanists. Nevertheless, the Committee were able to assemble a group of distinguished workers in the field from both disciplines who were willing to consider their specialisations in terms of the de Bary definition. Molecular and cell biologists were also invited to join our proceedings and the reader will see how their special contribution has thrown a new perspective on the study of symbiosis. But it is not only the theory of the origin of the eukaryotic cell that has generated the present considerable interest in symbiosis. As the reader will observe, the field is of great significance both for those who are concerned with the recognition and metabolic co-operation aspects of cell–cell interaction, and also those who are interested in organism–organism interactions and their ecological significance. The chapters that follow present a wealth of information relevant to all these topics. Inevitably, experimental facts are to the fore but semantics have not been ignored and we commend Professor M. P. Starr's first chapter as a notable contribution towards clarifying what is meant by the term symbiosis.

The success of the meeting owes much to the advice that the Symposium Committee received from Professor G. Bond, Dr A. D. Greenwood, Dr A. P. M. Lockwood, Dr B. C. Loughman, Dr N. Sunderland, Professor J. R. Postgate and particularly Professor D. C. Smith who guided our thinking along the right lines at the early stages of formulating the programme. We are also grateful to the University of Bristol and the Department of Zoology for the hospitality and facilities which were offered to the Society. Finally, a special word of thanks must go to the Local Secretary for the meeting, Dr Colin Mapes, whose administrative skills ensured that the meeting proceeded so smoothly.

D. H. JENNINGS
D. L. LEE
Editors of the twenty-ninth
symposium of the Society for
Experimental Biology

A GENERALIZED SCHEME FOR CLASSIFYING ORGANISMIC ASSOCIATIONS

By M. P. STARR

Department of Bacteriology, University of California,
Davis, California 95616, USA

INTRODUCTION

The terminology presently used for labelling organismic associations is confusing, parochial, and highly imprecise. Consider, for example, the shifts in the meaning of the term 'symbiosis' from the general sense in which it was first used (de Bary, 1879) as 'the living together of dissimilarly named organisms'. There was in de Bary's introduction of 'symbiosis' as a general term no connotation implied about the particular spatial or societal format of the 'living together' nor any entailment that there be benefit or harm (or neither) to either or both of the organisms. Neverthe-less, some later writers have used the term 'symbiosis' to mean a closely apposed and mutually beneficial association of organisms; that is, the special kind of association that others have called 'mutualism'. It is logically and semantically monstrous to give both a class and an included subclass the same name, yet that has been the fate of the term 'symbiosis'; the genesis of this unfortunate practice is traced by Hertig, Taliaferro & Schwartz (1937).

Another example of the terminological confusion: when van Beneden (1876) coined the term 'commensal', he referred only to the idea of 'a messmate', that is, an organism 'who is received at the table of his neighbor to partake with him of his day's fishing...his friend's superfluities'. There was a connotation intended of something akin to a permissive (but definitely not essential) gastronomic hospitality, an (at most) unilaterally beneficial and definitely non-harmful association grounded in nutrition. This is rather different from some of the meanings today attributed to the term 'commensalism'. For example, the meanings are 'often extended to include benefits other than nutritional' (Cloudsley-Thompson, 1965); such benefits (obscured sometimes by parochial and short-lived neologisms) may involve providing protection, affording shelter, serving to transport, being used as a hunting ground, warning of danger, warding off undesirable predators, etc. The original idea that commensalism has a tangentially unilateral benefit shifts to situations – all called 'commensalism' – in

[1]

which mutual benefit is claimed or to those 'in which neither of the partners profits at the expense of the other' (Jennings, 1974). The non-harmfulness intended in the original meaning drifts to cover cases in which so-called 'commensals' are claimed to 'do some harm to its host other than depriving it of some food' (Askew, 1971). The permissivity of commensalistic associations inherent in van Beneden's term 'messmates' wanders to a particular case in which neither so-called 'commensal' can 'survive without its partner' (Jones, 1967). Even the asserted spatial relationships range from van Beneden's 'being received at the table of its neighbor' and, equivalently, 'without the one living on the other' (Caullery, 1952) to arrangements in which so-called 'commensals' regularly occupy particular loci upon or within the bodies of their partners. Mind you, actual situations of all these sorts (and many more) exist in the great diversity of organismic associations – but my point here is that these associations cannot all be called 'commensalism' and still permit effectual intellectual discourse.

A similar situation obtains in the numerous and conflicting usages of the term 'parasite' to mean variously: an organism that 'lives on or in the body of the host', 'feeds on the cells, tissues, or body fluids of another . . . organism', 'is metabolically dependent on the host', 'adversely affects the other', 'causes disease', 'lives at the expense of the other', 'requires some vital factor . . . from another organism', 'use[s] other living animals as . . . environment and source of food', '[involves] overt exploitation', and combinations of these and still other meanings. A comparable exercise on the term 'predator' yields an assortment of meanings consisting of various mixtures of components such as 'feeds on', 'causes death of', 'consumes', 'destroys', 'eats', 'pursues', 'hunts', 'pounces on', 'kills', 'captures', 'uses as food', 'seizes', 'exploits'.

Two sorts of organismic associations have been long-term objects of my experimental interest: the associations between phytopathogenic bacteria and plants (Starr, 1946, 1959; Starr & Chatterjee, 1972), and the associations between *Bdellovibrio* and other bacteria (Stolp & Starr, 1963; Starr & Seidler, 1971; Starr & Huang, 1972; Starr, this volume). In an effort to bring some conceptual and terminological clarity into these experimental systems, a classification of organismic associations was attempted (Heise & Starr, unpublished; summarized in Starr & Chatterjee, 1972). The major thrust of the Heise–Starr classificatory scheme was the specification of organismic associations in terms of three continua dealing with 'locational–occupational', 'valuational', and 'degree of dependency' criteria. This scheme, even though useful for some purposes, suffered from various deficiencies, among which the following deserve mention here: well-worn terms, albeit explicitly redefined, were used to name various classes of the organismic associations and this practice caused much confusion because the old meanings and the new ones were quite disparate; the 'locational–occupational' continuum conjointly specified two kinds of criteria and this mixture did not always permit the desired conceptual clarity; and the three continua were admittedly inadequate for specifying with sufficient precision the limitless array of organismic associations. The original scheme was unacceptable to our peers for these and other reasons, and I no longer care to defend its specifics.

However, the basic approach – and, indeed, the undertaking itself – seems to me and to others (see Lewis, 1974) as a step in the right direction and I have continued my efforts to purify, extend, and find suitable terminology for an adequate and generally applicable classificatory scheme. To do so on a well-informed basis, a comprehensive analytical review was undertaken of the diverse literature on a great array of organismic associations. The systems examined include associations involving viruses, bacteria, eukaryotic protista, lower and higher plants, and invertebrate and vertebrate animals; related abiotic components also were considered. This extensive analysis has led to my present version of this classificatory scheme, which is summarized in Table 1. Where possible, I have tried to avoid ambiguous, previously abused terms in this scheme, and I have defined (or, at least, explained subject to the space limitations of the present enterprise) the terms that are used. I have tried to focus each continuum to the end that a single criterion be specified by each; this effort and other considerations have led to an increase in the number of continua, and further additions and refinements are contemplated.

SYMBIOSIS

'Symbiosis' is an eminently appropriate term for the superclass of most of the phenomena under consideration here (Table 1). But, as noted above, an unfortunate second usage, in which the meaning of the term 'symbiosis' has been limited to the special case of those organismic associations that are mutually beneficial, has crept in. After dithering for some time about whether to retain or to drop the term 'symbiosis' from this classificatory scheme, I have now come to the conviction that it is high time to reverse the semantic deterioration. I have therefore resolved, finally and quite firmly, not only to use and to foster the use of the term 'symbiosis', but also to have it mean pretty much what de Bary (1879) intended: '*Zusammenleben ungleichnamiger Organismen*'. '*Zusammenleben*', a noun which might be translated literally as 'living together', more conventionally means 'association'; '*ungleichnamiger*' is a plural adjective that means 'of dissimilarly (or differently) named'; '*Organismen*' is a genitive plural noun that means 'organisms'.

'*Organismen*'

Taking each of these three seminal words in turn, let us deal first with the word '*Organismen*'. Our knowledge and concepts about the nature of organisms have been greatly extended since de Bary wrote these words in 1879; for example, viruses and the molecular bases of genetics were not known a century ago. Hence, it is best to state explicitly that by the term 'organisms' is meant here all biological entities, both cellular and non-cellular, that show the basic, least-common-denominator attribute(s) of life; to me, these are the capacities to perpetuate itself and to respond to evolutionary forces. The viruses are included therein – as Luria & Darnell (1967) have done, and for the very same reasons they specify ('individuality, historical continuity, and evolutionary independence rather than functional independence' are stressed in their definition). For that matter, it seems to me entirely justifiable to count as 'organism', at least for the purposes of the present enterprise, any transmissable biological entity – such as viroid (Diener, 1974), plasmid or episome (Hayes, 1968; Meynell, 1973), and the like – which minimally has the capacity to 'commandeer' (Brown, 1970), in a self-perpetuating and evolutionarily responsive manner, the genetic apparatus and services of another biological entity.

The issue might be raised about the necessity for the organisms to be alive throughout the entire relationship in order for the association to be considered as a symbiosis. An association which consists entirely of a relationship between a living organism and a totally unlive entity (including an already dead organism) is not a symbiosis. There are fairly common examples of relationships of this kind in which a living organism merely consumes the carcass (remains) of another organism which was already dead when the relationship began; this sort of relationship is not a symbiosis – rather, it is a non-symbiotic association which is called 'scavenging' or 'saprotrophy'. On the other hand, there is, discernable at this conceptual interface, a significant distinction – almost as subtle as a Talmudic pilpul – that must now be made explicit. In my scheme, an association that *starts* with two living organisms counts as a symbiosis, even if subsequent to the initiation of the association a portion or all of one of the associants loses, reversibly or irreversibly, some or many of the attributes of life – or maybe all of them – depending on how one defines this difficultly-defined state of being alive.

A symbiosis is treated here as involving only two associants. Actual symbiotic associations often do involve more than two associants. Special cases of such plural associations are, for example, the relationships called 'hyperparasitism' and 'multiparasitism' (Askew, 1971) and others designated by a great array of societal terms. However, when the gamut of these plural organismic associations is scrutinized critically, no sharp demarcation can be discerned between the relatively simple ones (such as hyperparasitism and multiparasitism) and the generally more complex societal ones. The relationships pertaining to each pair of associants within a more complex plural association can be, and often are, dissected out and viewed separately. Unfortunately, the usual endproduct of dissection – namely, that the complex relationships cannot easily be reconstructed anatomically or conceptually – often follows. However, I am optimistic that the intellectual community will develop novel and adequate methods of anatomical and conceptual reconstruction, particularly if goaded by explicit statements of the significant need! Since some limit has to be set, it seems to me sensible to limit the term 'symbiosis' to those organismic associations which involve only two associants; symbiotic associations that involve more than two associants might collectively be termed 'plural symbioses'. If the simpler plural symbioses need to be designated and generalized, it should be possible to translate the terms 'hyperparasitism' to 'hypersymbiosis' and 'multiparasitism' to 'multisymbiosis'. By the same token, consideration herein of the formats and effects of the association will have to be limited to those which bear directly upon the two associants which are participating in the symbiosis being examined, though I am well aware that these direct effects have important derivatives pertaining to the larger populations and to the course of evolution.

'Zusammenleben'

Now let us turn to de Bary's 'Zusammenleben', which I have translated as 'association'. Because it is logically possible – albeit biologically and societally improbable – that two symbionts might associate without interacting (that is, acting on one another) or, somewhat equivalently in operational terms, without being known or interpreted by man to be interacting, the more general terms 'organismic association' and 'organismic associant' are used here in preference to 'organismic interaction' and 'organismic interactant', respectively. (If it is known that both of the symbionts are indeed acting upon one another, it would be formally correct to term that symbiosis an 'organismic interaction' and to call the symbionts 'interactants'.) 'Organismic association' is thus almost synonymous with 'symbiosis' in the sense this term is used here, and 'organismic associant' (or simply 'associant' or 'associate') is almost synonymous with 'symbiont' or its variant 'symbiote' (the latter spelling, although used infrequently except in French writings, is said to be the philologically correct one; Hertig *et al.*, 1937).

The reason for the caveat that the term 'organismic association' is 'almost' synonymous with the term 'symbiosis' is the necessity (in my view) that another criterion concerning the association be added to form the concept of symbiosis. What is necessary for an organismic association to be a symbiosis is that the organismic association must not be *casual*; rather, the association must be *significant* to the well-being (or to the 'unwell-being' or to both) of one or both of the organismic participants. In my view, symbioses may be transient to persistent (see Continuum C in Table 1 and text); hence, my denial of casualness does not extend to a requirement for particular temporal aspects (which are specified separately in terms of Continuum C). My requirement that a symbiosis be significant does not entail a particular degree of significance (this is specified separately in terms of Continuum D; see Table 1 and text). Nor does it entail any particular kind(s) or mode(s) of cause(s) and effect(s) on which the significance might be based (these are specified separately in terms of Continuum E and Continuum H, respectively; see Table 1 and text).

'Ungleichnamig'

The use by de Bary of the term '*ungleichnamig*' has caused me considerable difficulty. Nowhere in the 1879 pamphlet, nor in any of his other writings known to me (e.g. de Bary, 1887), does he present his reasons for limiting the term 'symbiosis' to associations between differently named organisms; that is, to associations between different species. It does seem important to make a distinction between interspecific associations and intraspecific associations. But, deviating from de Bary, I and many others (e.g. Hall, 1974; Lewis, 1973; Odum, 1971; Zuckerman & Weiss, 1973) prefer the term 'symbiosis' to cover intraspecific as well as interspecific organismic associations. If it becomes necessary to distinguish between the two concepts, the terms 'interspecific symbiosis' and 'intraspecific symbiosis' might be used. Sprent's (1963) similar distinction – inherent in his use of the terms 'heteroparasites' and 'homoparasites' – might be generalized and locutions such as 'heterosymbiont' and 'homosymbiont' could then be used.

Ecology and organismic associations

The organismic associations which are being considered here constitute a major subdivision of the general subject of relationships of organisms with the environment ('ecological relationships' in Table 1). These ecological relationships of organisms might be divided into two great

Table 1. *A generalized scheme for classifying organismic associations*[a]

Relationships of organisms with the environment (ecological relationships)
  I. With permanently abiotic components of the environment (geobiological relationships)
 II. With biotic components of the environment (organismic associations)

Criterional continua by conjoint use of which various organismic associations (including symbioses) can be specified
A. Based upon criteria pertaining to relative spatial relationships
   1. Exhabitational
   2. Inhabitational
B. Based upon criteria pertaining to the relative sizes of the associants
   1. Associants differ considerably in size ('anisosymbionts'; 'micro-' and 'macrosymbionts')
   2. Associants are more-or-less equal in size ('isosymbionts')
C. Based upon temporal (durability) criteria
   1. Persistent
   2. Transient
D. Based upon criteria pertaining to the degree of necessariness of one associant on the other
   1. Obligately dependent ('necessary'; 'obligate')
   2. Facultatively dependent ('contingent'; 'facultative')
E. Based upon criteria pertaining to the mode of the dependence of one associant on the other
   1. Dependent physically
   2. Dependent chemically (including bio-organic-chemically)
   3. Dependent organismically (e.g. physiologically, nutritionally, regulatively, metabolically, genetically, anatomically, behaviorally)
   4. Dependent societally
F. Based upon nutritional criteria, in terms of the viability of the fed-upon organism
   1. Biotrophic
   2. Necrotrophic
G. Based upon criteria pertaining to degree of associant specificity
   1. Highly specific (limited to particular associants)
   2. Essentially non-specific (permissive)
H. Based upon criteria pertaining to harmful or beneficial effects on each associant[b]
   1a. Antagonistic (agonistic) or harmful
   2a. Non-antagonistic or non-harmful ('neutral' – in one sense)
   1b. Beneficial
   2b. Non-beneficial ('neutral' – in another sense)
I. Based upon integrational criteria
   1. Integrates (forms a 'third entity')
   2. Separates

[a] Starr's 1974-H version. This classification has evolved from the scheme put forth by Heise & Starr (unpublished); that scheme is summarized in Starr & Chatterjee (1972).
[b] In Continuum H, the modes of the harmful and beneficial effects on each associant could be shown by using a terminology akin to the modal terms given in Continuum E.

groups: those with permanently abiotic components of the environment ('geobiological relationships' in Table 1) and those with biotic (that is, living organismic) components of the environment ('organismic associations' in Table 1). It is the latter element, primarily, which is dealt with here, but, as with most biological phenomena, such strict dichotomy – albeit didactically tidy – rarely holds up on careful scrutiny (see below). In the present case, there is an array of intermediate situations (between items I and II of Table 1), such as those in which the abiotic component is, at the time the relationship starts, the dead carcass of a formerly living organism. As already noted, this sort of relationship is properly termed 'scavenging' or 'saprotrophy' (in the sense of Lewis, 1973, 1974) and not 'symbiosis' or 'organismic association' (in the sense I am using these terms herein).

Meaning of the term 'symbiosis' as used herein

Summarizing the meaning of the term 'symbiosis' as it is being used herein: a symbiosis is an association between two organisms; the term 'organism' is taken very broadly to mean a cellular or non-cellular biological entity which has the capacities to perpetuate itself and to respond to evolutionary forces or which, minimally, can commandeer in a self-perpetuating and evolutionary responsive manner the genetic apparatus and services of another biological entity; the association in a symbiosis must be significant to the well-being (or 'unwell-being' or both) of at least one of the associants; the associants in a symbiosis may be of different taxa or of the same taxon.

CONTINUA

Biological phenomena are rarely abruptly discontinuous. Rather, when the fact-base is adequate, they often are found to exist in more-or-less smooth and continuous gradations extending from one extreme to the other. It is a common occurrence for biologists to perceive what looks at first like a tidy discontinuity in Nature, whether it be in taxonomy or in another genre, then to take a second look and see an intermediate group or state which obliterates or blurs the 'discontinuity'. So it is with the various aspects of organismic associations, including symbiosis. As related above, the earlier Heise–Starr scheme invoked the notion of multiple continua which individually focused on a particular aspect of organismic associations and collectively specified a particular kind of organismic association. The system presented in Table 1 makes use of such continua, albeit in many cases the continua are based on criteria different from or additional to those used in the Heise–Starr scheme. Only the limits are labelled for each continuum in Table 1; the precise position of a particular organismic association relative to these limits in each continuum can be specified quantitatively or qualitatively or both – depending on the fact-base and on the purpose behind the statement.

A vignette of each continuum currently in this classificatory scheme (Table 1) will now be presented. It must be emphasized: (a) that this classificatory scheme is still in a state of flux; (b) that it will undoubtedly continue to be refined by me and, I hope, by others; and (c) that this necessarily truncated version – really only a précis, and a preliminary one at that – lacks many of the formalities (full definitions, logical and semantic arguments and analyses, adequate examples which might also serve as ostensive definitions of the terms, etc.) which in my view are essential for the proper presentation of a conceptual and terminological scheme.

Continuum (A) based on relative spatial criteria

One of the first facts one usually learns about a symbiosis is the relative locations of the two symbionts, particularly whether one symbiont resides within the other (and precisely where inside) or outside the other (and precisely where outside). There is utter confusion in the existing terminology pertaining to the relative spatial arrangements of symbionts. This confusion stems largely from trying to include spatial connotations – together with many other connotations – in the meanings of such over-worked terms as 'parasite', 'predator', and 'commensal'. The originally asserted spatial criteria are often conveniently but confusingly ignored while dealing with other attributes of the symbiont, and again invoked, often with no apology or explanation for the semantic confusion thus engendered. It would not be difficult for me to cite many choice examples in this genre. A pox on such semantic and logical obfuscation!

When one symbiont is inside another, it is inhabiting the other; it can quite justifiably be called an 'inhabitant' or an 'inhabiting symbiont' or an 'endosymbiont' (special problems pertaining to the latter term are considered later in this volume, p. 113). The symbiont that is inhabited could be called just that: the 'inhabited (or entered) symbiont'. When symbionts are outside one another, they are exhabiting each other; they both can be referred to as 'exhabitants' or 'exhabiting (or exhabited) symbionts' or 'exosymbionts' or 'ectosymbionts'. If further spatial distinction of exhabitants is necessary or possible, the organism that is viewed as the substrate could be called the 'exhabited symbiont' and thus be distinguished from the other symbiont, the 'exhabiting symbiont', that is upon the substrate organism. What counts as 'inside' and 'outside' another organism is a separate and, in some cases, an undeniably knotty problem. Traditionally, the digestive tract of animals is considered to be 'outside', as are other invaginations which communicate with the exterior of the body. Rather than split such hairs, it seems much more useful to indicate explicitly where one symbiont is located in relation to the other by the device of adding modifiers to the terms 'exhabitant' and 'inhabitant'. This practice would lead to locutions which specify the relative locations as exactly as is possible or desirable: 'epidermal exhabitant', 'cytoplasmic (or nuclear or nucleolar or mitochondrial or intraperiplasmic) inhabitant', 'phloem inhabitant', 'leaf-surface exhabitant', etc. Many points about such locational matters are beautifully depicted in the excellent exercise on microbe–plant interfaces published recently by Bracker & Littlefield (1973). Those cases in which only part of one organism enters another (e.g. the haustoria of certain fungi penetrating plant tissues; aphid stylets inserted into the phloem of plants) illustrate the utility of treating the relative spatial relationships as a continuum. Here, there are situations intermediate between inhabitation and exhabitation; their positions in the continuum could be specified with whatever precision is warranted or desired.

Continuum (B) based on relative sizes of the symbionts

A great deal of fuss has been made in some components of the symbiosis literature about the relative sizes of the symbionts in an organismic association. It is not uncommon to find the distinction between 'predator–prey' and 'parasite–host' relationships made partly or even entirely on this basis. In such usage, usually, the 'predator' is larger than its 'prey' and the 'parasite' is smaller than its 'host'. A totally-contained inhabiting symbiont must, for purely geometric reasons, be at least a little smaller than the inhabited symbiont. Based solely on geometric considerations, an exhabiting symbiont can be larger than the exhabited (substrate) symbiont, but it need not necessarily be larger and certainly can be and often is smaller. Symbionts approximately equal in size might be termed 'isosymbionts'; where there is a definite difference in size, the term 'anisosymbionts' would be appropriate. In accordance with current practice (Alexander, 1971), the term 'microsymbiont' could be used for the smaller, and the term 'macrosymbiont' for the larger, of a pair of anisosymbionts. In any case, it seems useful to indicate the relative total sizes of the symbionts, as well as related size issues which might arise, such as the instances of partial inhabitation by portions of otherwise exhabiting larger organisms – such as the haustoria, suctoria, stylets, and other specialized penetrating structures of some symbionts.

Continuum (C) based on durability of the association

Some symbioses exist throughout the life-span of one or both of the symbionts; these are termed 'persistent'. Other symbioses are transient or temporary. In some cases, symbioses are limited to one phase of a complicated life-cycle. As long as the condition is met that the association be significant to the well-being (or 'unwell-being' or both) of at least one of the associating organisms, the relationship is (for me) a symbiosis, even if the association is relatively or absolutely brief in time. What is needed is the explicit specification of the temporal situation; probably some attention should be devoted to the proportion of each symbiont's normal life-span which is spent in the symbiotic association.

Continua (D, E, F) based on the degree and mode of functional dependence

Once the gross morphological elements of a symbiotic association are discerned and classified, the student of that association usually tries to determine how necessary that symbiosis is to each participant and the functional bases of the dependence of one symbiont or another. There are many ways in which this dependence might be expressed; what is presented here is a set of continua selected, on the basis of their general applicability, from among the several possibilities which have occurred to me.

Continuum (D) based on the degree of necessariness

It is relatively easy to gather information about the degree of dependence or necessariness of a symbiosis. What one does is to determine experimentally how well or how poorly each participant in a symbiosis lives with and without its symbiont. Some symbiotic associations are thought to be absolutely necessary; these are termed 'obligate symbioses' and the participant for which the symbiosis is obligate is called an 'obligately dependent symbiont' (usually abbreviated to 'obligate symbiont'). Other symbiotic associations are known to be contingent; that is, they are entered into or not entered into; these are termed 'facultative symbioses' and the participant for which the symbiosis is contingent is called a 'facultatively dependent symbiont' (usually abbreviated to 'facultative symbiont'). The notion of obligate symbiosis is sometimes a reflection of initial ignorance about the true nature of the association, which is later determined to be somewhat toward the facultative symbiosis end of the continuum. Other cases of obligate symbiosis have stood the test of continued scrutiny (see Ingram & Tommerup, 1973, for a particularly illuminating discussion of this point). The inferred degree of necessariness of a symbiosis may depend on the experimental or observational system, which should therefore be carefully specified. Where the basis of the functional dependence is not known or is irrelevant to the case in point, the terms 'obligate' (or, preferably, 'obligately dependent') and 'facultative' (or, preferably, 'facultatively dependent') can be used alone. Alternatively, where the basis of the functional dependence is known and is relevant to the matter under consideration, the terms 'obligate' and 'facultative' can be used as modifiers of the modal terms pertaining to functional dependence which are discussed in the next paragraph.

Continuum (E) based on the mode of functional dependence

Classifying a particular symbiosis into the complex mishmash of possible modes of functional dependence is not particularly easy, only partly because of the difficulty of determining experimentally exactly what functional need each symbiont has of the other. It is difficult, also, because there would inevitably be an opening of the Pandora's box of reductionism (Ghiselin, 1974), if I did not exercise a rather cowardly restraint at this point! What I want to do here, avoiding the quagmire of reductionism, is to classify the functional dependence of symbionts according to the integrative levels (Novikoff, 1945) and the modes of the events that occur at each level. Presented rather simplistically, these include physical (with the conventional subdivisions of physics), chemical (with its conventional subdivisions, including bio-organic-chemical), organismal (with numerous subdivisions, of which only a few are mentioned here: physiological, nutritional, regulative, metabolic, genetic, anatomical, behavioral), and societal (with many subdivisions, none of which I am yet prepared to lay out systematically). The categorial levels and modal classes (as listed here and in Table 1) are arranged in a manner rather different in format from the other continua in this system, in that the categories are based on an increasing integrative level as one moves from subatomic physics through chemistry to the realm of society. My claim that the presentation is simplistic is based on the realization that, for example, the categorial level simply labelled here 'organismal' might be 'magnified' into a series of integrative levels such as subcellular, organellar, cellular, tissual, organic, etc. The same could be said – but not amplified with any genuine skill here – about 'physical', 'chemical' and 'societal'. But, I am not attempting a high-magnification view of this topic! The terms applied to the categorial (integrative) levels and the modal classes (the subdivisions at each level) have implicit in their use their ordinary-language, everyday meanings. At each integrative level and within each modal class, the functional dependence of a particular symbiont could be specified with whatever precision is possible from the existing fact-base or desirable for the purpose in hand.

Continuum (F) based on viability of nutritional-source center

The functional dependence in a symbiosis is often grounded in nutrition and an array of terms pertaining to interorganismic nutritional relationships is in vogue. From among these terms, I have singled out one kind which seems to me most germane to my attempts to formulate a generally applicable scheme for classifying organismic associations. The particular choice stems in part from this consideration and also from the fact that Lewis (1973, 1974) lays out, very clearly, a distinction (derived from others) pertaining to the viability of the fed-upon organism in a symbiosis. What counts as 'living' and 'dead' is a subject that fills volumes; this is not the place to unpack that complicated issue. One could, however, specify as precisely as is possible where, in the continuum between life and death, the viability state of the organism or its components lies with, perhaps, some indication of the likely reversibility.

Biotrophy, necrotrophy, and saprotrophy are delineated by Lewis (1974) on the basis of 'the derivation of organic compounds', respectively, from 'living cells' of a symbiont, from 'dead cells' of a symbiont, and from the 'non-living environment'. Some modifications are necessary in order to generalize this important distinction, to make it applicable, for example, to derivation of any sort of nutrient (not only organic compounds) and to unicellular as well as multicellular symbionts (whereupon the plural 'living cells' and 'dead cells' move to 'living symbiont' and 'more-or-less dead symbiont', respectively). But, before I turn to these emendations, it is important to know that Lewis (1974) requires a 'permanent, or at least prolonged, intimate contact' in a symbiosis; his definition of the term 'symbiosis' thus includes temporal (durability) and locational elements, each of which is specified separately in my generalized scheme (see Continuum C and Continuum A, above and in Table 1). The definition of the term 'symbiosis' given by Lewis (1974) would translate in my terminology to 'a persistent symbiosis involving close apposition (inhabiting or exhabiting) of organisms'. My meaning of the term 'symbiosis' requires that the association be significant to the well-being (or 'unwell-being' or both) of at least one of the associants; this significance is specified separately, as to degree and mode, in terms of Continuum D, Continuum E, and Continuum H (see above, below, and Table 1).

Keeping in mind these differences in viewpoint, let us return to the delineation of biotrophy, necrotrophy and saprotrophy. A symbiosis must start with an association of two living organisms. I use the term 'biotrophy' to mean that both organisms remain living and associated, and that at least one of the symbionts derives nutrients from the other symbiont.

When a portion or even all of one symbiont dies partially or completely after the symbiosis has commenced, the relationship is still counted by me as a symbiosis as long as the association continues; then, if the more-or-less dead portion or entirety of one symbiont is used by the other symbiont as nutriment, the situation is in my view a necrotrophic symbiosis. The term 'saprotrophy' means to me a non-symbiotic feeding upon abiotic (non-living) substances in the environment (including the already dead carcasses of formerly living organisms). The examples of biotrophy given by Lewis (1973, 1974) remain biotrophy in my sense. We differ somewhat in the meaning of necrotrophic symbiosis, largely because of the aforementioned differences between our respective meanings of the term 'symbiosis' but also because Lewis's system was designed for application primarily to symbioses involving at least one multicellular organism. In this connection, a classical predation – say, of a lion stalking, capturing, killing, and consuming a gazelle – would be for me a good example of a necrotrophic symbiosis. Where an organism associates with – that is, encounters and consumes – the carcass of another creature only after it is dead, the relationship is not a necrotrophic symbiosis but, rather, it is a non-symbiotic scavenging or saprotrophic one (for example, the jackal consuming the remains of the lion's kill).

One must set conceptual limits somewhere and, after careful deliberation, that is where I draw the lines. But it is by no means a simple matter to delineate and generalize concepts so complex as these. I do not take lightly the realization that my 'cuts' at some of these conceptual interfaces differ somewhat from those made by my much respected and admired colleague, D. H. Lewis. However, I am somewhat less uneasy about suggesting these modifications after having tested in these generalized terms many of the examples of biotrophy, necrotrophy and saprotrophy given by Lewis (1973, 1974), and finding practically no discrepancies between our respective placements (except those having to do with classical predations, which Lewis considers not to be symbiotic because they are not sufficiently prolonged). The generalized meanings thus seem to have no significant drawbacks in practice, and they do offer the advantage of applicability also to intermicrobial associations, one of which is discussed in these terms elsewhere in this volume (p. 93).

Continuum (G) based on degree of associant specificity

There is an extensive literature (see Pearce & Lowrie, 1972; Meeuse, 1973) on what is usually called 'host specificity', but which relationship might be included under the more general term 'symbiont (or associant) specificity' so as to avoid the 'parasite'-related connotation of the word 'host'. Some symbioses show a fair degree of specificity; that is, they are limited to highly selected associants. In other cases the symbionts are much more permissive (promiscuous); that is, the symbioses are closer to being non-specific. Symbiont specificity can be designated in whatever detail it is known or is needed for the business at hand; for example, 'limited to the primates', 'enters only into F^- *Escherichia coli* strain K12', 'infects only the crucifers', etc.

Conjoint continua (H) based on harm and benefit criteria

Symbiosis usually – perhaps always – involves harm or benefit, or both. (It may be useful to remark that the notions of 'harm' and 'benefit' as used here are human constructs; that is, it is man's interpretation of certain events as harmful or beneficial that we are considering.) Sometimes the benefit or harm is one-sided, sometimes it is mutual, sometimes there is benefit to one symbiont and harm to the other, sometimes there seems to be neither benefit nor harm perceived or interpreted as such by the human observer (which, of course, is not the same as the actual non-occurrence of events that benefit or harm the symbionts). In the earlier (Heise–Starr) scheme, benefit and harm were treated as extremes in one continuum with a neutral (neither benefit nor harm) condition in the middle of the single continuum. It now seems to me logically more correct to treat 'harm' and the '(almost) absence of harm' (one sort of 'neutrality') as the respective ends of one continuum and to treat 'benefit' and the '(almost) absence of benefit' (another sort of 'neutrality') as the respective ends of the other, conjoined, continuum. The reason for twice using the parenthetical 'almost' in the previous sentence is that if there were truly no trace of either benefit or harm to at least one of the associants, then the association would not be a symbiosis (as I use the term) because it would not be significant to the well-being (or 'unwell-being' or both) of at least one of the associants. Another caveat might again be noted: the harmful or beneficial effects as treated here are the *direct* effects upon the organisms which are participating in the symbiosis; the indirect effects on populations and the course of evolution, although highly significant biologically, are outside the scope of the present treatment.

There are many possible modes of harm: 'physical', 'chemical', 'organismal', 'societal' (see footnote *b*, Table 1, and the discussion above, in connection with Continuum E, of integrative levels and modal classes). Specific examples of some of these modes of harm would be 'bites a chunk out of its symbiont' (physical); 'poisons' (physiological) 'its symbiont with a protein toxin' (chemical); 'frightens its symbiont by making a threatening gesture' (behavioral); etc. There are many modes of benefit, which also might be arranged into the same modal classes as is harm. Specific examples of some of these modes of benefit would be 'emits light' (physical) 'that serves as a sexual attractant device for its symbiont' (behavioral); 'provides for its symbiont a necessary vitamin' (chemical; nutritional); 'quiets the fear of its symbiont by stroking it' (behavioral); etc. Locutions such as 'benefited (and benefiting) symbiont' and 'harmed (and harming) symbiont' might be used, as might also 'beneficial symbiosis' (= 'beneficialism') or 'harmful (antagonistic or agonistic) symbiosis' (= 'harmfulness' or 'antagonism' or 'agonism'). Where it is known, the particular mechanism whereby the harm or the benefit is inflicted, as well as the outcome in each symbiont resulting from the harm or benefit, might also be specified in modal or other terms and also quantified. There are many kinds of beneficial and harmful symbiotic associations and whole volumes have been written about some of them, particularly those called 'infectious disease' and 'antibiosis' and 'mutualism'.

Continuum (I) based on integrational criteria

Most symbiotic associations result, at least temporarily, in something more than the sum of the parts. As Srere & Mosbach (1974) put it: the 'high degree of integration in symbiotic associations is demonstrated very clearly by the appearance of substances and structures that the individual components could not produce' by themselves. Most often, the two components in a symbiosis eventually separate. More rarely, the two symbionts merge to form a 'third entity', the kind of integration found in the lichens (Ahmadjian, 1967; Ahmadjian & Hale, 1973; Smith, 1963), insect mycetosomes (Hinde, 1971), hereditary symbionts of protozoa (Preer, Preer & Jurand, 1974) and – according to some authorities (Margulis, 1970), but not so according to others (Raff & Mahler, 1972) – in the integration during evolutionary time of prokaryote cells to form organelles of eukaryote cells. It seems desirable that symbioses be specified in terms of their positions in this continuum, based on the format, mode, degree, and ultimate resolution (separability) of the integration.

SUMMARY – AND A HOPEFUL PROGNOSIS

The major aim of this enterprise is to bring some clarity into the subject of organismic associations – a clarity that is now lacking, in part because the terms now in vogue have been stretched in meanings over the past century until they have no universally accepted meanings. The particular intellectual approaches used in this study were (a) to examine, analyze and index a large array of associations between all sorts of biological entities; (b) to attempt to discern from these examinations the salient criteria which might serve as 'unit characters' in classifying organismic associations; (c) to arrange each unit character in a qualitative criterional continuum, usually with only the end-stations labelled but with quantitative positioning within each continuum intended where warranted by the fact-base and the purpose; (d) to apply the set of continua to actual organismic associations to the end of specifying a particular kind of association; and (e) to test thus the 'universal applicability' (Lewis, 1974) of the classificatory scheme.

A number of cursory applications of this sort have been made, limited mainly by my lack of expert knowledge of the biological systems. Three more detailed analyses have been made of systems where there was expert knowledge. In one case (Starr & Chatterjee, 1972), the associations between phytopathogenic bacteria of the genus *Erwinia* and plants or animals were usefully classified in terms of the original Heise–Starr scheme. In another instance, Lewis (1974) used a modification of the earlier Heise–Starr scheme and found it quite applicable to classifying associations between plants and various micro-organisms. The third case, a detailed analysis by the present scheme of the associations between *Bdellovibrio* and other bacteria, given later in this volume (p. 93), shows that the various aspects of this particular interbacterial association can be specified quite tidily in terms of the criterional continua and, moreover, that there is considerable heuristic value in attempting to classify in these terms since an array of interesting and likely productive experimental approaches was suggested when points of factual ignorance came to light in attempting to apply the classificatory criteria.

Another benefit from the undertaking stems from the realization that each of the terms conventionally used today to name organismic associations bears an assortment of differing meanings and that considerable clarification can be attained by translating (or redefining) such conventional terminology into the terms of this classificatory scheme. Although construction and presentation of such a glossary cannot be part of the present enterprise, a few examples might be inserted here: (a) the term 'symbiosis' in de Bary's original sense would translate to 'interspecific symbiosis' or 'heterosymbiosis' in my terminology; (b) van Beneden's original meaning of 'commensalism' would translate to 'an exhabitational symbiosis, in which the exhabiting symbiont feeds on surplus food captured by the exhabited (substrate) symbiont'; (c) 'mutualism' would translate to 'a mutually beneficial symbiosis, which can be either inhabitational or exhabitational'; (d) classical 'predation' would translate to 'an exhabitational, antagonistic (often killing), nutritionally dependent (consuming for food), necrotrophic symbiosis, with behavioral (hunting, stalking, capturing) overtones'; (*e*) some instances of 'parasitism' would translate to 'an inhabitational symbiosis, usually nutritionally dependent (and either biotrophic or necrotrophic), sometimes metabolically dependent, often antagonistic', but there are so many variations in the meanings of 'parasitism' that an array of clarifying translations must be made to cover the exact meaning intended in each case.

Although consideration of complex organismic associations such as populations and societies have been deliberately avoided here in the interest of achieving manageable simplicity in this preliminary analysis and terminology, my initial probes (in progress) into these populational and societal areas show that the classificatory scheme can, with minimal wrenching, be applied to such systems at higher integrative levels. All in all, I am heartened not only by these findings but also by the recent shifts in the attitudes of my peers – from the shoulder-shrugging or even more negative responses exhibited by them toward our earlier (admittedly defective) efforts to the more affirmative expressions now displayed by some of my colleagues, including the increasingly frequent uses by the more adventurous ones (in conversations, correspondence, and on-going writings) of the present – still tentative and developing – conceptual system and its terminology.

There are many ways of cutting a conceptual pie. I have tried here to wield the knife rationally, but 'reason' is often colored by one's experiences and prejudices; these will inevitably differ among individuals and, with the passage of time, within an individual. There is a revealing note on the latter point in footnote *a* to Table 1; the classificatory scheme outlined there is the eighth version considered by me during 1974 – and there would undoubtedly have been a ninth and tenth had there not been a publication deadline. Given the natures of intellectual dialogue and personal development, this classificatory scheme does not purport to be the last word – mine or other's – on this subject. What is asserted here (including the meaning given to the term 'symbiosis') can best be viewed as being my last word – for just now!

ACKNOWLEDGEMENTS

Supported in part by a research grant (AI-08426) from the National Institute of Allergy and Infectious Diseases, US Public Health Services. I am deeply indebted to David H. Lewis and Michael T. Ghiselin for their conceptually alert and linguistically felicitous critiques of various versions of this work; however, these good friends should not be held accountable – in any manner whatsoever – either for the views I have espoused or for the words with which I have expressed these views.

REFERENCES

AHMADJIAN, V. (1967). *The Lichen Symbiosis*. Waltham, Massachusetts: Blaisdell Publishing Company.

AHMADJIAN, V. & HALE, M. E. [eds.] (1973). *The Lichens*. New York: Academic Press.

ALEXANDER, M. (1971). *Microbial Ecology*. New York: John Wiley & Sons, Inc.

ASKEW, R. R. (1971). *Parasitic Insects*. London: Heinemann Educational Books Ltd.

DE BARY, A. (1879). *Die Erscheinung der Symbiose*. Strassburg: Verlag von Karl J. Trübner.

—— (1887). *Comparative Morphology and Biology of the Fungi[,] Mycetozoa and Bacteria*. Oxford: Clarendon Press.

BRACKER, C. E. & LITTLEFIELD, L. J. (1973). Structural concepts of host–pathogen interfaces. In *Fungal Pathogenicity and the Plant's Response*, 159–318 (eds. R. J. W. Byrde and C. V. Cutting). London and New York: Academic Press.

BROWN, R. B. (1970). *General Biology*. New York: McGraw-Hill Book Co.

CAULLERY, M. (1952). *Parasitism and Symbiosis*. London: Sidgwick & Jackson.

CLOUDSLEY-THOMPSON, J. L. (1965). *Animal Conflict and Adaptation*. London: G. T. Foulis & Co.

DIENER, T. O. (1974). Viroids: The smallest known agents of infectious disease. *A. Rev. Microbiol.*, **28**, 23–39.

GHISELIN, M. T. (1974). *The Economy of Nature and the Evolution of Sex*. Berkeley, Los Angeles and London: University of California Press.

HALL, R. (1974). Pathogenism and parasitism as concepts of symbiotic relationships. *Phytopathology*, **64**, 576–577.

HAYES, W. (1968). *The Genetics of Bacteria and Their Viruses*, second edition. New York: John Wiley & Sons, Inc.

HERTIG, M., TALIAFERRO, W. H. & SCHWARTZ, B. (1937). The terms *symbiosis*, *symbiont* and *symbiote*. Report of the Committee on Terminology (American Society of Parasitologists). *J. Parasitol.*, **23**, 326–329.

HINDE, R. (1971). The fine structure of the mycetome symbiotes of the aphids *Brevicoryne brassicae*, *Myzus persicae* and *Macrosiphum rosae*. *J. Insect Physiol.*, **17**, 2035–2050.

INGRAM, D. S. & TOMMERUP, I. C. (1973). The study of obligate parasites *in vitro*. In *Fungal Pathogenicity and the Plant's Response*, 121–140 (eds. R. J. W. Byrde and C. V. Cutting). London & New York: Academic Press.

JENNINGS, J. B. (1974). Symbioses in the Turbellaria and their implications in studies on the evolution of parasitism. In *Symbiosis in the Sea*, 127–160 (ed. W. B. Vernberg). Columbia, South Carolina: University of South Carolina Press.

JONES, A. W. (1967). *Introduction to Parasitology*. Reading, Massachusetts: Addison-Wesley.

LEWIS, D. H. (1973). Concepts in fungal nutrition and the origin of biotrophy. *Biol. Rev.*, **48**, 261–278.

—— (1974). Micro-organisms and plants: The evolution of parasitism and mutualism. *Symp. Soc. Gen. microbiol.*, **24**, 367–392.

LURIA, S. E. & DARNELL, J. E., Jr (1967). *General Virology*, second edition. New York: John Wiley & Sons.

Margulis, L. (1970). *Origin of Eukaryotic Cells*. New Haven, Connecticut: Yale University Press.

Meeuse, A. D. J. (1973). Co-evolution of plant hosts and their parasites as a taxonomic tool. In *Taxonomy and Ecology*, 289–316 (ed. V. H. Heywood). London: Academic Press.

Meynell, G. G. (1973). *Bacterial Plasmids: Conjugation, Colicinogeny, and Transmissible Drug Resistance.* Cambridge, Massachusetts: M.I.T. Press.

Novikoff, A. B. (1945). The concept of integrative levels and biology. *Science, Wash.*, **101**, 209–215.

Odum, E. P. (1971). *Fundamentals of Ecology*, third edition. Philadelphia and London: W. B. Saunders Co.

Pearce, J. H. & Lowrie, D. B. (1972). Tissue and host specificity in bacterial infection. *Symp. Soc. gen. Microbiol.*, **22**, 193–316.

Preer, J. R. Jr, Preer, L. B. & Jurand, A. (1974). Kappa and other endosymbionts in *Paramecium aurelia. Bact. Rev.*, **38**, 113–163.

Raff, R. A. & Mahler, H. R. (1972). The non symbiotic origin of mitochondria. *Science, Wash.*, **177**, 575–582.

Smith, D. C. (1963). Experimental studies of lichen physiology. *Symp. Soc. gen. Microbiol.*, **13**, 31–50.

Sprent, J. F. A. (1963). *Parasitism*. Baltimore: Williams & Wilkins Co.

Srere, P. A. & Mosbach, K. (1974). Metabolic compartmentation: Symbiotic, organellar, multienzymic, and microenvironmental. *A. Rev. Microbiol.*, **28**, 61–83.

Starr, M. P. (1946). The nutrition of phytopathogenic bacteria. I. Minimal nutritive requirements of the genus *Xanthomonas. J. Bact.*, **51**, 131–143.

(1959). Bacteria as plant pathogens. *A. Rev. Microbiol.*, **13**, 211–238.

Starr, M. P. & Chatterjee, A. K. (1972). The genus *Erwinia*: Enterobacteria pathogenic to plants and animals. *A. Rev. Microbiol.*, **26**, 389–426.

Starr, M. P. & Huang, J. C.-C. (1972). Physiology of the bdellovibrios. *Adv. microbial Physiol.*, **8**, 215–261.

Starr, M. P. & Seidler, R. J. (1971). The bdellovibrios. *A. Rev. Microbiol.*, **25**, 649–678.

Stolp, H. & Starr, M. P. (1963). *Bdellovibrio bacteriovorus* gen. et sp. n., a predatory, ectoparasitic, and bacteriolytic microorganism. *Antonie van Leeuwenhoek*, **29**, 217–248.

van Beneden, P. J. (1876). *Animal Parasites and Messmates*. New York: D. Appleton & Company.

Zuckerman, A. & Weiss, D. W. (1973). Preface and introduction. In *Dynamic Aspects of Host–Parasite Relationships*, vol. 1, vi–xii (eds. A. Zuckerman and D. W. Weiss). New York and London: Academic Press.

SYMBIOTIC THEORY OF THE ORIGIN OF EUKARYOTIC ORGANELLES: CRITERIA FOR PROOF

By LYNN MARGULIS

Department of Biology, Boston University,
Boston, Massachusetts 02215, USA

INTRODUCTION

There are intrinsic limitations in proving any historical argument. Yet a specific historical theory may be extremely useful for the integration of many observations into a coherent whole. In this context I am presenting a brief summary of the serial endosymbiosis theory. That is, three classes of eukaryotic cell organelles are postulated to have originated by hereditary endosymbiosis (namely, mitochondria, flagella/cilia and photosynthetic plastids). I try to show that this concept provides an explanation of phenomena such as the genetic behavior of organelles, physiology of the mitotic process, phylogeny of the protists, gaps in the fossil record and in the living biota, and even allows the interpretation of certain atmospheric data. The reader will see that the serial endosymbiosis theory has considerable implications for the overall detailed classification of living organisms into five kingdoms. Furthermore, the theory generates many explicit hypotheses with experimental solutions and I have listed some of these.

I am also using this opportunity to comment upon a few recent papers presenting non-symbiotic views. They seem less useful for generating new knowledge than the serial endosymbiosis theory, even if the latter should prove to be an incorrect historical reconstruction.

Nature of historical arguments, intrinsic limitations

It is never possible to prove rigorously that a unique series of events did occur in any historical context. Evolutionary biologists are in the same logical predicament as historians: they deal with series of complex irreversible phenomena and can only present arguments based on the assumption that of all the plausible historical sequences one is more likely to be a correct description of the past events than another. The probability that a particular historical reconstruction is more accurate than another increases markedly if the number of present observables that can be explained increases. Such an evolutionary theory can be judged by its predictive power as well.

In this context it is the purpose of this paper to present succinctly a summary of some potential experimental studies generated by the concept of the symbiotic origin of organelles (Sagan, 1967; Margulis, 1970). These studies and their predicted conclusions do not follow necessarily from alternative monophyletic and/or partially symbiotic concepts of the origin of eukaryotic cells.

For those unfamiliar with the theory a brief oversimplified outline of it in its original formulation is presented in the next section, e.g. three classes of organelles are proposed to have originated endosymbiotically: mitochondria, flagella, cilia and other '(9 + 2) organelles', and photo-synthetic plastids. Although several modifications and extensions of the theory have been published (Goksøyr, 1967; Schnepf & Brown, 1971; Raven, 1970; Lee, 1972; Taylor, 1974), the extreme endosymbiotic view (Margulis, 1970) has the advantage of being holistic and relatively complete. That is, it provides the most comprehensive, explicit, and testable framework of necessarily interrelated evolutionary postulates.

The details and documentation of the statements in the following section are given in the references, as noted. Semipopular discussion and illustration of the role of symbiosis in the origin of eukaryotic cells, according to the theory, have been published (Margulis, 1971a, 1972a, b).

BRIEF OUTLINE OF THE SERIAL ENDOSYMBIOSIS THEORY OF THE ORIGIN OF EUKARYOTIC CELLS

All organisms on earth have anaerobic, fermentative, heterotrophic, prokaryotic ancestors. Selection pressures in the early Precambrian led to an extensive adaptive radiation among prokaryotes, primarily on the metabolic level. Among others, the following cell types evolved: myco-plasm-like fermenters (catabolizing glucose to pyruvate via anaerobic glycolysis using the Embden-Meyerhof metabolic pathway), spirochaetes, photosynthetic oxygen-eliminating prokaryotes (coccoid blue-green algae) and aerobic Gram-negative eubacteria that oxidized small organic acids via the Krebs cycle (Sagan, 1967; Margulis, 1970, 1971b, c).

The first step in the origin of eukaryotes from prokaryotes occurred when a fermentative anaerobe was invaded by Krebs cycle-containing Gram-negative eubacteria (protomitochondria): the stabilization of this once-predatory association led to the formation of the mitochondria-containing amoeboids, from which all other eukaryotes derive.† The

† Because pinocytosis and phagocytosis are virtually unknown in prokaryotes (Stanier, 1970) and because recently characterized *Bdellovibrio*-like organisms

nuclear and other endoplasmic membranes evolved autogenously after the steroid biosynthetic pathway was made available by the presence of the protomitochondria. Selection pressures for the formation of the nuclear membrane probably involved the segregation of newly synthesized DNA on endomembrane and/or the sequestering of nucleoplasm DNA to protect it from the more oxidizing conditions of the cytoplasm surrounding the mitochondria. The second symbiotic step was the acquisition by such amoeboids of highly motile anaerobic surface bacteria (thought to have been spirochaetes): this led (after many mutations and intracellular transfer of genes from bacteria to host) to the origin of amoeboflagellates, ciliates and other protists. The surface spirochaete bacteria evolved into the (9 + 2) homologues: basal bodies cilia, flagella and other microtubule-based structures. Mitosis evolved in many lines of organisms as the motile bacteria merged with the amoeboid host cell and the morphogenetic processes of the bacteria were eventually utilized in the formation of the 'achromatic apparatus' (mitotic spindle) of mitosis. (See Margulis, 1970, 1974b, and Pickett-Heaps, 1974, for detailed exposition of the possible steps involved in the origin of mitosis.)

The final symbiotic step involved the acquisition of photosynthesis by varied populations of eukaryotic heterotrophs. Ingestion, without digestion, of blue-green algae by various protist hosts under nutrient-poor conditions led to the establishment of stable, heritable, intracellular symbioses. The prokaryotic algae eventually became the obligatory symbiotic photosynthetic plastids in the origin of various lines of nucleated algae and, eventually, chlorophytes and archegoniate multicellular green plants. Hence all eukaryotes are at least digenomic (contain two independently derived genomes: host and mitochondria), most heterotrophic eukaryotes were originally trigenomic* (host, mitochondria and (9 + 2) homologue or flagella/cilia/basal body) and most photosynthetic eukaryotes were originally quadrigenomic* (host, mitochondria, (9 + 2) homologue and plastids).

I agree with Taylor's (1974) contention that the word 'cell' is inadequate; his suggestions ought to be followed. Words referring to cells should be modified to account for the number of membrane-bound, protein-synthesizing units, that is: monad* should be used for prokaryotes, dyad* for heterotrophic eukaryotes and triad* for photosynthetic

---

provide such a fine model for the penetration of bacterial hosts by aerobic Gram-negative eubacteria, I now feel the first prokaryotic-prokaryotic-symbiotic step was likely to have been by invasion, followed by modification and stabilization of *Bdellovibrio*-like behavior. See Starr & Seidler, 1971, and Starr, p. 93, for details of the *Bdellovibrio* story.

* For definitions and analytic discussions of these terms, please see Taylor, 1974 and Margulis, 1975.

eukaryotes. (Even if the spirochaete origin for the flagellar-mitotic system is correct, it is clear that the once-independent protein-synthesizing system of the motile bacteria must have become completely integrated into the complex eukaryotic host system. Thus Taylor's terminology is entirely consistent with the serial endosymbiotic theory.) Multicellular units, then (according to Taylor, 1974) become: polymonads* (e.g. filamentous and mycelial prokaryotes such as *Stigonema*, *Chondromyces* or *Actinomyces*); polydyads* (e.g. slime molds, ascomycetes, metazoa); and polytriads* (e.g. bryophytes, tracheophytes, brown and red seaweeds) and so forth.

USEFULNESS OF THE SERIAL ENDOSYMBIOTIC THEORY IN EXPLANATION

This section will review briefly examples of the usefulness of the theory in explaining many apparently unrelated phenomena and in making predictions.

Genetic behaviour of organelles and other cell biological phenomena

Three distinct patterns of inheritance can be detected in the ciliate protist *Paramecium aurelia*: nuclear (Mendelian), caryonidal (a specialized modified nuclear mode) and cytoplasmic. For many years the inheritance of the killer trait has been the standard example of the behavior of cytoplasmic genes (Sonneborn, 1959). The killer phenotype was found to depend on the presence of cytoplasmic kappa particles. Traits correlated with the presence of kappa particles are inherited independently of the nucleus yet kappa requires several Mendelian genes for its maintenance and is negatively affected by several other nuclear alleles. The similarity between the behavior of kappa and other cytoplasmically inherited factors now known to be associated with mitochondria and plastids, is striking. An obvious explanation is that both kappa and these organelles began as free-living prokaryotes and with time they have become hereditary endosymbionts more and more dependent on products of nuclear genes. The analysis of the paramecium nuclear gene products that affect kappa would be extremely useful for the elucidation of this analogy. In fact, the hypothesis that mitochondria and plastids originated as prokaryotic symbionts has generated some fine experiments (e.g. Pigott & Carr, 1972) and an immense unwieldy literature (for example, see Nass, M. M. K., 1969, 1971; Nass, S., 1969; Cohen, 1970, 1973, for reviews; Ebringer, 1972). Let us take just one limited but extremely exciting example of the usefulness of the endosymbiotic view for the explanation of behavior: the recently discovered sexual behavior of mitochondria and chloroplasts.

* See footnote, p. 23.

The evidence that recombination (e.g. formation of recombinant DNA correlated with alteration of linkage relations of markers) occurs between differently marked mitochondria entering a cross in yeast is now overwhelming (Thomas & Wilkie, 1968; Wilkie & Thomas, 1973; Mounolou, Jakob & Slonimski, 1967; see Gillham, 1974, 1975 for reviews). Comparable organellar sexual recombination phenomena are probably occurring after zygote formation in *Chlamydomonas* chloroplasts as well (Sager, 1972; Gillham, 1974). The best model for the nature of the recombination process in organelles may be a multiparental unidirectional bacterial merogenomic model (Gillham, Boynton & Lee, 1974; Gillham, 1974). However, a molecular mechanism for both mitochondrial recombination and suppressiveness associated with certain petite strains of yeast, based on analogy of phage excision and repair systems has recently been presented (Perlman & Birky, 1974). Thus the detailed comparison of the genetic behavior of organelles (mitochondria and plastids) with prokaryotic genetic systems has proved fruitful. The reader is strongly recommended to the two comprehensive and brilliant reviews of this literature by Gillham (1974, 1976). The probability that a prokaryote would evolve all major eukaryotic features (nuclear membrane, histone proteins, mitotic spindle) and subsequently sequester in its cytoplasm a functional, small ribosome, prokaryotic genetic system which happened to have features comparable to the recombination system of prokaryotes (as envisaged by Uzzell & Spolsky, 1974) seems very remote. On the other hand, that endosymbiotic bacteria have retained their DNA recombination mechanisms and lost many of their metabolic virtuosities (because selection acts against redundancies) is perfectly logical. Many precedents of bacterial hereditary endosymbiosis correlated with loss of autonomy are known; over a dozen different examples of hereditary bacteria-like symbionts (including kappa particles) have been reported in the species *Paramecium aurelia* alone (Beale, Jurand & Preer, 1969; Preer, Preer & Jurand, 1974; Preer, this volume).

Genetic interaction between organisms that are clearly recognizable symbiotic partners is not well understood. Because of the possible usefulness of such information in understanding the processes that have led to the evolution of organelles from free-living symbionts I have undertaken a review of this literature. There is no doubt that heritable modification of symbionts occurs with time after association. Although it can be strongly criticized, even evidence for gene transfer from bacteria to plant cell nuclei has been presented (Ledoux, Huart & Jacobs, 1972, 1974). The reader is referred to the details in a long, illustrated, forthcoming paper concerned with this problem (Margulis, 1975).

Protist phylogeny

The contradictory classification systems for the 'lower eukaryotes' by zoologists and botanists are notorious. The recognition of 'anastomosing phylogenies', i.e. that new groups such as lichens and chlorophyte algae are products of symbioses, has permitted great clarification of the relationships between these organisms. Single, consistent, phylogenies and taxonomic schemes may be drawn. Although symbiosis may not be a factor in the origin of most higher taxa, it has been decidedly significant in many cases (Fig. 1). (See Whittaker, 1969; Margulis, 1974a, c, for details and Leedale, 1974, for discussion.)

The microfossiliferous Precambrian

It has long been recognized that there is a major change in the nature of the fossil record between the Precambrian ($\simeq 3 - 0.6 \times 10^9$ years ago) and the Phanerozoic (0.6×10^9 years ago until the present) and there is little or no evidence for an environmental 'catastrophe' or discontinuity at the Precambrian border. The recognition that, fundamentally, the Precambrian Eon was the age of the prokaryotic microbes and the Phanerozoic Eon the age of eukaryotes clarifies this discrepancy (Margulis, 1969). Of course this is an oversimplification not addressed to the origin of shelled metazoans (Cloud, 1968) and the role of oxygen and other environmental variables shaping the selection pressures on the biota (Cloud, 1974; Palmer, 1974). However, the symbiotic theory explains nicely how oxygen-releasing photosynthesis preceded by hundreds of millions of years the origin of green plants (*sensu strictu*).

Integration of data from apparently unrelated fields: some examples

The recognition that microbes are products of neo-Darwinian evolution with selection acting mainly on metabolic rather than morphologic traits has aided in placing in a temporal sequence certain steps in the history of the atmosphere (Cloud, 1974; Lovelock & Margulis, 1974; Margulis & Lovelock, 1974; Walker, 1976).

The concept that regeneration of complex (9 + 2) microtubule-based flagella is a process homologous to the movement of chromosomes in mitosis has led to the discovery of mitotic arrest by melatonin and other specific actions of certain drugs (Banerjee & Margulis, 1973; Margulis, 1973). The awareness of the homology between the flagella/cilia system of eukaryotes has suggested a new theory of sensory transduction [primarily olfactory and auditory (Atema, 1973)].

Of course the concept that mitochondria and plastids were bacteria has led to widespread use of specific antibiotics (e.g. streptomycin and other

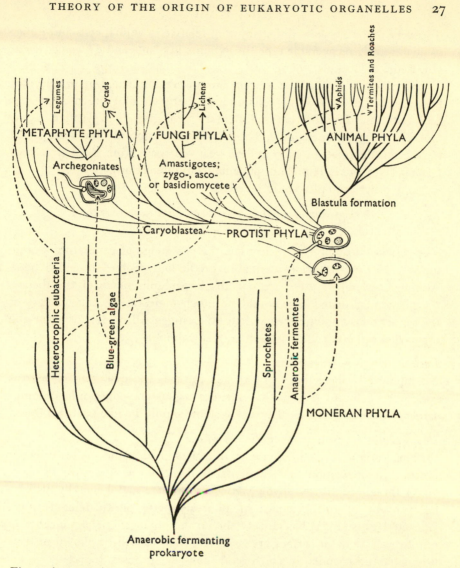

Fig. 1. A suggestion of the origin of some higher taxa by symbiosis (dashed lines indicate less conspicuous partners; capitals represent phyla, smaller print refer to lower taxa). The figure is drawn approximately against time; the lower phylogeny represents prokaryotes and the upper represents eukaryotes.

aminoglycosides, chloramphenicol, cycloheximide) to investigate the role of the organellar protein-synthetic system relative to the 'host' eukaryote (Ebringer, 1972). The assumption that plant cells are the products of hereditary endosymbiosis explains the presence of, for example, duplicate chloroplast and non-plastid metabolic pathways [e.g. for polyunsaturated fatty acids (Jacobson, Kannangara & Stumpf, 1973)]. It also clarifies the discovery of a very skewed distribution, relative to animal cells and fungi, of qualitatively recoverable types of mutants in plants and *Chlamydomonas* (Li, Redei & Gowans, 1967).

The recognition that Precambrian times were dominated by prokaryotes has already led to fruitful collaboration between blue-green algal mat ecologists and Precambrian sedimentary geologists (e.g. Walter, 1975; Golubic, 1973). For example, *Entophysalis*-like coccoid blue-green algae have been identified in 2500 million year old rocks from Belcher Island, Canada (Golubic & Hofmann, personal communication).

Many other examples could be cited.

Generation of explicit hypotheses with experimental solutions: some examples

The serial endosymbiosis theory has the advantage of generating a large number of experimentally verifiable predictions. Some are listed here. Although several of these are not absolute requirements of the theory, the phenomena listed here would be much more likely consequences of the serial endosymbiotic theory than other suggested models of eukaryote organelle origin (e.g. Raff & Mahler, this volume).

Lack of mitosis in blue-greens
No blue-green algae that show trends towards the evolution of mitosis will ever be found (e.g. with centrioles or histone-containing chromosomes). All blue-green algal sexual systems should be like bacterial ones.

Phylogenetic significance of mitotic variations
Striking variations on the theme of mitosis will be found in 'lower eukaryotes'. In opposition to statements in the classical literature (Wilson, 1925) these will be of phylogenetic significance and generally will be independent of the presence of photosynthetic plastid-related traits. [Pickett-Heaps (1974) has already shown the phylogenetic significance of mitotic variations in certain green algae.] Meiosis will be shown to have originated independently in several lines of protists (ciliates, hypermastigotes) just as homeothermy originated independently in several vertebrate lines.

The red algal thallus was a multicellular* heterotroph that acquired cyanelles
The photosynthetic plastids of red algae will have many metabolic pathways and primary amino acid sequences in proteins in common with blue-green algae; non-plastid cytoplasmic proteins in red algae will tend to be more like certain heterotrophic protists, primitive zygomycetes, or ascomycetes (amastigote-conjugating organisms).

* Or polydyad (Taylor, 1974).

Lack of automomy as the advanced state

Organisms containing well-developed Mendelian genetic systems (e.g. yeasts, metazoans, metaphytes) will tend to have *less* autonomous protein-synthetic systems in their mitochondria and photosynthetic plastids. Conversely certain 'lower eukaryotes' that have no eumitotic sexual systems will tend to contain organelles with greater autonomy and higher molecular weight unique sequence DNA. Attempts to 'culture' organelles will be more successful in organisms that lack or show idiosyncratic sexuality (e.g. euglenids, dinoflagellates, etc.). (See Margulis, 1974*b*, and Pickett-Heaps, 1974, for discussion of some of these forms.)

Bdellovibrio *symbionts*

Bdellovibrio-like bacteria will be found that maintain stable relationships with their 'prey'.

Tubulin homologies

The amino acid sequence of mitotic spindle tubulin proteins from all eumitotic organisms will show striking homologies with those from the $(9 + 2)$ flagellar and ciliary axonemes.

Tubulin re-utilized for mitotic spindle and flagella

Flagellar proteins (α and β tubulins) will be used directly in the formation of the mitotic spindle microtubules in many if not all eumitotic eukaryotes. This protein will not be found in blue-green algae. However, homologous tubulin proteins will be detected in red algae, fungi, and other eukaryotes that have mitosis but lack flagella in extant forms.

Eukaryote transformation

The phenomenon of gene transfer from prokaryotes or eukaryotic donors to eukaryote nuclei will be demonstrated. With respect to organelles, in any lineage (see Fig. 1) the particular genes transferred from the mitochondria and/or plastids will tend to be the same within that lineage and tend to be different outside that lineage.

In-vitro growth of organelles

Photosynthetic plastids and mitochondria will be cultured *in vitro*. The growth requirements for their culture will tend to be components of the protein-synthetic system and metabolites supplied to them by the in-vivo cycloheximide sensitive 80 S ribosomal system. Other organelles (those that presumably had an autogenous rather than an exogenous origin, e.g. Golgi, pigment granules, lysosomes, nuclei) will not be culturable.

Polygenomic control of metabolic pathways

Complex metabolic pathways providing end products of high selective advantage (e.g. certain alkaloids, toxins, steroids, lichenic acids) will tend to be products of more than one genome. That is, the nucleic acids coding for proteins in a pathway or polypeptides of certain proteins will reside in part in the organelle(s) as well as in the nucleus.

Motile protein: spirochaete-tubulin homology

Primary amino acid sequence of the eukaryotic flagella proteins (tubulins) will show greater similarities to axial filament proteins (or whatever protein is intrinsically motile in spirochaetes) than to any other prokaryotic proteins. Spirochaetes containing 25 nm microtubules (perhaps even showing colchicine-sensitivity) will be found. Maybe even extant spirochaetes showing (9 + 2) arrangement of microtubules (axial, fibers) or cell body will be discovered.

Protein homologies of the 80 S ribosome

By amino acid sequence analysis eukaryotic 80 S ribosomes will show protein homologies indicating that they have more than one prokaryotic ancestor (e.g. anaerobic fermenter and heterotrophic eubacteria, host and protomitochondrion respectively, or anaerobic fermenter and spirochaete, host and flagellum respectively, and so forth).

Hybridization between organelles and free-living micro-organism DNAs

Direct nucleic acid hybridization studies will show homologies between organisms and organelles as sketched in Fig. 1. The amount of unique-sequence transcribable DNA per organelle will tend to be an inverse function of the age of the association. This hypothesis can be applied to the nucleic acids of partners in clear-cut examples of symbiosis too.

Organellar genetics and physiology: models for infection

Understanding of the genetic interactions in symbioses and pathogenic relations will be greatly enhanced by an understanding of eukaryotic organellar genetics.

Anastomosing phylogenies

A consistent phylogeny at higher taxonomic levels (such as is available for tracheophytes and chordates) acceptable to botanists, zoologists and microbiologists (because of information generated by all these disciplines) will only be possible after acceptance of the symbiotic theory for the origin of organelles. Because of anastomosing relationships, the presence of many morphological traits visible only at the ultrastructural level, and convergent directions of ecological and metabolic differentiation among protists, the finalization of such a phylogeny may be difficult.

COUNTERARGUMENTS

Taylor (1974) has published a remarkably complete and well-argued paper in which several alternatives to the endosymbiotic view of the origin of mitochondria and plastids are explored; since he has already enumerated the logical problems with most autogenous or 'pinching off' hypotheses it is not necessary to repeat his arguments here. The reader is referred to his paper, 'Implications and extensions of the serial endosymbiosis theory of the origin of eukaryotes'. This section therefore will just address itself to subsequent work by two groups of authors: 'The differentiation of mitochondria by gene amplification hypothesis' (Perlman & Mahler, 1970; Raff & Mahler, 1972; Raff & Mahler, this volume) and the 'membrane sequester hypothesis' (Uzzell & Spolsky, 1974).

The possibility that mitochondria arose 'as a means of improvement of existing subcellular structures in the primordial (perhaps eukaryotic) cell' has been argued by Perlman & Mahler (1970) and Raff & Mahler (1972). They suggest that partial mitochondrial autonomy may be a relatively recent modification and that the present day 'mitochondrial genome had its origin in nuclear DNA and may have been amplified in a manner not unlike the amplification of ribosomal cistrons in developing oocytes' (see also Raff & Mahler, this volume). Uzzell & Spolsky distinguish between the 'evolutionary view' of the origin of organelles (i.e. the monophyletic origin of plastids and mitochondria in a single population of prokaryotes) and the 'creationist view' in their paper, 'Plastids and mitochondria as symbionts: A revival of special creation?'. They suggest that 'initially all nuclear and organellar structures were probably very similar. All three contained a complete genome; all may have had membrane bound respiratory and photosynthetic activities. With time, increasing differentiation of function occurred: nuclear membranes lost photosynthetic and respiratory activities, which became sequestered in mitochondria and plastids respectively; photosynthetic and respiratory activities were increased by invaginations of the inner organellar membranes; organellar genomes lost many duplicated genes; while the nuclear genome, through additional duplication (Ohno, 1970)* increased in size and differentiated tremendously' (p. 338 of Uzzell & Spolsky, 1974). These two groups of workers represent the only serious and modern attempts to defend the 'classical' view as outlined, say, by Allsopp (1969). Uzzell & Spolsky start with the

* See Uzzell & Spolsky, 1974, for Ohno reference.

very eukaryote they are trying to evolve (see quote)! Furthermore the probability of the differentiation and spatial sequestering of a system (mitochondria) much more similar to an ancient ancestor (bacteria) than a more current one (e.g. cycloheximide-sensitive, histone-containing eukaryotic one) under dubious selection pressures in organisms that neither left fossils nor have living counterparts seems to me remote. Primitive eukaryotes that lack mitochondria are known (for example *Pelomyxa palustris*). This suggests the origin of mitochondria and other features of the eukaryotic condition arose in populations of organisms ancestral to extant amoebae. Why would the mitochondrial system have retained prokaryotic features if it differentiated from membrane (i.e. 'pinched off') in an already eukaryotic cell? What mechanisms to sequester ribosomes, lipid-synthetic systems, redundant transfer RNAs and so forth can be imagined that would come out with a pattern so similar to the genetic system of free-living bacteria? That is, the entire contribution of molecular biology to cell evolution seems to have been overlooked.

However this criticism does not apply to the set of papers by Raff, Mahler and Perlman. These workers have chosen not to deal with photosynthetic plastids: in so doing they have failed to recognize some serious problems. Let us imagine two possibilities: that the mitochondria differentiated, as they suggest, for increased metabolic efficiency from mesosomal membranes in a respiring micro-organism. That this aerobic micro-organism also evolved the total photosynthetic apparatus (e.g. plastids) independently of blue-green algae and that this apparatus evolved to an end point so similar in many details, if not identical physiologically, is too much to believe. The alternative, more reasonable, hypothesis is that the aerobic organism that differentiated mitochondria by gene amplification acquired plastids symbiotically.

There are still bothersome problems with the mitochondria-non-symbiotic, plastid-symbiotic view. For example, one needs an *ad hoc* explanation of the total loss of the mitochondria as a unit system in entire groups of organisms such as trichomonads and *Pyrsonympha*. There is no precedence for the gradual intracellular segregation of an active protein-synthetic package as a unit which is then lost. There is much precedence for this phenomenon (it is called infection and cure) when the acquired entity is genetic. Why, too, have exactly the same mitochondrial enzyme systems been pinched off in organisms as distinct as *Chlorella*, *Sequoia*, *Homo* and *Trichoplax*? If this is an example of the retention of a primitive trait in these forms then why has comparable differentiation of a protein-synthetic read-out system not occurred and been retained in any other prokaryotes? (See Taylor, 1974, for further discussion of these points.)

The excellent point made by Perlman & Mahler (1970) that petite yeast retain many required non-oxidative functions is not an argument for the 'differentiation of mitochondria from the nucleus' hypothesis. It is simply consistent with the notion that mitochondria have indispensable functions for the rest of the cell in addition to supplying ATP by oxidative phos- phorylation and that in special cases some of these may have been trans- ferred to the hosts. This explains the relative rarity of eukaryotes which lack mitochondria. Those that do inevitably live in nutrient-rich environ- ments. Mitochondria tend to be indispensable; this is comparable to the observation that even heterotrophic higher plants, unable to synthesize chlorophyll and thus photosynthesize, retain a requirement for proplastids and etiolated plastids. That is, plastids perform essential functions other than photosynthesis in cells of green plants.

The argument leveled against the symbiosis view that the host cytoplasm contains primitive oxygen-mediated metabolic systems (superoxide dis- mutase, peroxisomes) and therefore aerobiosis could not have been the nature of the selection pressure for the protomitochondrion–host associa- tion is oversimplified on two counts. If the anaerobic hosts were strict anaerobes, entirely unable to tolerate molecular oxygen, they would never have been found in environments with aerobic respirers (Hall, 1973; Taylor, 1974). Furthermore, there are many symbioses known, probably the majority, where an aerobic organism houses another aerobic form. Indeed, photosynthetic diatoms are known that house photosynthetic blue-green algae (*Richelia intracellularis* in *Rhizoselenia*) (Norris, 1967). Nature is far more subtle than this. Symbioses involve a complex of metabolic, behavioral, topologic and other environment-dependent inter- actions that defy characterization by a single environmental variable such as aerobiosis, photosynthesis, acidity and so forth. *Euglena* will digest ('bleach') its plastids rather than die of phosphorus starvation just as unhappy giant clams (*Tridacna*) will expel their symbiotic algae. The hypothesis that heterotrophic amoeba evolved when fully aerobic eubacteria invaded anaerobic hosts leading to large aerobic complexes (dyads) is very reasonable. The anaerobes, just like many lactic acid bacteria today, are likely to have evolved oxygen-dissipating enzymes for their own protection long before.

IMPLICATIONS FOR THE OVERALL CLASSIFICATION OF LIVING ORGANISMS INTO FIVE KINGDOMS

Whittaker (1959, 1969), on ecological grounds, presented a sound scheme for the classification of all living organisms into five groups, one prokaryote and four eukaryote. The five kingdom system of Whittaker was originally published with only the inclusion of the highest taxa: kingdoms and phyla. By applying the concepts generated by the symbiotic theory Whittaker's scheme could be modified to form an overall classification consistent with information generated by genetic and developmental studies. The role of symbiosis in the origin of the higher taxa (based on this detailed classification, Margulis, 1974c) is schematically drawn in Fig. 1. This figure also includes other well-known groups that have traits strongly influenced by regular symbiotic associations. It is obvious that (although Wallin (1927) certainly overstated the case) 'symbionticism' *has* been a mechanism in the origin of new taxa; examples may be found on every level from species to phylum.

In summary, the basic modified Whittaker five kingdom scheme is as follows: monerans (all prokaryotes, or monads), protists (dyads and triads; diverse groups of asexual and eumitotic eukaryotes, ploidy levels and mitotic systems vary), fungi (dyads: amastigote* haploid or dikaryotic forms, zygo-, asco- and basidiospore formers that grow by absorptive nutrition), animals (dyads: diploid metazons that develop from blastulas), plants (archegoniate, embryophyte triad photoautotrophs).

This is not the place to detail the modified Whittaker classification but only to urge you to look up your organism in it. Is the system consistent with your own experience with the living forms? Your criticism is appreciated.

An alternative logical two kingdom system has been suggested by Taylor (1974; e.g. Prokaryote Kingdom and Eukaryote Kingdom); this and several other possibilities have been discussed by Leedale (1974). Following Whitehouse (1973), however, it seemed to me both logical and respectful of systematic tradition to keep the prokaryote (chromonemal)† versus eukaryote (chromosomal)‡ distinction at the superkingdom level. Thus the plant and animal kingdoms would remain in a restricted form supplemented by three other kingdoms: monerans, protists and fungi. For the first time the mycologists themselves have suggested the treatment of fungi as a kingdom (or at least a subkingdom; they have hedged a bit) in their most recent taxonomic treatise (Ainsworth, Frederick, Sparrow & Sussman, 1973) although they have not separated out flagellated 'lower' forms from the amastigote line. The prescient suggestion by Brooks (1924) that fungi be raised to kingdom status unfortunately has never been incorporated into the mycological literature.

* Those lacking flagella at all stages of the life-cycle.
† Uncomplexed DNA in nucleoids, not membrane bounded.
‡ Chromosomal DNA complexed with protein and RNA, membrane-bounded nuclei.

SUMMARY

The purpose of a scientific theory is to unite apparently disparate observations into a coherent set of generalizations with predictive power. Historical theories, which necessarily treat complex irreversible events, can never be directly tested. However they certainly can lead to predictions. The 'extreme' version of the serial endosymbiotic theory argues that three classes of eukaryotic organelles had free-living ancestors: mitochondria, basal bodies/flagella/cilia [(9 + 2) homologues] and photosynthetic plastids. Many lines of evidence support this theory and can be interpreted in relation to one another on the basis of this theory. Even if this theory should eventually be proved wrong it has the real advantage of generating a large number of unique experimentally verifiable hypotheses.

ACKNOWLEDGEMENTS

I am grateful to Max (F. J. R.) Taylor for his finely argued paper (1974) and the discussions generated by it. Many ideas here were developed in discussions with J. K. Kelleher, N. Gillham and R. Trench. I thank David Jennings, Robert Whittaker and Philip Perlman for useful critical comments on the manuscript. The sabbatical hospitality of both the Microbiology and Zoology Departments at the University of Washington, Seattle, Washington, was greatly appreciated. I acknowledge research support from NASA (NGR-004-025) and travel support to attend the Bristol Symposium which came from the Society for Experimental Biology and from the Boston University Graduate School.

REFERENCES

AINSWORTH, G. C., FREDERICK, G. C., SPARROW, K. & SUSSMAN, A. S. (1973). *The Fungi: An Advanced Treatise*, vol. IVA. New York and London: Academic Press.

ALLSOPP, A. (1969). Phylogenetic relationships of the procaryotes and the origin of the eucaryotic cell. *New Phytol.*, **68**, 591–612.

ATEMA, J. (1973). Microtubule theory of sensory transduction. *J. theoret. Biol.*, **38**, 181–190.

BANERJEE, S. & MARGULIS, L. (1973). Mitotic arrest by melatonin. *Expl. Cell Res.*, **78**, 314–318.

BEALE, G. H., JURAND, A. & PREER, J. R. Jr (1969). The classes of endosymbiont of *Paramecium aurelia*. *J. Cell Sci.*, **5**, 65–91.

BROOKS, F. T. (1924). Some present-day aspects of mycology. *Trans. Br. mycol. Soc.*, **9**, 14–32.

CLOUD, P. E. Jr (1968). Premetazoa evolution and the origin of Metazoa. In *Evolution and Environment* (ed. E. T. Drake). New Haven: Yale University Press. (1974). Evolution of ecosystems. *Am. Scient.*, **62**, 54–66.

COHEN, S. S. (1970). Are/Were mitochondria and chloroplasts microorganisms? *Am. Scient.*, **58**, 281–289.

— (1973). Mitochondria and chloroplasts revisited. *Am. Scient.*, **61**, 437–445.

EBRINGER, L. (1972). Are plastids derived from prokaryotic microorganisms? Action of antibiotics in chloroplasts of *Euglena gracilis. J. Gen. Microbiol.*, **71**, 35–52.

GILLHAM, N. W. (1974). Genetic analysis of the chloroplast and mitochondrial genomes. *A. Rev. Genet.* **8**, 347–391.

— (1976). *Organelle Heredity: An Inquiry into the Genetics of Chloroplasts and Mitochondria.* Holt, Reinhart & Winston (in press).

GILLHAM, N. W., BOYNTON, J. E. & LEE, R. W. (1974). Segregation and recombination of non-Mendelian genes in *Chlamydomonas. XIIIth Int. Congr. Genet.* **78**, 439–457.

GOKSØYR, J. (1967). Evolution of eucaryotic cells. *Nature, Lond.*, **214**, 1161.

GOLUBIC, S. (1973). The relationship between blue-green algae and carbonate deposits. In *The Biology of the Blue Green Algae*, 434–472 (eds. N. G. Carr and B. A. Whitton). Oxford: Blackwell Scientific Publications.

HALL, J. B. (1973). The nature of the host in the origin of the eukaryote cell. *J. theor. Biol.*, **38**, 413–418.

JACOBSON, B. S., KANNANGARA, C. G. & STUMPF, P. K. (1973). The elongation of medium chain trienoic acids to α-linolenic acid by a spinach chloroplast stroma system. *Biochem. biophys. Res. Comm.*, **52**, 1190–1198.

LEDOUX, L., HUART, R. & JACOBS, M. (1972). Fate and effect of exogenous DNA in *Arabidopsis thaliana. XIIIth Int. Congr. Cell Biol.*

— (1974). DNA mediated genetic correction of thiamineless *Arabidopsis thaliana. Nature, Lond.*, **249**, 17–21.

LEE, R. E. (1972). Origin of plastids and the phylogeny of algae. *Nature, Lond.*, **237**, 44–46.

LEEDALE, G. F. (1974). How many are the kingdoms of organisms? *Taxon*, **23**, 261–270.

LI, S. L., REDEI, G. P. & GOWANS, C. S. (1967). Phylogenetic comparison of mutation spectra. *Molec. gen. Genet.*, **100**, 77–83.

LOVELOCK, J. E. & MARGULIS, L. (1974). Homeostatic tendencies of the earth's atmosphere. *Origins of Life*, **5**, 93–103.

MARGULIS, L. (1969). New phylogenies of the lower organisms: Possible relation to organic deposits in Precambrian sediment. *J. Geol.*, **77**, 606–617.

— (1970). *Origin of Eukaryotic Cells.* New Haven: Yale University Press.

— (1971*a*). Origins of animal and plant cells. *Am. Scient.*, **59**, 230–235.

— (1971*b*). Early cellular evolution. In *Exobiology*, vol. 1, 342–368 (ed. C. Ponnamperuma). Amsterdam: North-Holland.

— (1971*c*). Microbial evolution on the early earth. *Molec. Evol.*, **1**, 480–484.

— (1972*a*). Symbiosis and evolution. *Scient. Am.*, **224**, 48–57.

— (1972*b*). Symbiosis and cell evolution. *Natuur en Techniek*, **40**, 395–407.

— (1973). Colchicine sensitive microtubules. *Int. Rev. Cytol.*, **34**, 333–361.

— (1974*a*). Classification and evolution of prokaryotes and eukaryotes. In *Handbook of Genetics*, vol. 1, 1–41 (ed. R. King). New York: Plenum Press.

— (1974*b*). On the origin and possible mechanism of colchicine-sensitive mitotic movement. *Biosystems*, **6**, 16–36.

— (1974*c*). The five kingdom classification and the origin and evolution of cells. *Evol. Biol.*, **7**, 45–78.

— (1975). Genetic and evolutionary consequences of symbiosis. *Expl. Parasitol. Rev.* (in press).

MARGULIS, L. & LOVELOCK, J. E. (1974). Biological modulation of the earth's atmosphere. *Icarus*, **21**, 471–489.

MOUNOLOU, J. C., JAKOB, H. & SLONIMSKI, P. P. (1967). Molecular nature of hereditary cytoplasmic factors controlling gene expression in mitochondria. In *Control of Nuclear Activity*, 413–431 (ed. L. Goldstein). Englewood Cliffs, New Jersey: Prentice-Hall.

NASS, M. M. K. (1969). Mitochondrial DNA: advances, problems and goals. *Science, Wash.*, **165**, 1128–1131.

    (1971). Properties and biological significance of mitochondrial DNA. In *Biological Ultrastructure: The Origin of Cell Organelles*, 41–63 (ed. P. S. Harris). Corvallis: Oregon State University Press.

NASS, S. (1969). The significance of the structural and functional similarities of bacteria and mitochondria. *Int. Rev. Cytol.*, **25**, 55–129.

NORRIS, R. E. (1967). *Algal consortisms in marine plankton*. Proc. of the Seminar on Sea Salt and Plants, 178–189. New Delhi: Catholic Press.

PALMER, A. R. (1974). Search for the Cambrian World. *Am. Scient.*, **62**, 216–224.

PERLMAN, P. S. & BIRKY, C. W. Jr (1974). Mitochondrial genetics in Bakers yeast: a molecular mechanism for recombinational polarity and suppressiveness. *Proc. natn. Acad. Sci. USA* (in press).

PERLMAN, P. S. & MAHLER, H. R. (1970). Formation of yeast mitochondria. III. Biochemical properties of mitochondria isolated from a cytoplasmic petite mutant. *Bioenergetics*, **1**, 113–138.

PICKETT-HEAPS, J. (1974). The evolution of mitosis and the eukaryotic condition. *Biosystems*, **6**, 37–48.

PIGOTT, G. H. & CARR, N. G. (1972). Homology between nucleic acids of blue-green algae and chloroplasts of *Euglena gracilis*. *Science, Wash.*, **175**, 1259–1261.

PREER, J. R. Jr, PREER, L. B. & JURAND, A. (1974). Kappa and other symbionts in *Paramecium aurelia*. *Bact. Rev.*, **38**, 113–163.

RAFF, R. A. & MAHLER, H. R. (1972). The non-symbiotic origin of mitochondria. *Science, Wash.*, **177**, 575–582.

RAVEN, P. H. (1970). A multiple origin for plastids and mitochondria. *Science, Wash.*, **169**, 641–646.

SAGAN, L. (Margulis). (1967). On the origin of mitosing cells. *J. theoret. Biol.*, **14**, 225–275.

SAGER, R. (1972). *Cytoplasmic Genes and Organelles*. New York: Academic Press.

SCHNEPF, E. & BROWN, R. M. Jr (1971). On relationships between endosymbiosis and the origin of plastids and mitochondria. In *Origin and Continuity of Cell Organelles*, 299–322 (eds. J. Reinert and H. Ursprung). Heidelberg: Springer-Verlag.

SONNEBORN, T. M. (1959). Kappa and related particles in *Paramecium*. *Adv. Virus Res.*, **6**, 229–338.

STANIER, R. Y. (1970). Some aspects of the biology of cells and their possible evolutionary significance. *Symp. Soc. gen. Microbiol.*, **20**, 1–38.

STARR, M. P. & SEIDLER, R. J. (1971). The *Bdellovibrios*. *A. Rev. Microbiol.*, **25**, 649–678.

TAYLOR, F. J. R. (1974). Implications and extensions of the serial endosymbiosis theory of the origin of eukaryotes. *Taxon*, **23**, 229–258.

THOMAS, D. Y. & WILKIE, D. (1968). Recombination of mitochondrial drug resistance factors in *Saccharomyces cerevisiae*. *Biochem. biophys. Res. Comm.*, **30**, 368–372.

UZZELL, T. & SPOLSKY, C. (1974). Mitochondria and plastids as endosymbionts: A revival of special creation? *Am. Scient.*, **62**, 334–343.

WALKER, J. C. G. (1976). *History of the Earth's Atmosphere*. New York: Macmillan (in press).

WALLIN, I. E. (1927). *Symbionticism and the Origin of Species*. London: Baillière, Tindall & Cox.

WALTER, M. (1975). *Stromatolites*. Amsterdam: Elsevier (in press).

WHITEHOUSE, H. K. L. (1973). *Towards an Understanding of the Mechanism of Heredity*. New York: St Martin's Press.

WHITTAKER, R. H. (1959). On the broad classification of organisms. *Q. Rev. Biol.*, **34**, 210–226.

     (1969). New concepts of the kingdoms of organisms. *Science, Wash.*, **163**, 150–159.

WILKIE, D. & THOMAS, D. Y. (1973). Mitochondrial genetic analysis by zygote cell lineages in *Saccharomyces cerevisiae*. *Genetics*, **73**, 367–377.

WILSON, E. B. (1925). *The Cell in Development and Heredity*. New York: Macmillan.

PARACOCCUS DENITRIFICANS: A PRESENT-DAY BACTERIUM RESEMBLING THE HYPOTHETICAL FREE-LIVING ANCESTOR OF THE MITOCHONDRION

BY P. JOHN AND F. R. WHATLEY

Botany School, South Parks Road, Oxford OX1 3RA

An attractive feature of the symbiotic theory of the evolutionary origin of the eukaryotic cell is that present-day representatives of the ancestral types called for in the theory can be found (see Margulis, this Symposium). Thus we know of aerobic bacteria (ancestors of mitochondria), blue-green bacteria (ancestors of chloroplasts), and *Pelomyxa palustris* (proto-eukaryote).

Two of the most significant features of present-day mitochondria are the respiratory chain and the ATPase, which together are responsible for oxidative phosphorylation. When these two features are examined in the widest variety of present-day bacteria for which information is available, *Paracoccus* (*Micrococcus*) *denitrificans* shows the greatest similarity to a mitochondrion. Thus the constitutive respiratory chain of *P. denitrificans* is almost identical with the mitochondrial respiratory chain (see John & Whatley, 1975). Similarly, the stoichiometry of oxidative phosphorylation, as indicated by measuring H^+/O ratios, is essentially the same as that of a mitochondrion (Scholes & Mitchell, 1970; Mitchell, 1972). Furthermore, a mitochondrial type of respiratory control has been demonstrated in *P. denitrificans* (John & Hamilton, 1970), though only rarely in other bacteria.

Finally, the cytochrome c_{550} isolated from *P. denitrificans* closely resembles mitochondrial cytochrome c spectrophotometrically, potentiometrically, physiologically (as shown by their cross-reactivity with cytochrome oxidases from mitochondria and *P. denitrificans*, see John & Whatley, 1975), and structurally (as shown by X-ray structure analysis, Timkovich & Dickerson, 1973). While none of these mitochondrial features is unique to *P. denitrificans* no other bacterium has been shown to possess as many mitochondrial features as *P. denitrificans*. However, *Rhodopseudomonas spheroides* when grown aerobically in the dark has a mitochondrial type of respiratory chain (Dutton & Wilson, 1974), but the stoichiometry of oxidative phosphorylation and the presence of respiratory control are not known in *Rps. spheroides*. It is probable that future research will reveal that *P. denitrificans* is not unique in the degree to which it resembles a mitochondrion, and it will then be viewed as a representative of a small group of aerobic bacteria (probably including *Rps. spheroides*) all of which will have an obvious affinity with the mitochondrion.

The essential similarity between *P. denitrificans* and a mitochondrion allows for a relatively simple evolutionary transition from an aerobic bacterium resembling *P. denitrificans* to a mitochondrion, as envisaged by the symbiotic theory. In this transition the ATPase and the constitutive components of the respiratory chain would be retained, the adaptive components of the respiratory chain would be lost, and the transport properties of the bacterial plasma membrane would be modified so that an effective integration of the metabolism of the organelle with that of the surrounding cell was attained. The adaptive components of the respiratory chain of *P. denitrificans* which are absent from a mitochondrion, are readily dispensed with in *P. denitrificans*, e.g. the synthesis of the respiratory nitrate reductase is repressed in the presence of oxygen (Fewson & Nicholas, 1961). The only new component necessary for an evolutionary transition from the plasma membrane of *P. denitrificans* to the inner mitochondrial membrane is the adenine nucleotide carrier, which appears to be absent from all bacteria and present in all mitochondria. Only after the acquisition of this carrier would ATP synthesized by the endosymbiont be available to the host.

REFERENCES

DUTTON, P. L. & WILSON, D. F. (1974). Redox potentiometry in mitochondrial and photosynthetic bioenergetics. *Biochim. biophys. Acta*, **346**, 165–212.

FEWSON, C. A. & NICHOLAS, D. J. D. (1961). Respiratory enzymes in *Micrococcus denitrificans*. *Biochim. biophys. Acta*, **48**, 208–210.

JOHN, P. & HAMILTON, W. A. (1970). Respiratory control in membrane particles from *Micrococcus denitrificans*. *FEBS Lett.*, **10**, 246–248.

JOHN, P. & WHATLEY, F. R. (1975). *Paracoccus denitrificans* Davis (*Micrococcus denitrificans* Beijerinck) as a mitochondrion. *Adv. Bot. Res.*, **4** (in press).

MITCHELL, P. (1972). Structural and functional organisation of energy-transducing membranes and their ion-conducting properties. *FEBS Symp.*, **28**, 353–370.

SCHOLES, P. & MITCHELL, P. (1970). Respiration-driven proton translocation in *Micrococcus denitrificans*. *J. Bioenerget.*, **1**, 309–323.

TIMKOVICH, R. & DICKERSON, R. E. (1973). Recurrence of the cytochrome fold in a nitrate-respiring bacterium. *J. molec. Biol.*, **79**, 39–56.

THE SYMBIONT THAT NEVER WAS:
AN INQUIRY INTO THE EVOLUTIONARY
ORIGIN OF THE MITOCHONDRION

By R. A. RAFF* and H. R. MAHLER†

Department of Zoology* and Department of Chemistry†
Indiana University, Bloomington, Indiana 47401, USA

Across his features and written into the very texture of his bones are the half-effaced signatures of what he has been, of what he is, or of what he may become.

Loren Eiseley,
The Night Country

VALIDITY OF THE MOLECULAR APPROACH

Regardless of the mode of origin of the eukaryotic cell and its organelles, whether by the symbiotic association model or by the plasmid model, both the nuclear–cytoplasmic and nuclear–mitochondrial systems of the eukaryotic cell have undergone over a billion years of co-evolution to their present forms. Because of the extreme antiquity of the events we are attempting to reconstruct, evidence relevant to the problem of the origin of eukaryotic organelles is difficult to obtain and even more difficult to interpret.

The almost complete lack of a fossil record bearing on these events requires us to find our clues by unraveling the nature of the contemporary systems available to us. The sheer mass of morphological, ultrastructural, ecological, taxonomic, biochemical and genetic data available are almost overwhelming and must be culled and sifted through by anyone attempting to study the problem. Various workers who have done this have, not surprisingly, emphasized particular lines of evidence (depending upon their predilections and disciplines) over others.

Unfortunately, this has resulted in a somewhat disjointed debate, since most of the evidence presented in favor of the symbiotic hypothesis is drawn from cytological studies of cells and from comparative studies of living intracellular symbionts, organelles and free-living prokaryotes (e.g. Margulis, 1970; Taylor, 1974), while we have concentrated on the molecular-biological aspects of the problem. While we agree that many lines of evidence have already proved valuable in arriving at an understanding of cellular evolution, we suggest that the appropriate comparative

[41]

molecular studies offer the greatest potential in comprehending the origin of eukaryotic cells and organelles.

Several types of molecular data provide useful evolutionary evidence. These include protein sequences, metabolic pathways, nucleic acid hybridization, and physical parameters of macromolecules and supramolecular structures. The most developed of these for evolutionary studies is the use of protein sequences, and indeed, the potential power of the molecular approach may be illustrated by the construction of phylogenies based upon amino acid sequence comparisons.

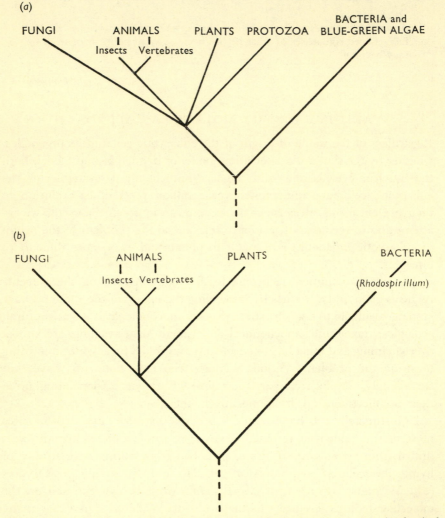

Fig. 1. Classical and molecular phylogenies of the kingdoms. (a) A classical phylogeny based upon Whittaker's (1969) five kingdom scheme. (b) A phylogeny of cytochrome c. See text for details.

A comparison of a classical and a molecular phylogeny of the kingdoms is illustrated in Fig. 1. The classical phylogeny follows the five kingdom scheme of Whittaker (1969). The animal kingdom is further divided into the deuterostome super-phylum and the protostome super-phylum (Barnes, 1963). These are respectively labelled vertebrates and insects for their most familiar representatives. This diagram is based upon the sum of the data conventionally used in construction of phylogenies (morphology, biochemistry, embryology, fossil record, etc.). The molecular phylogeny (for cytochrome c) is based on a phylogeny drawn by Dayhoff (1972) from sequence data. The relationships of the kingdoms and of the animal super-phyla based on protein sequence is essentially the same as that based on conventional data. Note that the cytochrome c phylogeny lacks the protozoa since no protozoan cytochrome c had been sequenced at the time this phylogeny was drawn (Dayhoff, 1972). Since then sequences of cytochrome c from two protozoans, *Crithidia* and *Euglena*, have been published (Pettigrew, 1972, 1973; Lin, Niece & Fitch, 1973). These differ very strikingly from any other eukaryotic cytochromes c, and if they were positioned into the dendrogram they would appear as an early offshoot of the eukaryotic stem.

Three general classes of molecular consequences may be predicted to have arisen from the evolutionary acquisition of an organelle such as the mitochondrion.

First are primitive features characteristic of the ancestral prokaryote that have been retained by the organelle.

Second are features which may have been characterized by rapid rates of evolution early in the history of the organelle and low rates of evolution subsequently. Simpson (1953) has pointed out that evolution may occur at an accelerated pace during evolutionary transitions between major adaptive zones. The origin of eukaryotic cells clearly was such a major evolutionary step which may have required a relatively short span of time. Since the innovations were subcellular and biochemical, this transition may have had a high rate of molecular evolution followed by a slowing of rates. Evidence that such variations occur in molecular evolution is available. For instance, α-lactoglobulin apparently had an increased rate of evolution subsequent to its divergence from lysozyme (Dickerson & Geis, 1969; Dickerson, 1971), and the acquisition of families of redundant nucleotide sequences in eukaryotic genomes may have occurred by saltatory events while subsequent sequence divergence has been gradual (Britten & Kohne, 1970).

Third are features indicating organellar divergence. This type of evolutionary change has been discussed in detail by Lewontin (1970) who has pointed out that the various genomes of a eukaryotic cell represent different units of selection and thus may evolve in divergent fashions.

Examples of all three categories appear to exist in the data we have gathered on mitochondria. Into the first category (retention of prokaryotic characteristics) fall such features as the amino acid sequence of mitochondrial superoxide dismutase and the inhibitor sensitivity spectrum of the mitochondrial ribosome. These features have been considered by some to constitute powerful arguments for the symbiotic theory, but as pointed out by Uzzell & Spolsky (1974) such conservative features only provide confirmation of the ultimate prokaryotic origin of eukaryotic systems and provide no proof of mode of origin. Other characters such as the structure of mitochondrial DNA may be best explained by the plasmid hypothesis since mitochondrial DNA strongly resembles bacterial plasmids.

There are probably a number of mitochondrial features that meet the second category (initial high rate of evolution, followed by low rates). Perhaps the best example is a member of the electron transport chain, cytochrome c, which has evolved slowly in eukaryotes, but very likely had an initial high rate since eukaryotic cytochromes c differ greatly from the prokaryotic species (Dickerson, 1971; Ambler, 1973; Mahler & Raff, 1975).

The third category (divergent mitochondrial evolution) provides some bizarre examples which indicate that very specialized evolutionary trends have occurred in mitochondria – trends that we feel are not consistent with the symbiotic model (see below). Two notable examples are mitochondrial RNA polymerase and the physical characteristics of mitochondrial ribosomes. The mitochondrial polymerases resemble neither nuclear nor bacterial species (which in fact resemble each other in several significant respects). Mitoribosomes, which have been characterized as bacterial in type because of their similar inhibitor sensitivities, differ among themselves and from bacterial ribosomes in size, and differ greatly from bacterial ribosomes in composition of rRNA and in lacking a 5 S RNA (which incidentally is found in cytoribosomes).

The existence of examples of molecular species fitting these categories indicates that mitochondrial evolution has been perhaps more complex and interesting than anyone has suspected.

THE SYMBIOTIC THEORY AND ITS CONSEQUENCES

According to the symbiotic theory of the origin of the eukaryotic cell, as the primitive anoxygenic atmosphere of the earth began to acquire free oxygen as a result of photosynthesis, prokaryotes that had utilized a wide variety of pathways for anaerobic energy metabolism were forced either to adapt to aerobic conditions or to become restricted to the limited anaerobic environment remaining. Prokaryotes exhibit a variety of anaerobic energy-generating pathways, and there is a diversity of anaerobic prokaryotes. Eukaryotes, however, possess only glycolysis as a pathway of anaerobic energy metabolism (Stanier, 1970): the ancestral proto-eukaryote thus likewise utilized glycolysis.

This proto-eukaryote evolved several adaptations that allowed it to escape from the selective pressure of free oxygen which was the determinant driving the evolution of advanced oxidative metabolic pathways in contemporary prokaryotes. By such innovations as larger cell size, intracellular translocation, advanced mechanisms of cell motility, and the ability to phagocytose, the proto-eukaryote became able to ingest prokaryotes as prey to provide substrates for glycolysis. Dependent upon and subsequent to these advances was the establishment of stable intracellular symbiotic relationships between the proto-eukaryote and certain ingested aerobic prokaryotes. Thus, the final step in the origin of the eukaryotic cell was the acquisition of oxygen-mediation mechanisms (photosynthesis and respiration) by discrete quantum steps. Symbiotic relationships involving prokaryotic symbionts housed in a eukaryotic cytoplasm exist among present-day organisms (Stanier, 1970; Preer, 1971). Interestingly, however, mutualistic endosymbiotic relationships in which the host is a prokaryote have apparently not been observed (Stanier, 1970). The symbiotic theory further requires that in the course of time the symbiotic association has become intimate to the point that most of the genetic information required for assembly of the organelle-symbiont has been transferred to the nuclear genome. Concomitantly the informational content of the organellar genome has been greatly reduced.

Two aspects of this hypothesis seem awkward. The model suggests that the proto-eukaryote possessed many advanced cellular adaptations, yet was primitive and inefficient metabolically. In the face of competition from other prokaryotic organisms possessing more efficient aerobic energy-yielding pathways foreshadowing the present-day patterns, such an organism should have found itself at a considerable disadvantage. Secondly, the integration of the symbiont to the extent of limited autonomy observed in the mitochondrion would have required a wholesale transfer of genes from the endosymbiont genome to the unrelated nuclear genome of the host. No mechanism has been put forward to account for such a massive transfer.

Unfortunately these aspects of the theory are not amenable to experimental investigation. However, other consequences emerge from this model for which pertinent experimental data do exist. These are outlined below, and will be discussed in detail in subsequent sections of this paper.

(1) By the symbiotic model, the eukaryotic cytoplasm should show evidence of a fundamentally anaerobic nature, since the anaerobic proto-eukaryote acquired its oxygen-mediating systems from the aerobic symbiont.

With the exception of some secondarily evolved anaerobic eukaryotic organisms that are ultimately dependent upon molecular oxygen, all eukaryotes are aerobes and utilize the efficient energy yielding pathways of the mitochondrion. Further, as will be substantiated below, all eukaryotes appear to be fundamentally aerobic not only in the possession of oxygen-detoxifying enzymes and mitochondria, but also in many details of biosynthesis, in which several important cellular components synthesized by anaerobic pathways in bacteria are synthesized in eukaryotes by reactions utilizing molecular oxygen. Thus, eukaryotic cells probably did not arise from their prokaryotic progenitors until after the atmosphere contained appreciable free oxygen. These ancestors were probably already adapted to the use of oxygen.

(2) According to the symbiotic model the limited organellar genomes of contemporary organisms as well as their protein synthetic systems are evolutionary relicts. We shall show that in fact these supposedly relict systems are of vital importance to organellar organization and function, and that the origin of organellar genetic systems probably lies in the necessity for in-situ protein synthesis on the inner membrane of a topologically closed organelle.

(3) A number of properties of present-day organelles are similar to those of prokaryotes: such characteristics are interpreted as being simply relicts in nature and have been taken to provide strong support for a symbiotic origin. We shall show that while many aspects of the mitochondrion are conservative, others are not, so that the degree of homology between prokaryotes and mitochondria is easily overstated.

AEROBIC ADAPTATIONS OF THE EUKARYOTIC CELL CYTOPLASM AND SURVIVAL OF PRIMITIVE FUNCTIONS

Several lines of evidence indicate that the eukaryotic cell cytoplasm is not simply an anaerobic system containing an aerobic respiratory organelle, but that in fact it possesses ancient and fundamental adaptations to an aerobic environment.

Enzymatic protection from oxygen toxicity

Some extremely interesting recent studies have dealt with the mechanism by which contemporary cells, both pro- and eukaryotic, protect themselves against the potential depradations of $[O_2 \cdot]^-$, the superoxide radical, generated in a variety of enzymatic and non-enzymatic reactions that utilize molecular oxygen. The reaction catalyzed is

$$2[O_2]^- + 2H^+ \rightleftharpoons H_2O_2 + O_2.$$

This scavenging of the extremely reactive radical – and of singlet oxygen – appears to constitute an essential defense mechanism evolved to permit survival of the aerobic cell in its intrinsically toxic environment (McCord, Keele & Fridovich, 1971; Fridovich, 1972; Lavelle, Michelson & Dimitrijevic, 1973).

The dismutase, and catalase, which destroys H_2O_2, are apparently ubiquitous in aerobes, and appear to be vital to the existence of organisms living in the presence of oxygen. This view is supported by the data of McCord *et al.* (1971), who assayed a number of species of prokaryotic aerobes, strict anaerobes and aerotolerant anaerobes for superoxide dismutase and catalase. All aerobes contained both enzymes, strict anaerobes contained neither, while aerotolerant anaerobes contained superoxide dismutase and lacked catalase. Yeast (a eukaryote) was found to contain both enzymes. On the basis of the evidence just cited, the speculation that the enzymes must have arisen quite early during evolutionary development, co-incident perhaps with the first organisms capable of producing oxygen, appears well founded. Bacteria, such as *Escherichia coli*, contain two types of enzyme; one, with iron at the active site, is localized in the periplasmic space and serves primarily as a defense against superoxide in the environment; the other, a mangano-zinc enzyme localized in the intracellular matrix, functions in the protection of the contents of this compartment against the superoxide generated within (Gregory, Yost & Fridovich 1973). Eukaryotic cells also contain two forms of the enzyme. One, a cupro-zinc enzyme (two atoms each per 33 000 daltons), is localized in the cell sap and in the intermembrane space of mitochondria (Weisiger & Fridovich, 1973*a*). The other enzyme resembles one of the bacterial enzymes in its metal co-factor and subunit mass (one Mn per 20 000 daltons) though not in the molecular weight (tetramers of molecular weight 80 000 versus dimers of molecular weight 40 000). Its localization is within the mitochondrial matrix, and like all other proteins found in this compartment is specified by nuclear genes and synthesized outside the mitochondria (Weisiger & Fridovich, 1973*b*).

The N-terminal sequence of four of these proteins have recently been obtained by Steinman & Hill (1973). Their results show a striking sequence homology between the two bacterial enzymes and that of chick liver mitochondria, and a complete lack thereof between the mitochondrial enzyme and the protein from beef erythrocyte cytoplasm. Therefore, the first three show a remarkably low degree of evolutionary divergence in this region of their protein structure, as a result, presumably, of constraints inherent in their function and/or localization. Whether similar constraints are operative for the remainder of the sequence in these molecules and for those of the other (Cu-Zn) class of dismutases can only be revealed by future studies. In any event, the close homology of the bacterial and mitochondrial classes of proteins argues in favor of a common ancestry within an early *aerobic* proto-eukaryote – but, other interpretations to the contrary (Weisiger & Fridovich, 1973*a*, *b*; Fridovich, 1974), cannot be used to test the validity of the two alternate theories. It represents the survival of an essential primitive function (Uzzell & Spolsky, 1974) subject to ponderous evolutionary constraints. In this respect, as well as in its small size and lack of membrane attachment, and its continued specification by the principal genome of contemporary eukaryotes, it bears a great deal of resemblance to other essential entities required for aerobic function such as cytochrome *c*. The cytosolic dismutases evidently have had a long, divergent evolutionary history from both the prokaryotic and the highly conservative mitochondrial species. Isolated from such diverse eukaryotes as fungi, plants and mammals they are in all cases cupro-zinc proteins. Unfortunately, not enough sequence data exists on these similar (though apparently not identical) enzymes to trace their evolutionary relationships to each other and to the prokaryotic mangano-enzyme (Steinman & Hill, 1973). Nevertheless, the evidence on hand strongly suggests that the basic cupro-zinc enzyme of the eukaryotic cell cytoplasm arose earlier in the history of the eukaryotic cell than the origin of the eukaryotic kingdoms, and thus probably represents an ancient adaptation of early eukaryotes to the presence of oxygen in the Precambrian atmosphere.

Catalase, the other major oxygen-detoxifying enzyme of eukaryotes, and various oxidases are packaged in specialized organelles called peroxysomes (De Duve & Baudhuin, 1966; De Duve, 1969; Avers, 1971). The obvious role for the catalase of these particles is in disposal of hydrogen peroxide. The oxidases and catalase of the peroxysome may also represent a primitive respiratory pathway allowing the cell to regenerate NAD^+ from $NADH + H^+$ and thus utilize non-fermentable substrates such as glycerol. Peroxysomes are found throughout the eukaryotic kingdoms in such evolutionary diverse organisms as *Tetrahymena*, yeasts, plants and mammals, and thus may, as originally pointed out by De Duve & Baudhuin (1966), represent an organelle evolved by primitive eukaryotes for protection from oxygen.

Metabolic requirements of anaerobic eukaryotes

Anaerobic prokaryotes exist in considerable diversity, but there are very few anaerobic eukaryotes. The largest group of anaerobic eukaryotes consists of the flagellated protozoa inhabiting the anaerobic environments of the intestinal tracts of animals. Examples of these are the rumen protozoa of cattle (Hungate, 1967) and the trichomonads which lack mitochondria (for example see Nielsen, Ludvik & Nielsen, 1966). There are also free-living examples such as the facultative anaerobic fungus *Aqualinderella* which is an obligate fermenter living upon submerged fruit in stagnant water (Emerson & Held, 1969). Stanier (1970) suggests that these organisms are not primitively anaerobic, but represent secondary adaptations to specialized niches. This is certainly the case for facultatively anaerobic animals, some of which are capable of surviving and producing ATP for prolonged periods under anaerobic conditions.

An example of a well-known facultative eukaryotic anaerobe is yeast, which can be cultured indefinitely under anaerobic conditions but only if these cells are provided with oleate and a steroid (Andreason & Stier, 1953, 1954). Oleate and steroids require the presence of oxygen for their biosynthesis and are synthesized by yeast cultured aerobically (Yuan & Bloch, 1961). Thus yeast, a primitive eukaryote capable of anaerobic growth, ultimately has an absolute requirement for oxygen (Keith, Wisnieski, Henry & Williams, 1972). The various other anaerobic eukaryotes discussed above probably have similar requirements which are met by their close association with aerobic eukaryotes. For example, *Aqualinderella*, like yeast, requires an unsaturated fatty acid and a sterol when cultured on artificial medium (Held, 1970).

Oxygen and biosynthetic patterns

Bloch (1962) and Goldfine & Bloch (1963) have drawn two significant generalizations from their detailed discussions of the relationship of molecular oxygen (O_2) to biosynthetic patterns. First, universal cellular components are not invariably synthesized by the same pathway in all cells. A number are synthesized by anaerobic pathways in some organisms and by aerobic pathways in others. Second, there are compounds for which there are only aerobic pathways: such compounds are consequently confined to aerobic organisms. Thus, there exist metabolic specializations superimposed upon the ancient anaerobic metabolic schemes evolved during the Precambrian prior to the existence of free oxygen in the atmosphere.

Alternative aerobic and anaerobic pathways exist for mono-unsaturated fatty acids, nicotinic acid, tyrosine, carotenoids and porphyrines. There are no anaerobic pathways at all for steroids and polyunsaturated fatty acids.

The mechanisms of synthesis of unsaturated fatty acids and steroids are of particular significance to our postulate that the eukaryotic cell is primitively adapted to the utilization of molecular oxygen. Eukaryotes and some advanced prokaryotes use the aerobic pathway for the synthesis of mono-unsaturated fatty acids. Eubacteria, whether aerobic or anaerobic, utilize an entirely different, anaerobic pathway (Bloch *et al.*, 1961; Bloomfield & Bloch, 1960). The anaerobic pathway for unsaturated fatty acids in eubacteria involves a modification of the elongation pathway as outlined in the reactions below for the synthesis of vaccenic acid (Bloch, 1969).

$$CH_3\text{—}(CH_2)_5\text{—}CH_2\text{—}COSR \xrightarrow{+C_2}$$

$$CH_3\text{—}(CH_2)_5\text{—}CH_2\overset{\overset{\displaystyle O}{\|}}{\text{—}C}\text{—}CH_2\text{—}COSR \xrightarrow{\text{Reduction}}$$

$$CH_3\text{—}(CH_2)_5\text{—}CH_2\overset{\overset{\displaystyle OH}{|}}{\underset{\underset{\displaystyle H}{|}}{\text{—}C}}\text{—}CH_2\text{—}COSR \xrightarrow{\text{Dehydration}}$$

$$CH_3\text{—}(CH_2)_5\text{—}CH = CH\text{—}CH_2\text{—}COSR \xrightarrow{\text{Further elongation}}$$

$$CH_3\text{—}(CH_2)_5\text{—}CH = CH\text{—}(CH_2)_9\text{—}COOH$$

By contrast the aerobic pathway for production of unsaturated fatty acids does not involve any modification of the elongation reactions, but begins with a saturated long-chain fatty acid. The essential reactions may be summarized as below for the synthesis of oleic acid.

$$CH_3\text{—}(CH_2)_7\text{—}CH_2\text{—}CH_2\text{—}(CH_2)_7\text{—}COSR \xrightarrow{O_2,\ NADPH}$$

$$\text{[oxy-derivative (exact structure unknown)]} \longrightarrow$$

$$CH_3\text{—}(CH_2)_7\text{—}CH = CH\text{—}(CH_2)_7\text{—}COSR$$

In eukaryotes this reaction occurs in the endoplasmic reticulum.

Bloch (1962) proposed that the change in pathway occurred during the evolution of advanced prokaryotes and was retained by the ancestral eukaryote. The proposed selective advantage was that aerobically synthesized fatty acids such as oleic acid serve as substrates for the production of certain polyunsaturated fatty acids such as linoleic and more highly unsaturated species. Polyunsaturated fatty acids are absent in both aerobic and anaerobic bacteria, but are universal in eukaryotes.

The final stage of steroid synthesis, the cyclization of squalene to lanosterol, similarly requires molecular oxygen and is localized in the microsomal membranes of eukaryotic cells (Tchen & Bloch, 1957; Willett *et al.*, 1967; Scallen, Dean & Schuster, 1968; Hayaishi & Nozaki, 1969).

Steroids, which are ubiquitous in eukaryotes, have been recently reported from prokaryotes as well (summarized by Bird *et al.*, 1971). In most prokaryotes in which they have been detected, they have been found in minute amounts, so that contamination may account for some of the reports. However, *Methylococcus* has been reported to contain amounts of steroids comparable to eukaryotes (Bird *et al.*, 1971). The lack of an anaerobic pathway for steroid synthesis as well as their universal occurrence in eukaryotes suggests that this pathway evolved early in eukaryotic cellular evolution.

Electron transport in bacteria

In bacteria, membrane-associated electron transport appears to have evolved quite early and thus may well represent an exceedingly primitive, perhaps even primordial function, already present in quite early anaerobic forms.

Bacterial cytochromes are more varied than are the mitochondrial cytochromes (for example, c-type cytochromes include not only c and c_1, but also c', c_2, c_3, c_4 and c_5), and there are several terminal oxidases [for example, a_1, a_2 (now called d), and o, as well as aa_3, the oxidase of mitochondria]. Some bacteria have only one oxidase, others have two or three (Gel'man, Lukoyanova & Ostrovskii, 1967). Multiple oxidases are particularly prevalent in cells capable of adapting to different oxidants. Mitochondrial cytochrome c interacts very poorly with most bacterial cytochrome oxidases and vice versa. Isolated bacterial c-type cytochromes are different from mammalian cytochrome c in primary sequence and in such properties as isoelectric point and redox potential, although they have a common prosthetic group (Kamen & Horio, 1970; Gel'man *et al.*, 1967; Smith, 1961). In common with mitochondrial cytochromes, bacterial cytochromes are membrane bound, and cytochrome c is the most readily extracted cytochrome (Gel'man *et al.*, 1967).

Also significant is the widespread occurrence of cytochromes in anaerobic bacteria in which they function in electron transport between organic substrates or molecular hydrogen and a variety of inorganic oxidants (Horio & Kamen, 1970; Newton & Kamen, 1961).

Certain contemporary, obligately aerobic, eubacteria such as *Micro-bacterium phlei* can lay claim to an electron transport system which in all relevant particulars is a virtual copy of that found in mitochondria (Brodie & Gutnick, 1971; Harold, 1972). This includes not only all the essential catalytic components, especially an oxidase containing cytochromes aa_3, but even such sophisticated features as multiple forms of cytochrome *b* and a variety of branch points. The tightly membrane-integrated cyto-chromes, which in mitochondria contain known gene products of the organelle, would therefore appear to be the most appropriate objects for further studies of possible evolutionary relationships.

Unless we entertain the possibility of parallel evolution, the facts just presented would argue for the likelihood that similar electron transport chains might already have been present in the postulated common ancestor of these bacteria and of mitochondria.

Non-mitochondrial electron transport systems of eukaryotes: mixed function oxidases

The most commonly considered electron transport chains are those of the respiratory organelle – the mitochondrion of eukaryotes or the cell mem-brane of prokaryotes. Nearly all present-day organisms (with the exception of the clostridia and lactobacilli) contain cytochromes (Horio & Kamen, 1970). This suggests that rather than being the exception among Pre-cambrian prokaryotes, cytochrome electron transport chains were probably the rule. When oxygen began to become available as an electron sink many organisms were able to modify their electron transport system to utilize oxygen as a terminal acceptor. We assume this process occurred in the cells ancestral to eukaryotes as well as in other phylogenies.

Eukaryotes and prokaryotes, however, are not limited to the cytochrome chain outlined above for electron transfer processes terminating with molecular oxygen. Both also utilize mixed function oxidases: these are largely microsomal in their location in the eukaryotic cell. These reactions can be formulated as (Ullrich, 1972)

$$RH + DoH_2 + O_2 \rightarrow ROH + Do + H_2O.$$

RH is the substrate and DoH_2 is a hydrogen donor, generally reduced nicotinamide adenine dinucleotide phosphate ($NADPH_2$). Oxygenases serve to introduce hydroxyl groups into hydrophobic compounds, and are involved in steroid and fatty acid metabolism as well as in the synthesis of such amino acids as tyrosine and hydroxproline, and in the degradation of hydrocarbons. The electron transport chains associated with such reactions vary in complexity, but all serve to carry electrons from NADPH or NADH via a flavoprotein to cytochrome P_{450} which interacts with molecular oxygen.

Another system involving a somewhat different set of components is involved in fatty acid desaturation (Oshino, Imai & Sato, 1971; Holloway & Wakil, 1970). Cytochrome b_5 functions in the desaturation scheme, but apparently does not function in most mixed function chains (Hrycay & O'Brien, 1974; Levin, Ryan, West & Lu, 1974).

Two of the components of the microsomal electron transport chains are cytochromes (cytochromes P_{450} and b_5) and thus of considerable evolutionary interest since all other eukaryotic cytochromes are located in the mitochondrial inner membrane.

Cytochrome P_{450} has been recently purified and crystallized from the bacterium *Pseudomonas putida* (Yu *et al.*, 1974), and partially purified from rat liver microsomes (Levin *et al.*, 1974). Cytochromes P_{450} from both sources have similar molecular weights 45 000–50 000. Unfortunately no sequence data is yet available for these proteins.

Cytochrome b_5 has been purified from mammalian liver microsomes (Ozols & Strittmatter, 1967; Nobrega & Ozols, 1971; Ozols, 1974), and the complete sequence of the calf liver protein has been determined. The sequence bears no resemblance to cytochrome c, but does have some homology to the α and β chains of hemoglobin, as well as to cytochrome b_2, a component of a *mitochondrial* enzyme (L-lactate dehydrogenase) in yeast.

Recently (Guiard & Lederer, 1973) an 80 amino acid long tryptic peptide of mitochondrial cytochrome b_2 bearing the heme binding site has been sequenced. This sequence is reported to show striking homology with the same region of microsomal cytochrome b_5, and the folding of the polypeptide backbone around the heme is claimed to be identical in these two cytochromes. Thus microsomal cytochrome b_5, which functions in reactions that appear to be primitively fundamental to eukaryotes, is apparently evolutionarily related to other non-mitochondrial proteins, the globins, and to a mitochondrial cytochrome, cytochrome b_2. We feel such a relationship is best explained by assuming that the proto-eukaryote carried the gene for a cytochrome ancestral to both the contemporary cytochrome b_2 and b_5 species. As both mitochondrial and non-mitochondrial electron transport systems evolved, the original gene may have duplicated and given rise to a pair of related genes which were subject to divergent evolution to produce the specialized mitochondrial and microsomal species.

The most significant point to be made from the available data is that these complex electron transport systems, which allow the cell to use molecular oxygen as a direct reactant in several metabolic schemes, are a part of the eukaryotic microsomal membrane systems or in some cases the outer mitochondrial membrane, which is closely related to and may have originated from the membrane of the endoplasmic reticulum. These systems are not confined to eukaryotes. We propose that these were also a part of the biochemical armamentarium of the ancestral proto-eukaryote which had thus both a respiratory chain (destined for enclosure in the mitochondrion) and a system for mixed function oxidase reactions which remained extra-mitochondrial.

Ferredoxins and iron–sulfur centers

Ferredoxins are members of a group of soluble, or easily solubilized, proteins of relatively modest size, active in electron transport by virtue of iron atoms co-ordinated to an unusual constellation of four cysteine residues by means of inorganic sulfur. The sites are called iron–sulfur centers (ISCs); proteins containing ISCs are called iron–sulfur proteins (ISPs).

Non-heme iron proteins have been reported to be present in sources as diverse as *Clostridium pasteurianum*, *Rhodospirillum rubrum*, *Pseudomonas putida*, blue-green algae, and the chloroplasts of all plants examined. Similar entities are present as well in the microsomal fraction of most animal cells and the mitochondria of adrenal glands. Their molecular weights equal 10 000–12 000 (occasionally 24 000, probably as the consequence of gene duplication) and they contain two ISCs, resulting in highly characteristic spectroscopic and magnetic properties.

Sequence data provided by Professor I. C. Gunsalus indicate that *putidaredoxin* from the aerobic eubacterium *Pseudomonas putida*, and the *adrenodoxin* from adrenal mitochondria show significant homology ($\sim$35 %). It is likely that this last species, which is probably localized in the outer mitochondrial membrane and constitutes an essential component of the hydroxylation (mixed function oxidase) system in this tissue, constitutes a representative example of other ISPs involved in analogous reactions by other metazoan cells. As mentioned earlier, generally these systems are found associated with the endoplasmic reticulum and are concentrated in the microsomal fraction after cell fractionation. Clearly, the homologies between the bacterial and the mammalian ISPs are at least as striking and significant as those previously established between bacteria and plants and even between photosynthetic bacteria and blue-green algae (see also Table 10.5 of Dayhoff, 1972).

MITOCHONDRIAL GENE EXPRESSION IS DIVERGENT FROM THAT OF PROKARYOTES

One of the more attractive features of the endosymbiont theory, and indeed the one that propelled it into the recent scientific consciousness and compelled its almost universal acceptance, is directly related to the discovery that all mitochondria contain their own genetic system. This is composed of a specific mitochondrial (mt) DNA as well as the means for its replication and, at least potentially, expression. Thus, the organelles possess a DNA-dependent RNA polymerase, and a system for protein synthesis, consisting of ribosomes, mRNA(s), tRNAs, aminoacyl tRNA ligases and the initiation, elongation and termination factors required for its function. The case for the genetic autonomy of mitochondria was strengthened by the conclusive demonstration that at least in organisms such as *Saccharomyces* and *Neurospora*, for which a pattern of extra-chromosomal, non-Mendelian inheritance had already been established, mtDNA could be identified as the genophore responsible. Related to this was the discovery that mitochondrial genomes were capable of undergoing not only mutation but also repair and mutual exchanges (recombination). However, the strongest inferences were based on the contention that all these mitochondrial entities, events and processes appeared not only to be conserved throughout the whole eukaryotic realm, from the most primitive to the most complex species, but also to be essentially bacteria-like in their properties. However, the recent rapid expansion of factual material has made possible a more critical re-evaluation of these two propositions. As will be shown below, neither of them appears tenable without modifications so profound as to make an alternative explanation, i.e. divergent evolution of the two genetic systems of the eukaryotic cell, the nucleo-cytosolic and the mitochondrial, at least equally plausible.

The mitochondrial genome

Mitochondrial DNA

Mitochondrial DNA (mtDNA) has been subject to considerable evolution, not only as concerns its size, but also its base composition. The implication is, therefore, that there have been corresponding evolutionary changes in the encoded sequences of proximal (RNA) and distal (protein) gene products. Three observations stand out: (i) The buoyant densities of mtDNA encompass a range of 1.684 to 1.715 g cm^{-3} – equivalent to the total span between about 20% and 80% G + C found in prokaryotes. (ii) While bacterial genomes vary in molecular weight between 2×10^8

$(3 \times 10^5$ nucleotide pairs) for the smallest, such as *Mycoplasma gallosepti-cum*, and 26×10^8 $(40 \times 10^5$ nucleotide pairs) for the largest, such as *E. coli*, even the largest mt genome probably does not exceed a value of 6×10^7 (or 7.5×10^4 nucleotide pairs). However, this is a size quite consistent with that of many of the larger bacterial plasmids (Table 1). Furthermore, there exists a class of small plasmids virtually identical in size to the mtDNA of animals (5×10^6). Related to these observations is the apparent complete lack of redundancy and gene duplication in 'normal' mito-chondrial genomes. (iii) At least for metazoa, nuclear and mitochondrial genomes appear to have become evolutionarily stabilized co-ordinately and convergently at a density value close to 1.700 (40 % G + C) for both, and a mass of the order of 10×10^6 daltons $(1.2 \times 10^4$ nucleotide pairs) for the mitochondrial component.

Similarities of mitochondrial DNA and plasmids

Let us briefly review and expand on the properties shared by mitochondrial and plasmid DNA (Tables 1, 2 and Meyer, 1973). These are (i) size (mass from $\sim 5-\sim 50 \times 10^6$; length $\sim 1-10$ μm); (ii) configuration: doubly covalently linked, supercoiled circles, easily transformed from this into relaxed circles, covalently linked in only one strand; (iii) arrest of replica-tion and resultant elimination (curing) by specific intercalating dyes in the acridinium and phenanthridinium series; (iv) ready accumulation of replicating forms; some, such as multiple catenanes, potentially of an

Table 1. *Properties of some mitochondrial and plasmid DNAs*

Type of DNA	Molecular weight ($\times 10^6$)	Circularity	Contour length (μm)
Plasmids			
Fertility factors			
F'	75	+	n.d.
F'Gal	51	+	n.d.
Golicinogenic factors			
E1	4	+	2.31
R1	10.6	+	5.5
	65	+	33
Mitochondrial DNAs			
Yeast	50	+	25
Sea urchin	10.4	+	5.4
Xenopus (Frog)	11.7	+	5.7
Human	9.6	+	5.0

Data from the compilations of Mahler (1973); Meyer (1973); Mahler & Raff (1975) and from Hartman, Ziegler & Comb (1971): n.d., not determined.

Table 2. *Characteristics of plasmids*

Characteristic	Observations
Molecular weights	1.5×10^6 to 1.0×10^8
Conformation	Circular duplex, supercoiled
Distribution	Eubacteria, photosynthetic bacteria
Amount of plasmid DNA per cell	Varies with plasmid and host; may be 1 to 30 copies per cell or up to 40 % of total cell DNA
Replication	Probable specific replication points on membrane; plasmid stability mutants linked with both plasmid and chromosome
Genomic content of various plasmids	Replication genes, sex factor, colicinogenic factors, variety of genes for antibiotic resistance, various genetic markers excised from bacterial chromosome (*lac, gal, tryp, cysB*)
Effects of acriflavine and other acridines, ethidium bromide	Plasmid eliminated from host

See Raff & Mahler (1972) for references.

aberrant variety; (v) replication requiring tight integration with, and potential regulation by, an attachment site on the appropriate membrane; (vi) presence of ribonucleotide tracts in the mature, vegetative form of the genophore (Williams, Boyer & Helinski, 1973; Wong-Staal, Mendelsohn & Goulian, 1973; Grossman, Watson & Vinograd, 1973); (vii) actual (plasmids) or potential (mtDNA) integration into the principal cellular genophore; (viii) partial or complete dissociation of replication of plasmid and mtDNA from the DNA of the principal genophore by the application of inhibitors of protein synthesis (Blamire, Cryer, Finkelstein & Marmur, 1972; Goebel, 1973; Kline, 1974) or temperature-sensitive mutations (Cottrell, Rabinowitz & Getz, 1973; Goebel & Schrempf, 1973); (ix) ability to confer antibiotic resistance to cells harboring an antibiotic resistant form of the subsidiary genome.

DNA-dependent RNA polymerases

All gene products, whether eventually translated into polypeptides or not, are in the first instance transcripts of DNA. We therefore turn our attention first to enzymes and processes responsible for transcription in general and in mitochondria in particular.

Bacterial RNA polymerases

The RNA polymerases of bacteria are large, multisubunit enzymes with molecular weights in the range of 400 000 to 500 000. The holoenzyme (*E. coli*) consists of four major subunits (α_2, β', β and σ), and frequently a fifth, ω (Burgess, 1971; Chamberlin, 1974).

The subunit structure of other prokaryotic RNA polymerases are similar with some variation in molecular weights. Properties of the polymerases are presented in Table 3.

The bacterial enzymes are strongly inhibited by the antibiotics strepto-lydigin, streptovaricin and rifampicin or the related compound rifamycin SV (rifampin) at concentrations of about one to two μg cm^{-3} or less (Hartmann, Honikel, Knüsel & Nüesch, 1967; Mizuno, Yamazaki, Nitta & Umezawa, 1968). These compounds bind to the β' subunit and block the initiation of transcription but not elongation (Hinkle, Mangel & Chamberlin, 1972).

Eukaryotic transcriptional systems

Eukaryotic cells contain several transcriptional systems that can be discriminated *in vivo* by their differential sensitivities to various inhibitors. The nucleolar system which produces rRNA is in most cells sensitive to low concentrations (< 1 μg cm^{-3}) of actinomycin D (Penman, Vesco & Penman, 1968). The nucleoplasmic system which produces heterogeneous nuclear RNA (HnRNA) and mRNA is sensitive to camptothecin (Perlman, Abelson & Penman, 1973), while the mitochondrial system is sensitive to ethidium bromide (Penman *et al.*, 1968; Zylber, Vesco & Penman, 1969; Mahler & Perlman, 1971; Mahler & Dawidowicz, 1973; Meyer, 1973).

The basis for these differential sensitivities has not yet been completely elucidated, but it has become clear that each system has its own species of RNA polymerase.

Eukaryotic nuclear polymerases

Eukaryotic nuclei (of animals, plants, fungi and protozoa) contain three or more RNA polymerase species (Lindell *et al.*, 1970; Kedinger *et al.*, 1970; Jacob, Sajdal & Munro, 1970; Roeder & Rutter, 1969, 1970*a, b*; Kedinger *et al.*, 1972; Mullinix, Strain & Bogorad, 1973; Ponta, Ponta & Wintersberger, 1972; Roeder, 1974). The major species are enzyme I or A, which is nucleolar in localization and functions in the synthesis of rRNA, and polymerase II or B, which is located in the nucleoplasm and apparently synthesizes mRNA. This enzyme, but not the nucleolar species, is inhibited by α-amanitin (Lindell *et al.*, 1970; Kedinger *et al.*, 1972). These enzymes are not inhibited by rifampicin, though derivatives of rifamycin bearing complex substituents do act as inhibitors in high concentrations (Adman, Schultz & Hall, 1972; DiMauro, Hollenberg & Hall, 1972; Meilhac, Tysper & Chambon, 1972).

These derivatives also inhibit *E. coli* RNA polymerase. Some such as AF/013 inhibit at very low concentrations (65 % inhibition at 0.4 μg cm^{-3}).

Table 3. *Subunit properties of purified DNA-dependent RNA polymerases*

Polymerase	Subunits MW	Molar ratios	Inhibitors — Sensitive to (>50% inhibition at given concentration)	Inhibitors — Insensitive to
Bacterial (*E. coli*)	β' 150 000–160 000	1	Rifampicin ⎫	α-amanitin
	β 145 000–150 000	1	Rifamycin SF ⎬ <0.5 μg cm^{-3}	
	σ 85 000– 90 000	1	Streptovaricin ⎪	
	α 40 000	2	AF/013 ⎭	
	ω 9 000– 12 000	2		
Eukaryotic				
Type I (A) (calf thymus nucleolar)	200 000	1	AF/013 12 μg cm^{-3}	Rifampicin
	126 000	1		Rifamycin SV
	51 000	1		Streptovaricin
	44 000	1		α-amanitin
	25 000	2		
	16 000	2		
Type II (B) (calf thymus nucleoplasmic)	∼200 000	1	α-amanitin 0.04 μg cm^{-3}	Rifampicin
	140 000	1	AF/013 12 μg cm^{-3}	Rifamycin SV
	34 000	1–2		Streptovaricin
	25 000	∼2		
	16 000	3–4		
Mitochondrial (*Neurospora*)	65 000		Rifampicin <6 μg cm^{-3}	α-amanitin
(yeast)	61 000		Rifampicin 38 μg cm^{-3} (crude enzyme)	Rifampicin (purified enzyme)
(*Xenopus*)	46 000		AF'013 20 μg cm^{-3}	Rifampicin
				Rifamycin SV
				α-amanitin
(rat)	65 000		Rifampicin < 10 μg cm^{-3}	α-amanitin
Chloroplast (maize)	220 000			α-amanitin
	150 000			Rifamycin SV
	several smaller proteins			

Comparable inhibition of animal RNA polymerase by AF/013 requires a concentration of greater than 12 μg cm^{-3} (Meilhac *et al.*, 1972). As in the case of inhibition of bacterial RNA polymerase by rifampicin, AF/013 inhibits initiation but not elongation by animal polymerase. However, the modes of action may not be identical since AF/013 inhibits primary binding of animal but not *E. coli* RNA polymerase to DNA (Meilhac *et al.*, 1972).

Nuclear polymerases resemble prokaryotic polymerases in being high molecular weight (380 000–450 000), multisubunit enzymes (Weaver, Blatti & Rutter, 1971; Gissinger & Chambon, 1972; Kedinger & Chambon, 1972; Ponta *et al.*, 1972; Mullinix *et al.*, 1973; Dezelee & Sentenac, 1973). The number and sizes of subunits have not yet been unequivocally decided, though there appear to be two large subunits present in 1:1 molar ratio and two to four small subunits. The subunit properties of the relatively well-characterized calf thymus enzymes are presented in Table 3.

These subunits are very similar to those of bacterial polymerase in their sizes, though it is not yet known if the subunits are functionally homologous.

Mitochondrial polymerase

Mitochondria purified from a variety of organisms have been found to contain RNA polymerase activity (Mahler, 1973). Highly purified polymerase has been prepared from the mitochondria of *Neurospora* (Küntzel & Schafer, 1971) and *Xenopus* (Wu & Dawid, 1972). These enzymes have about 5 % of the activity of purified *E. coli* polymerase. Both are single polypeptides of similar molecular weight (64 000 for *Neurospora* polymerase and 46 000 for *Xenopus*) which aggregate readily at low salt concentrations. Poly [d(A–T)] is the best template for both enzymes, with mitochondrial DNA about half as active. The native, covalently-closed, circular form of mitochondrial DNA is a better template than its nicked form (Wu & Dawid, 1972). The *Xenopus* enzyme is active with calf thymus DNA while the *Neurospora* enzyme is not. Both polymerases are insensitive to α-amanitin. However, there are significant differences in the sensitivities of these enzymes to rifampicin. The *Neurospora* polymerase is inhibited by this compound (6 μg cm^{-3}) while the *Xenopus* enzyme is unaffected by it at concentrations as high as 100 μgcm^{-3}, but is completely inhibited by more complex derivatives such as AF/013 (30 μg cm^{-3}). Wu & Dawid have also reported that the *Xenopus* enzyme is not stimulated by Mn^{2+} ions and is inhibited by salt concentrations above 0.05 mol l^{-1}. These characteristics set this polymerase well apart from the polymerases found either in *Xenopus* nuclei or bacteria.

Scragg (1971) has studied crude RNA polymerase from yeast mitochondria. Molecular weight estimates suggest that the main peak of activity has a molecular weight of 200 000. This apparently represents aggregates of a 61 000 molecular weight entity (Scragg, 1974). High concentrations of rifampicin are required for significant inhibition of the crude enzyme; the purified enzyme is apparently completely resistant.

Tsai, Michaelis & Criddle (1971) have also studied yeast mitochondrial polymerase which they find to be resistant to rifampin, streptovaricin and α-amanitin. Reid & Parsons (1971) have reported that rat liver mitochondrial RNA polymerase has a molecular weight of about 65 000 and is sensitive to 10 μg cm^{-3} of rifampicin.

Thus the various mitochondrial RNA polymerases that have been studied to date are easily distinguished from both nuclear and bacterial polymerases. The mitochondrial polymerases, in contrast to the other enzymes, are apparently composed of single subunits with molecular weights ranging from 46 000 to 65 000, and differ in inhibitor specificity, catalytic properties, and ion requirements.

The closest prokaryotic equivalent is the polymerase of bacteriophage T_7: it too is a single polypeptide (molecular weight of 70 000) resistant to the action of rifampicin.

The mitochondrial polymerases are also strikingly different from chloroplast polymerase, which has been reported to have a molecular weight of about 500 000 and to contain two large subunits of 220 000 and 150 000 molecular weight together with several smaller polypeptides (Bottomley, Smith & Bogorad, 1971; Smith & Bogorad, cited by Mullinix et al., 1973).

Site of synthesis of mitochondrial RNA polymerase

The genetic locus for mitochondrial polymerase has been investigated with petite strains of yeast that either altogether lack or have aberrant mitochondrial DNA. Mitochondria from such strains appear to lack the ability to synthesize proteins (Mahler, 1973). Though the issue is not completely resolved, evidence does exist to indicate that the mitochondrial RNA polymerase of yeast is specified by a nuclear gene. RNA polymerase activity has been found in petite strains (South, cited by Mahler, 1973; Wintersberger, 1970; Tsai et al., 1971). However, Tsai et al. (1971) and Scragg (1971) have reported finding no polymerase activity in petite strains completely lacking in mitochondrial DNA. Since the inner membranes of these mitochondria are aberrant, the lack of polymerase activity may have resulted from an inability to stimulate the synthesis of, or retain the enzyme in, the organelle. The experiments of Barath & Küntzel (1972), which demonstrate the stimulation of mitochondrial polymerase synthesis in the presence of ethidium bromide or chloramphenicol, which respectively inhibit mitochondrial transcription and translation, provide further support for a nuclear locus for the mitochondrial polymerase gene.

Mitochondrial messenger RNA

Polyadenylation of mt mRNA

Recent investigations (Perry et al., 1973; Darnell, Jelinek & Molloy, 1973) have disclosed a striking difference between well-characterized mRNAs of bacterial and eukaryotic origin. Most of the latter carry at their 3'–OH a sequence of polyriboadenylic acid residues that is apparently absent from bacterial mRNA. These are synthesized by a separate enzyme not requiring a template and are attached to the mRNA after the completion of its transcription. Two different classes of such post-transcriptionally polyadenylated RNA have been identified in metazoan cells (for references see Hirsch & Penman, 1973): one, carrying poly A stretches 150–200 nucleotides long, is constituted by mRNAs of nuclear origin destined for the cytoplasm and by the RNA of viruses of nuclear localization. Plant mRNA contains similar poly A tracts (Sagher, Edelman & Jakoh, 1974; Higgins, Mercer & Goodman, 1973). The second class, with approximately 80 to 110 adenylate residues, consists of the RNAs of viruses of cytoplasmic localization.

Yeast and fungal mRNA carries short (50–60 nucleotide) poly A tracts (Reed & Wintersberger, 1973; McLaughlin et al., 1973; Silver & Horgan, 1974).

Poly A-containing RNAs, likely candidates for a messenger function, have now been identified in HeLa, insect and yeast mitochondria. As in the nucleus, the attachment of poly A to mRNA occurs after transcription of the mRNA (Ojala & Attardi, 1973; Hirsch & Penman, 1973). The length of the poly A stretch is about 55 nucleotides. Thus mitochondrial mRNA appears to be similar in this respect to the mRNAs of lower eukaryotes and dissimilar to bacterial mRNA.

Mitochondrial ribosomal RNA

General properties of ribosomes

One of the mainstays of the symbiotic hypothesis for the origin of mitochondria and chloroplasts is their possession not only of DNA, but also of the machinery for its transcription and translation. Translation in these organelles is performed by ribosomes analogous both to those of prokaryotes and of the eukaryotic cytoplasm. The general physical properties of organellar ribosomes from three diverse groups of organisms are compared to the corresponding cytoplasmic ribosomes (cytoribosomes) and to bacterial ribosomes in Table 4.

The sedimentation properties of the mitochondrial ribosomes differ from species to species and from those of bacteria. The mitoribosomes of fungi have somewhat higher sedimentation coefficients than those of bacteria while those from the ciliate *Tetrahymena* are much higher, and those of animals lower.

Bacterial ribosomes are composed of about 60% RNA and 40% protein. The proportion of these two constituents in cytoribosomes has been claimed to be approximately equal, but may in fact be similar to that of bacteria (McConkey, 1974). On the other hand, the mitoribosomes of *Tetrahymena* and of higher animals have been reported to contain over 60% protein. The mitoribosomes of *Tetrahymena* are clearly in a class by themselves. These particles not only exhibit a very high sedimentation coefficient, but appear to be composed of two subunits equal in their sedimentation coefficients rather than a large and a small one as is customary in all other cases. However, the existence of two rRNAs of unequal size in these mitoribosomes indicates that the two subunits are probably functionally distinct. A detailed comparative discussion of rRNA and 5 S RNA will be presented below.

Base Composition of mitochondrial rRNAs

The DNA base compositions of organisms vary within the rather broad limits of about 22% G + C to about 75% G + C (Woese & Bleyman, 1972). These values may define the limits beyond which the overall DNA base composition of an organism may not evolve without losing its ability to perform a coding function. There may be two types of genetic code limit organisms – those that have as extreme a G + C content as is possible without generating atypical amino acid compositions in their proteins, and those in which the proteins have evolved such extremes in amino acid composition. It is unknown whether the latter type exists (Woese & Bleyman, 1972).

Ribosomes must exhibit some fundamental similarities. All carry out essentially the same reactions. All consist of two subunits – one large and one small, except for *Tetrahymena* mitoribosomes. Each subunit contains a distinct major species of ribosomal RNA as well as a set of proteins characteristic of that subunit. Ribosomal RNAs contain double helical regions that may control their conformation and appropriate sequences are apparently required for proper ribosome assembly (reviewed by Monier, 1972).

Table 4. *Properties of bacterial, mitochondrial and eukaryotic cytoplasmic ribosomes* [a]

Parameter	Bacteria	Ascomycetes Mitochondria	Ascomycetes Cytoplasm	Tetrahymena Mitochondria	Tetrahymena Cytoplasm	Animals Mitochondria	Animals Cytoplasm
Dimensions (nm)	20 × 16	26.5 × 21		27.5 × 23		37 × 24	~ 32 × 25
Sedimentation coefficient (S)							
Ribosome	70	72–78	80	80	80	50–60	80
Large subunit	50	50–58	60	55	60	33–45	60
Small subunit	30	35–40	40	55	40	25–35	40
Density (g cm^{-3})							
Ribosome	1.63	1.46–1.48	1.52	1.46	1.56	1.40–1.43	1.50
Large subunit	1.67			1.52	1.57		1.60–1.64
Small subunit	1.63			1.46	1.53		1.55
Average molecular weights ($\times 10^{-6}$)							
Ribosome	2.7	4.16	4.49			2.5	
Large subunit	1.8	2.47				1.7	
Small subunit	0.85	1.69				0.80	
Large subunit rRNA	1.09	1.30	1.30	0.82	1.30	0.53	1.4 –1.7
Small subunit rRNA	0.56	0.70	0.73	0.52	0.69	0.33	
Peptides of the							
Large subunit	0.60 (35)[b]	(30)				1.08 (~30)	0.90 (~35)
Small subunit	0.30 (20)	(23)				0.56 (~20)	0.70 (~20)
Presence of 5 S RNA in ribosome	Yes	No	Yes		Yes	No	Yes
Methylation of rRNA	Yes	Yes, low	Yes			Yes, low	Yes
Subunit exchange with bacterial ribosomes	Yes (various bacteria)	No					
Initiation and elongation factors active with bacterial ribosomes		Yes					

[a] See Mahler & Raff (1975) for references.
[b] Number in parenthesis is the number of distinct proteins detected in a given subunit.

The rRNA of the small subunit may also play a direct functional role in protein synthesis. When about 50 nucleotides from the 3′ end of the rRNA have been removed from *E. coli* 30 S subunits by certain colicins, the particles become inactive in protein synthesis (Senior & Holland, 1971; Bowman *et al.*, 1971). The 3′-terminal hexanucleotides of 18 S rRNAs from yeast, *Drosophila*, and rabbits have an identical sequence, $-UCAUUA_{OH}$, that is complementary to the known eukaryotic terminator codons, and therefore capable of playing a role in polypeptide chain termination by recognizing the terminator codons of the mRNA under translation (Dalgarno & Shine, 1973).

The rRNAs of the large and small ribosomal subunits are similar in their base composition (Amaldi, 1969; Lava-Sanchez *et al.*, 1972) but not in nucleotide sequence (reviewed by Monier, 1972). Bacterial rRNAs are quite uniform in base composition (50–55 % G + C), while eukaryotic cytoribosomal RNAs and mitoribosomal RNAs show considerable variations. These are plotted in Fig. 2 against the total DNA base compositions

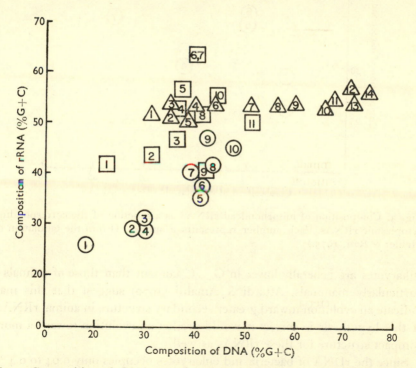

Fig. 2. Composition of rRNAs as a function of the corresponding DNA. Bacteria (△), eukaryotic nucleo–cytoplasm (□), mitochondria (○). The numbers in each symbol represent a particular species of that class. (From the tabulation of Mahler & Raff, 1975.)

of the cell or organelle from which they were derived. The constancy of rRNA composition in bacteria regardless of the DNA base composition is in accord with the hypothesis that in these organisms there exist functional constraints on rRNA composition that are different from those in eukaryotes.

An evolutionary trend is apparent in the base compositions of eukaryotic rRNAs (Fig. 3). Both the cytoplasmic and mitochondrial rRNAs of lower

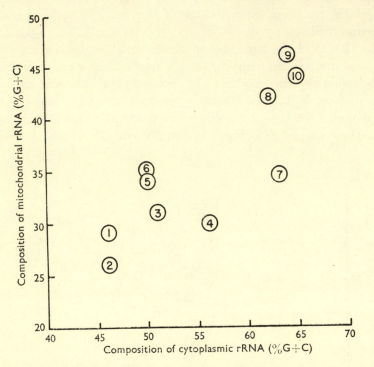

Fig. 3. Composition of mitochondrial rRNAs as a function of the corresponding cytoplasmic rRNAs. Each number represents a species. (From the tabulation of Mahler & Raff, 1975.)

eukaryotes are generally lower in G + C content than those of animals – particularly mammals. Attardi & Amaldi (1970) suggest that this may indicate an evolution toward greater secondary structure in animal rRNAs. If this hypothesis is correct it would predict an evolution toward a more compact structure for these rRNAs as well.

Since the rDNA of bacteria and eukaryotes occupies only 0.04 to 0.4% of the total genome (Attardi & Amaldi, 1970) one would not expect any correlation between the base compositions of rRNA and the total DNA of either the bacterial cell or the eukaryotic nucleus. However, the compositions of eukaryotic cytoribosomal RNA vary from 42 % G + C to 65 % and correlate roughly with the compositions of nuclear DNAs. Mitochondrial RNAs vary from 26 % G + C to 47 % G + C and correlate strongly with the compositions of mitochondrial DNAs (Fig. 3). Ribosomal DNA constitutes 2.5 to 7 % of the total mitochondrial DNA (Mahler, 1973).

Thus, mitoribosomal RNAs are very different from bacterial rRNAs in base compositions. Furthermore, there exists a correlation of base composition of mitochondrial RNAs not only with mitochondrial DNA, but also with those of cytoribosomal RNAs from the same species (Freeman, Mitra & Bartoov, 1973) (Fig. 3). These results suggest that whatever their nature, the rRNAs and perhaps the ribosomes in the mitochondria and the cytosol of the same cell are subject to similar functional or regulatory constraints. And as already mentioned this type of constraint on mtRNA is quite dissimilar from that governing the composition of the analogous entity in bacteria.

Evolution of size of ribosomal RNA

Polyacrylamide gel electrophoresis has been used for the determination of the molecular weights of a wide variety of rRNAs. The major rRNAs of the large and small subunits of prokaryotic ribosomes are not only nearly identical in compositions, but also in size. The molecular weight of the rRNAs of the large subunit equals 1.04 to 1.11×10^6 (Loening, 1968; Monier, 1972). The cytoribosomal RNAs of eukaryotes, on the other hand, show distinct, phylogenetically correlated, trends in molecular weights.

The rRNA of the small subunit with a few exceptions appears to be uniformly of a molecular weight of 0.7×10^6 (Loening, 1968; Monier, 1972). The molecular weights of large subunit rRNAs are about 1.3×10^6 in fungi, protozoa and plants (Loening, 1968). The large ribosomal RNAs of metazoans are distinctly larger than this. Invertebrates contain rRNAs with molecular weights of $1.40-1.44 \times 10^6$ (Loening, 1968; Sy & McCarthy, 1968; Raff, 1970). The values for lower vertebrates are about $1.5-1.6 \times 10^6$ (Loening, 1968), while mammals yield values of $1.7-1.9 \times 10^6$ by gel electrophoresis and equilibrium centrifugation (Loening, 1968; McConkey & Hopkins, 1969).

Molecular weights have been obtained for various species of mitochondrial rRNA by gel electrophoresis and by electron microscopy. It is clear that mitochondrial rRNAs of fungi are longer than the corresponding rRNAs from *E. coli*, while mammalian mitochondrial rRNAs are shorter. As shown in Table 5 the lengths agree with the molecular weight values determined by transport methods obtained under conditions that totally

Table 5. *Physical parameters of* E. coli *and mitochondrial rRNAs*

	Large		Small	
Species	Length (μm)	MW ($\times$ 10^{-6})	Length (μm)	MW ($\times$ 10^{-6})
E. coli	0.72–0.85	1.07	0.38–0.43	0.56
Fungi	0.91	1.30	0.46	0.70
Human	0.42–0.46	0.54	0.26	0.35

Data selected from the compilation of Mahler & Raff (1975).

denature the RNA so as to eliminate all conformational effects (Forrester, Nagley & Linnane, 1970; Groot, Aaij & Borst, 1970; Edelman *et al.*, 1971).

The data show clearly that mitochondrial rRNAs are not of the same size as those of bacteria, nor are their molecular weights positively correlated with those of the corresponding cytoplasmic rRNAs – in fact the contrary is true. The values of molecular weights indicate a divergent evolutionary trend in these RNAs quite unlike the apparent strong conservativeness of bacterial rRNAs. It is of particular interest to note that these values for the molecular weights of mitochondrial rRNAs indicate two opposing trends with molecules larger than the bacterial ones occurring in fungi, and smaller ones in animals. In neither case can they be considered to be typically prokaryotic.

Degree of methylation of rRNA

Methylation of rRNA is apparently important in ribosome assembly, and possibly in function.

Unfortunately, the nature and extent of methylation of mitochondrial rRNAs is still somewhat unclear. Dubin (1974) in a precise study of cultured hamster (BHK-21) cell mitochondrial rRNA has found that the 17 S rRNA contains 0.13 methyl groups per 100 nucleotides and the 13 S rRNA 0.37. Dubin draws the conclusion that hamster cell mitochondrial rRNA is significantly methylated, though to a considerably lesser extent than any other rRNAs. This low level of methylation may well play a functional role. The 13 S mitochondrial rRNA, but not the 17 S rRNA, contains N^6-dimethyl-adenine which appears to be restricted to or enriched in the small subunit rRNAs of bacteria and mammals (Klagsbrun, 1973). The existence of m_2^6Ap in the 13 S mitochondrial rRNA may be homologous to the $m_2^6Apm_2^6A$ regions of bacterial and eukaryotic cytoplasmic rRNAs. Thus, as Dubin (1974) points out, homology in a small region would indicate a remarkable degree of conservation in the face of extensive evolutionary changes in the rest of the molecule. The most convincing evidence for the occurrence and importance of methylation of mitochondrial rRNA is provided by some recent observations by Kuriyama

& Luck (1973*a*, 1974), who found that the deficiency of small ribosomal subunits in mitochondria of the *poky* mutant of *Neurospora crassa* is due to abnormal processing of the 32 S rRNA precursor. In *poky* the precursor and its resultant products are undermethylated, suggesting that abnormal processing in this mutant is due to faulty methylation.

5 S ribosomal RNA

The large ribosomal subunit of both prokaryotes and eukaryotes contains a small RNA (about 125 residues in length) corresponding to a molecular weight of 4×10^4 and a sedimentation coefficient of 5 S. This RNA has been isolated from ribosomes of such diverse sources as bacteria, lower eukaryotes, plants and animals, and appears to be a universal component of cytoribosomes (see Monier, 1972 for references). This RNA plays a structural role in the assembly of the large ribosomal subunits (Erdmann, Fahnestock, Higo & Nomura, 1971; Gray & Monier, 1971; Gray *et al.*, 1973).

Complete nucleotide sequences have been determined for 5 S RNAs from *E. coli*, *Pseudomonas fluorescens*, yeast, *Xenopus*, mouse, rabbit, rat and human cells (Brownlee, Sanger & Barrell, 1968; Forget & Weissman, 1969; Dubuy & Weissman, 1971; Hindley & Page, 1972; Brownlee, Cartwright, McShane & Williamson, 1972). The observed differences between these 5 S RNA sequences are presented in Table 6. The rate of

Table 6. *Fractional differences in 5 S RNA sequences*

	Human	Yeast	E. coli	P. fluorescens
Human	0	0.40	0.45	0.50
Yeast		0	0.47	0.57
E. coli			0	0.31
P. fluorescens				0

Data from Dayhoff (1972) and Kimura & Ohta (1973)

sequence divergence has apparently been slow enough for relationships to become discernible (Sankoff, Morel & Cedergren, 1973): while the sequences of human and yeast RNA are more similar to each other than to bacterial sequences and vice versa, considerable sequence homology has nevertheless been retained between the eukaryotic and bacterial 5 S RNAs.

Curiously, while mitoribosomes like all other ribosomes are composed of a large and a small subunit, each containing a characteristic high molecular weight rRNA, they apparently lack 5 S RNA. At least no molecule with such a sedimentation coefficient or a corresponding electro-phoretic mobility has been found in mitochondrial ribosomes of fungi or animals (Mahler, 1973). Since 5 S RNAs play a vital role in the assembly of ribosomes of all other types, it seems *a priori* peculiar that this function should have become dispensable for mitochondrial ribosomes. This is particularly so since chloroplast ribosomes contain 5 S RNA. Two major possibilities are that a 5 S analog exists, but is concealed under the 4 S tRNA peak (Lizardi & Luck, 1971; Gray & Attardi, 1973), or that the role of 5 S RNA is performed by a segment of the large rRNA (Lizardi & Luck, 1971).

Gray & Attardi (1973) have examined the low molecular weight RNAs of HeLa mitochondrial ribosomes by gel electrophoresis, and were able to partially resolve two components in the 4 S region of the gel. An RNA somewhat shorter than mitochondrial tRNA was found. There is one molecule of this RNA in each large ribosomal (45 S) subunit. Labeling of the RNA is inhibited by ethidium bromide indicating that it is a product of mitochondrial transcription. Dubin, Jones & Cleaves (1974) have studied an RNA of similar low molecular weight from hamster mitochondria. This RNA is present in approximately a 1 : 1 molar ratio with mitochondrial rRNA, is unmethylated and has a relatively low G + C content. Its synthesis is also sensitive to ethidium bromide. The function of these RNAs has not yet been determined. However, since the mitochondrial rRNAs in mammalian cells are considerably smaller than their prokaryotic or cytoplasmic counterparts and yet capable of functioning, it is not unreasonable that such a small RNA molecule could perform the function of the 5 S RNAs of other ribosomes. Since prokaryotic and eukaryotic 5 S RNAs exhibit sequence homology, it may well be possible to test for 5 S sequences in the small mitochondrial RNAs isolated by Gray & Attardi and by Dubin *et al.*, or in mitochondrial rRNA from other sources, by molecular hybridization techniques (Brown & Weber, 1968*a*, *b*; Brown, Wensink & Jordan, 1972).

In summary, while mitochondrial rRNAs have diverged from prokaryotic rRNAs, they have also diverged from one another in such properties as base composition and molecular weights. Furthermore, the ordinary (120 nucleotide) small RNA of the large subunit is absent in the mitochondrial particle. Thus, the loss or modification of this RNA may have occurred early in the evolutionary history of the mitochondrion and prior to the divergence of fungi and animals. This evidence supports, but of course by no means proves, the hypothesis that mitochondria had a common origin and that the protein synthetic machinery of the ancestral mitochondrion underwent radical modification early in the evolutionary history of the organelle (Fig. 4).

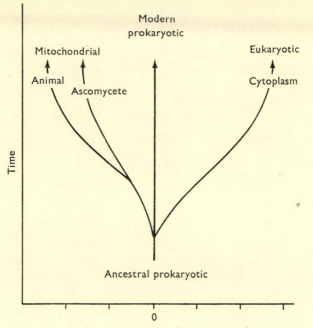

Fig. 4. Evolution of ribosomes: a simplified phylogeny of mitochondrial, eukaryotic and prokaryotic ribosomes. For simplicity, it is assumed that modern prokaryotic ribosomes are identical to those of Precambrian prokaryotes. Organellar (mitochondrial) and eukaryotic cytoplasmic ribosomes have diverged in various ways from the basic prokaryotic pattern. Mitochondrial ribosomes have diverged less than cytoplasmic ribosomes, but show significant differences from prokaryotic ribosomes and from each other. (From Raff & Mahler, 1972. Copyright © 1972 by the American Association for the Advancement of Science.)

A MODEL FOR THE NON-SYMBIOTIC ORIGIN OF THE MITOCHONDRION

The model

Contrary to the symbiotic model which assumes an anaerobic proto-eukaryote (i.e. an advanced prokaryote with characteristics of eukaryotes: a transitional form) that acquired a respiratory endosymbiont, we propose that the proto-eukaryote was an advanced, aerobic cell rather larger in size than is typical for prokaryotes. Among the changes in cellular organization which were concomitant with the trend to larger size (Stanier, 1970) was a large increase in respiratory membrane surface. This was initially accomplished by invagination of the inner cell membrane (Fig. 5*a*), and

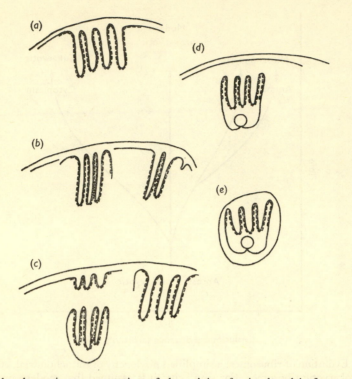

Fig. 5. A schematic representation of the origin of mitochondria from a simple prokaryotic respiratory organelle. The drawings present cross-sections of hypothetical cells representing various evolutionary stages. Blocks on the membrane represent respiratory assemblies. (*a*) Section of proto-eukaryote showing invaginated cell membrane possessing respiratory function. (*b*) As the proto-eukaryote becomes large, a more extensive respiratory surface becomes necessary, and is provided by blebbing off of respiratory membranes from the cell membrane. (*c*) Topologically closed respiratory organelles generated by blebbing. (*d*) Establishment of a stable plasmid (schematically represented by a circle) containing genes for ribosomal components and some elements of the respiratory membrane. (*e*) The final step in the evolution of the mitochondrion in the later acquisition of an outer membrane. (From Raff & Mahler, 1972. Copyright © 1972 by the American Association for the Advancement of Science.)

later by formation of membrane-bound vesicles generated from it (Fig. 5*b*, *c*). The respiratory organelles thus generated were topologically closed objects surrounded by a membrane providing a selective permeability barrier between the respiratory elements and the cytoplasm. This provided the basis for a more sophisticated performance and regulation of respiratory metabolism (Hughes, Lloyd & Brightwell, 1970; Hall, 1973), and presumably proved to be evolutionarily advantageous. However, the resulting segregation of the respiratory elements from the cytoplasm posed a problem to the cell since certain constituents of the respiratory chain (for example subunits of cytochrome oxidase) apparently require synthesis *in situ*.

While the membrane surrounding the respiratory elements was (like the present-day mitochondrial membrane) permeable to many proteins, including cytochrome *c* and enzymes of the respiratory and phosphorylation systems, it was impermeable to ribosomes or ribosomal RNA. Thus, these respiratory organelles required constant de-novo replacement from the cell membrane.

The continual re-formation and turnover of such a complex organelle would have been somewhat uneconomical. Thus a system for organelle maintenance based upon the presence of a system for protein synthesis on the inside of the organelle would have been selectively advantageous. The implantation of such an organellar protein-synthetic system with its concomitant organellar genome seems at first sight to present a formidable problem. However, genetic systems exist in contemporary eukaryotic and prokaryotic organisms which suggest that this should have been entirely feasible. We propose that the proto-eukaryotic cell implanted a protein synthesis system into the respiratory organelle by simply incorporating a stable plasmid containing the appropriate genes for ribosomal components (Fig. 5*d*).

A particularly well-studied analogous process occurs in the generation of multiple nucleoli during amphibian oogenesis. In that process, tandem multiple replicates of the chromosomal ribosomal RNA genes are released as circles of DNA. These are packaged into free nucleoli which produce the large amounts of rRNA required by the egg (Brown & Dawid, 1968; Brown & Blackler, 1972). The hypothetical respiratory organellar genome of the proto-eukaryote may not, of course, have been generated by the same mechanism; nevertheless, it serves as a useful example.

Elements of the model

Organelles of prokaryotes

While the organelles of eukaryotic cells are more complex (and more thoroughly studied) than those of prokaryotic cells, the latter nonetheless contain several types of membrane-bound organelles. Among these are the chlorobium vesicles and thylakoids of photosynthetic bacteria, the gas vesicles of photosynthetic bacteria, the carboxysomes of some chemo-autotrophs, and the mesosomes of bacteria (Stanier, 1970; Hughes *et al.*, 1970; Echlin, 1970; Walsby, 1972; Shively, Ball, Brown & Saunders, 1973). The functions of some of these membranous systems of prokaryotes are problematical. However, the cell membranes of bacteria contain the electron transport–oxidative phosphorylation systems of these organisms and membrane-bound replication sites for the cells' DNA. Thus the basic membranous equipment of the prokaryotic cell includes elements capable of evolutionary modification in the manner outlined in our model.

An early step in our model is the production of internal membranes possessing structural and functional specialization different from those of the outer cell membrane (Fig. 5). The existence of such prokaryotic membrane differentiation is best exemplified by the membranes of the photosynthetic bacteria (reviewed by Oelze & Drews, 1972). The bacterial photosynthetic apparatus is an integral part of the intracytoplasmic membrane system. These membranes form under anaerobic conditions in the light, and disappear if the cultures are made aerobic. The intracyto-plasmic membranes probably originate as modified extensions of the cell membrane. Under conditions which induce their formation there is an increase in lipids and in bacteriochlorophyll content co-incident with membrane proliferation. The intracytoplasmic membranes are similar to the plasma membrane in their components but differ from it in phospho-lipid ratio and in certain proteins and enzymic activities. An accelerated rate of membrane synthesis relative to that of the cell wall causes an invagination of the membrane which becomes differentiated from the plasma membrane by virtue of an active process of incorporation of photosynthetic components into the developing novel membranes. The complex internal membranes of these bacteria possess knob-like structures similar to those found on the cristae of mitochondria, where – as the mito-chondrial ATPase – they perform a key function in energy transduction. In the reverse process, under conditions in which the intracytoplasmic membranes are not needed, synthesis of photosynthetic units ceases, though synthesis of other membrane components continues, resulting in the dilution of photosynthetic components and the disappearance of intracytoplasmic membrane.

Photosynthetic membranes are not the only specialized intracytoplasmic membranes in prokaryotes. Some of the most complex membranes in non-photosynthetic bacteria are found in certain ammonia and nitrite-oxidizing organisms. Two of these, *Nitrosocystis* and *Nitrosolobus*, oxidize ammonia to nitrite and fix CO_2 as a carbon source (Watson, Graham, Remsen & Valois, 1971; Watson & Remsen, 1970). Both have extensive internal membranes. Those of *Nitrosolobus* seem to indicate physiological as well as morphological partitioning since glycogen-like material is confined to the outer compartments of the cell. Spectroscopic studies indicate the presence of cytochromes b, c and a_3. The intracytoplasmic membranes in the nitrite-oxidizer *Nitrococcus* are in the form of tubular invaginations of the cell membrane (Watson & Waterbury, 1971). These membranes appear studded, like those of photosynthetic bacteria, with particles approximately 10 nm in diameter similar to those found on mitochondrial cristae (Remsen, Valois & Watson, 1967). It is tempting to speculate that these complex membranous systems represent adaptations to the metabolic specializations of these organisms, similar to ones exhibited by certain contemporary mitochondria, i.e. those of facultative anaerobic yeasts (Linnane, Haslam, Lukins & Nagely, 1972; Perlman & Mahler, 1974). They certainly imply a great evolutionary flexibility for prokaryotic membranes.

Most of the above membrane systems are extensions of the cell membrane. It would be of interest to find examples of prokaryotes with more advanced organelles such as the one illustrated in Fig. 5.

Plasmids

A more detailed comparison of plasmid and mitochondrial DNAs has been made above. Here we will only summarize the properties of plasmids which support their evolutionary role as ancestors of the mitochondrial genome (Table 2).

Plasmids are similar in size to mitochondrial DNAs, and resemble the latter in being supercoiled circles, subject to elimination by acridines and ethidium. Like the primary chromosomes, plasmids are capable of autonomous self-replication in the bacterial cell. This process involves the products of both plasmid-linked and chromosome-linked genes, and may involve specific replication sites. Plasmids, however, contain not only genes required for their replication, but also a considerable variety of other genes as well.

Of particular significance to our model is the capability of plasmids for direct genetic interaction with the chromosomes by means of physical integration into the latter. The integration process seems to involve insertion of plasmid DNA, by means of the cell's recombination enzymes, into a region of the chromosome possessing some homology with a region of the plasmid. Integration is often reversible, and excision of the plasmid in some cases entails a concomitant excision of chromosomal genes so that a novel type of plasmid is generated (Fig. 6). These events are not particularly rare, and in natural populations subject to particular selection pressures, plasmids bearing advantageous genes very quickly become apparent. This property of plasmids has become medically significant since bacteria harboring plasmids bearing several genes for resistance to different antibiotics have been found. Plasmids provide an evolutionary mechanism of great flexibility since they are transmissible from one bacterium to another (Richmond & Wiedman, 1974). Thus there is a

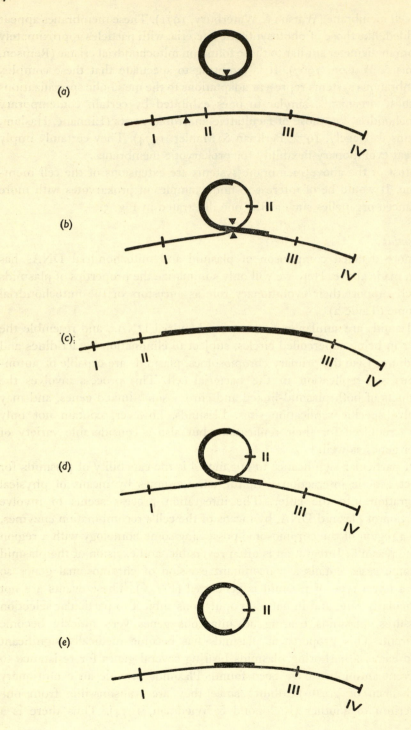

continuous flow of genetic information between chromosomes and plasmids. The incorporation of chromosomal genes into plasmids is not restricted to genes for antibiotic resistance or for certain metabolic enzymes, but also includes genes for rRNA and tRNA (Russell et al., 1970; Yu, Vermeulin & Atwood, 1970).

An interesting synthetic analog to the model has been reported by Morrow et al. (1974) who incorporated *Xenopus* rDNA into a plasmid *in vitro*. Recombinant plasmids were then introduced into *E. coli* cells, and were found to be stably replicated. *E. coli* minicells (which lack the main genophore) containing the plasmid synthesize RNA which hybridizes to *Xenopus* rDNA.

Given the existence of a selective pressure on the proto-eukaryote for the incorporation of a protein synthesis system into the primitive respiratory organelle, generation of a plasmid with the appropriate genes would constitute an efficient way to do so by exploiting the cell's already well-established genetic interactions between plasmid and chromosome.

Role of membrane-bound ribosomes

An integral part of our model is the postulate that certain elements of the respiratory machinery of the proto-eukaryote required synthesis *in situ* and thus, once a closed proto-mitochondrion evolved, an internal protein synthetic system became a necessity. There are several lines of evidence that support this proposal.

The strongest supporting data for our contention, that in-situ synthesis of certain components played a key role in mitochondrial origination, comes from the nature of the products synthesized on mitochondrial ribosomes. The proteins encoded by the mitochondrial genome and synthesized in the mitochondrion are few in number ($\sim$10) and are components of inner membrane activities such as the rutamycin sensitive ATPase, cytochrome oxidase and ubiquinone-cytochrome c reductase. Neupert & Ludwig (1971) and Bandow (1972) have provided direct evidence that the components of the outer mitochondrial membrane and

Fig. 6. Model for the reversible integration of a plasmid into a bacterial chromosome (model of Campbell, redrawn from Richmond, 1970). The roman numerals represent genes on the main chromosome. (a) Plasmid (heavy line) and chromosome (thin line) surviving independently in the same cell. (b) Apposition of homologous regions of plasmids and chromosome followed by single cross-over. (c) Integrated plasmid. (d) Apposition of homologous regions again followed by crossing-over. (e) Plasmid excised with incorporation of part of the chromosome carrying marker to yield a novel plasmid. By the proposed model such events occurred in the proto-eukaryote and the marker may have been, for example, a gene for rRNA. (From Raff & Mahler, 1972. Copyright © 1972 by the American Association for the Advancement of Science.)

of the intermembrane space are all synthesized on cytoplasmic poly-ribosomes while the mitochondrial system contributes only to the inner membrane. By and large these products of mitochondrial protein synthesis are extremely hydrophobic.

Beattie, Patton & Stuchell (1970) and Coote & Work (1971) found that the products of mitochondrial protein synthesis in mammalian cells were highly insoluble and were associated with the cristae. Kadenbach & Hadvary (1973) working with rat liver mitochondria found that the pro-ducts of endogeneous protein synthesis were generally insoluble in aqueous solvents, but that about one-third of the products were soluble in chloro-form-methanol (2 : 1). In agreement with such properties is the finding of Constantino & Attardi (1973) that HeLa mitochondria incorporate hydro-phobic amino acids but show a very low incorporation of charged amino acids.

The proteins synthesized by yeast mitochondria (Tzagoloff & Akai, 1972; Murray & Linnane, 1972) and by *Neurospora* mitochondria (Rowe, Lansman & Woodward, 1973) are likewise hydrophobic and largely soluble in chloroform-methanol. Sebald, Machleidt & Otto (1973) found that, in *Neurospora*, the components of cytochrome oxidase synthesized in the mitochondrion are higher in non-polar amino acids than the cyto-chrome oxidase components synthesized outside of the mitochondrion, and similar results have also been reported for the cytochrome oxidase (Poyton & Schatz, 1975) and ATPase (Sierra & Tzagoloff, 1973) of yeast.

The similarity of these results from mitochondria of mammals and ascomycetes, which are evolutionarily distant from each other, indicates that the peculiar properties of the proteins synthesized in the mitochondria are probably ancient and critical to mitochondrial function. Further evidence for the evolutionary necessity of an internal mitochondrial protein-synthetic system comes from the intimate association of mito-chondrial ribosomes with the inner mitochondrial membrane.

Cytoplasmic ribosomes occur in two forms, free and bound to mem-branes of the endoplasmic reticulum. Membrane-bound ribosomes are particularly prevalent in cells that synthesize proteins for export (Goldberg & Green, 1964; Campbell, 1965; Redman, Siekevitz & Palade, 1966; Redman & Sabatini, 1966; Palade, 1966; Redman, 1967, 1969; Takagi & Ogata, 1968; Kimmel, 1969). Free and membrane-bound ribosomes appear to be engaged in the synthesis of different proteins. Thus in liver, serum albumin is synthesized on bound ribosomes while ferritin is mainly a product of free polysomes (Ganoza & Williams, 1969; Hicks, Drysdale & Munro, 1969; Redman, 1969; Peters, 1962). Membrane-bound ribosomes are also present in non-secretory cells (Rosbash & Penman, 1971; Andrews

& Tata, 1971). Tata (1971) proposed that membrane-bound ribosomes may play a role in the topographical segregation within a cell of different groups of ribosomes synthesizing specific proteins. He also made the interesting suggestion that ribosomes on rough endoplasmic reticulum associated with mitochondria might be synthesizing mitochondrial proteins not synthesized by the protein synthetic system of the organelle.

A mechanism whereby the in-situ synthesis of membrane components might be accomplished by membrane-bound ribosomes is suggested by Adelman, Sabatini & Blobel (1973). This group found that membrane-bound ribosomes can be removed from microsomal membranes, without disrupting the membranes, by incubation with puromycin and high salt. Puromycin alone becomes linked to nascent chains, but high salt as well as puromycin was required for release of ribosomes from the membrane. Incorporation studies with [^{3}H]puromycin showed that nascent chains released from the ribosomes still remained bound to the membrane. Their model suggests that ribosomes are bound to membranes both by nascent chains and by a salt labile bond. Proteins synthesized by these ribosomes are then vectorially released into the membrane complex. Kellems, Allison & Butow (1974) found that about half of the puromycin-discharged nascent chains from cytoplasmic ribosomes attached to the outer surface of mitochondria remain associated with the mitochondria. These investigators suggested that in fact both cytoplasmic ribosomes on the outer mitochondrial membrane and mitochondrial ribosomes attached to the inner membrane are involved in the synthesis of mitochondrial proteins.

Several lines of evidence show that mitochondrial ribosomes are bound to the inner membrane (Bunn, Mitchell, Lukins & Linnane, 1970; Linnane et al., 1972; Towers, Kellerman, Raison & Linnane, 1973; Villa, Morimoto & Halvorson, 1974). Kuriyama & Luck (1973b) using the poky mutant of Neurospora have very convincingly demonstrated that functional mitochondrial ribosomes are membrane bound. These mitochondria carry out protein synthesis albeit at a lower rate than wild type. The mitochondria of this mutant contain large to small ribosomal subunits in a mass ratio of 10 : 1 in contrast to functional ribosomes which should exhibit a 2 : 1 mass ratio. Kuriyama & Luck sonicated mitochondria from the poky mutant, and found that most of the mitoribosomes remained associated with the resultant membrane-bound vesicles. When these were subjected to puromycin plus high salt the released ribosomal subunits were found to have the ratio of 2 : 1 predicted for active ribosomes. These results provide the strongest evidence for the direct attachment of active mitochondrial ribosomes to the inner membrane. Since nascent chains are involved in binding of mitochondrial ribosomes to the membrane, it is very likely that these represent in-situ synthesis of inner membrane components. Chloroplast ribosomes are also apparently bound to the thylakoids (Chua, Blobel, Siekevitz & Palade, 1973).

The nature of the products of mitochondrial protein synthesis and the membrane-bound nature of the mitochondrial ribosomes strongly suggest a requirement for in-situ synthesis of the former. The exact nature of this requirement is not yet known, but it is of interest to note that specific protein–protein interactions between membranes and nascent chains of membrane-bound ribosomes have been described. A striking example is provided by Diegelmann, Bernstein & Peterkofsky (1973) who prepared microsomes from chick embryo connective tissue. These membrane-bound ribosomes were found to carry out the cell-free synthesis of collagen, including hydroxylation of proline in the nascent chains. Polyribosomes released by detergent treatment of such preparations were still competent to synthesize collagen nascent chains, but proline in these chains was not hydroxylated. These results suggest that procollagen chains synthesized on membrane-bound polyribosomes enter the membrane in such a fashion as to interact with prolyl hydroxylase in the membrane.

Thus the outer mitochondrial membrane is synthesized by cytoplasmic ribosomes translating mRNAs transcribed in the nucleus. The inner membrane is composed partly of nuclear-encoded proteins which are synthesized in the cytoplasm and then enter the mitochondrion. However, some proteins of the inner membrane are highly insoluble and must be synthesized, released and integrated *in situ*. This small group of proteins required the evolution of an organellar genome and system of protein synthesis. This system has apparently been in many respects evolutionarily conservative and retains many primitive (i.e. prokaryotic) traits, but its origin did not require a symbiotic event since all of the necessary genetic and biochemical mechanisms for the production of two divergent protein synthesis systems was already in existence in the ancestral proto-eukaryote.

Genes for the expression of the mitochondrial genome

Mitochondrial systems for DNA replication, RNA and protein synthesis are quite distinct from the corresponding nuclear–cytoplasmic systems. Yet with the exception of rRNAs and tRNAs, the components of these systems are almost exclusively encoded by nuclear genes (Table 7). This closely resembles the situation for most mitochondrial enzymes including the highly conservative superoxide dismutase.

According to the symbiotic model this state of affairs arose by a transfer of genes from the endosymbiont to the nucleus. We, on the other hand, propose that the genes involved resided in the nucleus from the first, and that the plasmid contained only the minimal number of genes needed for its replication and function after its sequestration into a closed organelle.

The most suggestive evidence comes from the peculiarly divergent properties of gene expression in mitochondria. As shown in Tables 1, 2 and 7, the transcriptive systems of mitochondria have several features not found in either nuclear or bacterial transcription. Most strikingly, mitochondrial RNA polymerases are composed of a single polypeptide chain (MW 45 000–60 000) instead of being a large complex composed of several polypeptide chains as are the polymerases of both the nucleus and bacteria. At least in yeast this unique polymerase is encoded in the nucleus.

The components of the mitochondrial protein-synthesis system exhibit a mixture of conservative and divergent properties (Tables 4 and 7, Figs. 3, 4 and 5). Ribosome size and properties of the mitochondrial rRNAs are highly divergent from the prokaryotic pattern: significantly

Table 7. *Intracellular location of genes for proteins of mitochondrial DNA-replicative, RNA-synthetic and protein-synthetic systems*

Protein	Function and properties	Gene location
DNA polymerase (mammalian, ascomycetes)	Polymerase distinct from nuclear enzyme	Nuclear
RNA polymerase (*Neurospora*)	Rifampicin-sensitive and amanitin-insensitive, therefore, bacteria-like; single polypeptide: 64 000 daltons (smallest known polymerase)[a]	Probably nuclear[c]
(Yeast)	Sensitive to ethidium bromide	Nuclear
(HeLa)	Symmetrical transcription[b] sensitive to ethidium bromide	?
Ribosomal proteins (*Neurospora*)	30 in large subunit; 23 in small subunit; all distinct from cytoplasmic ribosome proteins	Nuclear
Erythromycin-resistance factor (yeast)	Probably a modified RNA	Mitochondrial
Polypeptide chain initiation factors (*Neurospora*)	Bacteria-like	Probably nuclear[c]
Polypeptide chain elongation factors (yeast)	Factors required for mitochondrial protein synthesis can substitute for bacterial elongation factors with bacterial ribosomes	Nuclear

[a] Bacterial polymerase consists of several polypeptide chains.
[b] This is unique; bacterial and nuclear transcription are asymmetrical.
[c] Transcription stimulated by ethidium bromide. See Raff & Mahler (1972) and Mahler & Raff (1975) for references.

these divergent rRNAs are encoded in the mitochondrial DNA. On the other hand, ribosomal proteins and elongation factors are encoded in nuclear genes, yet the elongation factors are bacteria-like and can substitute for bacterial factors with bacterial ribosomes.

We suggest that this peculiar melange of prokaryote-like and uniquely mitochondrial properties are an evolutionary product of two different inputs. First, both nuclear–cytoplasmic and nuclear–mitochondrial systems of interactions arose in an advanced prokaryotic cell (proto-eukaryote) and thus the divergent evolution of these two systems utilized the same starting material. In several respects the nuclear–cytosolic system has been the more conservative. Second, the regulatory requirements of the two systems are very likely different, and the simplest way to manage these controls is to utilize separate components. Thus there are different mitochondrial and nuclear polymerases, although the genes for both are probably nuclear. Precedent for this hypothesis comes from the situation in the nucleus in which there are separate nucleoplasmic and nucleolar polymerases with different transcriptional functions. Further, in metabolic pathways in which synthetic and degradative pathways (e.g. for fatty acids) utilize similar reactions, completely different sets of enzymes are employed.

Finally, we wish to stress that our approach entails looking upon the mitochondrion not as an imperfect organism but rather as the most striking example of a complex supra-molecular structure.

ACKNOWLEDGEMENTS

This article is contribution 974 from the Department of Zoology and contribution 2518 from the Department of Chemistry, Indiana University. The research was supported by USPHS Grant HD 6902 to R.A.R. and Grant GM 12228 to H.R.M. H.R.M. is a recipient of PHS Research Career Award GM 05060.

We also thank the American Association for the Advancement of Science and Charles Scribner's Sons for permission to use copyrighted material.

REFERENCES

ADELMAN, M. R., SABATINI, D. D. & BLOBEL, G. (1973). Ribosome-membrane interaction. *J. Cell Biol.*, **56**, 206–229.

ADMAN, R., SCHULTZ, L. D. & HALL, B. D. (1972). Transcription in yeast: separation and properties of multiple RNA polymerases. *Proc. natn. Acad. Sci. USA*, **69**, 1702–1706.

AMALDI, F. (1969). Non-random variability in evolution of base compositions of ribosomal RNA. *Nature, Lond.*, **221**, 95–96.

AMBLER, R. P. (1973). Bacterial cytochromes c and molecular evolution. *Syst. Zool.*, **22**, 554–565.

ANDREASON, A. A. & STIER, T. J. B. (1953). Anaerobic nutrition of *Saccharomyces cerevisiae*. I. Ergosterol requirement for growth in a defined medium. *J. cell. comp. Physiol.*, **41**, 23–36.

— (1954). Anaerobic nutrition of *Saccharomyces cerevisiae*. II. Unsaturated fatty acid requirement for growth in a defined medium. *J. cell. comp. Physiol.*, **43**, 271–281.

ANDREWS, T. M. & TATA, J. R. (1971). Protein synthesis by membrane-bound and free ribosomes of secretory and non-secretory tissues. *Biochem. J.*, **121**, 683–694.

ATTARDI, G. & AMALDI, F. (1970). Structure and synthesis of ribosomal RNA. *A. Rev. Biochem.*, **39**, 183–224.

AVERS, C. J. (1971). Peroxisomes of yeast and other fungi. *Sub-cell. Biochem.*, **1**, 25–37.

BANDOW, W. (1972). Membrane separation and biogenesis of the outer membrane of yeast mitochondria. *Biochim. biophys. Acta*, **282**, 105–122.

BARATH, Z. & KÜNTZEL, H. (1972). Induction of mitochondrial RNA polymerase in *Neurospora crassa*. *Nature New Biol.*, **240**, 195–197.

BARNES, R. D. (1963). *Invertebrate Zoology*. Philadelphia: W. B. Saunders.

BEATTIE, D. S., PATTON, G. M. & STUCHELL, R. N. (1970). Studies *in vitro* on amino acid incorporation into purified components of rat liver mitochondria. *J. biol. Chem.*, **245**, 2177–2184.

BIRD, C. W., LYNCH, J. M., PIRT, F. J., REID, W. W., BROOKS, C. J. W. & MIDDLEDITCH, B. S. (1971). Steroids and squalene in *Methylococcus capsulatus* grown on methane. *Nature, Lond.*, **230**, 473.

BLAMIRE, J., CRYER, D. R., FINKELSTEIN, D. R. & MARMUR, J. (1972). Sedimentation properties of yeast nuclear and mitochondrial DNA. *J. molec. Biol.*, **67**, 11–24.

BLOCH, K. (1962). Oxygen and biosynthetic patterns. *Fedn Proc.*, **21**, 1058–1063.

— (1969). Enzymatic synthesis of mono-unsaturated fatty acids. *Accts Chem. Res.*, **2**, 193–202.

BLOCH, K., BARONOWSKY, P., GOLFINE, H., LENNARZ, W. J., LIGHT, R., NORRIS, A. T. & SCHEUERBRANDT, G. (1961). Biosynthesis and metabolism of unsaturated fatty acids. *Fedn Proc.*, **20**, 921–927.

BLOOMFIELD, D. K. & BLOCH, K. (1960). The formation of $\triangle^9$-unsaturated fatty acid. *J. biol. Chem.*, **235**, 337–345.

BOTTOMLEY, W., SMITH, H. J. & BOGORAD, L. (1971). RNA polymerases of maize: partial purification and properties of the chloroplast enzymes. *Proc. natn. Acad. Sci. USA*, **68**, 2412–2416.

BOWMAN, C. M., DAHLBERG, J. E., IKEMURA, T., KONISHY, J. & NOMURA, M. (1971). Specific inactivation of 16 S ribosomal RNA induced by colicin E3 *in vivo*. *Proc. natn. Acad. Sci. USA*, **68**, 964–968.

BRITTEN, R. J. & KOHNE, D. E. (1970). Repeated segments of DNA. *Scient. Am.*, **222**, 24–31.

BRODIE, A. F. & GUTNICK, D. L. (1971). Electron transport and oxidative phosphorylation in microbial systems. In *Electron and Coupled Energy Transfer Systems*, vol. I, part *B*, 599–681 (eds. T. S. King and M. Klingenberg). New York: M. Dekker.

BROWN, D. D. & BLACKLER, A. W. (1972). Gene amplification proceeds by a chromosome copy mechanism. *J. molec. Biol.*, **63**, 75–83.

BROWN, D. D. & DAWID, I. B. (1968). Specific gene amplification in oocytes. *Science, Wash.*, **160**, 272–280.

BROWN, D. D. & WEBER, C. S. (1968a). Gene linkage by RNA–DNA hybridization. I. Unique DNA sequences homologous to 4 S RNA, 5 S RNA and ribosomal RNA. *J. molec. Biol.*, **34,** 661–680.

——— (1968b). Gene linkage by RNA–DNA hybridization. II. Arrangement of the redundant gene sequences for 28 S and 18 S ribosomal RNA. *J. molec. Biol.*, **34,** 681–697.

BROWN, D. D. & WENSINK, P. C. & JORDAN, E. (1972). A comparison of the ribosomal DNA's of *Xenopus laevis* and *Xenopus mulleri*: the evolution of tandem genes. *J. molec. Biol.*, **63,** 57–73.

BROWNLEE, G. G., CARTWRIGHT, E., McSHANE, T. & WILLIAMSON, R. (1972). The nucleotide sequence of somatic 5 S RNA from *Xenopus laevis*. *FEBS Lett.*, **25,** 8–12.

BROWNLEE, G. G., SANGER, F. & BARRELL, B. G. (1968). The sequence of 5 S ribosomal ribonucleic acid. *J. molec. Biol.*, **34,** 379–412.

BUNN, C. L., MITCHELL, C. H., LUKINS, H. B. & LINNANE, A. W. (1970). Biogenesis of mitochondria. XVII. A new class of cytoplasmically determined antibiotic resistant mutants in *Saccharomyces cerevisiae*. *Proc. natn. Acad. Sci. USA*, **67,** 1233–1240.

BURGESS, R. R. (1971). RNA polymerase. *A. Rev. Biochem.*, **40,** 711–740.

CAMPBELL, P. N. (1965). The biosynthesis of proteins. *Prog. Biophys.*, **15,** 3–38.

CHAMBERLIN, M. J. (1974). Bacterial DNA-dependant RNA polymerase. In *The Enzymes*, vol. X, 333–374 (ed. P. D. Boyer). New York: Academic Press.

CHUA, N.-H., BLOBEL, G., SIEKEVITZ, P. & PALADE, G. E. (1973). Attachment of chloroplast polysomes to thylakoid membranes in *Chlamydomonas reinhardtii*. *Proc. natn. Acad. Sci. USA*, **70,** 1554–1558.

COOTE, J. L. & WORK, T. S. (1971). Proteins coded for by mitochondrial DNA of mammalian cells. *Eur. J. Biochem.*, **23,** 564–574.

CONSTANTINO, P. & ATTARDI, G. (1973). Atypical pattern of utilization of amino acids for mitochondrial protein synthesis in HeLa cells. *Proc. natn. Acad. Sci. USA*, **70,** 1490–1494.

COTTRELL, S., RABINOWITZ, M. & GETZ, G. S. (1973). Mitochondrial deoxyribonucleic acid synthesis in a temperature-sensitive mutant of deoxyribonucleic acid replication of *Saccharomyces cerevisiae*. *Biochemistry*, **12,** 4374–4378.

DALGARNO, L. & SHINE, J. (1973). Conserved terminal sequences in 18 S rRNA may represent terminator anticodons. *Nature New Biol.*, **245,** 261–262.

DARNELL, J. E., JELINEK, W. R. & MOLLOY, G. R. (1973). Biogenesis of mRNA: genetic regulation in mammalian cells. *Science, Wash.*, **181,** 1215–1221.

DAYHOFF, M. O. (1972). *Atlas of Protein Sequence and Structure*, vol. **5.** Silver Springs, Maryland: National Biomedical Research Foundation.

DE DUVE, C. (1969). Evolution of the peroxisome. *Ann. N.Y. Acad. Sci.*, **168,** 369–381.

DE DUVE, C. & BAUDHUIN, P. (1966). Peroxisomes (microbodies and related particles). *Physiol. Rev.*, **46,** 323–357.

DEZELEE, S. & SENTENAC, A. (1973). Role of DNA–RNA hybrids in eukaryotes. *Eur. J. Biochem.*, **34,** 41–52.

DICKERSON, R. E. (1971). The structure of cytochrome c and the dates of molecular evolution. *J. molec. Evol.*, **1,** 26–45.

DICKERSON, R. E. & GEIS, I. (1969). *The Structure and Action of Proteins*. New York: Harper and Row.

DIEGELMANN, R. F., BERNSTEIN, L. & PETERKOFSKY, B. (1973). Cell-free collagen synthesis on membrane-bound polysomes of chick embryo connective tissue and the localization of prolyl hydroxylase on the polysome–membrane complex. *J. biol. Chem.*, **248**, 6514–6521.

DiMauro, E., Hollenberg, C. P. & Hall, B. D. (1972). Transcription in yeast: a factor that stimulates yeast RNA polymerases. *Proc. natn. Acad. Sci. USA*, 2818–2822.

Dubin, D. T. (1974). Methylated nucleotide content of mitochondrial ribosomal RNA from hamster cells. *J. molec. Biol.*, **84**, 257–273.

Dubin, D. T., Jones, T. H. & Cleaves, G. R. (1974). An unmethylated '3 S_E' RNA in hamster mitochondria: a 5 S RNA-equivalent? *Biochem. biophys. Res. Comm.*, **56**, 401–406.

Dubuy, B. & Weissman, S. M. (1971). Nucleotide sequence of *Pseudomonas fluorescens* 5 S ribonucleic acid. *J. biol. Chem.*, **246**, 747–761.

Echlin, P. (1970). Photosynthetic apparatus in prokaryotes and eukaryotes. *Symp. Soc. gen. Microbiol.*, **20**, 221–248.

Edelman, M., Verma, I. M., Herzog, R., Galun, E. & Littauer, U. Z. (1971). Physico-chemical properties of mitochondrial ribosomal RNA from fungi. *Eur. J. Biochem.*, **19**, 372–378.

Emerson, R. & Held, A. A. (1969). *Aqualindeansella fermentus* gen. sp. n., a phycomycete adapted to stagnant waters. II. Isolation, cultural characteristics, and gas relations. *Am. J. Bot.*, **56**, 1103–1120.

Erdmann, V. A., Fahnestock, S., Higo, K. & Nomura, M. (1971). Role of 5 S RNA in the functions of 50 S ribosomal subunits. *Proc. natn. Acad. Sci. USA*, **68**, 2932–2936.

Forget, B. G. & Weissman, S. M. (1969). The nucleotide sequence of ribosomal 5 S ribonucleic acid from KB cells. *J. biol. Chem.*, **244**, 3148–3165.

Forrester, I. T., Nagley, P. & Linnane, A. W. (1970). Yeast mitochondrial ribosomal RNA: a new extraction procedure and unusual physical properties. *FEBS Lett.*, **11**, 59–61.

Freeman, K. B., Mitra, R. S. & Bartoov, B. (1973). Characteristics of the base composition of mitochondrial ribosomal RNA. *Sub-cell. Biochem.*, **2**, 183–192.

Fridovich, I. (1972). Superoxide radical and superoxide. *Accts chem. Res.*, **5**, 321–326.

(1974). Evidence for the symbiotic origin of mitochondria. *Life Sci.*, **14**, 819–826.

Ganoza, M. C. & Williams, C. A. (1969). *In vitro* synthesis of different categories of specific proteins by membrane-bound and free ribosomes. *Proc. natn. Acad. Sci. USA*, **63**, 1370–1376.

Gel'man, N. S., Lukoyanova, M. A. & Ostrovskii, D. N. (1967). *Respiration and Phosphorylation of Bacteria*. New York: Plenum Press.

Gissinger, F. & Chambon, P. (1972). Animal DNA-dependent RNA polymerases 2. Purification of calf thymus AI enzymes. *Eur. J. Biochem.*, **28**, 277–282.

Goebel, W. (1973). Extrachromosomal DNA in bacteria. *Angew. Chemie*, **12**, 517–524.

Goebel, W. & Schrempf, H. (1973). Possible involvement of DNA polymerase I in excision of RNA from *Col. E_1* DNA *in vivo*. *Nature New Biol.*, **245**, 39–41.

Goldberg, B. & Green, H. (1964). An analysis of collagen secretion by established mouse fibroblast lines. *J. Cell Biol.*, **22**, 227–258.

Goldfine, H. & Bloch, K. (1963). Oxygen and biosynthetic reactions. In *Control Mechanisms in Respiration and Fermentation*, 81–103 (ed. B. Wright). New York: Ronald.

Gray, P. N. & Attardi, G. (1973). An attempt to identify a presumptive 5 S RNA-equivalent RNA species in mitochondrial ribosomes. *Abst. 13th meeting Am Soc. Cell Biol.*, 120a.

GRAY, P. N., BELLMARE, G., MONIER, R., GARRETT, R. A. & STÖFFLER, G. (1973). Identification of the nucleotide sequences involved in the interaction between *E. coli* 5 S RNA and specific 50 S subunit proteins. *J. molec. Biol.*, **77**, 133–152.

GRAY, P. N. & MONIER, R. (1971). Formation of a complex between 23 S RNA, 5 S RNA and proteins from *E. coli* 50 S ribosomal subunits. *FEBS Lett.*, **18**, 145–148.

GREGORY, E. M., YOST, F. J. Jr & FRIDOVICH, I. (1973). Superoxide dismutases of *Escherichia coli*: intracellular localization and functions. *J. Bact.*, **115**, 987–991.

GROOT, P. H. E., AAIJ, C. & BORST, P. (1970). Variations with temperature of the apparent molecular weight of rat liver mitochondrial RNA determined by gel electrophoresis. *Biochem. biophys. Res. Comm.*, **41**, 1321–1327.

GROSSMAN, L. I., WATSON, R. & VINOGRAD, J. (1973). Sensitivity of mitochondrial DNA to alkali and ribonuclease H. *Fedn Proc. Abst.*, **1747**, 529.

GUIARD, B. & LEDERER, F. (1973). Structure of the heme binding site of baker's yeast L-lactate dehydrogenase (cytochrome b₂). *9th Int. Cong. Biochem. Abstr.*, **2n7**, 92.

HALL, J. B. (1973). The nature of the host in the origin of the eukaryote cell. *J. theoret. Biol.*, **38**, 413–418.

HAROLD, F. M. (1972). Conservation and transformation of energy by bacterial membranes. *Bact. Rev.*, **36**, 172–230.

HARTMAN, J. F., ZIEGLER, M. M. & COMB, D. G. (1971). Sea urchin embryogenesis. I. RNA synthesis by cytoplasmic and nuclear genes during development. *Devl Biol.*, **25**, 209–231.

HARTMANN, G., HONIKEL, K. O., KNÜSEL, F. & NÜESCH, J. (1967). The specific inhibition of the DNA-dependent RNA synthesis by rifamycin. *Biochem. biophys. Acta*, **145**, 843–844.

HAYAISHI, O. & NOZAKI, M. (1969). Nature and mechanisms of oxygenases. *Science, Wash.*, **164**, 389–396.

HELD, A. A. (1970). Nutrition and fermentative energy metabolism of the water mould *Aqualinderella fermentans*. *Mycologia*, **62**, 339–358.

HICKS, S. J., DRYSDALE, J. W. & MUNRO, H. N. (1969). Preferential synthesis of ferritin and albumin by different populations of liver polysomes. *Science, Wash.*, **164**, 584–585.

HIGGINS, T. J. V., MERCER, J. F. B. & GOODMAN, P. B. (1973). Poly (A) sequences in plant polysomal RNA. *Nature New Biol.*, **246**, 68–70.

HINDLEY, J. & PAGE, S. M. (1972). Nucleotide sequence of yeast 5 S ribosomal RNA. *FEBS Lett.*, **26**, 157–160.

HINKLE, D. C., MANGEL, W. F. & CHAMBERLIN, M. J. (1972). Studies of the binding of *E. coli* RNA polymerase to DNA. IV. The effect of rifampicin on binding and on RNA chain initiation. *J. molec. Biol.*, **70**, 209–220.

HIRSCH, M. & PENMAN, S. (1973). Mitochondrial polyadenylic acid-containing RNA: localization and characterization. *J. molec. Biol.*, **80**, 379–391.

HOLLOWAY, P. W. & WAKIL, S. J. (1970). Requirement for reduced diphosphopyridine nucleotide cytochrome b_5 reductase in stearyl coenzyme A desaturation. *J. biol. Chem.*, **245**, 1862–1865.

HORIO, T. & KAMEN, M. D. (1970). Bacterial cytochromes. II. Functional aspects. *A. Rev. Microbiol.*, **24**, 399–428.

HRYCAY, E. G. & O'BRIEN, P. J. (1974). Microsomal electron transport. II. Reduced nicotinamide adenine dinucleotide–cytochrome b₅ reductase and cytochrome P-450 as electron carriers in microsomal NADH-peroxidase activity. *Arch. Biochem. Biophys.*, **160**, 230–245.

HUGHES, E. D., LLOYD, D. & BRIGHTWELL, R. (1970). Structure, function and distribution of organelles in prokaryotic and eukaryotic microbes. *Symp. Soc. gen. Microbiol.*, **20**, 295–322.

HUNGATE, R. E. (1967). *The Rumen and its Microbes*. New York: Academic Press.

JACOB, S. T., SAJDAL, E. M. & MUNRO, H. N. (1970). Different responses of soluble whole nuclear RNA polymerase and soluble nucleolar RNA polymerase to divalent cations and to inhibition by α-amanitin. *Biochem. biophys. Res. Comm.*, **38**, 765–770.

KADENBACH, B. & HADVARY, P. (1973). Demonstration of two types of protein synthesized in isolated rat liver mitochondria. *Eur. J. Biochem.*, **32**, 343–349.

KAMEN, M. D. & HORIO, T. (1970). Bacterial cytochromes. I. Structural aspects. *A. Rev. Biochem.*, **39**, 673–700.

KEDINGER, C., GISSINGER, F., GNIAZDOWSKI, M., MANDEL, J-L. & CHAMBON, P. (1972). Animal DNA-dependent RNA polymerases. I. Large-scale solubilization and separation of A and B calf-thymus RNA-polymerase activities. *Eur. J. Biochem.*, **28**, 269–276.

KEDINGER, C. & CHAMBON, P. (1972). Animal DNA-dependent RNA polymerases. III. Purification of calf-thymus BI and BII enzymes. *Eur. J. Biochem.*, **28**, 283–290.

KEDINGER, C., GNIAZDOWSKI, M., MANDEL, J-L., GISSINGER, F. & CHAMBON, P. (1970). α-Amatin: a specific inhibitor of one of two DNA-dependent RNA polymerase activities from calf thymus. *Biochem. biophys. Res. Comm.*, **38**, 165–171.

KEITH, A. D., WISNIESKI, B., HENRY, S. & WILLIAMS, J. C. (1972). Membranes of yeast and *Neurospora*: lipid mutants and physical studies. In *Biological Membranes of Eucaryotic Microbes*, 259–321 (ed. J. Erwin). New York: Academic Press.

KELLEMS, R. D., ALLISON, V. F. & BUTOW, R. A. (1974). Cytoplasmic type 80 S ribosomes associated with yeast mitochondria. II. Evidence for the association of cytoplasmic ribosomes with the outer mitochondrial membrane *in situ*. *J. biol. Chem.*, **249**, 3297–3303.

KIMMEL, C. B. (1969). On the RNA in cultured myeloma cells producing immunoglobulin. *Biochim. biophys. Acta*, **182**, 361–374.

KIMURA, M. & OHTA, T. (1973). Eukaryote–prokaryote divergence estimated by 5 S ribosomal RNA sequences. *Nature New Biol.*, **243**, 199–200.

KLAGSBRUN, M. (1973). An evolutionary study of the methylation of transfer and ribosomal ribonucleic acid in prokaryote and eukaryote organisms. *J. biol. Chem.*, **248**, 2612–2620.

KLINE, B. C. (1974). Mechanism and biosynthetic requirements for F plasmid replication in *Escherichia coli*. *Biochemistry, N.Y.*, **13**, 139–146.

KÜNTZEL, H. & SCHÄFER, K. P. (1971). Mitochondrial RNA polymerase from *Neurospora crassa*. *Nature New Biol.*, **231**, 265–269.

KURIYAMA, Y. & LUCK, D. J. L. (1973*a*). Ribosomal RNA synthesis in mitochondria of *Neurospora crassa*. *J. molec. Biol.*, **73**, 425–437.

(1973*b*) Membrane-associated ribosomes in mitochondria of *Neurospora crassa*. *J. Cell Biol.*, **59**, 776–784.

(1974). Methylation and processing of mitochondrial ribosomal RNA in *poky* and wild-type *Neurospora crassa*. *J. molec. Biol.*, **83**, 253–266.

LAVA-SANCHEZ, P. A., AMALDI, F. & LAPOSTA, A. (1972). Base composition of ribosomal RNA and evolution. *J. molec. Evol.*, **2**, 44–55.

LAVELLE, F., MICHELSON, A. M. & DIMITRIJEVIC, L. (1973). Biological protection by superoxide dismutase. *Biochem. biophys. Res. Comm.*, **55**, 350–357.

LEVIN, W., RYAN, D., WEST, S. & LU, A. Y. H. (1974). Preparation of partially purified lipid-depleted cytochrome P-450 and reduced nicotinamide adenine dinucleotide phosphate cytochrome c reductase from rat liver mitochondria. *J. biol. Chem.*, **249**, 1747–1754.

LEWONTIN, R. C. (1970). The units of selection. *A. Rev. Ecol. Syst.*, **1**, 1–18.

LIN, D. K., NIECE, R. L. & FITCH, W. M. (1973). The properties and amino-acid sequence of cytochrome c from *Euglena gracilis*. *Nature, Lond.*, **241**, 533–535.

LINDELL, T. J., WEINBERG, F., MORRIS, P. W., ROEDER, R. G. & RUTTER, W. J. (1970). Specific inhibition of nuclear RNA polymerase II by α-amanitin. *Science, Wash.*, **170**, 447–449.

LINNANE, A. W., HASLAM, J. M., LUKINS, H. B. & NAGLEY, P. (1972). The biogenesis of mitochondria in micro-organisms. *A. Rev. Microbiol.*, **26**, 163–198.

LIZARDI, P. M. & LUCK, D. J. L. (1971). Absence of a 5 S RNA component in the mitochondrial ribosomes of *Neurospora crassa*. *Nature New Biol.*, **229**, 140–142.

LOENING, U. (1968). Molecular weights of ribosomal RNA in relation to evolution. *J. molec. Biol.*, **38**, 355–365.

McCONKEY, E. H. (1974). Composition of mammalian ribosomal subunits: A re-evaluation. *Proc. natn. Acad. Sci. USA*, **71**, 1379–1383.

McCONKEY, E. H. & HOPKINS, J. W. (1969). Molecular weights of some HeLa ribosomal RNA's. *J. molec. Biol.*, **39**, 545–550.

McCORD, J. M., KEELE, B. B. Jr. & FRIDOVICH, I. (1971). An enzyme-based theory of obligate anaerobiosis: the physiological function of superoxide dismutase. *Proc. natn. Acad. Sci. USA*, **68**, 1024–1027.

McLAUGHLIN, C. S., WARNER, J. R., EDMONDS, M., NAKAZATO, H. & VAUGHN, M. H. (1973). Polyadenylic acid sequences in yeast messenger ribonucleic acid. *J. biol. Chem.*, **248**, 1466–1471.

MAHLER, H. R. (1973). Biogenetic autonomy in mitochondria. *CRC crit. Rev. Biochem.*, **1**, 381–460.

MAHLER, H. R. & DAWIDOWICZ, K. (1973). Autonomy of mitochondria of *Saccaromyces cerevisiae* in their production of messenger RNA. *Proc. natn. Acad. Sci. USA*, **70**, 111–114.

MAHLER, H. R. & PERLMAN, P. S. (1971). Mitochondriogenesis as analyzed by blocks on mitochondrial translation and transcription. *Biochemistry, N.Y.*, **10**, 2979–2990.

MAHLER, H. R. & RAFF, R. A. (1975). The evolutionary origin of the mitochondrion: a non-symbiotic model. *Int. Rev. Cytol.* (in press).

MARGULIS, L. (1970). *Origin of Eukaryotic Cells*. New Haven, Conn.: Yale University Press.

MEILHAC, M., TYSPER, A. & CHAMBON, P. (1972). Animal DNA-dependent RNA polymerases. 4. Studies on inhibition by rifamycin derivatives. *Eur. J. Biochem.*, **28**, 291–300.

MEYER, R. R. (1973). On the evolutionary origin of mitochondrial DNA. *J. Theoret. Biol.*, **38**, 647–663.

MIZUNO, S., YAMAZAKI, H., NITTA, K. & UMEZAWA, H. (1968). Inhibition of DNA-dependent RNA polymerase reaction of *E. coli* by an antimicrobial antibiotic, streptovaricin. *Biochim. biophys. Acta*, **157**, 322–332.

MONIER, R. (1972). Structure and function of ribosomal RNA. In *The Mechanisms of Protein Synthesis and its Regulation*, 353–394 (ed. L. Bosch). Amsterdam: North-Holland.

MORROW, J. G., COHEN, S. N., CHANG, A. C. Y., BOYER, H. W., GOODMAN, H. M. & HELLING, R. B. (1974). Replication and transcription of eukaryotic DNA in *E. coli*. *Proc. natn. Acad. Sci. USA*, **71**, 1743–1747.

MULLINIX, K. P., STRAIN, G. C. & BOGORAD, L. (1973). RNA polymerases of maize. Purification and molecular structure of DNA-dependent RNA polymerase II. *Proc. natn. Acad. Sci. USA*, **70**, 2386–2390.

MURRAY, D. R. & LINNANE, A. W. (1972). Synthesis of proteolipid protein by yeast mitochondria. *Biochem. biophys. Res. Comm.*, **49**, 855–862.

NEUPERT, W. & LUDWIG, G. D. (1971). Sites of biosynthesis of outer and inner membrane proteins of *Neurospora crassa* mitochondria. *Eur. J. Biochem.*, **19**, 523–532.

NEWTON, J. W. & KAMEN, M. D. (1961). Cytochrome systems in anaerobic electron transport. In *The Bacteria*, vol. II, 397–423 (eds. I. C. Gunsalus and R. Y. Stanier). New York: Academic Press.

NIELSEN, M. H., LUDVIK, J. & NIELSEN, R. (1966). On the ultrastructure of *Trichomonas vaginalis* Donné. *J. Microsc.*, **5**, 229–250.

NOBREGA, F. G. & OZOLS, J. (1971). Amino acid sequence of tryptic peptides of cytochromes b_5 from microsomes of human, monkey, porcine and chicken liver. *J. biol. Chem.*, **246**, 1706–1717.

OELZE, J. & DREWS, G. (1972). Membranes of photosynthetic bacteria. *Biochim. biophys. Acta*, **265**, 209–239.

OJALA, D. & ATTARDI, G. (1973). Expression of the mitochondrial genome in HeLa cells. XIX. Occurrence in mitochondria of polyadenylic acid sequences, 'free' and covalently linked to mitochondrial DNA-coded RNA. *J. molec. Biol.*, **82**, 151–175.

OSHINO, N., IMAI, Y. & SATO, R. (1971). A function of cytochrome b_5 in fatty acid desaturation by rat liver microsomes. *J. Biochem.*, **69**, 155–167.

OZOLS, J. (1974). Cytochrome b_5 from microsomal membranes of equine, bovine and porcine livers. Isolation and properties of preparations containing the membraneous segment. *Biochemistry*, **13**, 426–433.

OZOLS, J. & STRITTMATTER, P. (1967). The homology between cytochrome b_5, hemoglobin, and myoglobin. *Proc. natn. Acad. Sci. USA*, **58**, 264–267.

PALADE, G. E. (1966). Structure and function at the cellular level. *J. Am. med. Assn*, **198**, 815–825.

PENMAN, S., VESCO, C. & PENMAN, M. (1968). Localization and kinetics of formation of nuclear heterodisperse RNA, cytoplasmic heterodisperse RNA and polyribosome-associated messenger RNA in HeLa cells. *J. molec. Biol.*, **34**, 49–69.

PERLMAN, S., ABELSON, H. T. & PENMAN, S. (1973). Mitochondrial protein synthesis: RNA with the properties of eukaryotic messenger RNA. *Proc. natn. Acad. Sci. USA*, **70**, 350–353.

PERLMAN, P. S. & MAHLER, H. R. (1974). Derepression of mitochondria and their enzymes in yeast: regulatory aspects. *Arch. Biochem. Biophys.*, **162**, 248–271.

PERRY, R. P., GREENBERG, J. R., KELLEY, D. E., LATORRE, J. & SCHOCHETMAN, G. (1973). Messenger RNA: its origin and fate in the mammalian cell. In *Gene Expression and its Regulation*, 149–168 (ed. F. T. Kenny). New York: Plenum Press.

PETERS, T. Jr (1962). The biosynthesis of rat serum albumin. II. Intracellular phenomena in the secretion of newly formed albumin. *J. biol. Chem.*, **237**, 1186–1189.

PETTIGREW, G. W. (1972). The amino acid sequence of a cytochrome c from a protozoan *Crithidia oncopelti*. *FEBS Lett.*, **22**, 64–66.

—— (1973). The amino acid sequence of cytochrome c from *Euglena gracilis*. *Nature Lond.*, **241**, 531–533.

PONTA, H., PONTA, U. & WINTERSBERGER, E. (1972). Purification and properties of DNA-dependent RNA polymerases from yeast. *Eur. J. biochem.*, **29**, 110–118.

POYTON, R. O. & SCHATZ, G. (1975). Cytochrome c oxidase from Baker's yeast. III. Physical characterization of isolated subunits and chemical evidence for two different classes of polypeptides. *J. biol. Chem.* **250**, 762–766.

PREER, J. R. Jr (1971). Extrachromosomal inheritance: hereditary symbionts, mitochondria, chloroplasts. *A. Rev. Genet.*, **5**, 361–406.

RAFF, R. A. (1970). Molecular weights of *Hydra* and planarian ribosomal RNA. *Curr. Mod. Biol.*, **3**, 250–252.

RAFF, R. A. & MAHLER, H. R. (1972). The non-symbiotic origin of mitochondria. *Science, Wash.*, **177**, 575–582.

REDMAN, C. M. (1967). Studies on the transfer of incomplete polypeptide chains across rat liver microsomal membranes *in vitro*. *J. biol. Chem.*, **242**, 761–768.

(1969). Biosynthesis of serum proteins and ferritin by free and attached ribosomes of rat liver. *J. biol. Chem.*, **244**, 4308–4315.

REDMAN, C. M. & SABATINI, D. D. (1966). Vectorial discharge of peptides released by puromycin from attached ribosomes. *Proc. natn. Acad. Sci. USA*, **56**, 608–615.

REDMAN, C. M., SIEKEVITZ, P. & PALADE, G. E. (1966). Synthesis and transfer of amylase in pigeon pancreatic microsomes. *J. biol. Chem.*, **241**, 1150–1158.

REED, J. & WINTERSBERGER, E. (1973). Adenylic acid-rich sequences in messenger RNA from yeast polysomes. *FEBS Lett.*, **32**, 213–217.

REID, B. D. & PARSONS, P. (1971). Partial purification of mitochondrial RNA polymerase from rat liver. *Proc. natn. Acad. Sci. USA*, **68**, 2830–2834.

REMSEN, C. C., VALOIS, F. W. & WATSON, S. W. (1967). Fine structure of the cytomembranes of *Nitrosocystis oceanus*. *J. Bact.*, **94**, 422–433.

RICHMOND, M. H. (1970). Plasmids and chromosomes in prokaryotic cells. *Symp. Soc. gen. Microbiol.*, **20**, 249–294.

RICHMOND, M. H. & WIEDMAN, B. (1974). Plasmids and bacterial evolution. *Symp. Soc. gen. Microbiol.* **24**, 59–85.

ROEDER, R. G. (1974). Multiple forms of deoxyribonucleic acid-dependent ribonucleic acid polymerase in *Xenopus laevis*. *J. biol. Chem.*, **249**, 241–248.

ROEDER, R. G. & RUTTER, W. J. (1969). Multiple forms of DNA-dependent RNA polymerases in eukaryotic organisms. *Nature, Lond.*, **224**, 234–237.

(1970a). Multiple ribonucleic acid polymerases and ribonucleic acid synthesis during sea urchin development. *Biochemistry, N.Y.*, **9**, 2543–2553.

(1970b). Specific nucleolar and nucleoplasmic RNA polymerases. *Proc. natn. Acad. Sci. USA*, **65**, 675–682.

ROSBASH, M. & PENMAN, S. (1971). Membrane-associated protein synthesis of mammalian cells. I. The two classes of membrane-associated ribosomes. *J. molec. Biol.*, **59**, 227–241.

ROWE, M. J., LANSMAN, R. & WOODWARD, D. O. (1973). Characterization of the products of *Neurospora* mitochondrial protein synthesis. *Fedn Proc.* **32**, *Abst.*, **2407**, 641.

RUSSELL, R. L., ABELSON, J. N., LAUDY, A., GEFTER, M. L., BRENNER, S. & SMITH, J. D. (1970). Duplicate genes for tyrosine transfer RNA in *E. coli*. *J. molec. Biol.*, **47**, 1–13.

SAGHER, D., EDELMAN, M. & JAKOH, K. M. (1974). Poly (A)-associated RNA in plants. *Biochim. biophys. Acta*, **349**, 32–38.

SANKOFF, D., MOREL, C. & CEDERGREN, R. J. (1973). Evolution of 5 S RNA and the non-randomness of base replacement. *Nature New Biol.*, **245**, 232–234.

SCALLEN, T. J., DEAN, W. J. & SCHUSTER, M. W. (1968). Enzymatic conversion of squalene to cholesterol by an acetone powder of rat liver microsomes. *J. biol. Chem.*, **243**, 5202–5206.

SCRAGG, A. H. (1971). Mitochondrial DNA-directed RNA polymerase from *Saccharomyces cerevisiae* mitochondria. *Biochem. biophys. Res. Comm.*, **45**, 701–706.

——— (1974). A mitochondrial DNA-directed polymerase from yeast mitochondria. In *The Biogenesis of Mitochondria*, 47–57 (eds. A. M. Kroon and C. Saccone). New York: Academic Press.

SEBALD, W., MACHLEIDT, W. & OTTO, J. (1973). Products of mitochondrial protein synthesis in *Neurospora crassa*. *Eur. J. Biochem.*, **38**, 311–324.

SENIOR, B. W. & HOLLAND, I. B. (1971). Effect of colicin E3 upon the 30 S ribosomal subunit of *E. coli*. *Proc. natn. Acad. Sci. USA*, **68**, 959–963.

SHIVELY, J. M., BALL, F., BROWN, D. H. & SAUNDERS, R. E. (1973). Functional organelles in prokaryotes: polyhedral inclusion (carboxysomes) of *Thiobacillus neapolitanus*. *Science, Wash.*, **182**, 584–586.

SIERRA, M. F. & TZAGOLOFF, A. (1973). Assembly of the mitochondrial membrane system. Purification of a mitochondrial product of the ATPase. *Proc. natn. Acad. Sci. USA*, **70**, 3155–3159.

SILVER, J. C. & HORGAN, P. A. (1974). Hormonal regulation of presumptive mRNA in the fungus *Achlya ambisexualis*. *Nature, Lond.*, **249**, 252–254.

SIMPSON, G. G. (1953). *The Major Features of Evolution*. New York: Columbia University Press.

SMITH, L. (1961). Cytochrome systems in aerobic electron transport. In *The Bacteria*, vol. II, 365–396 (eds. I. C. Gunsalus and R. Y. Stanier). New York: Academic Press.

STANIER, R. Y. (1970). Some aspects of the biology of cells and their possible evolutionary significance. *Symp. Soc. gen. Microbiol.*, **20**, 1–38.

STEINMAN, H. M. & HILL, R. L. (1973). Sequence homologies among bacterial and mitochondrial superoxide dismutases. *Proc. natn. Acad. Sci. USA*, **70**, 3725–3729.

SY, J. & McCARTHY, K. S. (1968). Ribosomal RNA of *Arbacia punctulata*. *Biochim. biophys. Acta*, **166**, 571–574.

TAKAGI, M. & OGATA, K. (1968). Direct evidence for albumin biosynthesis by membrane bound polysomes in rat liver. *Biochem. biophys. Res. Comm.*, **33**, 55–60.

TATA, J. R. (1971). Ribosomal segregation as a possible function for the attachment of ribosomes to membranes. *Sub-cell. Biochem.*, **1**, 83–89.

TAYLOR, F. J. R. (1974). Implications and extensions of the serial endosymbiosis theory of the origin of eukaryotes. *Taxon*, **23**, 229–258.

TCHEN, T. T. & BLOCH, K. (1957). On the conversion of squalene to lanosterol *in vitro*. *J. biol. Chem.*, **226**, 921–930.

TOWERS, N. R., KELLERMAN, G. M., RAISON, J. K. & LINNANE, A. W. (1973). The biogenesis of mitochondria. 29. Effects of temperature induced phase changes in membranes on protein synthesis by mitochondria. *Biochim. biophys. Acta*, **299**, 153–161.

TSAI, M. J., MICHAELIS, G. & CRIDDLE, R. S. (1971). DNA-dependent RNA polymerase from yeast mitochondria. *Proc. natn. Acad. Sci. USA*, **66**, 473–479.

TZAGOLOFF, A. & AKAI, A. (1972). Assembly of the mitochondrial system. VIII. Properties of the products of mitochondrial protein synthesis in yeast. *J. biol. Chem.*, **247**, 6517–6523.

ULLRICH, V. (1972). Enzymatic hydroxylations with molecular oxygen. *Angew. Chemie, Int. Edn*, **11**, 701–712.

UZZELL, T. & SPOLSKY, C. (1974). Mitochondria and plastids as endosymbionts: a revival of special creation? *Am. Scient.*, **62**, 334–343.

VILLA, V. D., MORIMOTO, H. & HALVORSON, H. O. (1974). An improved method for the isolation of the membrane-bound mitochondrial ribosomes in yeast. *J. Bact.* (in press).

WALSBY, A. E. (1972). Structure and function of gas vacuoles. *Bact. Rev.*, **36**, 1–32.

WATSON, S. W., GRAHAM, L. B., REMSEN, C. C. & VALOIS, F. W. (1971). A lobular, ammonia-oxidizing bacterium, *Nitrosolobus multiformis nov. gen. nov. sp. Arch. Mikrobiol.*, **76**, 183–203.

WATSON, S. W. & REMSEN, C. C. (1970). Cell envelope of *Nitrocystis oceanus. J. ultrastruct. Res.*, **33**, 148–160.

WATSON, S. W. & WATERBURY, J. B. (1971). Characteristics of two marine nitrite oxidizing bacteria, *Nitrospina gracilis nov. gen. nov. sp.* and *Nitrococcus mobilis nov. gen. nov. sp. Arch. Microbiol.*, **77**, 203–230.

WEAVER, R. F., BLATTI, S. P. & RUTTER, W. J. (1971). Molecular structure of DNA-dependent RNA polymerases (II) from calf thymus and rat liver. *Proc. natn. Acad. Sci. USA*, **68**, 2994–2999.

WEISIGER, R. A. & FRIDOVICH, I. (1973a). Superoxide dismutase. *J. biol. Chem.*, **248**, 3582–3592.

WEISIGER, R. A. & FRIDOVICH, I. (1973b). Mitochondrial superoxide dismutase. *J. biol. Chem.*, **248**, 4793–4796.

WHITTAKER, R. H. (1969). New concepts of kingdoms of organisms. *Science, Wash.*, **163**, 150–160.

WILLETT, J. D., SHARPLESS, K. B., LORD, K. E., VAN TAMELEN, E. E. & CLAYTON, R. B. (1967). Squalene-2, 3-oxide, an intermediate in the enzymatic conversion of squalene to lanosterol and cholesterol. *J. biol. Chem.*, **242**, 4182–4191.

WILLIAMS, P. H., BOYER, H. W. & HELINSKI, D. R. (1973). Size and base composition of RNA in supercoiled plasmid DNA. *Proc. natn. Acad. Sci. USA*, **70**, 3744–3748.

WINTERSBERGER, E. (1970). DNA-dependent RNA polymerase from mitochondria of a cytoplasmic 'petite' mutant of yeast. *Biochem. biophys. Res. Comm.*, **40**, 1179–1184.

WOESE, C. R. & BLEYMAN, M. A. (1972). Genetic code limit organisms – do they exist? *J. molec. Evol.*, **1**, 223–229.

WONG-STAAL, F., MENDELSOHN, J. & GOULIAN, M. (1973). Ribonucleotides in closed circular mitochondrial DNA from HeLa cells. *Biochem. biophys. Res. Comm.*, **53**, 140–148.

WU, G.-J. & DAWID, I. B. (1972). Purification and properties of mitochondrial deoxyribonucleic acid dependent ribonucleic acid polymerases from ovaries of *Xenopus laevis. Biochemistry, N.Y.*, **11**, 3589–3593.

YU, C.-A., GUNSALUS, I. C., KATAGIRI, M., SUHARA, K. & TAKEMORI, S. (1974). Cytochrome P-450$_{CAM}$. I. Crystallization and properties. *J. biol. Chem.*, **249**, 94–101.

YU, M. T., VERMEULIN, C. W. & ATWOOD, K. C. (1970). Location of the genes for 16 S and 23 S ribosomal RNA in the genetic map of *E. coli. Proc. natn. Acad. Sci. USA*, **67**, 26–31.

YUAN, C. & BLOCH, K. (1961). Conversion of oleic acid to linoleic acid. *J. biol. Chem.*, **236**, 1277.

ZYLBER, E., VESCO, C. & PENMAN, S. (1969). Selective inhibition of the synthesis of mitochondria-associated RNA by ethidium bromide. *J. molec. Biol.*, **44**, 195–204.

BDELLOVIBRIO AS SYMBIONT: THE ASSOCIATIONS OF BDELLOVIBRIOS WITH OTHER BACTERIA INTERPRETED IN TERMS OF A GENERALIZED SCHEME FOR CLASSIFYING ORGANISMIC ASSOCIATIONS

By M. P. STARR

Department of Bacteriology, University of California,
Davis, California 95616, USA

INTRODUCTION

The classificatory scheme

A generalized scheme for classifying organismic associations has been sketched out earlier in this volume.* This classificatory scheme stems from a broadly based analysis of many kinds of associations between organisms of all sorts: viruses, bacteria, eukaryotic protista, lower and higher plants, and invertebrate and vertebrate animals. This analysis and other considerations have led me to the conviction that many organismic associations can be covered quite adequately by the term 'symbiosis' in the sense of de Bary (1879), as emended slightly, made more explicable, and updated by me to yield the summarizing statement: 'a symbiosis is an association between two organisms; the term "organism" is taken very broadly to mean any cellular or non-cellular biological entity which has the capacities to replicate itself and to respond to evolutionary forces or which, minimally, can commandeer in a self-perpetuating and evolutionarily responsive manner the genetic apparatus and services of another biological entity; the association in a symbiosis must be significant to the well-being (or "unwell-being" or both) of at least one of the associants; the associants in a symbiosis may be of different taxa or of the same taxon' (this volume, p. 8).

In my classificatory scheme – which has evolved from an earlier, aborted effort (Heise & Starr, unpublished; summarized in Starr & Chatterjee, 1972) – the significant properties of a particular organismic association are specified in terms of a set of continua which act as criteria. Each

* For details of the classificatory scheme and the terminology which I use in this article, the reader should turn to my other article on p. 1 of this volume.

continuum in the set focuses on a particular aspect of organismic associations. The continua in my present, still tentative, classificatory scheme are based on the following criteria: A. relative spatial relationships (exhabitational–inhabitational); B. relative sizes (isosymbionts–anisosymbionts); C. temporal (durability) factors (persistent–transient); D. degree of necessariness (obligate–facultative); E. modes of dependence; F. viability of nutritional source (biotrophic–necrotrophic); G. degree of specificity; H. harmful or beneficial effects (and modes thereof); I. factors pertaining to integration (integrates–separates). (For further details, see the Table 1 and the text of my earlier article.) The properties of a particular organismic association can be specified in terms of their qualitative and quantitative positions within each continuum – with whatever precision the fact-base and purpose warrant.

A 'textual footnote' intended to alleviate concern about the word 'symbiosis'

As I said in my earlier article, many biologists have, perhaps quite justifiably in view of the confusing usages, developed linguistic hang-ups pertaining to the word 'symbiosis'. One of my purposes here is to facilitate communication; I do not wish to cause semantic pain and thus withdrawal from any dialogue! The labels 'symbiosis' and 'symbiont' are not inherently essential to the present enterprise. What is essential here is to make classificatory distinctions that enable particular kinds of organismic associations to be clearly specified and usefully delineated from other kinds. Hence, if it reduces the load on one's semantic pain-centers, the less committal words 'organismic association' (or 'association') and 'organismic associant' (or 'associant') could be mentally substituted in most places where the words 'symbiosis' and 'symbiont' appear in this essay.

Raison d'être and plan

As I have explained in my first article, one motive for initiating and carrying out this conceptual analysis and synthesis was to bring some terminological clarity into the systems of organismic associations which have been long-term objects of my experimental interest. The time has come to attempt application of the classificatory scheme to such systems. The associations of phytopathogenic bacteria with plants (Starr, 1959; Starr & Chatterjee, 1972), much too diverse to yield the compact analysis needed for the present enterprise, will have to be dealt with at another time and place. Hence, I have turned here to another system, also under intensive study by us for some time, the associations of *Bdellovibrio* with other bacteria (Stolp & Starr, 1963; Starr & Seidler, 1971; Snellen &

Starr, 1974). A brief summary of the salient facts about *Bdellovibrio* and this interbacterial association will be presented first, followed by an attempt to apply my classificatory scheme to this system.

THE ASSOCIATIONS OF SYMBIOSIS-COMPETENT BDELLOVIBRIOS WITH OTHER BACTERIA

The discovery of bdellovibrios and other prefactory matters

The bacteria which now comprise the genus *Bdellovibrio* (Stolp & Starr, 1963) were discovered 'accidentally' in 1962 by Heinz Stolp (Stolp & Petzold, 1962; Stolp, 1973) when he was attempting to isolate bacterial viruses (bacteriophages; phages). Stolp had inoculated soil filtrates onto the confluent cellular growths (lawns) of the bacterium *Pseudomonas phaseolicola* being cultivated, in the conventional double-layer method, on nutrient agar in Petri plates. Following the usual 24 hour incubation period, the lawns showed no cleared zones (bacteriophage plaques) where the bacteria would have been lysed by phages in the otherwise opaque bacterial lawns. For some undetermined reason, Stolp did not discard the plates – as one would expect a phage worker to do because phages develop only in young, growing bacteria – but he re-examined them after another 24 hours had passed. The later examination was crucial for, at this time, plaques were evident in the *Pseudomonas* lawns; these plaques continued to increase in size for about a week. Because of the delay in starting plaque formation and the continuing increase in plaque size, Stolp concluded that conventional bacteriophages were probably not the cause of the plaques. He then – quite exceptionally for a phage researcher – examined material from the plaques in a phase-contrast microscope.

A large number of rapidly motile tiny microbes, in addition to a few cells of the larger pseudomonad, were demonstrated by these microscopic observations of material removed from the developing plaques. The small organisms collided with the much larger pseudomonad cells, adhered to their surfaces, and seemed to cause them to lyse. The lytic agent was clearly not a phage: the agent did not pass through a filter with very small pore size (200 nm) through which phages can pass, nor did plaques form on a streptomycin-containing medium with a lawn of streptomycin-resistant pseudomonad cells, whereas phages do yield plaques under these conditions.

A comprehensive study (Stolp & Starr, 1963) of these unusual microbes led to the conclusion that they are indeed unique bacteria. Based on the ability to enter into this peculiar interbacterial association, and the bacterio-lytic and other properties including the relatively small size of these bacteria, a new genus, *Bdellovibrio*, was established (Stolp & Starr, 1963), with the specific epithet, *bacteriovorus*, for the single species then recognized. The generic name reflects the organism's mode of initial attachment to other bacteria and its shape ('*Bdello-*' is derived from the Greek word for a leech and '*-vibrio*' refers to its comma shape), while '*bacteriovorus*' refers to the fact that *Bdellovibrio* seemed to devour the bacteria with which it was associated.

It might be useful to examine the epistemological implications of this early work on *Bdellovibrio*. In view of the presently known ubiquity of *Bdellovibrio* in soils and waters the world over, it is quite likely that many bacteriologists had 'seen-without-seeing' bdellovibrios, because they had no prior knowledge or concept of such creatures. Then came the 'chance' observation in 1962 by Stolp of the late-forming plaques in bacterial lawns. His microscopic examination, as already related, was crucial because it provided the first 'seeing' glimpse of these highly motile, tiny, vibrioid bacteria which attached to other bacteria and seemed to cause them to lyse. Taking into account that another unusual and (at the time) little-known bacterium, *Caulobacter*, was known to attach sometimes to other bacteria, the early notion (Stolp & Petzold, 1962) that *Bdellovibrio* might be some sort of *Caulobacter*-like creature is understandable. However, this notion was soon corrected. Comparative study (Stolp & Starr, 1963) of several strains of *Bdellovibrio* in 1962–1963 extended the factual base; the conceptual base, unfortunately, hardened into an erroneous notion – namely, 'ectoparasitism' – which conceived of *Bdellovibrio* as being able to act antagonistically upon its symbiont only from outside the symbiont. Despite numerous clues – now patently clear from hindsight – it took another two years before the error was corrected and the ability of *Bdellovibrio* to enter other bacterial cells and to develop in this intramural (intraperiplasmic) locus was established experimentally (Starr & Baigent, 1966; Scherff, DeVay & Carroll, 1966).

The account which follows presents very briefly the salient facts about *Bdellovibrio*, emphasizing those which bear on its association with other bacteria. Further details and other aspects are treated in recent review articles (Shilo, 1969, 1973*a*; Starr & Seidler, 1971; Starr & Huang, 1972; Stolp, 1973; Varon, 1974). This first part of the presentation deals only with the significant associations of *Bdellovibrio* cells with the cells of other bacteria. That is to say, we shall consider here only those aspects of the life of the bdellovibrios which are symbiotic (in the sense given by the statement in the Introduction). Bdellovibrios which are capable of entering into symbiotic associations with other bacteria will be termed herein 'symbiosis-competent' ('S-C'; conventionally: 'host-dependent' or 'H-D' or 'parasitic' or 'predatory'). Non-symbiotic aspects of *Bdellovibrio* life, and terminological points pertaining thereto, are handled separately in a subsequent section on p. 104. The associations of symbiosis-competent bdellovibrios with other bacteria will be dissected into a series of approximately chronological stages; however, the exact timing is in many cases unknown or varies with the experimental system. Hence, the form of presentation should be viewed primarily as a convenient didactic device, rather than as an accurate chronology.

'*Recognition*' *and* '*chemotaxis*'

The initial association of *Bdellovibrio* swarmers (the tiny, highly motile, usually vibrioid cells) with other bacteria may involve some sort of 'recognition' of a suitable associant cell. This 'recognition' (if indeed it exists) might be mediated by a 'chemotaxis'. However, diligent efforts over several years in various laboratories including ours have not yielded experimental results in accordance with the reasonable expectations stemming from this hypothesis. These expectations are that there should be massive, selective locomotion of bdellovibrios toward congenial bacteria (potential symbionts) or extracts prepared therefrom used as 'bait' in the Adler (1966, 1973) capillary-tube procedure and various modifications thereof. Although we have sometimes found in our own work statistically valid increases in numbers of bdellovibrios moving toward congenial bacteria, as compared with uncongenial bacteria, used as bait, generally rather small numbers of bdellovibrios swam even to the congenial bacterial bait; moreover, the results in replicate experiments have sometimes been contradictory. Hence, we have been disinclined to publish our ambiguous and still unexplainable findings. A recent publication (Straley & Conti, 1974) reports similar 'weak chemotactic responses' of bdellovibrios with bacterial preparations, remarks that 'possible explanations are being explored' and then turns to the chemotaxis of bdellovibrios toward yeast extract – a subject which is not yet directly relevant to the present issue of trying to decide whether or not there is 'recognition' by *Bdellovibrio* of a congenial bacterial symbiont and a 'directed motility' of the *Bdellovibrio* toward it, controlled by some sort of 'chemotaxis'. So, although the hypothesis is tempting, there is little yet to support it – with the possible exception of some (also still perplexing) work on 'symbiont specificity' to which we will now turn.

'*Symbiont specificity*'

The solid facts about the specificity of the symbiotic associations of bdellovibrios with other bacteria ('symbiont specificity' in my sense;

conventionally: 'host specificity') are not numerous and what has been reported is still fairly confusing. A given strain of *Bdellovibrio* is generally reported to enter into associations only with a fairly limited array of rather closely related bacteria. Some *Bdellovibrio* strains are reportedly limited to a single symbiont strain; others will enter into associations with many different strains belonging to one or more bacterial genera. Symbiont specificities of bdellovibrios are reported or summarized in several works (Stolp & Petzold, 1962; Stolp & Starr, 1963; Uematsu & Wakimoto, 1970; Starr & Seidler, 1971; Stolp, 1973; Taylor, Baumann, Reichelt & Allen, 1974). In some cases, the reported differences in ability of a given *Bdellovibrio* strain to attack various bacteria extended to cases wherein the attacked and unattacked bacteria were practically indistinguishable on other grounds even by an experienced bacteriologist! The bases and indeed the full extent of this symbiont specificity remain largely unknown; moreover the observed 'specificity' will often vary with the assay condition. Up to now, only Gram-negative bacteria have been conclusively shown to serve as symbionts for the presently known bdellovibrios (but see Burger, Drews & Ladwig, 1968, and Gromov & Mamkaeva, 1972, for associations of purported bdellovibrios with Gram-positive bacteria and with green algae of the genus *Chlorella*).

An interesting start has been made in understanding one possible basis of symbiont specificity, in a study (Varon & Shilo, 1969a) wherein *Salmonella* and *Escherichia* mutants with various blocks in their syntheses of cell wall lipopolysaccharides were used as prospective symbionts for *Bdellovibrio*. The bdellovibrios attached better to rough strains having the complete lipopolysaccharide core but lacking the type-specific O antigens than they did either to the smooth wild type (having the type-specific O antigens) or to the extremely rough strains with defects in the lipopolysaccharide core.

Attachment

Regardless of the uncertainty about whether there is 'recognition' of a suitable associant cell and whether there is a 'chemotaxis', there is no question about the existence of a violent collision of a motile *Bdellovibrio* swarmer with a prospective associant cell, followed by attachment. The bdellovibrio swarmer, which is provided with an unusual sheathed flagellum (Seidler & Starr, 1968), has been estimated to swim at the astonishing speed of 100 cell-lengths per second (Stolp, 1967 *a*, *b*). The bdellovibrio swarmer may shove the larger bacterial cell, with some ten to twenty times the mass of the *Bdellovibrio* cell, over a distance corresponding to several cell-lengths. Loss of motility or flagellar sheath integrity, induced in the *Bdellovibrio* by any of several procedures (Stolp & Starr, 1963; Varon & Shilo, 1968; Abram & Davis, 1970; Dunn, Windom, Hansen & Seidler, 1974), results in total inability of the *Bdellovibrio* to attach to other bacteria and thus to enter into a detectable symbiotic association with them.

When an actively motile preparation of symbiosis-competent bdellovibrio is added to a suspension of susceptible ('congenial') bacterial cells, attachments by the bdellovibrio cells to the cells of the other bacterial organism (the 'associant' or the 'symbiont'; conventionally: the 'host' or the 'prey') begin immediately. The bdellovibrio attaches by its anterior, aflagellated end – which has some interesting and unique structures (Abram & Davis, 1970). Often, more than one bdellovibrio attaches to a single symbiont cell. The kinetics of attachment and various other factors pertaining to attachment are presented in the still unsurpassed pioneering study by Varon & Shilo (1968). The most recent work on the subject (Dunn *et al.*, 1974) reports the isolation and many interesting properties of temperature-sensitive 'attachment mutants' of *Bdellovibrio*. These mutants could be further exploited as a powerful experimental tool in the clarification of still unresolved questions about attachment.

The early stages of contact may be non-specific (Dunn *et al.*, 1974): for example, symbiosis-competent bdellovibrios sometimes collide with and attach to non-susceptible bacteria or even to the glass coverslips of microscopic slide preparations. The attachment to an uncongenial bacterial cell and even to a congenial symbiont cell may be reversible in the earliest stages; that is, the attachment may be aborted and the bdellovibrio may swim away to another bacterial cell. It is not known whether these sorts of attachments have any functional, anatomical, or behavioral relationship to the usual symbiotic connection.

Penetration by bdellovibrios into cells of other bacteria

After a successful *Bdellovibrio*-symbiont attachment has been established, the next noticeable morphological aspect of the association involves the penetration of the symbiont cell by the *Bdellovibrio* cell. This penetration requires the formation (by mechanisms which are still not known with certainty) in the symbiont cell wall of a rather undersized pore through which the bdellovibrio eventually either squeezes by its own locomotor activity (Starr & Baigent, 1966; Stolp, 1973) or possibly is pulled 'passively' into the other bacterial cell (Abram, Castro e Melo & Chou, 1974); perhaps both penetration procedures operate. The fate of this pore (hole) in the symbiont's cell wall seems not to be known: is it repaired or plugged and, if so, how?

Although not all of the following notions are based on equally solid experimental evidence (and, indeed, some of them may have been fairly effectively refuted), the pore formation and penetration have been claimed to result from one or more of the following events: ballistic damage to the cell wall in the initial violent collision of the *Bdellovibrio* with the symbiont cell (Stolp, 1973), enzymatic action by the *Bdellovibrio* on the symbiont's cell wall (Huang & Starr, 1973*a*; Fackrell & Robinson, 1973; Engelking & Seidler, 1974), autolysis mechanisms induced in the symbiont by the *Bdellovibrio* (Stolp & Starr, 1965), damage to the cell wall by the mechanical drilling or swivelling motion of the attached *Bdellovibrio* prior to penetration (Starr & Baigent, 1966; Stolp, 1973), and other less explicitly stated factors. What are thought to be rebuttals of some of these notions have appeared. When the motility of the *Bdellovibrio* is retarded by the use of viscous media (thus reducing the ballistic impact) both attachment and penetration are claimed (Abram *et al.*, 1974) still to occur. Loss of *Bdellovibrio* motility immediately after attachment (thus making a drilling motion impossible) has been reported (Horowitz, Kessel & Shilo, 1974) and also vehemently denied (Stolp, 1967*a*, *b*, 1973). Abram *et al.* (1974) state that 'the motion and the impact between parasite and host are neither sufficient nor absolute prerequisites for attachment and penetration'. Then these authors deny on other grounds that the rotational or swivelling or other active motility of the *Bdellovibrio* plays any significant role in its penetration – which latter process they claim is a 'passive act of the parasite', firmly bonded to the symbiont's protoplast. This – by its retraction or that of the cell wall – pulls the *Bdellovibrio* into the periplasmic space. The notion of autolysis induction in the symbiont as a factor in penetration has been said (Shilo, personal communication) to be ruled out by the experiments summarized in Table 3 of Varon & Shilo (1968) with streptomycin-resistant *E. coli* and streptomycin-sensitive *Bdellovibrio* in which 'the inhibitory effect of streptomycin on invasion was expressed fully'.

Although I expect some significant findings about penetration to follow the use of temperature-sensitive 'penetration mutants' (Dunn *et al.*, 1974), my own view at the moment is that making the pore in the symbiont's cell wall may possibly be aided by the purported ballistic damage and the swivelling or drilling motion of the *Bdellovibrio*, but that there is now both morphological and biochemical evidence to indicate that the penetration process is facilitated or mediated by enzymatic action of the bdellovibrio and by the purported retraction of the symbiont's cell wall or its cytoplasmic membrane to which latter structure the penetrating bdellovibrio is said to be firmly bonded. Clearly this entire situation demands further clarification.

Regardless of the mechanismic bases for pore formation and penetration, when these processes have been effected the *Bdellovibrio* cell becomes lodged in the periplasmic region between the symbiont's cytoplasmic membrane and cell wall. This is one of the rare points on which all *Bdellovibrio* workers seem to be in agreement: the cytoplasmic membrane of the symbiont cell is not breached (except to the extent of the localized lesions reported by Snellen & Starr, 1974) and the *Bdellovibrio* cell occupies an intramural (intrategumental; intraperiplasmic) position within the symbiont cell. Re-stating this important point in other words, the *Bdellovibrio* cell is not inside the protoplast of the symbiont cell, although it is within a space (the periplasmic region) between the cell wall and the cytoplasmic membrane of the symbiont cell.

Early effects of the symbiosis

Symbiont cells that ordinarily are motile have been reported to stop swimming within moments after their initial contacts with *Bdellovibrio* swarmers; precisely how the bdellovibrio brings about this particular dysfunctioning of its symbiont is not yet known, but it may well be a concomitant of the elimination of the symbiont's respiratory potential (Rittenberg & Shilo, 1970). Much more often reported, with many *Bdellovibrio*-symbiont combinations, is the subsequent conversion of the attacked cell into a swollen, often spherical body at an early stage following attachment and before complete penetration is microscopically apparent. Although I have been chided rather strongly (Rittenberg, personal communication) about my use of the words 'before complete penetration' in the previous sentence (he says that 'the bdelloplast forms after complete penetration'), I can only reply that my views on the matter are supported unequivocally by many observations, including those depicted in Figure 1 of Starr & Baigent (1966) and Figure 9 of Seidler & Starr (1969a). These swollen or spherical bodies are probably not identical with spheroplasts – because, unlike some spheroplasts, they are not osmotically sensitive (Starr & Seidler, 1971; Starr & Huang, 1972) – hence, many *Bdellovibrio* workers prefer to use the terms 'spherical (or swollen) body' or 'bdelloplast' (rather than 'spheroplast') in referring to them. When the symbiont cell is a short rod, the entire cell is converted into such a bdelloplast; in a longer symbiont cell and at low multiplicities of infection, the swelling is more localized and a kind of multiple 'ballooning' is often seen. The formation of these bdelloplasts and balloons has been related circumstantially to enzymatic action of the *Bdellovibrio* on the rigid layer (murein) of the symbiont cell wall (Huang & Starr, 1973a).

Local damage has been reported to occur on the symbiont's cell surface (even though the *Bdellovibrio* cell may subsequently have become detached) in the form of holes and pitted areas in that surface of the symbiont's cell wall which had been in contact with the bdellovibrio (Shilo, 1969). Most recently, localized damage to the symbiont's cytoplasmic membrane in the vicinity of the entered *Bdellovibrio* has been shown in ultrastructural studies of the association (Snellen & Starr, 1974).

In addition to the ultrastructural membrane and wall damage, the cessation of motility, and the formation of bdelloplasts and other swollen bodies, there are signs of other serious dysfunctionings of the symbiont's cells in the earlier stages of their association with *Bdellovibrio*. Attachment of the *Bdellovibrio* cell to the cell wall of its symbiont has been shown to be sufficient to cause cessation of protein and nucleic acid synthesis in the symbiont, even before penetration has become microscopically apparent (Varon, Drucker & Shilo, 1969; Shilo, 1973*a*, *b*). The cytoplasmic membrane of the entered symbiont cell also functions abnormally (becomes less selectively permeable, but in what appears to be a controlled fashion) already in the earliest stages of its association with the *Bdellovibrio* (Rittenberg & Shilo, 1970; Crothers & Robinson, 1971). This controlled leakage has recently been related circumstantially to the aforementioned localized ultrastructural damage effected by the bdellovibrio on the cytoplasmic membrane of the entered symbiont (Snellen & Starr, 1974), but it must be emphasized that there is as yet no direct experimental evidence about actual leakage at these localized damaged sites. The respiratory potential of the symbiont is essentially eliminated within a few minutes after *Bdellovibrio* attachment (Rittenberg & Shilo, 1970). Many of these detrimental early effects on the symbiont may, quite unexpectedly, be reversible, as is suggested by the exciting finding of Mielke and Stolp (personal communication) that they could 'cure *Bdellovibrio* infections' by means of deoxycholate; further details on this point are given in a subsequent section (pp. 116–17).

Development of Bdellovibrio *inside its symbiont*

Once in the periplasmic locus, the small and vibrioid *Bdellovibrio* cell begins its intramural growth phase and, as a second stage in its characteristic dimorphism, elongates into a C- or helical-shaped cell. The helical (more usually, but incorrectly in a geometric sense, called 'spiral') cell may be ten or more times longer than the entering vibrioid swarmer cell; its size seems to be directly proportional to the size of the symbiont cell within which it is developing. During this intramural growth phase, there is a progressive – but meticulously regulated (Matin & Rittenberg, 1972; Hespell, Rosson, Thomashow & Rittenberg, 1973) – disorganization and dissolution and utilization, by the *Bdellovibrio*, of the symbiont's cytoplasm, nucleoplasm, and other components. The digested symbiont components are converted at very high efficiency into *Bdellovibrio* components (Hespell *et al.*, 1973; Hespell, Thomashow & Rittenberg, 1974). The rather large, helical bdellovibrio cell undergoes multiple constriction prior to its segmentation into several vibrioid daughter (progeny swarmer) cells. At about this time, the typical sheathed flagellum (Seidler & Starr, 1968) develops on each progeny swarmer (Burnham, Hashimoto & Conti, 1968) and the swarmers can then be seen, by phase-contrast microscopy, swimming within the swollen and usually heavily disorganized ('ghosted') bdelloplast.

The final stage: release of Bdellovibrio progeny

The means by which the *Bdellovibrio* progeny swarmers leave the ghosted symbiont cell are not entirely clear. The juvenile swarmers are rapidly motile within the ghosted symbiont cell; perhaps they are able to break down its remnants mechanically by virtue of this motility. In many instances, the destruction of the symbiont cell is so extensive that no significant structural barrier remains to retard departure of the *Bdellovibrio* progeny ('swarmers') and reinitiation of another symbiotic cycle when a congenial associant is found. In other cases, the remnants of the symbiont, after the *Bdellovibrio* swarmers have left, appear to be fairly substantial.

Exhabitational symbiosis by bdellovibrios

Rittenberg (personal communication) has recently informed me about the interesting associations of bdellovibrios with very small bacteria, *Bordetella pertussis*, approximately the same size as *Bdellovibrio*. Here something akin to an exhabitational symbiosis sometimes occurs, possibly because the relatively equal sizes of the associants precludes complete entrance of the *Bdellovibrio*. Another tendency toward an exhabitational symbiotic mode has recently been related to me by Abram (personal communication). By manipulation of the ionic and other environmental conditions, penetration (but not attachment) of the *Bdellovibrio* is inhibited and the attached bdellovibrio develops in this exhabitational locus. However, in both of these exhabitational cases, it was not clear whether or not the symbiont's cell wall is breached (I expect it is) and whether or not the bdellovibrio bonds tightly to the symbiont's cytoplasmic membrane (I expect it does, if the purported bonding in the inhabitational mode is confirmed and generalized). If my expectations are realized, then the spatial situations in these cases are neither purely exhabitational nor entirely inhabitational, and they must be classed in intermediate positions within the exhabitational–

inhabitational continuum (see earlier article) in the same way as the haustoria of some fungi.

These expectations seem indeed to have been realized by the latest news from Abram (personal communication) wherein she describes a *Bdellovibrio–Acinetobacter* system in which there is simultaneously an array of relative spatial arrangements ranging from the usual intramural inhabitation, through partial inhabitation (only the anterior end of the *Bdellovibrio* enters), to full exhabitation. In many or all of these relative spatial arrangements, the *Bdellovibrio* has been reported to go through its characteristic dimorphic developmental cycle; fairly substantial *Bdellovibrio* progeny yields were observed in some. But her generosity in sharing this exciting news with me should be partially compensated by allowing her to disclose the full details of this fantastic story elsewhere!

Important, indeed essential, for the present enterprise is that these findings by Rittenberg and Abram (separate personal communications) add the occurrence in *Bdellovibrio*-symbiont systems of exhabitational symbiosis, and various intermediates between inhabitational and exhabitational symbiosis, to the better-known, intraperiplasmic, inhabitational kind. These findings support the utility of arranging criteria about organismic associations into continua. The same might be said about the continuum extending from symbiosis to non-symbiosis: some of the cases related here and some of those discussed elsewhere in this essay in connection with non-symbiotic development of bdellovibrios, may involve relationships intermediate between symbiosis and some sorts of non-symbiosis ('saprotrophy' or 'scavenging' – in my terminology).

NON-SYMBIOTIC DEVELOPMENT OF BDELLOVIBRIOS

Consideration of non-symbiotic aspects of *Bdellovibrio* might be thought outside the scope of this exploration of a symbiosis. However, the non-symbiotic ('N-S'; conventionally: 'host-independent' or 'H-I' or 'saprophytic' or 'non-parasitic' or 'axenic') and symbiosis-competent ('S-C'; conventionally: 'host-dependent' or 'H-D' or 'parasitic' or 'predatory') phases in *Bdellovibrio* can alternate, and an understanding of their relationships is highly relevant to an understanding of the nature of the bdellovibrio symbiosis.

Isolation of non-symbiotic forms

Stolp & Petzold (1962) – in the very first report on the creature now known as *Bdellovibrio* – had demonstrated the existence of such non-symbiotic strains. Upon plating *Bdellovibrio*-symbiont lysates containing relatively few ($< 10^8$ cm^{-3}) symbiosis-competent cells onto a variety of symbiont-free nutrient agar media, no growth of *Bdellovibrio* colonies – that is, no non-symbiotic development – was observed. However, when the lysates were concentrated substantially (to $> 10^9$ S-C *Bdellovibrio* cells cm^{-3}) and plated onto symbiont-free nutrient medium, a few yellow colonies of N-S bdellovibrios appeared after five to six days. Apparently such N-S *Bdellovibrio* cells were relatively rare; in those trials, none was detectable in populations of 10^8 S-C *Bdellovibrio* cells cm^{-3} and only a few in populations of over 10^9 cells cm^{-3}.

Non-symbiotic derivatives (probably of differing sorts) were isolated from several other S-C *Bdellovibrio* strains by similar inoculation of concentrated suspensions of the S-C strains onto nutrient agar media or by a single differential filtration of a S-C lysate followed by plating onto a nutrient medium (Stolp & Starr, 1963; Shilo & Bruff, 1965). Another procedure which has been described (Seidler & Starr, 1969*b*) for the routine isolation of N-S bdellovibrios relies on the prior isolation of spontaneous streptomycin-resistant mutants of S-C bdellovibrios, followed by their propagation on streptomycin-sensitive symbionts and subsequent transfer to a nutrient selection medium containing streptomycin. The streptomycin prevents the growth of residual symbiont cells in the lysate and thus counter-selects S-C bdellovibrios and favors N-S bdellovibrios.

By means of these procedures, N-S derivatives have been isolated by several groups from every one of the different S-C strains so studied. Since these S-C cultures belong to the several, presently recognized species of the genus *Bdellovibrio* (Seidler, Starr & Mandel, 1969; Seidler, Mandel & Baptist, 1972; Gloor, Klubek & Seidler, 1974; Burnham & Robinson, 1974), and since they had been isolated from different geographical regions and habitats (Taylor *et al.*, 1974), it seems safe to conclude that S-C bdellovibrios, in general, have the potential for yielding N-S derivatives. It is very important to emphasize that there are probably numerous bases for the loss of the symbiotic capacity, and that the term 'non-symbiotic' as used here covers a variety of non-symbiotic sorts, only some of which are usefully delineated in what I have written. Shilo (1973*a*) and Varon, Dickbuch & Shilo (1974) have made a good start at sorting out this situation, which, however, requires the further attempt at elucidation given in the next section.

Terminology pertaining to non-symbiotic (and symbiotic) bdellovibrios

Partly as an act of contrition for my role in aiding and abetting the terminological confusion pertaining to non-symbiotic and symbiotic bdellovibrios, let me now try to clarify the terminology. A bdellovibrio able to enter into a symbiotic association with other bacteria is herein termed 'symbiosis-competent'. This competence – the overall capacity of *Bdellovibrio* to enter into a significant association with other bacteria – involves two intertwined, frequently confused, elements: (*a*) the physical element, which reflects the relative spatial arrangements of the *Bdellovibrio* and its bacterial symbiont – either when entering into and living within other bacterial cells or living in locations upon or only partly within the other bacterial cells; and (*b*) the element which reflects the dependence and the effects, and modal aspects thereof, of the *Bdellovibrio* upon its symbiont, while they are associating. For the physical element (*a*), I now use (see earlier article) the terms 'inhabitational symbiosis' or 'exhabitational symbiosis', with specific modifiers (for example, 'intramural' or 'intra-periplasmic'). The dependence, effects, and modality element (*b*) is dealt with by an array of explicit terms such as 'nutritionally and possibly regulatively but not energetically dependent', 'digests the symbiont's nucleoplasm and cytoplasm', 'the respiratory potential of the symbiont is reduced', 'localized lesions appear on the symbiont's cytoplasmic membrane', etc.; a later section treats this subject in more detail. However, I should point out in the present context that the ambiguous term 'host-dependent' which I have helped to perpetrate and perpetuate, could refer to the *Bdellovibrio* strain being dependent upon another organism (the 'symbiont'; conventionally: the 'host') either (*c*) while in a symbiosis or (*d*) while living non-symbiotically (axenically) but still dependent upon various cell-free extracts and products of organisms which are usually, although not necessarily, those which can serve as symbionts. For situation (*c*), I would prefer to use the term 'symbiosis-dependent'. Situation (*d*) is actually non-symbiotic but still dependent upon the products of the former symbiont for axenic cultivation.

Non-symbiotic bdellovibrios clearly are of several sorts. (*a*) There might be permanently non-symbiotic ('P-N-S') creatures – none of these is presently known, in part because of the conceptual and methodological approaches in vogue and which are discussed elsewhere in this article. (*b*) There are those N-S clones which have been derived at rather low frequencies from S-C clones and which can revert to the S-C form, the kind of N-S strains first isolated by Stolp & Petzold (1962) and perhaps some of those isolated by Stolp & Starr (1963), and termed by them 'saprophytic'. (*c*) There are those occurring at somewhat higher frequencies in S-C clones (also able to revert to S-C forms) and which were isolated and studied in some detail by Seidler & Starr (1969*b*) who termed them (together with the 'saprophytic' sort noted in the previous sentence) 'host-independent'. (*d*) There are the S-C forms which are being cultivated non-symbiotically in complex preparations derived from former symbionts (Horowitz *et al.*, 1974). (*e*) Finally, there are those N-S clones which derive from or form all or part of the population of facultatively symbiotic ('F-S'; conventionally: 'facultatively parasitic' or 'F-P') strains. These are exemplified by strain A3.12 of *Bdellovibrio starrii* (studied in some detail by Shilo & Bruff, 1965), strain UKi2 of *Bdellovibrio stolpii* (studied in some detail by Burnham *et al.*, 1970; Diedrich, Denny, Hashimoto & Conti, 1970; and Varon, Dickbuch & Shilo, 1974), and several of the strains reported by Uematsu & Wakimoto (1970, 1971). The F-S condition may be general for all *Bdellovibrio* strains, if the recent reports of Shilo (1973*b*) and Varon *et al.* (1974) can indeed be generalized. The latter work suggests certain operational definitions for some of these sorts; although their classification and terminology differ somewhat from mine, the concepts are about the same. The exact mechanisms whereby some of these N-S forms might arise remain unknown.

Nutritional requirements for non-symbiotic growth

In this context, the few available studies on the nutritional requirements for axenic growth of these N-S, F-S and S-C forms might be cited here (Shilo & Bruff, 1965; Scidler & Starr, 1969*b*; Reiner & Shilo, 1969; Ishiguro, 1973, 1974; Rittenberg, 1973; Horowitz *et al.*, 1974; Gloor & Stolp, unpublished). One major achievement in this area has been the cultivation of S-C bdellovibrios axenically in the presence of complex substances ('host extract') which were derived from the cells of a former symbiont, *Escherichia coli*, and partially characterized (Horowitz *et al.*, 1974). In this notable study, the S-C bdellovibrios went through something akin to the characteristic dimorphic life-cycle when cultivated axenically in the presence of 'host extract'. In another direction, a 'growth initiation factor' has been isolated from several sorts of microbes (not necessarily only those which serve as suitable symbionts) and shown to be required for the axenic growth of N-S strains freshly derived from S-C clones, but not to be required for the growth of those N-S clones which had been cultivated for some time axenically (Ishiguro, 1973, 1974).

What is emerging from the imaginative studies of my colleagues in Jerusalem, Vancouver, Los Angeles and Hamburg is elucidation of the nutritional bases for axenic cultivation of N-S, F-S and S-C bdellovibrios and, directly or circumstantially, the possible nutritional bases for their symbiotic development. But the facts provided by these significant experimental programs cannot be used for slipping into implicit (almost subliminal) denials that bdellovibrios are (or can be) symbionts. Most microbes can be cultivated axenically, but no other bacteria are known which have the peculiar capacity that bdellovibrios have of entering into and developing within their associates; a few other bacteria share with *Bdellovibrio* the exhabitational symbiotic capacity. It is very important to learn the nutritional and other functional bases for a symbiosis, but learning them does not constitute a ground for any claims other than (*a*) that the symbiosis is facultative and (*b*) that the nutritional and other bases of the symbiosis can be simulated under axenic conditions by the culture medium and other components which are specified. Knowing how to cultivate, for example, monkey kidney cells axenically in cell cultures does not permit denying the existence either of kidneys or of monkeys (each of which are cellular assemblages at integrative levels increasingly higher than the cell cultures). So it is with the emerging facts about the nutritional needs for axenic cultivation of bdellovibrios; these are indeed essential facts with important relevance for understanding this cellular assemblage which constitutes an interbacterial symbiosis, but they are by no means applicable to denying the existence of the symbiosis!

Interconversion of non-symbiotic and symbiotic forms

Populations of N-S bdellovibrios, exposed to bacteria which can serve as suitable symbiont cells, can revert to populations of S-C bdellovibrios. The precise mechanism of this shift also remains unknown. The potential of several N-S isolates for regaining the symbiotic capacity (becoming again S-C) has been examined in a semiquantitative manner by determining the ability of such strains to grow on living symbiont cells (Seidler & Starr, 1969*b*). Non-symbiotic *Bdellovibrio* cells (the 'H-I' kind derived by these workers from S-C strains) were added to cell suspensions of the previously congenial symbiont, and examined periodically for several days. In every N-S *Bdellovibrio* of this kind so investigated, bdellovibrios were re-selected with symbiotic capabilities seemingly identical to those of the original S-C cultures.

On the other hand, the direct formation, by bdellovibrio clones which can be grown axenically, of isolated plaques on congenial symbiont lawns has up to now been observed only with a few *Bdellovibrio* strains – different in nature from the N-S ('H-I') sort studied by Seidler & Starr (1969*b*) – notably, strain UKi2 of *Bdellovibrio stolpii* and strain A3.12 of *Bdellovibrio starrii* and several strains isolated by Uematsu & Wakimoto (1970, 1971). These latter sorts have been termed 'facultatively parasitic' (Diedrich *et al.*, 1970; Burnham *et al.*, 1970; Varon *et al.*, 1974). The literature on this subject contains many conceptual obscurities as well as some determined efforts to clarify (for the latter, see Stolp, 1973; Shilo, 1973*b*; Varon *et al.*, 1974). From the facts now available, the situation called 'facultatively parasitic' ('F-P') or 'host-independent-dependent' ('H-I-D') or, in the terminology I now prefer, 'facultatively symbiotic' ('F-S') might be characterized as one in which a *Bdellovibrio* population develops with more-or-less equal competence in the presence or absence of living symbiont cells, and thus can shift from a symbiotic to a non-symbiotic mode and vice versa. In this sense, it is unknown whether each individual cell of such F-S bdellovibrio populations is indeed facultatively symbiotic, or whether the F-S populations consist of more-or-less equal mixtures of S-C and N-S individuals. The former situation is usually assumed by others to be the case, but there is as yet little hard evidence to support this view – and certainly not with the vehemence shown by some proponents of this notion.

The symbiotic capacity of some N-S and F-S *Bdellovibrio* strains dwindles after successive axenic transfers through symbiont-free nutrient media (Stolp & Starr, 1963; Shilo & Bruff, 1965; Seidler & Starr, 1969b), though in one case, the symbiotic capacity of a F-S strain was retained after many such transfers (Diedrich *et al.*, 1970). But, in all cases, it has been invariably possible (although sometimes difficult) to re-isolate S-C forms from N-S or F-S cultures (Starr & Seidler, 1971; Varon *et al.*, 1974). The point emphasized here is that, like many other symbiotic organisms, some N-S or F-S bdellovibrios may lose part of the symbiotic capacity when cultivated axenically (in the absence of living symbionts), making it increasingly difficult to re-select for symbiotic forms. It thus appears that only certain *Bdellovibrio* strains behave as though they are F-S, and that even in these strains the N-S property seems to be a transient character of the population because there is no irrevocable selection upon repeated axenic cultivation in symbiont-free nutrient media.

Symbiosis, epistemology, and taxonomy

Only S-C and F-S bdellovibrios have been isolated from Nature. All known N-S strains are laboratory derivatives, of various sorts, from these S-C and F-S forms. This situation probably stems in part from the fact that the actual or potential capacity to enter into an intramural symbiosis with other bacteria has been and is being treated – in a taxonomic sense – by *Bdellovibrio* workers as a necessary and sufficient condition for a small, aerobic, vibrioid bacterium to be a bdellovibrio. I state only as an historical fact that the capacity to be or to become an intramural symbiont of other bacteria has been treated by *Bdellovibrio* workers as having an 'epistemological primacy', meaning thereby that two related matters were in their minds: '(a) in the course of deciding whether or not an unidentified organism is a *Bdellovibrio*, an experienced student of *Bdellovibrio* would early direct his efforts to determining whether this symbiotic trait is present; and (*b*) he would be more reluctant to accept as a *Bdellovibrio* an organism which lacked the symbiotic capacity than he would if it lacked some other trait, say, the vibrio shape' (Heise and Starr, unpublished; cited by Starr & Seidler, 1971). I quite agree with a perceptive remark made about the latter (*b*) statement by Abram (personal communication), who pointed out that nowadays we would certainly accept as a possible *Bdellovibrio* a bacterial organism which had no known symbiotic capacity but which did show the characteristic dimorphic developmental cycle (alternation of large, immotile, helical cells and small, highly motile, vibrioid cells). I sometimes muse about the likely consequences if the unusual dimorphism of the bdellovibrios had been uncovered instead of their also unusual symbiotic capacity; one of them which comes irreverently to mind is that I would not be writing this article for a volume entitled 'Symbiosis'!

The aforementioned epistemological primacy and the associated experimental methodology (generally the procedures of Stolp & Starr, 1963) may have biased the isolation of naturally occurring bdellovibrios to yield only the S-C and F-S kinds. The same philosophical and methodological factors may have led to another sort of biasing: the bacteria isolated solely on the basis of the symbiotic capacity – or its operational equivalent, the formation of plaques on lawns of living bacteria – may be taxonomically a quite diverse assemblage which transcends the single genus *Bdellovibrio* and its three presently named species (Seidler *et al.*, 1972; Gloor *et al.*, 1974; Burnham & Robinson, 1974). Such unrecognized diversity may account for some of the conflicting statements in the literature, including those which appear here! It is high time for a critical look at this matter, an undertaking which would need the polyphasic methodology and integrated co-operation of the international community of *Bdellovibrio* workers united into a formal 'Working Party on *Bdellovibrio* and Relatives'. I pledge my services to such an enterprise.

The on-going studies with mutant strains (Dunn *et al.*, 1974; Varon *et al.*, 1974) are permitting separation of various elements comprising the symbiotic capacity of *Bdellovibrio*, and are already leading to important clarifications of the nature and interrelationships of S-C, N-S and F-S strains. Perhaps they will also lead to rationales for seeking and finding N-S and even P-N-S bdellovibrios from Nature – if indeed they exist in natural habitats. In this connection Shilo (1973*b*) has analyzed, with his usual sagacity, the likely selective advantages the S-C forms would have over the others.

IS *BDELLOVIBRIO* A 'PARASITE' OR A 'PREDATOR' OR BOTH OR NEITHER?

Having completed my attempt to lay out in a reasonably objective and tolerably compact manner the salient facts about *Bdellovibrio* and its association with other bacteria, it is now time to consider the question which heads this section. There has been a considerable quantity of heated but sterile bickering – to which I have contributed – about whether *Bdellovibrio* is a 'parasite' or a 'predator' or both or neither; examples can be found in the writings of practically all the significant experimental workers on *Bdellovibrio*. I do not propose to prolong this essentially non-productive and ill-focused quarrel except to say, somewhat tautologically and paradoxically, that by some definitions and usages of the term 'parasite' *Bdellovibrio* is a parasite and by others it is not a parasite. Following a similar process, by some definitions and usages of the term 'predator' *Bdellovibrio* is a predator and by others it is not a predator. These rather assertive statements are based on a lengthy and very instructive exercise in which scores of definitions and usages of each of these two terms were gleaned from a wide array of textbooks, monographs, review articles, research reports, dictionaries and encyclopedias. These definitions and usages (in this broadly based literature) were then 'translated' into the terms of my generalized classificatory scheme, tabulated, and analyzed. There is an extremely high variance in the meanings attributed to each term; moreover, there is a substantial overlap between the meanings given for the two terms. Presentation here – even in a highly abridged but still meaningful version – of the results of this gleaning, translation, and analysis is precluded because of space limitations. Detailing these findings publicly will simply have to await the availability of another suitable forum.

Actually, some entomologists (Doutt, 1964) use a term 'parasitoid' to label an association which comes close to fitting the situation of *Bdellovibrio* and its symbiont. A parasitoid is an entomophagous insect that places its immature forms inside another living organism (an insect in this case), which latter creature is eventually consumed and thereby killed by the developing inhabitant. So, my final contribution in this genre is the dictum that *Bdellovibrio* is a parasitoid of other bacteria!

But my intention here is to do something that I consider to be more useful; namely, to drop the 'parasite' and 'predator' and 'parasitoid' terminology, to analyze the associations of bdellovibrios with other bacteria in terms of the generalized classificatory scheme summarized here in the Introduction and detailed in the text and Table 1 of my earlier article, and to apply the terminology of this scheme to the association. One caveat should be noted: I have limited my consideration here and in the classificatory scheme to the *direct* effects of the two associants on one another. I realize, of course, that there can be significant subsidiary effects on the rest of the population which is not directly antagonized, and refer with no further comment to studies on such indirect and possibly beneficial populational effects in *Bdellovibrio*-symbiont systems (Guélin & Cabioch, 1974; Keya & Alexander, 1975).

BDELLOVIBRIOS ARE INHABITING (AND PERHAPS EXHABITING), ANTAGONISTIC SYMBIONTS OF OTHER BACTERIA

Bdellovibrios are indeed symbionts

It seems to me self-evident that the relationship between what I have been calling symbiosis-competent bdellovibrios and certain other bacteria is an organismic association, a symbiosis in the sense this term is used in my classificatory scheme. But perhaps my reasons for holding this view so strongly should now be stated explicitly. It goes without argument that two organisms are involved, that they are both alive at least at the start of the relationship, that they become associated and that the association is certainly significant to the well-being or 'unwell-being' of both organisms. Hence, the relationship is a symbiosis in the sense that the term 'symbiosis' is used here. (At this point, it might be useful to re-read the Introduction.)

Bdellovibrios ordinarily are inhabiting symbionts, but they can be exhabiting symbionts

That symbiosis-competent bdellovibrios ordinarily are inhabiting symbionts of other bacteria should by now be clear from the usual relative locations of the two associants. For a very brief period, during the act of attaching to the outsides of the other bacteria which are about to be entered, the bdellovibrios might be thought of as exhabiting associants. As already noted, there may be many effects on the associant's cell during this very brief exhabitational (attachment) period which precedes penetration. The rare aborted attachments referred to previously are also exhabitational, but probably without further significance unless there has been some mechanical or enzymatic damage. Also exhabitational are the unusual associations, involving very small or otherwise peculiar symbionts or manipulated ionic environments, alluded to in a preceding section (p. 103). In any case, associations in this mode are still relatively little known. Some of them are very brief and probably without much significance (unless they proceed to the usual inhabitational or the unusual exhabitational symbiosis).

Once the *Bdellovibrio* cell penetrates the cell wall of its symbiont, however, it certainly is an inhabiting symbiont. The *Bdellovibrio* cell lodges and develops between the cell wall and the cytoplasmic membrane of the entered symbiont and, although within the cell, it is not inside the symbiont's protoplast. This same sort of spatial relationship is found in other inhabiting symbioses, for example in the cases of the penetrating haustoria of some phytopathogenic (Ehrlich & Ehrlich, 1971) and lichenizing (Smith, 1963; Ahmadjian & Hale, 1973) fungi.

A digression concerning the term 'endosymbiont'

In laying out the distinctions between prokaryotes and eukaryotes, Stanier (1970) declares that 'stable endosymbiosis' does not occur in prokaryotes and argues most eloquently to the effect that bdellovibrios are not 'in the strict sense' endosymbionts of other prokaryotes. His argument is that endosymbiosis requires an ability to 'internalize objects of cellular dimensions by endocytosis' and that the 'real barrier to the establishment of endosymbioses in prokaryotes lies at the level of the cytoplasmic membrane'. However, in another more general work, a superb textbook co-authored by Stanier at about the same date (Stanier, Doudorff & Adelberg, 1970), the meaning of the term 'endosymbiont' is given as an organism that 'grows in the cells and tissues of its host', where the term 'host refers to the larger of two symbionts'; nothing is said explicitly in the cited portion of this textbook about a requirement for endocytosis and the examples of endosymbioses given in that work may not in all cases actually involve a recognizable endocytosis. Returning for a glance at *Bdellovibrio*, the recent report of Abram *et al.* (1974) might again be noted in the present context. Here, there is the purported involvement of the symbiont's cytoplasmic membrane and cell wall in retracting and thereby bringing the bdellovibrio (which is said to be firmly bonded to the symbiont's cytoplasmic membrane) through the pore in the symbiont's cell wall and into its periplasmic space. Is this a primitive endocytosis?

The term 'endosymbiosis' had been used previously by Buchner (1965, and earlier) with the meaning of a 'well-regulated and essentially undisturbed co-operative living' in which one partner 'shelters another within its body'. Buchner's and both of Stanier's meanings of 'endosymbiosis' all include the concept of inhabitation; the meaning in Stanier's 1970 review, but not his 1970 textbook, requires the occurrence of endocytosis; Buchner's meaning seems to require mutual benefit or, at least, mutual non-harmfulness ('essentially undisturbed co-operative living'). In my classificatory scheme the concept of relative location is separated from other concepts (benefit/harm; modes and formats of dependence; etc.). It seems unarguable to me that the inhabiting bdellovibrios are endosymbionts in the broad sense that they are symbionts which are indeed within another organism. Explicitness demands the term 'intramural' (or 'intraperiplasmic') to designate the precise spatial relationship of the inhabiting *Bdellovibrio* to its inhabited symbiont. Although the term 'endosymbiont' is etymologically felicitous as a synonym for 'inhabiting symbiont', the word might be retained for one (but which one?) of its special usages – if that indeed is and remains the considered judgement and consensus of the workers in the field. However, my views on the matter should by now be clear: the word 'endosymbiont' could be an exact synonym for 'inhabiting symbiont'!

Relative sizes of the symbionts

In terms of relative size, the *Bdellovibrio* swarmer is usually substantially smaller than its associant (they are, thus, anisosymbionts in my terminology); the smaller swarmer (the microsymbiont) is somewhere in the neighborhood of one-third to one-twentieth the mass of the usual bacterial cell (the macrosymbiont) that it enters. The mass of the fully developed intracellular helical ('spiral') form (and that of the combined progeny swarmers derived therefrom) is about one-half that of the associant cell which it has consumed. The interesting exhabitational associations of bdellovibrios with other very small bacteria, approximately the same size as *Bdellovibrio* (isosymbionts in my terminology), have already been noted in the section on 'Exhabitational symbiosis by bdellovibrios' (p. 103).

Temporal aspects

The durability of this association extends over most of the life-spans of the two associated bacteria. In general terms, an individual bacterial cell has a life-span of a few hours; when this time runs out, the cell usually has divided into two daughter cells or it has died. The time-schedule for this symbiotic association (Seidler & Starr, 1969*a*; Varon & Shilo, 1969*b*) might, in an equally simplified generalization, be given as one in which a young or at most middle-aged *Bdellovibrio* swarmer enters a young-to-old symbiont cell and, within a few hours at the most, the symbiont cell has been consumed and the several *Bdellovibrio* progeny swarmers have emerged. Viewed thus, simplistically and in generalized kinetic terms, and taking the normal life-spans of the associants into account, the symbiosis might be termed 'persistent'.

Degree of necessariness of the bdellovibrio symbiosis

Symbiosis-competent bdellovibrios generally tend toward the obligate end of the degree of necessariness continuum; sometimes they behave as though they are facultative symbionts. The often obligate dependence is clearly indicated by studies such as those of Hespell *et al.* (1973, 1974) which show that the life of a S-C progeny swarmer is quite brief unless it 'finds' and enters another symbiont cell. As already related, non-symbiotic derivatives (possibly of differing sorts) can be selected from (probably) all symbiosis-competent and facultatively symbiotic *Bdellovibrio* populations, and S-C bdellovibrios can be selected from N-S and F-S *Bdellovibrio* populations. The nutritional, regulative, and other features which might be involved in these shifts are just beginning to be clarified (Shilo, 1973*a*, *b*; Ishiguro, 1973, 1974; Horowitz *et al.*, 1974; Varon *et al.*, 1974). Still unclear is whether an individual bdellovibrio cell can be, at the same time, both S-C and N-S; that is, whether the F-S strains consist of populations which are more-or-less equal mixtures of S-C and N-S cells, or whether each individual cell is actually a facultative symbiont (Shilo & Bruff, 1965; Diedrich *et al.*, 1970; Varon *et al.*, 1974). I hope that raising this issue again in the present context may stimulate performance of the necessary experimental work which would permit a clear decision to be made between these alternatives.

Nutritional and regulative, but not energetic, dependence

A symbiosis-competent *Bdellovibrio*, while in a symbiosis, is dependent upon its associant. The major modes of dependence presently known must be classified as nutritional and possibly also regulative (Shilo, 1973*b*). It is fairly certain that the dependence is not energetic in the sense that the chlamydiae are energetically dependent (Storz & Page, 1971). The bdellovibrios have the capacities for obtaining energy from both oxidative and substrate-linked phosphorylations (Simpson & Robinson, 1968; Shilo, 1973*b*; Hespell *et al.*, 1973; Gadkari and Stolp, unpublished); though, while in a symbiosis, the substrates are of course coming from the material of the symbiont. During the intramural phase, the *Bdellovibrio* consumes the associant's cell material in a meticulously regulated fashion (as already outlined). Exactly what is going on in this regard in the various sorts of exhabitational symbiosis (Rittenberg and Abram; separate personal communications) is still not known to me. Whether modes of dependence other than nutritional and regulative might justifiably be claimed or denied to exist cannot be determined from the present fact-base.

Biotrophy versus necrotrophy

Regarding nutritional dependence of symbionts, I have adapted (that is, modified – with apologies to my mentor, D. H. Lewis) in my classificatory scheme the biotrophy/necrotrophy distinction (Lewis, 1973, 1974; Luttrell, 1974). By my definition, a symbiosis starts with two living organisms and remains a symbiosis even if one of the associants subsequently becomes more-or-less dead. If both organisms remain alive and there is a nutritional dependence of one symbiont on the other, I apply the term 'biotrophy'. If, after the symbiosis begins, a part or all of one of the symbionts dies and serves as nutriment for the other, I apply the term 'necrotrophy'. The term 'saprotrophy' means to me 'a non-symbiotic feeding upon'. Delineation of the three terms is drawn in this way because of my 'scientific tact' – which is inherently neither better nor worse than that of others; merely, sometimes, different!

Generally, symbiosis-competent bdellovibrios are at first biotrophic symbionts – in the sense that their associations start with attack upon, attachment to, and entrance into living bacterial cells. These 'living bacterial cells' soon lose a sufficient component of livingness so that the rest of the association must be thought of as tending toward necrotrophy. By some criteria the entered symbiont becomes 'dead' fairly soon; for example, while in contact with the live *Bdellovibrio*, it can no longer grow and divide, its respiratory potential is essentially eliminated, and its nucleic acid and protein syntheses have stopped. However, it still functions in a 'live' fashion with regard to another criterion that is very important to the bdellovibrio; that is, the symbiont's cytoplasmic membrane retains meticulous control over passage of substances between its periplasmic and cytoplasmic regions. There is localized ultrastructural damage to the symbiont's cytoplasmic membrane in the vicinity of the inhabiting *Bdellovibrio* (Snellen & Starr, 1974); a crucial question which remains to be answered is whether these damaged membrane sites actually are the loci of the controlled leakage into and out of the periplasmic space. In this context, it is indeed significant that the symbiont's cytoplasmic membrane retains much of its ultrastructural and functional integrity and is not totally destroyed, at least until the *Bdellovibrio* progeny swarmers are mature. In fact, if the integrity of the symbiont's cytoplasmic membrane, and possibly of its cell wall, is disturbed (by, for example, calcium deprivation), the intramural symbiotic development of the *Bdellovibrio* is severely retarded (Seidler & Starr, 1969a; Huang & Starr, 1973b).

While on the subject of the symbiont being 'dead' or 'alive', the exciting recent findings of Mielke and Stolp (personal communication) should be kept in mind. These workers claim that 'bdelloplasted' cells of *Escherichia coli*, which already had bdellovibrios within their periplasmic spaces, could be 'cured' of the *Bdellovibrio* infections by deoxycholate (to which substance *Bdellovibrio* is very sensitive and *E. coli* quite resistant). Such cured *E. coli* cells are reported to be able, once again, to divide and to form colonies on nutrient agar. In the system they used, wherein practically all of the *E. coli* cells were seen by them microscopically to have been entered by bdellovibrios, 40 % of such *E. coli* cells could once again grow into colonies after the deoxycholate had killed the inhabiting bdellovibrios. This report, if confirmed, suggests that many of the early dysfunctions of the symbiont caused by the *Bdellovibrio* are – quite surprisingly – reversible, and that the irreversible death of the symbiont caused by the *Bdellovibrio* occurs some time after the periplasmic inhabitation has been achieved. In the present context, then, the symbiosis may remain biotrophic for a somewhat longer time than I had originally suspected, before the shift to the necrotrophic format occurs.

The interface between symbiotic and non-symbiotic development of bdellovibrios

Intertwined with consideration of the continuum between biotrophy and necrotrophy is another issue which still needs to be examined. I can best express this issue in the form of a question: what sorts of associations exist between *Bdellovibrio* and other bacteria at the highly relevant interface between symbiosis and non-symbiosis? Perhaps I can get closer to the issue by asking the question in an operational form: what are the formats of the associations when bdellovibrios are placed in contact with more-or-less dead cells of other bacteria? There is a continuum between life and death; the ultimate ends of the continuum are essentially unarguable, but the intermediate states have elicited volumes of writings.

I have already reported, in perhaps tedious detail, the natures of the associations of bdellovibrios with living bacterial symbionts. Similarly, the non-symbiotic cultivation of bdellovibrios in various non-cellular brews has been described. And, I have already noted that some clones of ordinarily symbiosis-competent bdellovibrios can live non-symbiotically by scavenging on previously autoclaved (definitely dead) bacterial cells, provided certain special environmental conditions are met (Crothers, Fackrell, Huang & Robinson, 1972; Huang & Starr, 1973*b*). The symbiotic situation with less drastically damaged, perhaps still more-or-less living bacterial cells, remains somewhat ambiguous – as we shall now see.

Varon & Shilo (1968, 1969*b*) showed that ultraviolet-treated *Escherichia coli* cells and those heated for 15 minutes at 70 °C (but not those heated at 98 °C or 120 °C) were still adequate associates for a S-C *Bdellovibrio* strain, yielding essentially the same number of *Bdellovibrio* progeny as did untreated *E. coli* cells. However, these otherwise excellent papers are silent on a point crucial to the present issue; namely, whether there was a sufficient structural integrity – a sufficient component of 'livingness' – in the more mildly treated cells to allow intramural development of the bdellovibrios. Abram (personal communication) has repeated the Varon & Shilo experiment, but supplemented with microscopic observations. In her trials, over 90 % of the *E. coli* cells were killed (in the sense that they could no longer form colonies) by heating at 60 °C for 15 minutes, yet over 80 % of the cells were seen by phase-contrast microscopy to be entered by the bdellovibrios; the developmental cycles of the bdellovibrios in both heated and unheated *E. coli* cells were about the same, and the yield of *Bdellovibrio* progeny was only slightly lower in the heated than in the unheated *E. coli* cells. A recent report (Ross, Robinow & Robinson, 1974) provides some microscopic evidence that S-C bdellovibrios may enter into and develop within *Spirillum serpens* cells that have been heated at 56 °C for 90 minutes. It may be the case here, and in the aforementioned experiments of Varon & Shilo (1968, 1969*b*) and of Abram (personal communication), that a sufficient component of the associant cell's structural integrity – and, hence, some aspects of its 'livingness' – are maintained in this relatively mild exposure to heat, so that attachment, penetration, and possibly controlled intramural growth of the bdellovibrios are still possible, whereas this is not the case with autoclaved cells (Varon & Shilo, 1969*b*; Huang & Starr, 1973*b*; Lambina *et al.*, 1974; but see the contrary claims of Ross, Robinow & Robinson, 1974, pertaining to their Figure 5).

The fact-base is still inadequate for a clean decision to be made concerning the precise point in this series at which this interbacterial relationship can justifiably be considered no longer to be a symbiosis (in the sense the term is used here), but rather to be some sort of (scavenging) non-symbiosis. Are there truly inhabitational, saprotrophic, non-symbiotic relationships which can be initiated between bdellovibrios and the fully dead carcasses of other bacteria, something like the relationships between some fly maggots and the dead carcasses of animals? Perhaps the foregoing explication of these conceptual issues will have the desired heuristic effect: initiation of ingenious experimental programs which might provide the answers.

Bdellovibrios are antagonistic symbionts, which are benefited by the symbiosis

In my terminology, the bdellovibrio–symbiont association is clearly antagonistic (harmful; agonistic) as far as the inhabited symbiont is concerned; it is also clearly beneficial nutritionally, and possibly in other ways, to the *Bdellovibrio*. *Bdellovibrio* may be physically (mechanically) harmful and, without stretching the meanings of the terms too much, also biochemically and physiologically harmful ('enzymatically toxigenic') and pathogenic to its symbiont. The factual situation regarding these claims of physical (mechanical) and biochemical-physiological (enzymatic) harmfulness has already been discussed extensively. I would simply note here that moving from the concept of being biochemically-physiologically harmful to the concept of being toxigenic has ample precedent in other areas of pathology. Now, let me try to justify the notion that *Bdellovibrio* might even be considered to be pathogenic to its symbiont.

An antagonistic symbiont which causes 'infectious disease' in its symbiont is called a 'pathogen' (see Hall, 1974). The complicated concept of infectious disease (Smith, 1934; Burnet, 1962) might be generalized, rather simplistically, as requiring three criteria: (*a*) abnormal physiological functioning, (*b*) which dysfunctioning is caused by a symbiont (a 'parasite' in conventional terminology), and (*c*) which dysfunctioning usually proceeds in a characteristic pattern, a syndrome. That *Bdellovibrio* is undeniably a symbiont, in my terminology, should now be clear. That, as inhabiting symbiont, it causes 'abnormal physiological functioning' which 'proceeds in a syndrome' needs yet to be re-stated here. Among the abnormal functionings of its bacterial symbiont caused by the *Bdellovibrio*, the following can be reiterated: (*a*) the motility of the symbiont ceases; (*b*) protein and nucleic acid syntheses by the symbiont halts; (*c*) the respiratory potential of the symbiont is eliminated; (*d*) the rigidity of the symbiont's cells is reduced (bdelloplast formation; alteration of the rigid murein layer); (*e*) the symbiont's cytoplasmic membrane is damaged locally and the physiologically abnormal leakage that is known to occur might be happening at such damaged sites; (*f*) disorganization and disintegration of all components of the entered cell eventually occur. The dysfunctioning proceeds in a characteristic pattern which is quite regular and typical. (The foregoing order may be roughly the correct sequence, insofar as is presently known; however, I quite agree with the comments of some colleagues to the effect that we cannot yet order the sequence with any great precision.) Although it may be disquieting to some, *Bdellovibrio* could be viewed as a pathogen which causes in its symbiont an infectious disease – a disease which is, moreover, manageable up to a point by 'chemotherapy' as is suggested by the aforementioned work of Mielke & Stolp (personal communication). On top of these manifestations of antagonism, the *Bdellovibrio* consumes its symbiont from within – in much the same format as is shown by the entomophagous parasitoid insects.

SUMMARIZING – PLUS A FEW LOOSE ENDS

In summary, the association between bdellovibrios and other bacteria is a symbiosis in the sense that I use the term. Non-symbiotic bdellovibrios, capable of saprotrophic growth in axenic cultures, have been derived from symbiosis-competent clones. Some of these non-symbiotic bdellovibrios, as well as symbiosis-competent ones, are still dependent for axenic cultivation on complex substances which can be extracted from the former symbiont. While it is in a symbiosis, a symbiosis-competent *Bdellovibrio* behaves like an obligately to facultatively dependent, persistent, fairly specific, biotrophic becoming necrotrophic, antagonistic (possibly physically harmful, toxigenic and pathogenic), generally (intramurally) inhabiting and possibly exhabiting symbiont of other bacteria. The bdellovibrios are benefited nutritionally and possibly regulatively. There are possibly some behavioral overtones ('recognition' of a congenial symbiont and locomotion toward it mediated by 'chemotaxis') but the fact-base on this point is quite incomplete. Intriguing cases of plural symbiosis – the kind usually called 'hyperparasitism', but which I have renamed 'hypersymbiosis' – involving bdellovibrios, *Bdellovibrio* phages, and symbiont bacteria, are now known (Althauser, Samsonoff, Anderson & Conti, 1972; Kessell & Varon, 1973; Varon, 1974).

The symbiosis generally can be characterized quite tidily in the terms of the criterional continua of my classificatory scheme that is being used here. Where there is uncertainty regarding the applicability of a criterional term, the uncertainty is due to a lack of factual knowledge. Identifying explicitly the factual deficiency surely has heuristic value in that it may stimulate the necessary experimental work which, in turn, might overcome the factual deficiency and then permit proper categorization.

ACKNOWLEDGEMENTS

For thoughtful critiques of various versions and sections of this work and for generous disclosures of research-in-progress, I am deeply indebted to many individuals: D. Abram, A. K. Chatterjee, D. M. Coder, S. C. Rittenberg, R. J. Seidler, M. Shilo, J. E. Snellen and H. Stolp. Although I have not always heeded the generously proffered and much appreciated advice of these colleagues and friends, every one of them has helped me in this enterprise. My grateful thanks go to each individual; however, none is to be held accountable for what I say here.

REFERENCES

ABRAM, D., CASTRO E MELO, J. & CHOU, D. (1974). Penetration of *Bdellovibrio bacteriovorus* into host cells. *J. Bact.*, **118**, 663–680.

ABRAM, D. & DAVIS, B. K. (1970). Structural properties and features of parasitic *Bdellovibrio bacteriovorus*. *J. Bact.*, **104**, 948–965.

ADLER, J. (1966). Chemotaxis in bacteria. *Science, Wash.*, **153**, 708–716.

(1973). A method for measuring chemotaxis and use of the method to determine optimum conditions for chemotaxis by *Escherichia coli*. *J. gen. Microbiol.*, **74**, 77–91.

AHMADJIAN, V. & HALE, M. E. [eds.] (1973). *The Lichens*. New York: Academic Press.

ALTHAUSER, M., SAMSONOFF, W. A., ANDERSON, C. & CONTI, S. F. (1972). Isolation and preliminary characterization of bacteriophage for *Bdellovibrio bacteriovorus*. *J. Virol.*, **10**, 516–524.

DE BARY, A. (1879). *Die Erscheinung der Symbiose*. Strassburg: Verlag von Karl J. Trübner.

BUCHNER, P. (1965). *Endosymbiosis of Animals with Plant Organisms*. New York: John Wiley & Sons, Inc.

BURGER, A., DREWS, G. & LADWIG, R. (1968). Wirtskreis und Infektionscyclus eines neu isolierten *Bdellovibrio bacteriovorus*-Stammes. *Arch. Mikrobiol.*, **61**, 261–279.

BURNET, M. (1962). *Natural History of Infectious Disease*, third edition. London: Cambridge University Press.

BURNHAM, J. C., HASHIMOTO, T. & CONTI, S. F. (1968). Electron microscopic observations on the penetration of *Bdellovibrio bacteriovorus* into Gram-negative bacterial hosts. *J. Bact.*, **96**, 1366–1381.

BURNHAM, J. C., HASHIMOTO, T. & CONTI, S. F. (1970). Ultrastructure and cell division of a facultatively parasitic strain of *Bdellovibrio bacteriovorus*. *J. Bact.*, **101**, 997–1004.

BURNHAM, J. C. & ROBINSON, J. (1974). *Bdellovibrio*. In *Bergey's Manual of Determinative Bacteriology*, eighth edition, 212–214 (eds. R. E. Buchanan and N. E. Gibbons). Baltimore: Williams & Wilkins Company.

CROTHERS, S. F., FACKRELL, H. B., HUANG, J. C.-C. & ROBINSON, J. (1972). Relationship between *Bdellovibrio bacteriovorus* 6-5-S and autoclaved host bacteria. *Can. J. Microbiol.*, **18**, 1941–1948.

CROTHERS, S. F. & ROBINSON, J. (1971). Changes in the permeability of *Escherichia coli* during parasitization by *Bdellovibrio bacteriovorus*. *Can. J. Microbiol.*, **17**, 689–697.

DIEDRICH, D. L., DENNY, C. F., HASHIMOTO, T. & CONTI, S. F. (1970). Facultatively parasitic strain of *Bdellovibrio bacteriovorus*. *J. Bact.*, **101**, 989–996.

DOUTT, R. L. (1964). Biological characteristics of entomophagous adults. In *Biological Control of Insect Pests and Weeds*, 145–167 (eds. P. DeBach and E. I. Schlinger). London: Chapman & Hall.

DUNN, J. E., WINDOM, G. E., HANSEN, K. L. & SEIDLER, R. J. (1974). Isolation and characterization of temperature-sensitive mutants of host-dependent *Bdellovibrio bacteriovorus* 109D. *J. Bact.*, **117**, 1341–1349.

EHRLICH, M. A. & EHRLICH, H. G. (1971). Fine structure of the host–parasite interfaces in mycoparasitism. *A. Rev. Phytopathol.*, **9**, 155–184.

ENGELKING, H. M. & SEIDLER, R. J. (1974). The involvement of extracellular enzymes in the metabolism of *Bdellovibrio*. *Arch. Mikrobiol.*, **95**, 293–304.

FACKRELL, H. B. & ROBINSON, J. (1973). Purification and characterization of a lytic peptidase produced by *Bdellovibrio bacteriovorus* 6-5-S. *Can. J. Microbiol.*, **19**, 659–666.

GLOOR, L., KLUBEK, B. & SEIDLER, R. J. (1974). Molecular heterogeneity of the bdellovibrios: Metallo and serine proteases unique to each species. *Arch. Mikrobiol.*, **95**, 45–56.

GROMOV, B. V. & MAMKAEVA, K. A. (1972). [Electron microscope examination of *Bdellovibrio chlorellavorus* parasitism on cells of the green alga *Chlorella vulgaris*.] *Tsitologiia*, **14**, 256–260.

GUÉLIN, A. & CABIOCH, L. (1974). Caractères dynamiques de l'interaction entre le microprédateur *Bdellovibrio bacteriovorus* et la bactérie-hôte en fonction de leurs densités initiales respective. *C. r. hebd. Séanc. Acad. Sci. Paris*, **278**, 1293–1296.

HALL, R. (1974). Pathogenism and parasitism as concepts of symbiotic relationships. *Phytopathology*, **64**, 576–577.

HESPELL, R. B., ROSSON, R. A., THOMASHOW, M. F. & RITTENBERG, S. C. (1973). Respiration of *Bdellovibrio bacteriovorus* strain 109J and its energy substrates for intraperiplasmic growth. *J. Bact.*, **113**, 1280–1288.

HESPELL, R. B., THOMASHOW, M. F. & RITTENBERG, S. C. (1974). Changes in cell composition and viability of *Bdellovibrio bacteriovorus* during starvation. *Arch. Mikrobiol.*, **97**, 313–327.

HOROWITZ, A. T., KESSEL, M. & SHILO, M. (1974). Growth cycle of predacious bdellovibrios in a host-free extract system and some properties of the host extract. *J. Bact.*, **117**, 270–282.

HUANG, J. C.-C. & STARR, M. P. (1973*a*). Possible enzymatic bases of bacteriolysis by bdellovibrios. *Arch. Mikrobiol.*, **89**, 147–167.

(1973*b*). Effects of calcium and magnesium ions and host viability on growth of bdellovibrios. *Antonie van Leeuwenhoek*, **39**, 151–167.

ISHIGURO, E. E. (1973). A growth initiation factor for host-independent derivatives of *Bdellovibrio bacteriovorus*. *J. Bact.*, **115**, 243–252.

(1974). Minimal nutritional requirements for growth of host-independent derivatives of *Bdellovibrio bacteriovorus* strain 109 Davis. *Can. J. Microbiol.*, **20**, 263–265.

KESSELL, M. & VARON, M. (1973). Development of bdellophage VL-1 in parasitic and saprophytic bdellovibrios. *J. Virol.*, **12**, 1522–1533.

KEYA, S. O. & ALEXANDER, M. (1975). Regulation of parasitism by host density: The *Bdellovibrio-Rhizobium* interrelationship. *Soil Biol. Biochem.* (in press).

LAMBINA, V. A., AFINOGENOVA, A. V., KONOVALOVA, S. M., PECHNIKOV, N. V., FEDOROVA, A. M., FICHTE, B. A. & SKRYABIN, G. K. (1974). [On the character of parasitism of *Bdellovibrio bacteriovorus* Stolp et Starr gen. et sp. nov.] *Akad. Nauk.* [*S.S.S.R.*], *Institut Biochimii i Fiziologii Mikroorganizmov, Seriia Biologischeskaia*, **1**, 81–88.

LEWIS, D. H. (1973). Concepts in fungal nutrition and the origin of biotrophy. *Biol. Rev.*, **48**, 261–278.

(1974). Micro-organisms and plants: The evolution of parasitism and mutualism. *Symp. Soc. gen. Microbiol.*, **24**, 367–392.

LUTTRELL, E. S. (1974). Parasitism of fungi on vascular plants. *Mycologia*, **66**, 1–15.

MATIN, A. & RITTENBERG, S. C. (1972). Kinetics of deoxyribonucleic acid destruction and synthesis during growth of *Bdellovibrio bacteriovorus* strain 109D on *Pseudomonas putida* and *Escherichia coli*. *J. Bact.*, **111**, 664–673.

REINER, A. M. & SHILO, M. (1969). Host-independent growth of *Bdellovibrio bacteriovorus* in microbial extracts. *J. gen. Microbiol.*, **59**, 401–410.

RITTENBERG, S. C. (1973). Aspects of the physiology and biochemistry of intraperiplasmic development of *Bdellovibrio*. *Symp. First Internat. Congr. Bacteriol.* [*Jerusalem, September, 1973*], *Abstracts*, **1**, 108.

RITTENBERG, S. C. & SHILO, M. (1970). Early host damage in the infection cycle of *Bdellovibrio bacteriovorus*. *J. Bact.*, **102**, 149–160.

ROSS, E. J., ROBINOW, C. F. & ROBINSON, J. (1974). Intracellular growth of *Bdellovibrio bacteriovorus* 6-5-S in heat-killed *Spirillum serpens* VHL. *Can. J. Microbiol.*, **20**, 847–851.

SCHERFF, R. H., DEVAY, J. E. & CARROLL, T. W. (1966). Ultrastructure of host–parasite relationships involving reproduction of *Bdellovibrio bacteriovorus* in host bacteria. *Phytopathology*, **56**, 627–632.

SEIDLER, R. J., MANDEL, M. & BAPTIST, J. N. (1972). Molecular heterogeneity of the bdellovibrios: Evidence of two new species. *J. Bact.*, **109**, 209–217.

SEIDLER, R. J. & STARR, M. P. (1968). Structure of the flagellum of *Bdellovibrio bacteriovorus*. *J. Bact.*, **95**, 1952–1955.

(1969*a*). Factors affecting the intracellular parasitic growth of *Bdellovibrio bacteriovorus* developing within *Escherichia coli*. *J. Bact.*, **97**, 912–923.

(1969*b*). Isolation and characterization of host-independent bdellovibrios. *J. Bact.*, **100**, 769–785.

SEIDLER, R. J., STARR, M. P. & MANDEL, M. (1969). Deoxyribonucleic acid characterization of bdellovibrios. *J. Bact.*, **100**, 786–790.

SHILO, M. (1969). Morphological and physiological aspects of the interaction of *Bdellovibrio* with host bacteria. *Curr. Top. Microbiol. Immunol.*, **50**, 174–204.

(1973*a*). *Bdellovibrio bactereovorus* as a model for the study of bacterial endoparasitism. In *Dynamic Aspects of Host–Parasite Relationships*, vol. I, 1–12 (eds. A. Zuckerman and D. W. Weiss). New York and London: Academic Press.

(1973*b*). Rapports entre *Bdellovibrio* et ses hôtes. Nature de la dépendance. *Bull. Inst. Pasteur*, **71**, 21–31.

SHILO, M. & BRUFF, B. (1965). Lysis of Gram-negative bacteria by host-independent ectoparasitic *Bdellovibrio bacteriovorus* isolates. *J. gen. Microbiol.*, **40**, 317–328.

SIMPSON, F. J. & ROBINSON, J. (1968). Some energy producing systems in *Bdellovibrio bacteriovorus* strain 6-5-S. *Can. J. Biochem.*, **46**, 865–873.

SMITH, D. C. (1973). Experimental studies of lichen physiology. In *Symbiotic Associations*. *Symp. Soc. gen. Microbiol.*, **13**, 31–50.

SMITH, T. (1934; reprinted 1963). *Parasitism and Disease*. New York and London: Hafner Publishing Co. Inc.

SNELLEN, J. E. & STARR, M. P. (1974). Ultrastructural aspects of localized membrane damage in *Spirillum serpens* VHL early in its association with *Bdellovibrio bacteriovorus* 109D. *Arch. Mikrobiol.*, **100**, 179–195.

STANIER, R. Y. (1970). Some aspects of the biology of cells and their possible evolutionary significance. In *Organization and Control in Prokaryotic and Eukaryotic Cells*. *Symp. Soc. gen. Microbiol.*, **20**, 1–38.

STANIER, R. Y., DOUDOROFF, M. & ADELBERG, E. A. (1970). *The Microbial World*, third edition. Englewood Cliffs, New Jersey: Prentice-Hall, Inc.

STARR, M. P. (1959). Bacteria as plant pathogens. *A. Rev. Microbiol.*, **13**, 211–238.

STARR, M. P. & BAIGENT, N. L. (1966). Parasitic interaction of *Bdellovibrio bacteriovorus* with other bacteria. *J. Bact.*, **91**, 2006–2017.

STARR, M. P. & CHATTERJEE, A. K. (1972). The genus *Erwinia*: Enterobacteria pathogenic to plants and animals. *A. Rev. Microbiol.*, **26**, 389–426.

STARR, M. P. & HUANG, J. C.-C. (1972). Physiology of the bdellovibrios. *Adv. microbial Physiol.*, **8**, 215–261.

STARR, M. P. & SEIDLER, R. J. (1971). The bdellovibrios. *A. Rev. Microbiol.*, **25**, 649–678.

STOLP, H. (1967*a*). *Bdellovibrio bacteriovorus* (*Pseudomonadaceae*). Parasitische Befall und Lysis von *Spirillum serpens*. Film E-1314. Göttingen: Institut für den wissenschaftlichen Film.

(1967*b*). Lysis von Bakterien durch den Parasiten *Bdellovibrio bacteriovorus*. Film C 972. Göttingen: Institut für den wissenschaftlichen Film. (Begleittext in *Publ. Inst. Wiss. Film*, Göttingen, Bd. A-II, 695–706, 1969.)

(1973). The bdellovibrios: Bacterial parasites of bacteria. *A. Rev. Phytopathol.*, **11**, 53–76.

STOLP, H. & PETZOLD, H. (1962). Untersuchungen über einen obligat parasitischen Mikroorganismus mit lytischer Aktivät für *Pseudomonas-Bakterien*. *Phytopathol. Z.*, **45**, 364–390.

STOLP, H. & STARR, M. P. (1963). *Bdellovibrio bacteriovorus* gen. et sp. n., a predatory, ectoparasitic, and bacteriolytic microorganism. *Antonie van Leeuwenhoek*, **29**, 217–248.

STOLP, H. & STARR, M. P. (1965). Bacteriolysis. *A. Rev. Microbiol.*, **19**, 79–104.

STORZ, J. & PAGE, L. A. (1971). Taxonomy of the chlamydiae: Reasons for classifying organisms of the genus *Chlamydia*, family *Chlamydiaceae*, in a separate order, *Chlamydiales* ord. nov. *Int. J. Syst. Bacteriol.*, **21**, 332–334.

STRALEY, S. C. & CONTI, S. F. (1974). Chemotaxis in *Bdellovibrio bacteriovorus*. *J. Bact.*, **120**, 549–551.

TAYLOR, V. I., BAUMANN, P., REICHELT, J. L. & ALLEN, R. D. (1974). Isolation, enumeration, and host range of marine bdellovibrios. *Arch. Mikrobiol.*, **98**, 101–114.

UEMATSU, T. & WAKIMOTO, S. (1970). Biological and ecological studies on *Bdellovibrio*. 1. Isolation, morphology, and parasitism of *Bdellovibrio*. *Ann. phytopathol. Soc. Japan*, **36**, 48–55.

UEMATSU, T. & WAKIMOTO, S. (1971). Biological and ecological studies on *Bdellovibrio*. 3. Growth of *B. bacteriovorus* in media composed of living and of autoclaved bacterial cells. *Ann. phytopathol. Soc. Japan*, **37**, 91–99.

VARON, M. (1974). The bdellophage three-membered parasite system. *CRC Crit. Rev. Microbiol.*, **4**, 221–241.

VARON, M., DICKBUCH, S. & SHILO, M. (1974). Isolation of host-dependent and nonparasitic mutants of the facultative parasitic *Bdellovibrio* UKi2. *J. Bact.*, **119**, 635–637.

VARON, M., DRUCKER, I. & SHILO, M. (1969). Early effects of *Bdellovibrio* infection on the synthesis of protein and RNA of host bacteria. *Biochem. biophys. Res. Comm.*, **37**, 518–525.

VARON, M. & SHILO, M. (1968). Interaction of *Bdellovibrio bacteriovorus* and host bacteria. I. Kinetic studies of attachment and invasion of *Escherichia coli* B by *Bdellovibrio bacteriovorus*. *J. Bact.*, **95**, 744–753.

(1969*a*). Attachment of *Bdellovibrio bacteriovorus* to cell wall mutants of *Salmonella* spp. and *Escherichia coli*. *J. Bact.*, **97**, 977–979.

(1969*b*). Interaction of *Bdellovibrio bacteriovorus* and host bacteria. II. Intracellular growth and development of *Bdellovibrio bacteriovorus* in liquid cultures. *J. Bact.*, **99**, 136–141.

THE HEREDITARY SYMBIONTS OF *PARAMECIUM AURELIA*

By J. R. PREER, JR

Department of Zoology, Indiana University,
Bloomington, Indiana 47401, USA

INTRODUCTION

For many years after its discovery by Sonneborn (1938) the killer trait in *Paramecium aurelia* was thought to be due to an ordinary genetic determinant. It was not until 1946, three years after the demonstration that the killer trait was due to the cytoplasmic factor kappa, that the suggestion was made that the phenomenon was due to an extrinsic organism. Lindegren (1946) noted that the facts available at that time were consistent with the view that kappa was a virus. Altenburg (1946) suggested that kappa might be a relative of the symbiotic algae infecting *Paramecium bursaria*, and asserted that his suggestion was also in accord with the information then available. Later Altenburg (1948) cited facts which he believed supported his view: (1) The green algae of *P. bursaria* and kappa are present in about the same number (Preer, 1946). (2) Kappa is large enough to be seen in the light microscope (Preer, 1948). (3) Kappa may be eliminated from paramecia by differential growth (Preer, 1946) and by high temperature (Sonneborn, 1946). (4) Kappa is, under special conditions, infective (Sonneborn, 1948). Of course these facts which gave courage to proponents of the theory that kappa was an organism were in no way decisive and in fact fitted the notion that kappa was a genetic cytoplasmic factor quite satisfactorily. Further observations since that time, however, have left no doubt that kappa and similar particles such as mu, lambda, etc. are bacterial in nature and that Lindegren and Altenburg were correct in their belief that kappa had an external origin.

Observations bearing on the bacterial relations of the endosymbionts have been made by the author, Dippell, van Wagtendonk, Soldo, Smith, Sonneborn, Kung, Stevenson, Jurand, L. B. Preer, Baker, Gibson and Williams. Their work has been considered in detail in a recent review of the endosymbionts of *P. aurelia* (Preer, Preer & Jurand, 1974). The main points are summarized in Table 1. Aside from a general similarity in morphological and chemical properties, the most convincing evidence that the endosymbionts are bacteria consists of Kung's (1970, 1971) finding that

Table 1. *Evidence that the endosymbionts are bacteria*

1. Sizes and shapes typical of bacteria
2. Electron microscope shows cell walls, no nuclear membrane, no mitochondria
3. Delta, lambda, and mu have flagella
4. Amounts of protein, lipid, DNA and RNA like bacteria
5. DNA base composition different from that of *Paramecium*
6. Ribosomal RNAs of mu and alpha have sedimentation coefficients of bacteria, not *Paramecium*
7. Ribosomal RNAs of mu hybridize with DNA of *Escherichia coli*, not *Paramecium*
8. Mu contains a DNA-dependent RNA polymerase sensitive to rifampicin and actinomycin D
9. Mu contains diaminopimelic and probably muramic acids
10. Kappa respires, utilizes glucose, contains enzymes of the glycolytic and citric acid cycles
11. Kappa has cytochromes like *E. coli* and unlike *Paramecium*
12. Reports (disputed) of the culture of lambda and mu
13. Kappa infected with defective lysogenic phages inducible by ultraviolet light

kappa respires, has the enzymes of the glycolytic and oxidative pathways and contains uniquely bacterial cytochromes; Stevenson's (1967) finding that mu has diaminopimelic acid, a substance found only in bacterial cell walls; the discovery by Jurand & Preer (1969) that lambda and sigma have flagella whose structure is unique to bacteria; Baker's (1970, 1971) demonstration that the ribosomal RNA of mu is like that of bacteria and unlike that of *Paramecium*; and finally, the finding that kappa has defective phages inducible by ultraviolet light (Preer, Rudman, Preer & Jurand, 1974).

This paper will deal with three topics: first, taxonomic and evolutionary considerations; second, recent work on the bacteriophages and associated structures found in kappa; and last, certain aspects of the physiology of killing which are currently under investigation.

TAXONOMY AND EVOLUTION

Recently the endosymbionts of *P. aurelia* were described in detail and given binomial names (Preer, Preer & Jurand, 1974) (Table 2). The assignment of names to organisms which have not been cultured *in vitro* is feasible partly because of our increasing knowledge of the endosymbionts and partly because of the fact that systematists of the bacteria now recognize that the single most reliable taxonomic character is the DNA base ratio. This fact has led to a complete upheaval in bacterial taxonomy as revealed in the new eighth edition of *Bergey's Manual* (1974), and is of great importance in assigning the symbionts of *P. aurelia* to groups.

Table 2. *The bacterial endosymbionts of Paramecium aurelia*

	Common Name	Length (single forms) (μm)	Prelethal effects in sensitives	Distribution in Species	DNA base ratio G+C (%)	Remarks
Caedobacter taeniospiralis	Kappa	1.0–2.5	Spinning, humps, vacuoles, paralysis	2, 4	35–43	Contains R bodies and defective phages
Caedobacter conjugatus	Mu	1.0–4.0	Mate-killing	1, 2, 8	35–37	
Caedobacter minutus	Gamma	0.5–1.0	Vacuoles	8	38	Surrounded by extra membranes
Caedobacter falsus	Nu	1.0–1.5	Non-killer	2, 5	36	
Cytophaga caryophila	Alpha	1.0–6.0	Non-killer	2	36	Found in macronucleus
Lyticum flagellatum	Lambda	2.0–4.0	Rapid lysis	4, 8	49	Flagellated
Lyticum sinuosum	Sigma	2.0–10.0	Rapid lysis	2	45	Flagellated
Tectobacter vulgaris	Delta	1.0–2.0	Non-killer	1, 2, 4, 6, 8	—	Sparsely flagellated

(The data on which this table is based are summarized in Preer, Preer & Jurand (1974). Tau (Stevenson, 1970) is a non-killer in my hands and is indistinguishable from nu. Although delta has been reported to kill in the past, no strains of delta now on hand kill.)

The symbionts are all Gram-negative rods and with a single exception have been assigned to new genera. The exception is the macronuclear symbiont, alpha, which is remarkably similar in morphology (long, tapered rods) and DNA base ratio (36 % G + C) to the gliding bacteria of the genus *Cytophaga* (33–42 % G + C). Alpha is the new species *Cytophaga caryophila*. The recently described micronuclear symbiont, omega, in *Paramecium caudatum* (Ossipov, 1973) would appear to be a close relative of alpha.

The new genus *Caedobacter* consists of small, Gram-negative, non-spore-forming, non-flagellated rods with a G + C content of 35–43 %. They are aerobic or facultatively anaerobic and are diffcult or impossible to culture free of paramecia. All are related, as shown by DNA hybridization studies (Behme, 1969). *Caedobacter taeniospiralis* is kappa, distinguished by the presence of defective phages and tightly wound, ribbon-like inclusions called refractile or R bodies. Kappa-bearing paramecia liberate toxins into the medium in which they live. These toxins kill sensitive strains of paramecia. Kappa renders the paramecia which bear it resistant to the toxin. *Caedobacter conjugatus* is mu, the agent of mate-killing. Mate-killers liberate no toxins but kill their mates at conjugation. *Caedobacter minutus* is a very small form which liberates a powerful toxin into the medium in which it lives, able to kill sensitive paramecia. *Caedobacter falsus*, or nu, is similar to the other forms but produces no toxins. DNA hybridization studies (Behme, 1969) show that the species of *Caedobacter* are closely related to each other. They seem to resemble most closely the Gram-negative facultatively anaerobic rods such as *Proteus* and *Haemophilus* or the aerobic, oxidase-negative coccobacillus *Acinetobacter*.

Lyticum flagellatum (lambda) and *Lyticum sinuosum* (sigma) appear to be closely related forms. They are heavily flagellated, large rods producing a very rapid lysis of sensitive paramecia. According to Behme (1969), sigma is 45 % G + C and lambda is 49 % (density in caesium chloride of 1.708). Soldo & Godoy (1973a) have reported a very different value (1.686) for lambda DNA. The meaning of this discrepancy is not clear, but Allen, Byrne & Cronkite (1971) report that Soldo's strain of lambda-bearing paramecia also has a difference in esterases that sets it apart from their strain of lambda-bearers. It is suggested that although the original strains came from the same source, they may now be different. According to Behme (1969) the DNAs from the two species of *Lyticum* hybridize strongly with each other but only weakly with *Caedobacter*.

Delta, or *Tectobacter vulgaris*, is a sparsely flagellated, sometimes motile form with a rather heavy cell wall. It is widely distributed in *P. aurelia* and is often found with other symbionts such as kappa and mu. Although it has been reported to be a weak killer, no strains now on hand kill.

Paramecia are essential for the maintenance of the symbionts. Do the symbionts benefit their hosts? In the laboratory it is easily shown that sensitives are rendered resistant and converted to killers when infected by kappa, but whether killing and sensitivity play a role in nature is unknown. Recently Grun (personal communication) has pointed out that an invader which only protects its host from itself can hardly be thought of as conferring much advantage to the invaded. The fact that lambda killers, unlike lambda-free strains, do not require folic acid in the laboratory (Soldo & Godoy, 1973*b*) is of unknown significance in nature. The bacterial endosymbiont, omicron, recently described in *Euplotes aediculatus* by Heckmann (1974), however, clearly provides a benefit to its host. Fauré-Fremiet's (1952) earlier conclusion that symbiont-infected *Euplotes* was killed by relatively low doses of penicillin because the essential symbionts were eradicated was not thoroughly convincing. Heckmann's identical observations on the action of penicillin, however, were followed by showing that potentially doomed penicillin-treated *Euplotes* can be rescued by reinfection with omicron after penicillin treatment and before onset of death. His finding is an important one, and demonstrates again the difficulty in distinguishing between infectious agents and cell organelles and their hypothetical intermediates. The distinction would apparently have to be made on the basis of the extent of the metabolic and genetic systems retained by the endosymbiont.

BACTERIOPHAGES AND ASSOCIATED STRUCTURES

Since the finding of virus-like structures in kappa by Preer & Preer (1967) a number of studies by workers in my laboratory, described in detail by Preer, Preer & Jurand (1974), have led to the following picture. All strains of kappa are infected by one of several different, yet possibly related bacteriophages. These phages all appear to be defective. In form they vary from typical icosahedral protein envelopes each filled with about 0.8×10^{-16} g of DNA (Preer, Preer, Rudman & Jurand, 1971) to abortive unfilled heads or incompletely assembled capsid protein and free, covalently-closed, circular, 14 μm molecules of DNA (Dilts, 1974). Except for certain mutant strains of kappa, called pi, all individual kappa particles contain the phage. In most of the kappa particles the phage is in a latent, presumably integrated form. Occasionally, however, induction occurs spontaneously in some individual kappas, the DNA and its protein products increase and the kappa in which induction has occurred is killed. In one strain of killers induction may be brought about with exposure of killers to ultraviolet light

(Preer, Rudman, Preer & Jurand, 1974). Two remarkable phenomena accompany this induction. One is the synthesis of an R body in each induced kappa. An R body is a long (15 μm), narrow (0.4 μm), thin (13 nm) protein ribbon wound into a roll, yet with the capacity to unroll suddenly into a long twisted or tubular coiled ribbon. The other phenomenon accompanying induction is the acquisition of toxicity. The toxic action of killers results from the fact that induced kappas bearing R bodies are liberated by the killers into the medium in which they live, and are taken up by sensitive paramecia. Since killing has only been obtained with subcellular fractions containing rolled-up R bodies, the R body is thought to play an important, but as yet unknown, role in the process of killing sensitive paramecia.

Recently Singler (1974) has made a very significant discovery concerning relations among the phages, R bodies, and killing activity. She injected highly purified strain 562 phages into a rabbit and obtained an antiserum which agglutinated the phages. It also agglutinated isolated R bodies, while control serum agglutinated neither. This finding suggests that R bodies contain a protein constituent of the phage coat. If this conclusion is correct it is likely that R body protein is specified by the phage genome. R bodies may thus be likened to polyheads of T4, tubular aggregations of phage coat protein misassembled into sheets rather than phage capsid. Different structural types of R bodies have been found in different strains of killers (Preer, Jurand, Preer & Rudman, 1972). All of the facts are consistent with the view that their R bodies differ because of differing strains of phages. Indeed many of the phages differ in their morphology and their DNA.

Furthermore, Singler found that although the serum does not agglutinate whole 562 kappa, it does inhibit killing when serum, kappa, and sensitives are mixed simultaneously. Control serum had no effect. These facts support the conclusion that phages are involved in killing, a conclusion already reached on the basis of the correlation between presence of phages and killing. The findings further suggest that a molecule of the phage coat, presumably specified by the phage genome, is the toxic agent, or is necessary in some way for killing. The relation may not be a simple one, for isolated phages do not kill sensitive paramecia even when microinjected into the cytoplasm (Koizumi & Singler, unpublished) nor is there any evidence for increase of kappa or phage DNA in sensitive paramecia which are being killed. Instead the activity of chymotrypsin in affecting the nature of the toxic reaction has been taken as evidence that the toxin is protein. Perhaps the antibodies interfere with the reaction by inhibiting the action of R bodies in delivering the toxin to its site of action, or perhaps it reacts with the toxin itself. It is, of course, not ruled out that the toxin is the R body. Although some isolated R bodies are active in killing, the inactivity of highly purified R bodies has made this possibility appear less likely.

THE PHYSIOLOGY OF KILLING

General aspects

The discussion will be limited to killing by the kappa-bearing killers, about which most is known [see Preer, Preer & Jurand (1974) for a detailed consideration of the general aspects of the physiology of killing]. In brief, the situation is as follows. Toxic action on sensitive paramecia is due to those kappa particles which have been induced and contain R bodies. These kappa particles find their way into the medium in which paramecia live, probably by way of the cytopyge, and are taken up from the medium by sensitives. The route of entrance of kappa into sensitives will be considered presently. The prelethal symptoms depend on both the strain of kappa and the strain of sensitive being attacked. The immediate symptoms include in different cases a change in the rate and direction of ciliary beat, a reduction and cessation of the ability of the cells to form food vacuoles, interference with the operation of the contractile vacuole, and breakdown of food vacuole membranes, as well as other membranes at or near the outer surface of the paramecia. All of the effects may be ascribed to direct attack on the integrity of membranes or membrane related functions. The effects are very rapid. Cessation of food vacuole formation (judged by the inability to take up a latex marker and form latex-containing vacuoles) begins in about 12 minutes at 27 °C when concentrated suspensions of stock 7 kappa act on stock 7 kappa-free sensitives (7.s) and when stock 51 kappa acts on stock 51 sensitives (51.s).

The toxin has not been isolated from the R body-containing kappa. Simple physical or chemical disruption of the kappa particles of some strains such as 51 leads to complete loss of activity. In other strains such as 7, free R bodies very highly active in killing may be obtained following breaking of kappa, but disruption or unrolling of the R bodies leads to inactive preparations. In no case has a soluble extract contained toxicity. Thus, one of the requirements for killing seems to be an intact R body capable of unrolling. This is not the only requirement, however, for, depending upon the strain, isolated R bodies are either inactive when isolated or easily inactivated afterwards.

The basis for the remarkable specific resistance induced in paramecia by each strain of kappa is still almost completely unknown. It is known that kappas are taken into killers and that R bodies unroll in food vacuoles of certain killers, as well as sensitives, but killers are not affected by the toxin.

Theories concerning the uptake of toxin

As part of a larger program concerned with the physiology of killing and mechanisms of resistance, we have tried recently to understand and control the factors which govern the uptake of the toxin by sensitives. Two views of the route of uptake of the toxin by sensitives have been held (Preer, Preer & Jurand, 1974). The first is that the toxin gains entrance into paramecia by being taken into the gullet and becoming incorporated into food vacuoles (Jurand, Rudman & Preer, 1971). In support of this view are the facts that the vacuoles of exposed sensitives contain kappas when observed in the electron microscope and that none is seen to be absorbed elsewhere. It has also been found that in certain situations (in certain gullet-free mutants, during fission, autogamy, and conjugation) paramecia are resistant to killing; one concludes that killing is dependent upon the presence of a functional ingestatory apparatus. So if ingestion cannot occur, sensitives cannot take in toxic kappa particles and are protected. Proponents of this view note that the unwinding of R bodies in vacuoles and disruption of food vacuole membranes during killing is in accord with the hypothesis. It is also observed that whole kappas and isolated R bodies, both of which kill, are both readily taken into vacuoles. On the other hand, presumably similar combining molecules would have to lie on the outside of each to effect surface penetration, yet the surfaces of kappa and of R bodies are very different.

The second view is that absorption and penetration of toxic kappas occur at the cell surface. In support of this view it might be argued that only one toxic particle is required to kill a sensitive since only a few are adsorbed and low numbers could easily be overlooked in the electron microscope. Furthermore, it has been observed that sensitives destined to die because they have taken in kappa are saved if they are allowed to conjugate (Sonneborn, 1959). Hence the protective action of conjugation is not due to the prevention of uptake of the toxin. Butzel & Pagliara (1962), noting that inhibitors of protein synthesis also protect, suggest than an active metabolism in sensitives is necessary for their death. Since metabolism may lag during certain stages of the life cycle, resistance during these times is explained. Butzel, Brown & Martin (1960) point out that the protective action of lecithin is not brought about by an interference with incorporation into food vacuoles, for treating during or *after* exposure is effective.

If the toxin gains entrance by way of the food vacuoles and not at other sites, a thorough knowledge of food vacuole formation should lead to predictions which are decisive in settling the matter, as well as providing a background necessary for understanding the factors determining rate of uptake. Accordingly we will now look briefly at the factors influencing uptake of particles by *Paramecium*.

Rate of formation of food vacuoles in Paramecium

We have found that the rate of formation of food vacuoles is, within broad limits, independent of the kind and nature of particles in the medium. We have exposed paramecia cultured at one doubling per day (see Sonneborn, 1970, for general methods) to various concentrations of small particles of latex, carmine, bacteria (*Aerobacter aerogenes*, the organism on which the paramecia were cultured), yeast, and carbon. After five minutes paramecia were fixed in Perenyi's fixative, mounted on slides, stained with acetocarmine, and examined with phase microscopy. The mean number of labelled vacuoles per paramecium for latex, bacteria, and yeast is recorded in Fig. 1. Despite considerable variability, no evidence for dependence upon kind or concentration of particles is seen. It should, of course, be emphasized that

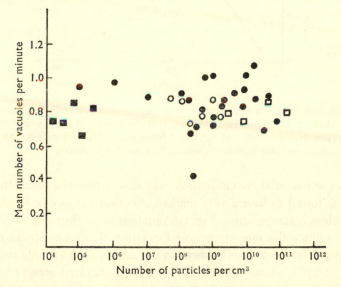

Fig. 1. Relation between mean number of labelled vacuoles per paramecium per minute and particle concentration. Paramecia of stock 7.s were placed into an appropriate suspension of particles, kept at 25–27 °C for 5 min, fixed in Perenyi's fixative, and stained in acetocarmine. The mean number of vacuoles in 20 paramecia was ascertained for each mixture using the phase contrast microscope. ●, 0.557 μm latex; □, 0.188 μm latex; ○, bacteria; ■, yeast.

the rates observed are those found during the first five minutes after the paramecia were placed into the marking medium, and that this operation itself may influence the rate. Indeed it has often been reported that the rate of food vacuole formation in very low concentrations of particles is much reduced (see Wichterman, 1953, for a review). If well-fed, rather than starved, paramecia are tested in this fashion, almost the same, perhaps slightly lower, rates are obtained. Poor cultural conditions such as extreme

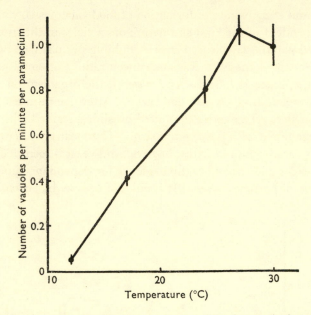

Fig. 2. Relation between temperature and rate of food vacuole formation. Stock 7.s cultured at 27 °C; 20 min equilibration at the new temperature; vacuoles marked with 0.557 μm diameter latex at 10^8 cm^{-3} for 5 min. The points represent means of 20 paramecia and the bars represent 95 % confidence limits.

starvation or bacterial contamination may also reduce the rate. One factor which was found to have a very marked effect on the rate of formation of food vacuoles is temperature (Fig. 2). Agitation of a culture on a shaker was found to reduce the rate of vacuole formation. Different stocks of paramecia may show marked differences in the rate of food vacuole formation; stock 51 at 24 °C showed a rate of 1.12 ± 0.04 (standard error) in one test while stock 7 was measured at 0.80 ± 0.03 at the same temperature. Other tests have confirmed that 51.s is consistently more rapid than 7.s at 24–27 °C. The rate of food vacuole formation during the cell cycle is very constant (Berger & Kimball, 1964), except for brief periods during fission, autogamy, and conjugation when it ceases altogether (Wichterman, 1953;

Dryl & Preer, 1967). The rate of food vacuole formation, while varying from animal to animal in mass cultures, proves nevertheless to be remarkably unaffected by the particulate environment in which the paramecium finds itself.

Rate of uptake of particles

The rate at which a given particle is taken up from the medium is dependent not only on the rate of food vacuole formation but also on the number of particles entering each food vacuole. Hence the rate of food vacuole formation and the rate of uptake of particles may be, and in fact often are, very different. While the rate of food vacuole formation is virtually independent of particle concentration, the rate of particle uptake is strongly dependent on particle concentration. For example, the number of yeast cells taken up per animal by stock 7.s in one experiment is given in Fig. 3. These results are typical of many others in showing a strong dependence of the number of particles ingested on the concentration of particles in the medium.

It has also been found that the absolute number of particles taken in in a given interval of time, varies much more markedly in different experiments than does the rate of food vacuole formation. It has been discovered that the major factor producing this variation is the number of particles of other kinds present in the medium. Thus if paramecia of stock 7.s are placed into culture medium (bacterial concentration of 10^8 per cm^3) which contains

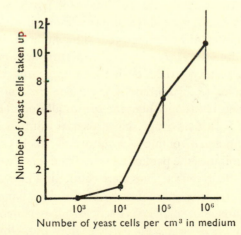

Fig. 3. Results of an experiment showing the effect of the concentration of particles on their rate of uptake. The number of yeast cells taken up per paramecium in 5 minutes at 22 °C by stock 7.s is plotted. The points represent means of 20 paramecia and the bars represent 95 % confidence limits

latex spheres with a mean diameter of 0.557 μm, the vacuoles are well marked by the latex in concentrations of 10^8 cm^{-3} and higher; at 10^7 cm^{-3} and lower the number of latex spheres getting in is too low for adequate marking. If, on the other hand, the same experiment is run in culture fluid from which the bacteria have been removed by centrifugation, vacuoles are well marked in concentrations of latex as low as 10^5; only at 10^4 and lower do too few latex get in. In other words, by eliminating the bacteria, the latex particles may be reduced in concentration 1000 times from 10^8 to 10^5 and approximately the same number of latex spheres will enter the vacuoles. Experiments of this kind have been repeated with numerous combinations of different particles. Particles of all kinds when in sufficient concentration tend to exclude particles of other kinds, and the effects are very large indeed.

Is it possible that *Paramecium* takes in particles in the same ratio in which they are found in the medium? Clearly not, for selection amongst different kinds of particles can be demonstrated. For example, if one mixes in equal concentrations latex particles of a mean diameter of 0.557 μm and 1.011 μm, the ratio of the two types found in the food vacuoles varies from experiment to experiment, and, depending upon the number and kinds of other particles present is often very far from 1:1. Thus it is evident that all kinds of particles are not taken up with the same efficiency. This fact has been reported in the past several times (see Wichterman, 1953). There is some disagreement as to whether selection is made simply on the basis of size and shape or whether chemical composition is also important.

Can one then describe the rate of intake of particles on the basis of competition between different kinds of particles, each with a characteristic and constant efficiency of being ingested? The situation is more complex, as shown by the following experiment. A mixture was made which contained 5×10^7 cm^{-3} latex particles of diameter 0.557 μm, 5×10^7 cm^{-3} latex particles of diameter 1.011 μm, and 0.625 g 100 cm^{-3} of a finely divided precipitate of denatured (boiled) bovine serum albumin (BSA) which had been homogenized with a syringe. Now a serial dilution series was made (full, $\frac{1}{2}$, $\frac{1}{4}$, etc.); an equal volume of a culture of stock 51.s was added to each; and after 5 minutes the paramecia were fixed and the numbers of each kind of latex particle in the vacuoles were counted. The results are given in Table 3. Note that as the concentration of the three kinds of particles is increased, the numbers of small latex particles taken in increases enormously, while the numbers of large latex particles does not. Obviously the competitive advantage of large and small latex spheres depends upon the concentrations of BSA and of large and small latex spheres, for their ratio is constant. A sample (0.06 g 100 cm^{-3}) of small clay particles (bentonite) or a homogenate of 64 000 paramecia cm^{-3} may be substituted for the 0.625 g 100 cm^{-3} of BSA with essentially the same results. Thus particle competition is complex and its rules and mechanisms are not fully known.

Table 3. *The mean number (± standard error) of latex particles of two different diameters taken up in 5 mins. by groups of 30 paramecia in medium containing particles of precipitated protein*

Concentration of particles	1.011 μm latex	0.557 μm latex
Full	0.3 ± 0.1	12.8 ± 2.1
1/2	0.2 ± 0.1	7.8 ± 1.1
1/4	1.9 ± 0.5	11.8 ± 1.5
1/8	2.9 ± 0.7	7.6 ± 0.9
1/16	1.8 ± 0.6	3.1 ± 0.7
1/32	3.5 ± 0.5	1.8 ± 0.3
1/64	2.0 ± 0.8	0.9 ± 0.3
1/128	2.0 ± 0.3	1.1 ± 0.3

(The absolute numbers of all 3 kinds of particles in the medium was varied, but their ratio one to another remained constant.)

Non-specific protection of sensitives by particulates in the culture medium

This background now enables us to make an important prediction with respect to killing by kappa, provided that the route of entry of the toxin is by way of the food vacuoles: any particle, even though it be entirely inert, which can be ingested by sensitive paramecia should be able to inhibit killing if present in sufficient quantity to compete with kappa.

A quantitative measure of the action of killers is generally obtained by exposing sensitives to a dilution series of kappa and counting the number of affected sensitives at later times. This method is sometimes not convenient, for the corpses produced by some killers quickly disintegrate, while other killers produce lethal effects only after a long period of time. In the work reported here sensitives have been exposed to kappa and then either (1) examined 60 minutes later for the frequency of animals which have ceased to form food vacuoles when challenged for five minutes with latex or other marking particles, or (2) washed free of kappa after a few minutes and isolated into culture fluid and scored after 24–48 hours for reduced fission rate (stock 7 kappa) or death (stock 51 kappa). Except for differences in sensitivity the two kinds of tests gave very similar results. The results of typical tests on stock 7 are given in Table 4. It is noted that a concentration of stock 7 kappa of 1.3×10^6 cm^{-3} is strong enough to give very strong toxic action. Nevertheless the toxic action of stock 7 kappa at that concentration could be completely prevented by adding concentrated

Table 4. *Stock 7 killing*

[Stock 7.s sensitives were exposed to isolated kappa (Preer, Hufnagel & Preer. 1966) for 1 hour at 27 °C and washed free by dilution with culture fluid and light centrifugation. Vacuoles were marked by adding 0.760 μm latex to a concentration of 10^8 cm^{-3} at room temperature. Fission rate and survival determined by isolating paramecia into culture fluid.]

Concentration of kappa particles cm^{-3}	Mean number of vacuoles per 5 mins (20 cells)	Mean number of fissions in 22 h (20 cells)	% Survival after 7 days (20 cells)
2.0×10^8	0.50	0.00	0
1.0×10^8	0.00	0.00	0
5.0×10^7	0.15	0.00	0
2.5×10^7	0.60	0.00	0
1.3×10^7	1.20	0.00	5
6.1×10^6	2.15	0.00	10
3.1×10^6	3.00	0.10	45
1.5×10^6	3.50	0.20	90
7.7×10^5	4.25	0.95	75
3.8×10^5	4.75	2.05	100
1.9×10^5	4.50	2.40	95
0.0	4.70	2.70	100

Table 5. *Protection by particles in the medium*

[The table gives the concentration of particles required to protect completely stock 7.s from the toxic effects (reduced fission rate) by 1.3×10^6 cm^{-3} isolated kappa of stock 7.]

	Total solids % (w/v)	Particles cm^{-3}
Bentonite	0.02	
Precipitated BSA	0.04	
Bacteria from culture medium	0.09	4×10^8
0.188 μm latex	0.25	7×10^{11}
0.557 μm latex	0.55	6×10^{10}

bacteria, heat-denatured homogenized BSA, bentonite, or latex (Table 5). Native BSA (2.5 %) and proteose-peptone (1.25 %) were completely ineffective even though used at a much higher concentration than the particles. The action of stock 51 on food vacuole formation and killing was also inhibited by all particles tested: bentonite, boiled BSA, and latex. Small particles, from homogenates of killers and sensitives, of a size less than that of mitochondria, bacteria, kappa, cilia, and trichocyts proved highly effective in protecting 51.s against stock 51 killing. Supernatants centrifuged for 60 minutes at 40 000 g, however had no protective action whatever. These findings strongly support the view that the route of entrance of the toxin is via the food vacuole and that particles of various sorts in high concentration compete with and exclude kappa. It should be emphasized that no study of the toxic reaction produced by killers can be properly evaluated without considering the question of whether food vacuoles are being formed at the normal rate and whether kappa is being excluded by an excess of ingestable particles. Particles in the medium producing even a very faint turbidity can have a very strong protective effect against the toxin of killer paramecia.

The feeding effect

The 'feeding effect' has been known for many years. Sonneborn (personal communication) found that high concentrations of homogenates of stock 51 killers kill less strongly than more dilute concentrations. Both stocks 51 and 7 killers show the same effect, and the effect holds not only for killing, but also for the inhibition of food vacuole formation and, in the case of stock 7, reduction in fission rate too. Do high concentrations of kappa, *per se*, inhibit killing or does some factor in the homogenate protect if present in sufficiently high concentration? The data indicate the latter, for purified kappa or simply the sedimentation obtained from the light centrifugation (5000 g for 2 min) of a homogenate containing kappa do not show protection in high concentrations (Table 6). It has also been shown that the effect can be at least partially restored by adding to the 5000 g precipitate,

Table 6. *The feeding effect*

[Stock 51 sensitives were exposed for one hour to dilutions of a stock 51 killer homogenate or to the sediment of a homogenate centrifuged at 5000 g for 2 min. The paramecia were then exposed to 0.790 μm latex at 10^8 cm^{-3} for 5 minutes. The mean number of latex-containing vacuoles is recorded (20 paramecia).]

Concentration of homogenate (cells cm^{-3})	Whole homogenate	5000 g sediment
64 000	4.20	0.00
32 000	2.10	0.00
16 000	0.20	0.00
8 000	0.20	0.00
4 000	0.00	0.00
2 000	0.00	0.00
1 000	0.00	0.00
500	0.10	1.00
250	0.85	1.40
125	2.50	2.20
63	3.80	4.80
0	5.20	

the 5000–40 000 g precipitate. The particle-free supernatant from 40 000 g for 60 minutes has no protective effect. Therefore the feeding effect can be eliminated by removing the very small particles sedimenting between 5000 and 40 000 g and partially restored by adding them back. Microscopical examination of this fraction reveals only ribosomes and small membranes; larger particles such as bacteria, kappa, trichocysts, cilia, large fragments of the body wall, kinetodesmal fibers, crystals, intact food vacuoles have all been eliminated. The same fraction from sensitives is as effective as the 5000–40 000 g fraction from killers in providing protection at high concentrations.

The activity of the small particles in a homogenate in producing this effect suggests an explanation based on particle competition. Even though the ratio of particles of all kinds is the same in all dilutions, perhaps the small particles compete more effectively with kappa at high concentrations. Indeed the experiments on particle competition cited above suggest that the relative competitive advantage of small particles increases greatly as absolute concentrations increase (Table 3). If this explanation for the feeding effect is correct, then it should be possible to duplicate this effect with artificial mixtures of particles. The addition of various kinds of small particles in the right concentrations to a 5000 g sediment should afford protection against the killing action of kappa, and simply by diluting the mixtures, killing should be obtained. In fact it was possible to obtain this result with bentonite, precipitated BSA, and 0.188 μm latex, all very small particles relative to kappa (Table 7). Note that stock 51.s was partially protected in the full strength mixtures, but all were killed when the mixtures (including kappa) were diluted to 1/16 of their initial values. Thus the predictions were fully confirmed by the data. It is likely therefore that the protection afforded sensitives in high concentrations of homogenates of

Table 7. *The feeding effect with artificial mixtures*

[Percent of 20 sensitives surviving after exposure to small particles plus a partially purified suspension (5000 g sediment) of 51 kappa. Treatment was for 1 hour at full and 1/16 concentrations.]

		Bentonite	Precipitated BSA	0.188 μm latex 0.10%
Small particles:		0.12%	0.63%	2.7×10^{11} cm^{-3}
5000 g killer sediment:		64 000	64 000	64 000
Concentration of mixture	Full	80%	50%	70%
	1/16	0%	0%	0%

killers is due to the fact that small particles in the homogenate when present in high concentrations successfully exclude kappa from its normal site of entry, the food vacuole.

SUMMARY AND CONCLUSIONS

A large body of evidence now supports the view that the endosymbionts of *P. aurelia* are bacteria. Recently they have been described as eight new species of Gram-negative aerobic or facultatively anaerobic rods. All but one have been assigned to new genera. No clear evidence that they provide a benefit to their host in nature has been obtained. It is interesting that Heckmann (1975) has recently shown that the bacterial symbionts of Euplotes are essential to the life of their hosts.

The endosymbiont kappa (*Caedobacter taeniospiralis*) contains defective phages, whose induction results in the production of the refractile (R) body, the acquisition of toxicity to sensitive paramecia, and death of the kappa particle. Recent work by Singler has shown that the phage coats have antigenic sites in common with R bodies, suggesting that the R body is specified by the phage genome and represents a misassembly of the phage coat protein. She has also found that antiserum made against the isolated phages inhibits killing, demonstrating a role of phage or phage-related proteins in killing.

The uptake of toxic kappa particles by sensitive paramecia has been thought to be either through the food vacuoles or by adsorption to the surface of sensitive paramecia. A study of the feeding behavior of paramecia reported here for the first time indicates that uptake is by means of food vacuoles. Vacuoles are formed at a fairly constant rate of about one vacuole per minute at 27 °C. This rate, within broad limits, is not influenced by number and kinds of particles encountered and is relatively resistant to differing physiological conditions. It is inhibited to varying degrees by large quantities of bacterial food, by unfavourable cultural conditions, and by agitation. It is strongly reduced at certain stages of the cell cycle (during fission, autogamy and conjugation when feeding stops altogether) and by lowered temperatures. However, the numbers of particles of a given kind ingested in a given time interval is greatly increased by raising the concentration of that kind of particle and greatly reduced by raising the concentrations of other particles present. Particles are not taken in in strict proportion to the number in the medium: selection occurs. Moreover, the ratio of different particles taken in may vary with the absolute concentration of the particles in the medium, even though their ratio in the medium is held constant.

The prediction that all kinds of ingestible particles in high concentration should be able to give virtually complete protection to sensitives from killing was tested. It was shown to be true for latex spheres of different diameters, for bacteria, bentonite (clay) particles, precipitated bovine serum albumin and particles in homogenates of paramecia.

A phenomenon of kappa killing known as the 'feeding effect' is seen when sensitives are exposed to homogenates of killers; dilute homogenates kill more strongly than more concentrated homogenates, even though the ratio of particles must be the same in all concentrations. It is postulated that the phenomenon results from the unusual kinetics of particle competition. Small particles in homogenates tend to exclude larger particles (such as kappa) more readily in high than in low concentration. It was shown that mixtures of partially purified kappa and small particles (latex, bentonite, precipitated protein) in a carefully chosen ratio could be used to duplicate the feeding effect. It is concluded that the competitive effects of inert particles on killing are consistent only with the view that kappa kills sensitives by entering the food vacuoles.

ACKNOWLEDGEMENTS

Contribution number 971 from the Department of Zoology, Indiana University. This work was supported by grants from the National Science Foundation (GB 27609) and the National Institutes of Health (GM 20038). The author wishes to acknowledge the assistance of L. B. Preer and B. M. Rudman.

REFERENCES

ALLEN, S. L., BYRNE, B. C. & CRONKITE, D. L. (1971). Intersyngenic variations in the esterases of bacterized *Paramecium aurelia*. *Biochem. Genet.*, **5**, 135–150.

ALTENBURG, E. (1946). In Discussion of paper by T. M. Sonneborn: Experimental control of the concentration of cytoplasmic genetic factors in *Paramecium*. *Cold Spring Harb. Symp. quant. Biol.*, **11**, 236–255.

(1948). The role of symbionts and autocatalysts in the genetics of the ciliates. *Am. Natur.*, **82**, 252–264.

BAKER, R. (1970). Studies on the RNA of the mate-killer particles of *Paramecium*. *Heredity*, **25**, 657–662.

(1971). Studies on the RNA of the mate-killer particles of *Paramecium aurelia*. Ph.D. thesis, University of East Anglia, England.

BEHME, R. (1969). Deoxyribonucleic acid relationship among symbionts of *Paramecium aurelia*. Ph.D. thesis, Indiana University, Bloomington, USA.

BERGER, J. D. & KIMBALL, R. F. (1964). Specific incorporation of precursors into DNA by feeding labelled bacteria to *Paramecium aurelia*. *J. Protozool.*, **11**, 534–537.

Bergey's Mannual of Determinative Bacteriology (1974). Eighth edition (eds. R. E. Buchanan and N. E. Gibbons). Baltimore: Williams and Wilkins.

BUTZEL, H. M., BROWN, L. H. & MARTIN, W. B. (1960). Effects of detergents upon killer-sensitive reactions in *Paramecium aurelia*. *Physiol. Zool.*, **33**, 213–224.

BUTZEL, H. M. & PAGLIARA, A. (1962). The effect of biochemical inhibitors upon the killer-sensitive system in *Paramecium aurelia*. *Expl. Cell Res.*, **27**, 382–395.

DILTS, J. A. (1974). Evidence for covalently closed circular DNA in the kappa endosymbiont of stock 51 of *Paramecium aurelia*. *Genetics*, **77** (*Suppl.*), s17–18.

DRYL, S. & PREER, J. R. Jr (1967). The possible mechanism of resistance of *Paramecium aurelia* to kappa toxin from killer stock, 7, syngen 2, during autogamy, conjugation and cell division. *J. Protozool.*, **14** (*Suppl.*), 33–34.

FAURÉ-FREMIET, E. (1952). Symbiontes bactériens des ciliés du genre *Euplotes*. *C. r. hebd. Séanc. Acad. Sci., Paris*, **235**, 402–403.

HECKMANN, K. (1975). Omikron, ein essentieller Endosymbiont von *Euplotes aediculatus*. *J. Protozool.*, **22**, 97–104.

JURAND, A. & PREER, L. B. (1969). Ultrastructure of flagellated lambda symbionts in *Paramecium aurelia*. *J. gen. Microbiol.*, **54**, 359–364.

JURAND, A., RUDMAN, B. M. & PREER, J. R. (1971). Prelethal effects of killing action by stock 7 of *Paramecium aurelia*. *J. exp. Zool.*, **177**, 365–387.

KUNG, C. (1970). The electron transport system of kappa particles from *Paramecium aurelia* stock 51. *J. gen. Microbiol.*, **61**, 371–378.

— (1971). Aerobic respiration of kappa particles from *Paramecium aurelia*. *J. Protozool.*, **18**, 328–332.

LINDEGREN, C. C. (1946). In Discussion of paper by T. M. Sonneborn: Experimental control of the concentration of cytoplasmic genetic factors in *Paramecium*. *Cold Spring Harb. Symp. quant. Biol.*, **11**, 236–255.

OSSIPOV, D. V. (1973). Specific infectious specificity of the omega-particles, micronuclear symbiotic bacteria of *Paramecium caudatum*. *Cytologia*, **15**, 211–217.

PREER, J. R. Jr (1946). Some properties of a genetic cytoplasmic factor in *Paramecium*. *Proc. natn. Acad. Sci. USA*, **32**, 247–253.

— (1948). The killer cytoplasmic factor, kappa. *Am. Natur.*, **82**, 35–42.

PREER, J. R. Jr, HUFNAGEL, L. & PREER, L. B. (1966). Structure and behavior of R bodies from killer paramecia. *J. ultrastr. Res.*, **15**, 131–143.

PREER, J. R. Jr & PREER, L. B. (1967). Virus-like bodies in killer paramecia. *Proc. natn. Acad. Sci. USA*, **58**, 1774–1781.

PREER, J. R. Jr, PREER, L. B. & JURAND, A. (1974). Kappa and other endosymbionts in *Paramecium aurelia*. *Bact. Rev.*, **38**, 113–163.

PREER, J. R. Jr, PREER, L. B., RUDMAN, B. M. & JURAND, A. (1971). Isolation and composition of bacteriophage-like particles from kappa of killer paramecia. *Molec. gen. Genet.*, **111**, 202–208.

PREER, L. B., JURAND, A., PREER, J. R. Jr & RUDMAN, B. M. (1972). The classes of kappa in *Paramecium aurelia*. *J. Cell Sci.*, **11**, 581–600.

PREER, L. B., RUDMAN, B. M., PREER, J. R. Jr & JURAND, A. (1974). Induction of R bodies by ultraviolet light in killer paramecia. *J. gen. Microbiol.*, **80**, 209–215.

SINGLER, M. J. (1974). Antigenic studies on the relationship of the viral capsid, R body, and toxin of kappa in stock 562, *Paramecium aurelia*. *Genetics*, **77** (*Suppl.*), s60–61.

SOLDO, A. T. & GODOY, G. A. (1973a). Molecular complexity of *Paramecium* symbiont lambda deoxyribonucleic acid: evidence for the presence of a multicopy genome. *J. Molec. Biol.*, **73**, 93–108.

(1973*b*). Observations on the production of folic acid by symbiont lambda particles of *Paramecium aurelia* stock 299. *J. Protozool.*, **20**, 502.

SONNEBORN, T. M. (1938). Mating types in *P. aurelia*: diverse conditions for mating in different stocks; occurrence, number and interrelations of the types. *Proc. Am. phil. Soc.*, **79**, 411–434.

(1946). Experimental control of the concentration of cytoplasmic genetic factors in *Paramecium. Cold Spring Harb. Symp. quant. Biol.*, **11**, 236–255.

(1948). Symposium on plasmagenes, genes and characters in *Paramecium aurelia. Am. Natur.*, **82**, 26–34.

(1959). Kappa and related particles in *Paramecium. Adv. Virus Res.*, **6**, 229–356.

(1970). Methods in paramecium research. In *Methods in Cell Physiology*, **4**, 241–339. New York: Academic Press.

STEVENSON, I. (1967). Diaminopimelic acid in the mu particles of *Paramecium aurelia. Nature, Lond.*, **215**, 434–435.

(1970). Endosymbiosis in some stocks of *Paramecium aurelia* collected in Australia. *Cytobios*, **2**, 207–224.

WICHTERMAN, R. (1953). *The Biology of Paramecium*. Blakiston: New York.

SYMBIOSIS IN *PARAMECIUM BURSARIA*

By MARLENE W. KARAKASHIAN

Max-Planck-Institut für Zellbiologie, Wilhelmshaven,
Federal Republic of Germany

Paramecium bursaria is a ciliated protozoan which normally lives in association with several hundred intracellular *Chlorella*. These algae grow and divide within the cytoplasm of the host cell at a rate compatible with that of the host's growth: they do not normally increase beyond the capacity of the host cell to hold them nor do they disappear in later descendants of the cell line. The algae also persist in the cytoplasm throughout conjugation and the subsequent reorganization of the ex-conjugant cell (Siegel, 1960). Thus, the symbiotic relationship between the organisms is hereditary.

Under the light microscope, infected paramecia look green and often seem packed with symbionts (Plate 1*a*). However, electron micrographs (Vivier, Petitprez & Chive, 1967; Karakashian, Karakashian & Rudzinska, 1968) of sectioned material show that algae may be widely separated by host cytoplasm (Plate 1*b*). Close inspection of an infected cell reveals that the *Chlorella* are situated within individual vacuoles (Plate 2*a*, *d*). The cytoplasm surrounding the perialgal vacuoles does not display any unusual structural features, such as have been observed by Jones, Yeh & Hirsch (1972) in macrophages infected with the parasite *Toxoplasma gondii*. In macrophages, additional membranes of endoplasmic reticulum are laid down around the vacuole containing the parasite and there is a tendency for host mitochondria to accumulate in the vicinity.

As befits hereditary units, the alga–vacuole complexes replicate under normal conditions. A number of intracellular division stages have been described (Karakashian *et al.*, 1968). The alga grows, its vacuole enlarging to accommodate it, and eventually enters division to produce autospores (Plate 2*a*, *d*). Upon completion of algal division, the old cell wall breaks into fragments which transiently adhere to the young cells and then detach, coiling into distinctive scrolls. At about this time, the perialgal vacuole begins to constrict in several places so as to segregate the young algal cells again into isolated units. The cell wall scrolls from the preceding generation are eliminated from the replicated perialgal vacuoles and later appear in digestion vacuoles (Plate 2*b*, *d*).

[145]

The alga–vacuole complexes are resistant to digestion by the host cell throughout their vegetative growth and replication (Plate 2*a*, *d*; Karakashian & Karakashian, 1973). This resistance appears to be conferred upon the perialgal vacuole, for most of the algae are digested if they are isolated from a homogenate and fed immediately to another paramecium. This point will be taken up in more detail later.

Evidently, *P. bursaria* has acquired a novel class of cell organelles, which may be removed from the cell without injury to it and which may be cultured separately. These organelle-like units are quickly re-established whenever aposymbiotic animals ingest large numbers of previously symbiotic algae, regardless of the time the algae and paramecia have been cultured independently.

In view of the experimental advantages of this system, it is surprising that the *P. bursaria–Chlorella* association has not attracted more active interest both from those interested in symbiosis as a phenomenon and from laboratories concerned primarily with the molecular biological aspects of cell recognition and cell–organelle relationships. This neglect greatly simplifies the task of summarizing what is known about symbiosis in *P. bursaria*, but it also means that substantial progress in solving basic questions has remained slow.

This review will focus on three primary problem areas which have been investigated within the last 15 years, chiefly by Stephen Karakashian and myself. These areas are: the benefit of the association to each partner, symbiont entry into the host cytoplasm and the specificity of the relationship. Earlier work on symbiosis in *P. bursaria* (Oehler, 1922; Pringsheim, 1928) has been discussed in previous publications (Karakashian, 1963; 1970*b*).

BENEFIT OF THE SYMBIOTIC ASSOCIATION
TO EACH PARTNER

In contrast to the impressive detail known for certain other symbiotic systems (Smith, Muscatine & Lewis, 1969), relatively little is known about the benefits of symbiosis to the ciliate host *P. bursaria*. This information consists largely of observations on the growth of infected and aposymbiotic animals under various lighting and feeding regimens (Siegel, 1960; Karakashian, 1963; Pado, 1965). Rather less is known about the value of symbiosis to the algal partners in any association (Taylor, 1973), but the outlook for learning more about the benefit to the *P. bursaria* algae may be more promising than for other symbiotic strains, because they can be isolated, cloned and cultured axenically.

Benefit to the host

The symbiotic association provides no apparent benefit to infected paramecia under the optimal growth conditions prevailing in the laboratory (Siegel, 1960; Karakashian, 1963). This may be seen by comparing the fission rates of infected and aposymbiotic isolates under standard laboratory

Table 1. *Mean daily fission rate of infected and aposymbiotic paramecia re-isolated daily in undiluted bacterized medium*[a]

Culture[b]	Fission rate[c]
3–25, LD	1.78 ± 0.26
3W, LD	1.78 ± 0.16
3–25, DD	1.90 ± 0.15
3W, DD	1.81 ± 0.12

[a] Data from Karakashian (1963).

[b] 3–25 is a culture of stock 3 paramecia infected with algae from stock 25; 3W is an aposymbiotic clone of stock 3 paramecia; LD – 10 hours light, 14 hours dark daily; DD – continuous darkness. All cultures were at 25 °C.

[c] There were five replicate lines in each group. One standard deviation is given for each mean.

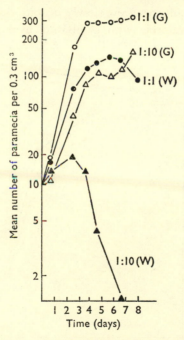

Fig. 1. Effect of decreasing bacterial concentration on growth of infected (G) and aposymbiotic (W) stock 3 paramecia under diurnal illumination (10 h light, 2300–3300 lx, 14 h dark) at 25 °C. Points represent average cell numbers in samples of six microcultures, each of which was initiated at 0 time by placing 10 individuals in 0.3 cm^{-3} of either undiluted (1 : 1) or a 1 : 10 dilution of bacterized medium (Karakashian, 1963).

conditions (Table 1). However, symbionts increasingly enhance the growth of infected cells relative to aposymbiotic controls as the initial supply of bacteria becomes depleted and assure the survival of infected paramecia when bacterial food is gone (Fig. 1; Karakashian, 1963). As the figure also shows, the beneficial effect of having symbionts is accentuated in diluted (1 : 10) bacterized medium.

Pado's (1965) report that the division rate of *P. bursaria* in bacterized medium depends on the prevailing light intensity stands in apparent contradiction to the above conclusion. However, his findings may be explained by the fact that he employed a culture technique in which the bacteria were not replenished as the paramecia divided.

The growth enhancement is a direct effect of the symbionts and not mediated by bacteria which are normally present in the ciliate cultures (Karakashian, 1963; Pado, 1965). This was shown by comparing the growth and survival of sterile cultures of infected animals maintained under diurnal illumination or continuous darkness (Fig. 2). Dark-grown cultures could not be maintained for more than a week under these conditions, whereas illuminated cultures grew slowly despite the absence of bacteria.

The growth of paramecia with symbionts in a medium filtered free

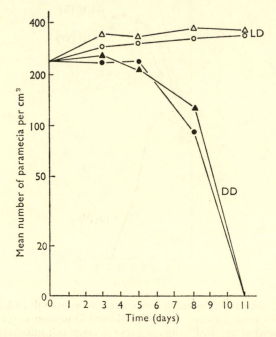

Fig. 2. Growth of stock 32–32 in a bacteria-free inorganic salts medium in LD (10 h light, 14 h dark) or DD (continuous darkness). Triangles and circles each represent replicate cultures (Karakashian, 1963).

of bacteria depends on the intensity of illumination provided (Fig. 3; Karakashian, 1963), suggesting that the host growth rate is governed by the rate of algal photosynthesis under these conditions. Pado (1965) reported that the growth rate of sterile cultures also depends on the amount of light available.

Host growth in a filtered medium is also correlated with the size of the intracellular algal population (Karakashian, 1963; Pado, 1965). As shown in Fig. 4, aposymbiotic paramecia of stocks 3 and 8 grow equally well in a bacterized medium. However, when they were infected with algae from another paramecium stock (25), and placed in a medium filtered free of bacteria, their growth rates diverged sharply and in a manner consistent with the average number of algae present in each cell (3–25: 602 algae; 8–25: 419 algae). How much of this growth difference can be attributed to a differential ability of these two hosts to exploit the strain 25 symbionts and how much is caused simply by the differences in the numbers of algae present is not easy to decide. This ambiguity is present in all studies dealing with the relative effectiveness of various host–alga combinations,

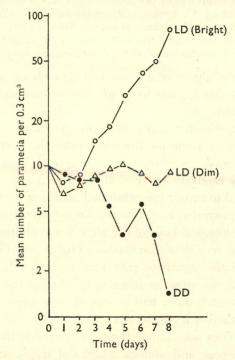

Fig. 3. Effect of light intensity on the growth of infected (3–25) and aposymbiotic (3W) paramecia in medium initially filtered free of bacteria. Bright LD, 4500–4800 lx; dim LD, 800–1350 lx; DD, continuous darkness. For details of microculture method, see legend of Fig. 1 (Karakashian, 1963).

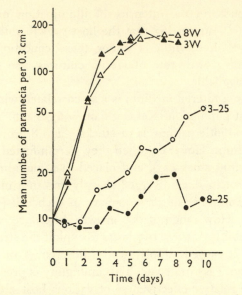

Fig. 4. Comparison of growth of two stocks of paramecia under 10 h light 2300–3300 lx, 14 h dark at 25 °C. Aposymbiotic paramecia (3W, 8W) were introduced into an undiluted bacterized medium at 0 time. Infected paramecia (3–25, 8–25) were placed in a medium filtered free of bacteria at the same time. For details of microculture method, see legend of Fig. 1 (Karakashian, 1963).

since the number of algae present in a cell depends on culture conditions (Pado, 1965) and also appears to be an association-specific characteristic (Karakashian, 1963).

The above experiments show that the symbiotic association in *P. bursaria* has adaptive value for the host, since the bacterial densities at which the symbionts' contribution to host growth becomes insignificant would occur only rarely under natural conditions and selective pressure would be expected to favour the continued maintenance of symbionts. To my knowledge, aposymbiotic *P. bursaria* do not occur in natural habitats.

The exact physiological basis of the algae's contribution to the ciliate is still unclear. The results of Karakashian (1963) and Pado (1965, 1967) suggest that algal photosynthetic rate is an important determinant of host growth rate. If so, the question arises as to whether the release of photosynthate occurs continuously and is superfluous when adequate bacteria are available or whether it is subject to regulation by the host cell. Present information also does not exclude the possibility that the host 'farms' its algae to some degree and obtains additional nutrients by periodically harvesting part of the intracellular population.

P. bursaria symbionts are potentially capable of transferring a large proportion of their photosynthate to the host in the form of a sugar. As shown by Muscatine, Karakashian & Karakashian (1967), 15–86 % of the ^{14}C fixed by diverse symbiotic strains is released into the medium, primarily in the form of maltose (Table 2). The extent of release is affected by the

Table 2. *Distribution of ^{14}C after photosynthesis*[a]
(30 min, 15 000 lx)

		Extracellular distribution of labelled photosynthate		
	Percent found extracellular relative to	Percent extracellular found as:		
Source of algae	total fixed	Maltose	Glucose	Glycolic acid and unknowns
P. bursaria				
NC64A	86.0	97.1	1.2	1.7
32B1	29.1	99.1	0.3	0.6
130C	15.3	97.0	0.0	3.0
3H	84.5	95.2	4.8	0.0
34A	46.2	99.4	0.5	0.1
42	86.7	99.2	0.7	0.1
41K	86.5	99.1	0.4	0.5
Free-living				
20 (*C. ellipsoidea*)	6.5	0	0	100.0
GrB1 (*C. vulgaris*)	1.1	0	0	100.0
250 (*C. protothecoides*)	0.9	0	0	100.0
490 (*C. miniata*)	7.6	0	0	100.0
262 (*C. vulgaris*)	0.4	0	0	100.0
263 (*C. vulgaris*)	0.5	0	0	100.0
326 (*Selenastrum minutum*)	3.8	0	0	100.0
397 (*C. vulgaris*)	1.0	0	0	100.0
398 (*C. vulgaris*)	0.6	0	0	100.0

[a] From Muscatine *et al.* (1967).

pH of the medium, some strains being more sensitive to pH than others. None of the free-living strains of *Chlorella* tested in this study liberate carbohydrates; most excrete only traces of organic acids. Nevertheless, there is no substitute for a direct demonstration of the transfer of organic materials from the symbionts to the host cytoplasm *in situ*. This demonstration is long overdue, especially since it is not technically difficult.* Experiments should be designed so as to reveal quantitative differences in the transfer of materials under various conditions of illumination and starvation. Complementary electron micrographic studies of the intracellular digestive activity of infected paramecia in sterile illuminated cultures will also be needed to clarify finally the basis for symbiont enhancement of host growth.

* See Note added in proof, p. 171.

Benefit to the symbiont

Paramecium symbionts cannot be cultured on the minimal medium suitable for growing many free-living strains of *Chlorella* (Karakashian, unpublished). The medium presently used for their culture, a rich broth containing proteose-peptone, yeast extract and glucose, is adequate for culturing most of the *P. bursaria* algae tested. However, symbiotic algae grow less rapidly in this medium than all of the free-living strains examined, suggesting the medium could still be improved. It is not known, for example, whether all of the ingredients are necessary or even desirable or whether alterations in the proportions of some nutrients would improve symbiont growth.

Because of this lack of work on the culture medium, there has been no systematic comparison of the growth requirements of various symbiotic strains. Differences in requirements certainly exist, since not all strains can be cultured in the present medium nor do those in culture grow equally well. The nutritional requirements of dark-grown cultures would be of particular interest, since it is in darkness that the host's contribution to algal growth is most apparent. Symbiotic algae grow and divide many times in the host cytoplasm (up to 50 divisions in some associations) when paramecia are maintained for long periods in darkness with excess food (Siegel, 1960; Karakashian, 1963); this is the best evidence to date that the symbiosis can be of some value to the algae. Whether the association also benefits the algae under more natural growth conditions is not known.

Pado (1967) has studied the photosynthetic and respiratory activity of the symbionts *in situ* and shown that the photosynthetic production of oxygen depends on the light available and on the algal number and their chlorophyll content. He reported also that the respiration rate of the symbionts is very high in comparison to the host cell, suggesting that the algae must consume large amounts of organic materials in darkness.

Preliminary examination of the photosynthetic capacity of some axenically cultured symbiont strains (Larner and Karakashian, unpublished) revealed that their photosynthetic rates are less dependent on carbon dioxide tension than the free-living *Chlorella* examined. On the other hand, the photosynthetic efficiency of the *Chlorella* strain studied *in situ* by Pado (1967) compares well with values reported for free-living *Chlorella*. A thorough study of the photosynthetic properties of a variety of symbiotic strains seems warranted. These studies might best be conducted on algae freshly isolated from the host cytoplasm, since it is known that the chloroplast morphology changes drastically in axenic culture (Karakashian, 1970a).

Pado (1965) has studied the intracellular growth of one strain of symbiotic algae as a function of light intensity and during adaptation to new light conditions. His results illustrate the extremely sensitive regulation of symbiont growth rate under diverse culture conditions. *P. bursaria* stocks differ with respect to the ease with which they can be rid of their symbionts by rapid growth of the infected animals in darkness (Siegel, 1960). This implies that either some host strains provide more nutrients to their symbionts than others or that the symbionts themselves have intrinsic differences in intracellular growth rate. This could easily be analysed further in reconstituted associations of selected paramecia and algal strains. As noted previously, the intracellular algal population is maintained at a higher level in some associations than in others. This limitation appears to be imposed by the host (Fig. 4; Pado, 1965), but how this is achieved is unknown. Observations by Pringsheim (1928) and Karakashian (1963) indicate this control sometimes fails when foreign algae are introduced into the host.

In summary, *P. bursaria* with symbionts grow and survive better than aposymbiotic animals in environments deficient in food for the host. Indirect evidence suggests that algal photosynthate provides the bulk of host nutrients under these conditions. The symbionts, on the other hand, are able to extract nourishment from their host when it is well fed and they are deprived of light. Given the number of algae present and their high metabolic activity, the host contribution of nutrients to the symbionts must be substantial.

SYMBIONT ENTRY INTO THE HOST CYTOPLASM

There are two ways in which a potential symbiont can enter the host cell: by being ingested and incorporated into a food vacuole or, less frequently, by moving across the cytoplasmic bridge which may form during sexual union with a symbiont-bearing partner. Experiments designed to reveal direct symbiont uptake through the host pellicle yielded negative results (Siegel, 1960).

The most common means of entry, through the host's food vacuoles, is also the most hazardous, because the potential symbionts must run the gauntlet of the host's digestive system. Many must do so successfully, however, for new associations form quickly in all aposymbiotic paramecia exposed to large numbers of symbiotic algae over a period of hours. This ability to escape digestion is a symbiont property, since few free-living strains of *Chlorella* survive the entry process (Karakashian & Karakashian, 1965).

Recent studies have shown that the conversion of a phagosome into a digestive vacuole does not require the presence of nutritive material

(Karakashian & Karakashian, 1973). Digestive enzyme activity can be detected in phagosomes containing latex (Plate 2*b*), carmine (Plate 2*c*) or clay particles under conditions in which bacteria were not ingested. Thus, if symbiotic algae in phagosomes are not digested, they must be able to interfere with the normal course of host digestion.

Active algal interference with host digestion has been demonstrated by examining the host's response to mixtures of live and dead algae. Heat-killed algae are ingested as readily as live ones of the same strain and can be observed in the cell interior if they are stained beforehand with the vital stain Congo Red. Aposymbiotic cells exposed to 50:50 mixtures of live and dead cells form phagosomes which have approximately equal proportions of each type. As indicated in Fig. 5, dead algae disappear from the host cytoplasm much more rapidly when they are ingested alone than when they enter a phagosome with live algae. In the absence of live algae, all of the Congo Red-stained algae were digested within 24 hours. When live

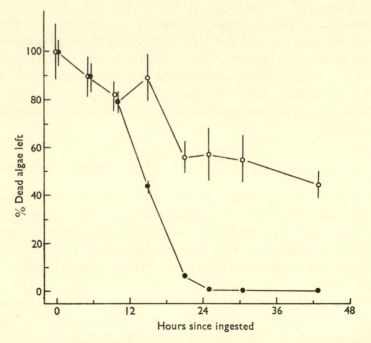

Fig. 5. Retention of heat-killed algae by aposymbiotic paramecia when they were ingested with and without live symbiotic algae. Groups of washed and starved paramecia were placed in suspensions of heat-killed algae alone or in 50:50 mixtures of heat-killed and live algae. After 2½ min, they were washed and set aside in salt solution for the times indicated on the abscissa and then examined microscopically. Open circles: dead algae left in animals exposed to 50:50 mixtures; closed circles: dead algae left in animals exposed to dead algae alone. Range bars denote the standard error (Karakashian & Karakashian, 1973).

algae were present, only about 50 % of the dead algae had been digested after 43 hours. Apparently, live algae interfere with the host digestive process so as to prevent their own digestion and, under the conditions of the experiment described, incidentally delayed the consumption of any dead algae which were in the same food vacuole. The limited protection afforded to the dead algae is presumed to be due to a gradual segregation of the live algae into their individual vacuoles following which they would no longer exert a protective effect. The means by which symbionts interfere with the digestive process is unknown, but it could involve either an inhibition of the host's digestive enzymes or the prevention of enzyme release into the phagosome. In the one instance where a similar phenomenon has been carefully analysed, it was shown that *Toxoplasma gondii* prevents the fusion of enzyme-bearing lysosomes with the vacuole surrounding the parasite (Jones & Hirsch, 1972).

Not all symbiotic algae escape digestion. Detailed studies of the aposymbiotic host's response to algae taken up in brief ($2\frac{1}{2}$ min) exposure periods showed that most of the algae are killed (Karakashian & Karakashian, 1973). In a typical experiment (Fig. 6), only about 20 % of the algae remained undigested ('healthy') 24 hours after ingestion. Of these survivors, most were located in a few heavily infected paramecia. The number of these heavily infected animals varies from a few percent to as many as 30 % of the sample in different experiments. The basis for the difference between experiments is unknown.

The fate of ingested algae may depend not only on their innate digestion-suppressing ability, but also on the manner in which they are packaged during ingestion. Preliminary observations indicate that large groups of algae swept into a single phagosome are more likely to escape digestion than those which enter phagosomes in smaller numbers. If this can be verified, it might account for some of the variability in the susceptibility of individual paramecia. It is also possible that a minimum number of algae must be present in a phagosome in order to assure that their collective action is both sufficient and rapid enough to protect them.

The bimodal variation commonly encountered in the host's response to newly ingested algae implies that the fates of algae in the several phagosomes of an individual host are linked to some extent. This would be the case if the overall digestive efficiency of the cell is subject to variation. If it were reduced, for example, more algae would be likely to survive entry, even if they entered in small groups. The digestive efficiency of a related ciliate, *Tetrahymena pyriformis*, is known to be correlated with the cells' nutritional state (Ricketts, 1971*a*, *b*). Following two days of starvation, *Tetrahymena* cells form food vacuoles at a slow rate and have a low initial

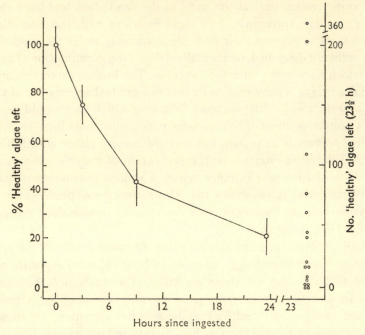

Fig. 6. Retention of symbiotic algae by aposymbiotic paramecia. Groups of washed and starved paramecia were placed in suspensions of washed symbiotic algae for $2\frac{1}{2}$ min following which they were again washed and set aside for microscopic examination at the times indicated on the abscissa. The average number of 'healthy' algae in each individual declined from 285 at 0 hours to 60 at $23\frac{1}{2}$ hours. The distribution of algae left in the individual cells at $23\frac{1}{2}$ hours after ingestion (right part of figure) reveals a clear dichotomy in infectivity. Range bars denote the standard error (Karakashian & Karakashian, 1973).

digestive enzyme content. If this were true for *P. bursaria*, too, minor differences in the nutritional state of the individual animals might be adequate to shift the balance towards or against the survival of most of the algae ingested by an individual cell.

Evidence for marked variations in the digestion-suppressing activity of symbiotic algae is still lacking. Axenically cultured symbionts display the same pattern of loss through digestion as freshly isolated symbionts (Karakashian, unpublished). Hirshon (1969) reported that symbionts are more rapidly ingested if they have been illuminated more than an hour prior to ingestion. Less than half as many symbionts were ingested if they did not receive illumination prior to and during the exposure period. The fate of the ingested algae was not followed, but one would predict, on the basis of their numbers alone, that the algae ingested in the dark might be more vulnerable to digestion than those which enter during the light. Conceivably, there might also be a diurnal fluctuation in the digestion-suppressing activity of the algae. Such differences should be revealed by comparing the symbionts' ability to protect co-ingested heat-killed algae at different times of day and night.

Symbiont entry during reciprocal nuclear exchange at conjugation is a relatively rare event, occurring in less than 10 % of conjugating pairs. However, treatment of conjugants with a suitable antiserum increases the incidence of appreciable cytoplasmic exchange to 100 % (Siegel, 1960). The fate of algae introduced into a new host by this means has never been studied in detail, but it seems probable that the algae enter with their perialgal vacuoles around them and are immediately able to colonize the new host. Thus, cytoplasmic exchange at conjugation may provide a means of introducing new types of symbionts without the complication of many of them being digested on entry. If these new symbionts could be recognized in some way, it should be possible to study symbiont competition within the host environment. It would be interesting to know, for example, whether the host is able to select more beneficial symbionts from a mixture of types in its cytoplasm.

The injection of symbiotic algae into aposymbiotic hosts should also be feasible, given modern microinjection techniques. Should it be possible to effect symbiont entry in this way, we would have a means of directly studying specificity restrictions which are imposed after the initial entry process.

Our current interpretation of what is occurring during the first few hours following the ingestion of potential symbionts may be summarized as follows:

(1) Each host cell at any given moment is in some modifiable state of readiness to take up and digest materials. This state could be considered its digestive efficiency and encompasses a number of interrelated characteristics, such as the supply of membrane components, the supply of digestive enzymes and the transport and delivery of these enzymes.

(2) Each algal cell has an innate, but small capacity to suppress the digestive processes of the host cell, probably by altering the fusion properties of the phagosome membrane. This capacity may vary, depending on the strain of algae and the environmental conditions, but it is not affected by axenic culture of the algae.

(3) Following their ingestion, there is a rate competition between the normal digestive processes of the host and the digestion-suppressing activity of the algae. Any circumstance which increases the collective digestion-suppressing activity of the algae within a given phagosome favours the algae surviving entry as does any condition which reduces the digestive efficiency of the host cell.

SPECIFICITY OF THE SYMBIOTIC ASSOCIATION

The relative ease with which a paramecium can be rid of its symbionts plus the fact that its aposymbiotic descendants and former symbionts can be cultured separately (Karakashian, 1963) make it feasible to study the formation and effectiveness of reconstituted associations.

In order to facilitate comparisons among strains, a standard procedure has been devised for measuring the compatibility of experimental combinations (Karakashian & Karakashian, 1965). This entails exposing paramecia to a standard density of algae for one week in a bacterized medium, after which a sample of 20–30 cells is washed free of extracellular algae and placed in a salt solution for 48 hours before microscopic examination for the presence of algae. This procedure is simple and sensitive enough to identify successful and non-successful combinations. The reproducibility of this assay method is shown for three cultures (Table 3).

Table 3. *Reproducibility of assay for infectibility of aposymbiotic clones*

Culture	Proportion infected in sample of 30				Mean proportion infected with 95 % confidence limits
I	0.07	0.13	0.20		0.13 ± 0.20
2	0.00	0.07	0.00		0.02 ± 0.10
3	0.27	0.10	0.20	0.20	0.19 ± 0.11

None of the cultures tested were significantly different in susceptibility to infection by the GrB1 strain of *Chlorella vulgaris*; all would be considered poorly susceptible. Assays made on cultures highly susceptible to infection by this strain of algae indicated the variability of the results is comparable and not sufficient to confuse the two classes of cultures.

A number of strains of algae of both symbiotic and free-living origin have been screened with respect to their ability to infect several stocks of *P. bursaria* (Bomford, 1965; Karakashian & Karakashian, 1965). Table 4 illustrates that algae of symbiotic origin (isolated from *P. bursaria*) are universally infective while free-living strains vary from relatively infective (397) to poorly infective (GrB1). Diverse paramecia stocks may react quite differently to the same strain of free-living algae (stocks 3W and 41W, for example). Thus, both the host organism and the alga contribute to the specificity of the association.

Weis & Berry (1974) have recently subjected symbiotic algae isolated from several stocks of *P. bursaria* to taxonomic analysis. They tentatively conclude that each host stock contains a single and different algal species and that the algae of different host syngens are from distinct subgenera. This high degree of specificity is not apparent in reinfection studies and, if

Table 4. *Infectivity of diverse algae in selected paramecia stocks*

Algae	Paramecia			
	3W	34W	41W	42W
Symbiotic				
Chlorella sp., 3H	+	+	+	+
Chlorella sp., 41K	+	+	+	+
Chlorella sp., NC64A	+	+	+	+
Free living				
Chlorella vulgaris, 397	o	(+)	+	+
Chlorella vulgaris, 263	o	+	+	o
Chlorella vulgaris, GrB1	o	o	+	o

+ = Infects most individuals of stock clone; (+) = infects some individuals of stock clone; o = infects no individuals of stock clone.

borne out by further studies, suggests that either very little superinfection occurs in nature or the host is capable of discriminating between fundamentally similar algal types.

It should be emphasized in this context that symbiotic strains have not yet been compared with respect to their infectivity following brief (2–3 min) exposure times. Conceivably, subtle differences in uptake of algae or packaging into food vacuoles would appear under these more stringent assay conditions. Symbiotic strains might also differ with respect to their ability to avoid attack by the digestive enzymes of the ciliate. This could be measured by comparing the extent to which different symbiotic strains delay digestion of heat-killed algae incorporated along with them.

Twenty-five free-living strains of algae were tested for infectivity in *P. bursaria* in one extensive survey (Karakashian & Karakashian, 1965). Ten of these strains proved to be significantly infective in at least one host stock. Occasionally, a free-living algal infection is very successful, at least in terms of the numbers of algae which survive the entry process. These infections are always less stable and far less able to support the growth of the host cell than infections involving symbiotic algae. They also do not lead to the loss of the many crystalline inclusions normally present in the cytoplasm of aposymbiotic cells; these inclusions usually disappear in cells infected with symbiotic algae (Karakashian, unpublished). The nature of these crystals and their significance for the symbiosis is unknown.

Infective strains of free-living algae are not confined to any particular species of *Chlorella* nor indeed to this genus. However, relatively few non-*Chlorella* species have been tested. Strain differences in the infectivity of a given algal species is not uncommon and this fact, together with the observation that host stocks also differ in infectability, makes it difficult to assign much weight to apparently incompatible combinations. The marked infectivity of a number of free-living *Chlorella* strains does suggest, however, that this genus is prone to enter into intracellular associations with paramecia (Karakashian & Karakashian, 1965).

Genetic analysis of host differences in susceptibility to free-living algae*

The striking differences in the susceptibility of host stocks 3W and 41W to infection by several strains of *Chlorella vulgaris* (Table 4) prompted an investigation of F_1 exconjugant clones derived from a cross between these stocks. Eleven pairs of F_1 clones were tested for their ability to accept the GrB1 strain of *C. vulgaris* (Table 5). These clones displayed a spectrum of

Table 5. *Susceptibility of progeny of stock 3 × stock 41W*

Pair number (A = stock 3 cytoplasm)[a] (B = stock 41 cytoplasm)	Number of infected cells (sample size – 15)	Number of algae per paramecium		
		1–50	51–100	> 100
3 (parent)	0			
41 (parent)	13	6	7	
354A	8	8		
354B	9	9		
357A	3	3		
357B	0			
363A	1	1		
363B	2	2		
367A	5	4	1	
367B	9	9		
371A	12	11	1	
371B	13	12	1	
393A	9	8	1	
393B	3	3		
421A	7	6	1	
421B	2	2		
445A	0			
445B	2	2		
448A	4	4		
448B	2	2		
473A	4	2	1	1
473B	3	3		
478A	15	11	3	1
478B	1	1		

[a] It should be noted that the stock 3 cells used in the cross were green due to the presence of symbiotic algae. Hence stock 3 cytoplasm was easily identified by the presence of the stock's naturally-occurring symbionts in one of the exconjugants resulting from each conjugating pair. Aposymbiotic sublines of the infected exconjugant were subsequently obtained by starving cells in darkness.

* These and other previously unpublished data included in this review were obtained in collaboration with S. J. Karakashian.

susceptibilities to the GrB1. More significant, however, was the occasional lack of uniformity between the two isogenic clones of an exconjugant pair, e.g. between 393A and 393B, 421A and 421B, and 478A and 478B. These differences rule out the direct involvement of nuclear genes as the basis for the susceptibility differences observed. Furthermore, in all three cases where the exconjugant pair of clones differed, the more susceptible of the two was the cytoplasmic descendant of the stock 3 or *unsusceptible* parent. Thus, a cytoplasmically-inherited susceptibility factor is not responsible for the parental stocks' difference in infectability either.

Eventual loss of host susceptibility to GrB1 algae

Repeated tests of the susceptibility of stocks 41W and 478A–W (the latter an F1 strain derived from the cross of stocks 3 and 41W) to GrB1 infection established that there is a pronounced drop in host susceptibility after a period of several months (Table 6). Tests of additional aposymbiotic clones of stock 41 paramecia revealed that this loss of susceptibility is not exceptional. Indeed, it is not possible to maintain clones in a highly susceptible state for more than a limited period of time. These data also show that the loss of infection susceptibility by a culture (as measured by the decreased proportion of infectable cells in test samples) does not mean that some individuals in that culture are not still highly infectable. Occasionally, a single infected cell in an assay sample contained as many as 200 GrB1 algae.

Table 6. *Susceptibility of paramecium stocks 41 and 478A to infection by GrB1 algae*

Date exposed	Proportion infected (in 15)	Number of algae per paramecium		
		1–50	51–100	100
Stock 41				
1–12–64	0.87	6	6	1
1–19–64	1.00	9	6	
1–24–64	0.80	9	3	
1–26–64	0.93	8	6	
1–31–64	0.80	9	1	2
2–2–64	0.07	1		
2–7–64	0.47	7		
4–7–64	0.00			
Stock 478A				
1–26–64	1.00	11	3	1
2–28–64	0.80	6	3	3
3–6–64	1.00	3	5	7
3–16–64	0.73	4	4	3
4–19–64	0.93	8	2	4
6–24–64	0.20	1		2
8–5–64	0.07	1		

Specificity of loss in susceptibility to GrB1 algae

Although susceptibility to GrB1 infection is eventually lost by an aposymbiotic clone, it remains infectable by other strains of free-living algae (Table 7). In addition, a change in susceptibility to GrB1 is not necessarily

Table 7. *Infectivity of diverse algae in paramecia differing in susceptibility to GrB1 infection*

| | | Proportion of animals infected | | |
Algae		478A	41Wa[c]	41Wb[c]
Chlorella vulgaris,	GrB1	0.15[a]	0.30[b]	0.03[b]
	261	1.00	–	–
	262	0.45	–	–
	263	0.95	0.67	0.53
	397	0.00	0.23	0.13
	398	0.90	0.97	0.26
Chlorella miniata	490	–	1.00	0.37
Selenastrum min.	326	–	0.40	0.87

[a] In a sample of 20.
[b] In a sample of 30.
[c] 41Wa and 41Wb were derived from a single culture and maintained at 19 °C and 29 °C, respectively.

matched by a similar change in susceptibility to the other strains tested. It seems that the whole spectrum of responses to free-living algal infections undergoes changes with time. Present data do not permit any conclusions as to whether the host responds similarly to certain groups of strains (490 and 398, for instance) or whether its infectability is regulated uniquely for each free-living strain.

The specificity of the response to an individual strain is further demonstrated by experiments in which a GrB1-resistant host culture was exposed simultaneously to GrB1 algae and *Selenastrum minutum* (326), an alga with a distinctive morphology. Table 8 shows that individual paramecia clearly discriminated between GrB1 algae, which most cells did not retain, and the *Selenastrum*, which was infective whether GrB1 was present in the infection mixture or not.

Loss of susceptibility in relation to duration of aposymbiosis

An attempt was made to determine whether there is a connection between the GrB1 susceptibility of a culture and the number of fissions it has undergone since the elimination of its normal symbionts. This was done by testing the GrB1 susceptibility of multiple cultures of a susceptible clone at various fission intervals. The results (Table 9) show that all lines eventually

Table 8. *Host selection of infective algae from mixtures of algae*

	Number of paramecia			
Composition of infection mixture	Uninfected	Infected with GrB1 only	Infected with 326 only	Infected with both
GrB1 only	27	3	–	–
326 only	11	–	19	–
GrB1 + 326	11	0	15	4

Table 9. *Host susceptibility to GrB1 infection at various fission ages following loss of naturally occurring symbionts*

	Number of fissions since loss of symbionts				
Line	10^a	19	38	58	78
S 2	0.83^b	0.93	–	0.90	0.07
S 4	0.83	0.83	0.43	0.00	–
S 9	0.83	0.73	0.00	–	–
S 10	0.83	0.90	0.67	0.00	–
S 13	0.83	–	0.90	0.00	–
S 15	0.83	–	0.00	–	–

[a] Susceptibility of line which was subcultured for experiment at age of 10 fissions.

[b] Numbers in each column represent proportion of infected paramecia found in a sample of 30.

lose their susceptibility to GrB1 algae, but the time in fissions at which the change occurs differs in each. Thus, lines S9 and S15 were resistant to GrB1 infection by 38 fissions, while line S13 was still 90 % infectable at that same fission age. Line S2, on the other hand, remained highly susceptible through at least 20 more fissions. It may be concluded that subcultures vary considerably in the rate at which they lose susceptibility to GrB1 and that this loss is not closely correlated with the clone's aposymbiotic age measured in fissions.

Relationship between GrB1 susceptibility and method of obtaining aposymbiotic lines

The aposymbiotic lines used in this study were obtained by either allowing paramecia to starve in darkness at 29 °C. for three or four weeks, or by allowing paramecia to divide in darkness at 25 °C. at their maximal rate (about two fissions per day for this stock). The starvation procedure relies for its success on the fact that survivors of the treatment have usually rid themselves of symbionts. The isolation line technique depends on the fact that the maximum host fission rate is slightly greater than that of the

Table 10. *GrB1 susceptibility of aposymbiotic lines following starvation or rapid growth at 25 °C*

Treatment	Proportion of infectable animals in sample of 30 from culture				
	1.00–0.75	0.74–0.50	0.49–0.25	0.24–0.01	0.00
41W paramecia made aposymbiotic by starvation in DD at 29 °C	0	0	0	3[a]	3
41W paramecia made aposymbiotic by rapid growth in DD at 25 °C	2	1	0	6	2
41W paramecia, previously aposymbiotic, starved 8–29 days in DD at 29 °C	0	0	0	6[b]	9
41W paramecia, previously aposymbiotic, grown rapidly for more than 50 fissions in DD at 25 °C	4	1	1	2	0

[a] Number of independently treated cultures in this category.
[b] Four of these cultures were starved 14–20 days; two were starved 29 days.
DD = continuous darkness.

symbionts when the latter are deprived of light. Table 10 summarizes the GrB1 susceptibility of aposymbiotic lines obtained by both of these methods. It is apparent that the probability of finding a highly susceptible line is not high by either method, but it is much improved by allowing the paramecia to outgrow their symbionts in darkness.

This conclusion is supported by the results of experiments in which GrB1-resistant aposymbiotic lines were subjected to the conditions used to rid infected cells of their symbionts (Table 10). Starvation of resistant cultures in darkness does little to improve their susceptibility to GrB1 algae. If such cultures are subjected to rapid growth in darkness, however, there is an appreciable increase in susceptibility to infection in many cases.

Restoration of susceptibility to resistant aposymbiotic lines

A partial restoration of susceptibility to GrB1 infection can be brought about by several treatments: rapid growth at 25 °C in darkness, as indicated above, infection with symbiotic algae, and removal of these algae again after a period of some weeks and conjugation with a non-susceptible clone of another stock of *P. bursaria*. Susceptibility has also been observed to increase 'spontaneously' in cultures which were maintained in the laboratory stock collection for a year between tests of their responses to GrB1 algae.

The effect of growth at one fission per week at diverse temperatures is summarized in Table 11. Growth at 19 °C has a salutary effect on the GrB1 susceptibility of cultures maintained at this temperature, but the effect may be temporary. Most clones kept at 29 °C either lose or do not regain sus-

Table 11. *Effect of growth (1 fission/week) at diverse temperatures on GrB1 susceptibility of 41W*

Culture	19 °C	25 °C	29 °C
A		LD 11–30 (0.30)	
		12–13 (0.33)	
	DD 12–21 (0.10)	DD 12–21 (0.17)	DD 12–21 (0.07)
		1–3 (0.23)	
	1–11 (0.63)	1–11 (0.27)	1–11 (0.33)
	1–25 (0.60)	1–25 (0.43)	1–25 (0.00)
	2–7 (0.17)	2–7 (0.00)	
	2–22 (0.17)	2–21 (0.00)	2–22 (0.07)
	3–29 (0.13)	3–29 (0.10)	3–29 (0.00)
B		LD 11–16 (0.03)	
		12–20 (0.03)	
	DD 1–3 (0.03)	DD 1–3 (0.03)	DD 1–3 (0.03)
	1–18 (0.03)	1–18 (0.00)	1–18 (0.00)
	2–7 (0.03)	2–7 (0.00)	2–7 (0.00)
	3–5 (0.67)	3–5 (0.07)	3–5 (0.00)
	3–7 (0.30)		
	3–19 (0.07)	3–19 (0.00)	3–19 (0.00)
	4–4 (0.00)	4–4 (0.00)	4–4 (0.00)
			4–11 (0.03)
		4–18 (0.00)	
		4–25 (0.03)	
C		LD 12–27 (0.83)	
		1–10 (0.00)	
	DD 1–17 (0.00)	DD 1–17 (0.07)	DD 1–17 (0.00)
	2–1 (0.03)	2–1 (0.03)	2–1 (0.00)
	2–15 (0.03)	2–15 (0.03)	
	3–19 (0.13)	3–19 (0.00)	3–19 (0.00)
	4–4 (0.50)	4–4 (0.00)	4–4 (0.00)
D		LD 1–10 (0.50)	
		1–24 (0.27)	
	DD 2–1 (0.40)	DD 2–1 (0.07)	DD
	2–15 (0.47)	2–15 (0.27)	2–15 (0.23)
	3–5 (0.70)	3–5 (0.57)	3–5 (0.13)
	4–4 (0.67)	4–4 (0.33)	4–4 (0.47)

First numbers in each column represent the dates on which the test of GrB1 susceptibility was begun. Numbers in parentheses represent the proportion of infected paramecia found in a sample of 30 animals removed from the culture for assay purposes. The cultures placed at 19 °C and 29 °C were derived from the culture maintained at 25 °C on the date of the last LD assay at 25 °C, at which time all cultures were placed in DD.

LD and DD as in Table 1.

ceptibility, suggesting this temperature is not favourable for maintaining a susceptible state. Slow growth at 25 °C yields variable results. Usually susceptibility diminishes or remains low with time, but temporary improvements in the response to GrB1 algae may be observed. Cultures growing at the rate of one fission per week are in a semi-starved state which is greatest at 29 °C.

Numbers of intracellular algae in susceptible paramecia

A highly susceptible culture has been defined as one having a high proportion of paramecia which are infectable by the test strain of algae. A loss of susceptibility is seen therefore as a drop in this infectable proportion until eventually no infectable animals are found. In the studies described here, even a single healthy GrB1 alga present at the end of the 48 hour test period qualified that cell for inclusion in the infectable category. It is of considerable interest therefore to review the sizes of the GrB1 populations in all of the infectable individuals examined.

Table 12. *Distribution of intracellular GrB1 in infectable paramecia*

Proportion infectable in culture	Number of cells examined	Number of algae in each paramecium[a]				
		1–25	26–50	51–75	76–100	> 100
1.00–0.75	165	22 (39.2)	33 (31.2)	21 (22.3)	23 (19.3)	66 (52.8)
0.74–0.50	163	31 (38.7)	30 (30.9)	29 (22.1)	21 (19.1)	52 (52.2)
0.49–0.25	149	36 (35.4)	22 (28.2)	15 (20.1)	20 (17.5)	56 (47.7)
0.24–0.01	188	69 (44.6)	41 (35.6)	25 (25.4)	14 (22.0)	39 (60.2)
Totals	665	158	126	90	78	213

[a] Numbers in parentheses represent expected intracellular populations calculated from totals in all infectable animals. The distributions fail a $\chi^2 h$ test of independence.

A compilation of the available data (Table 12) indicates there is not a random distribution of GrB1 populations among the infectable cells. More than 30% of all the cells examined were green to the eye and contained in excess of 100 algae each. This visible minority was detected in all four classes of susceptible cultures and is reminescent of the type of results obtained after paramecia are briefly exposed to symbiotic algae (Fig. 6).

The distributions of GrB1 algae among infectable paramecia from highly susceptible and poorly susceptible cultures differ markedly from each other. The differences are so apparent it need hardly be mentioned that the distributions fail a χ^2-test of independence. In the highly susceptible category, there is a deficiency of small algal populations and an excess of large ones. The converse is true among individuals from cultures with a low susceptibility to GrB1 algae.

It is reasonable to assume that similar processes operate during the entry of symbiotic and free-living algae. If so, the GrB1 results suggest that the susceptibility level of a culture is inversely related to the digestive efficiency of most individuals in it. For some reason, most paramecia in highly susceptible cultures have a low digestive efficiency and retain many algae, while in poorly susceptible cultures, most paramecia have a high efficiency and kill most of the algae they ingest. The many low-level infections in individuals from poorly susceptible cultures could be produced by the long exposure period used for these assays. Even cells with a high digestive efficiency might not be able to kill all of the algae they ingest in the course of a week.

Observations on the effectiveness of transmission of GrB1 algae during fissions of the host suggest that drastic changes in the ciliate's ability to retain the algae may occur within a single interfission interval, leading to a segregation of infected and uninfected daughter cells (Table 13). Segregation occurs in lineages having either a high or a low number of algae present, an indication that the algae are not simply diluted out. The paramecia used to initiate each lineage were selected because they were heavily infected with GrB1 algae. As each paramecium divided, its daughter cells were separated. After a line had undergone four fissions, its 16 fission products were sacrificed and scored for their content of algae. In lineage 2, a segregation of algae must have occurred at the third fission; in lineage 3, it may be inferred that segregation occurred at the first fission of the infected isolate. Segregation occurred at both the first and second fission of lineage 4 and not at all in lineages 5 and 6. In the latter cases, the cells were cultured in an undiluted medium, which meant there was always an excess of food available to the host.

These studies of GrB1 transmission show that the successful entry of free-living algae into the cytoplasm of the host does not guarantee their continued survival there. The algae's situation appears particularly precarious when the host's food supply is limited. It is possible that their retention by the host up to this point is not due to their having an innate ability to protect themselves, but, rather, to other constraints on the host's digestive processes, constraints which are abruptly lifted in some cells especially when the host is in a suboptimal nutritional condition.

Basis for changes in GrB1 susceptibility

A working hypothesis which guided much of this study on the GrB1 susceptibility of aposymbiotic paramecia was that the possession of symbionts somehow alters the physiology of the host cell in such a way that it remains tolerant of many kinds of algal infections for some time after the removal of its normal symbionts. It can be imagined that the symbionts continually produce an agent which reduces the digestive efficiency of the

Table 13. *Distribution of GrBi algae among first 16 fission products of 478A – GrBi isolates*

Average number of algae per initial cell	Line	Medium for host growth	Number of algae in fission products															
			1	2	3	4	5	6	7	8	9	10	11	12	13	14	15	16
Not determined	1	10% lettuce infusion	211	152	214	376	77	—	290	—	222	226	361	—	—	467	500	194
Not determined	2	10% lettuce infusion	113	100	—	—	83	70	83	—	151	—	0	0	225	127	—	73
555	3	10% lettuce infusion	61	73	68	55	75	33	66	59	0	0	0	0	0	0	—	—
555	4	10% lettuce infusion	0	0	0	0	85	66	59	47	0	0	0	0	0	0	0	0
226	5	100% lettuce infusion	6	22	13	20	20	35	18	19	30	28	27	32	17	26	34	34
226	6	100% lettuce infusion	28	29	37	51	27	34	18	23	34	36	51	39	21	15	29	32

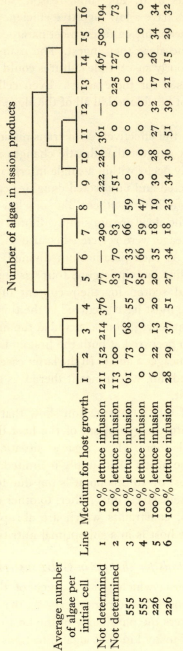
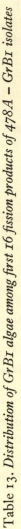

PLATE I

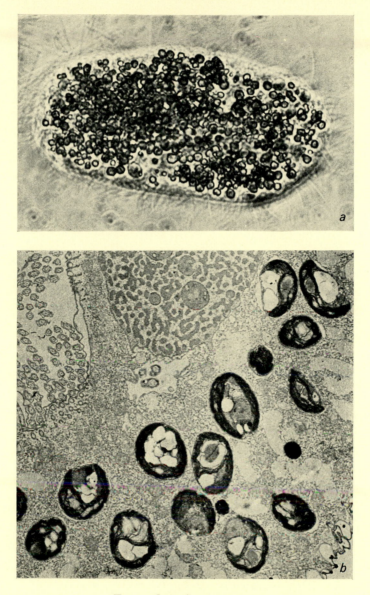

For explanation see p. 173

PLATE 2

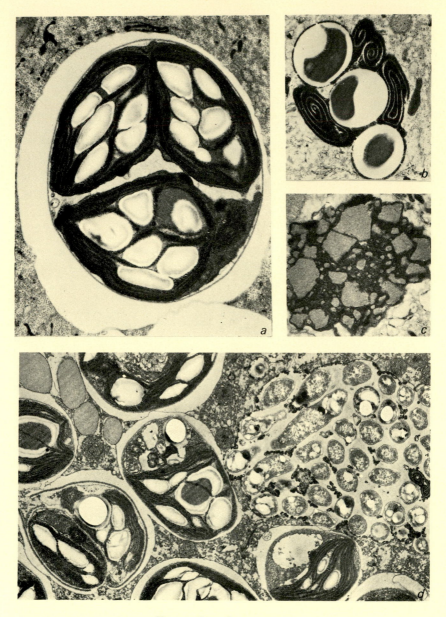

For explanation see p. 173

host and which is still present in the host after its symbionts are eliminated. If this agent were somewhat unstable and its rate of destruction dependent on the growth conditions of the host, it would eventually be diluted out or destroyed in all members of an aposymbiotic clone, but not necessarily simultaneously in all cells of the clone.

Comparable phenomena have been described for the several bacterial symbionts of *Paramecium aurelia* (reviewed by Preer, Preer & Jurand, 1974). The maintenance of each of these symbionts by their respective host is dependent upon the presence of specific host nuclear genes. When such a maintenance gene is removed by inducing heterozygotes to undergo autogamy (self-fertilization), the symbionts are eventually lost by the descendents of the exautogamous individuals which inherit the incorrect genotype. However, this loss is not immediate and there is great interclonal variability in the stage at which the symbionts disappear. They are retained for many fissions in some clones and vanish rapidly in others. The segregation of GrB1-free cells during fission strongly resembles the segregation of kappa-free animals following the removal of the K gene needed for the maintenance of kappa (Beale & McPhail, 1967). Beale & McPhail also noted that starvation and retardation of growth accelerates the loss of kappa symbionts. They proposed that the nuclear gene K is responsible for the production of a long-lived (perhaps replicating) cytoplasmic factor which is essential for the support of kappa. According to their interpretation, loss of kappa is directly associated with the disappearance of this factor from the cell and only indirectly with the loss of the K gene.

In another striking parallel to the GrB1 susceptibility pattern described above, Gibson (1973) reported that the bacterial symbionts alpha, lambda and mu can be successfully injected into host cells which no longer have the appropriate maintenance genes, providing the transfer is made no later than 20–30 fissions following gene replacement. Symbionts injected after this time are not retained by the host cells. Thus, *P. aurelia* cultures also undergo a change in susceptibility to infection with time, a change in susceptibility to reinfection by their normal symbionts.

The loss of the putative symbiont-support factor in *P. aurelia* is irreversible in individuals which no longer have the appropriate gene. The loss of susceptibility to GrB1 infection is not irreversible in *P. bursaria*. However, if the hypothetical factor in *P. bursaria* acts by altering the digestive efficiency of the host cell in some way, it is likely that its effect would be partially simulated by other conditions which also alter the ciliates' digestive efficiency. Thus, at the present time there is no reason to discard the hypothesis that symbionts have an effect on host metabolism which may endure for some time after they have been removed, especially since specific features of this interpretation, those related to the relative digestive efficiency of susceptible aposymbiotic cells, can be experimentally tested.

Further speculation must await a better elucidation of the events occurring during the infection of a host cell by normal algal symbionts. If the fate of digested algae is indeed a consequence of the outcome of a rate competition between the digestive processes of the paramecium and a digestion-suppressing activity of the algae, the course of future work seems clear. In addition to learning more about the ingestion and packaging into phagosomes of normal symbionts, it would be important to have similar information about infective free-living strains. Reliable data are needed on how environmental conditions affect the digestive efficiency of the host cell, particularly on the effect of host growth rate and culture temperature on the paramecium's ability to form food vacuoles and make, store and release digestive enzymes into phagosomes. Finally, it will be important to learn whether infective free-living algae also have some capacity to interfere with the normal course of host cell digestion. All of these questions are amenable to experimental study with techniques now at hand.

CONCLUDING REMARKS

The aim of this paper was to review the state of our knowledge about symbiosis in *Paramecium bursaria*. What has been learned is that the inter-action between the host cells and its symbionts is complex, highly integrated and very modifiable by environmental conditions. It is a mutually beneficial association, but the benefit shifts from one partner to the other in response to environmental pressure. It is a stable association and yet the opportunity remains to create, test and consolidate new partnerships. Finally, it is a specific association, with the specificity expressed at multiple levels: at the time of symbiont entry into the host cell, during their intracellular dispersal and growth and in the kind and degree of metabolic exchanges occurring between the partners.

SUMMARY

Paramecium bursaria normally appears green due to several hundred symbiotic Chlorella which are dispersed throughout its cytoplasm. The symbionts are situated within individual vacuoles and these alga–vacuole complexes grow and divide at a rate compatible with that of the paramecium. The symbiotic units also persist through conjugation and the subsequent reorganization of the host. Studies of the benefit of the symbiosis to the ciliate hosts have shown that they are able to grow and survive better than aposymbiotic animals in environments deficient in bacteria. The symbionts are also able to extract nourishment from the host when it is well fed and they are deprived of light. The biochemical nature of these exchanges has not been determined. Potential symbionts usually enter the host in food vacuoles. If they are ingested in sufficient numbers, they are able to interfere with the normal course of host digestion, perhaps by preventing the release of digestive enzymes into the food vacuole. All natural symbionts of *P. bursaria* appear able to reinfect aposymbiotic cells. Some free-living strains of *Chlorella* and related algae are also infective, but these associations are relatively unstable and provide little evident benefit to the host. Host susceptibility to infection by certain strains of free-living algae is invariably lost with time. This loss is specific and often rapid, but it does not occur simultaneously in subcultures derived from the original susceptible culture. The basis for these susceptibility changes is still unknown, but they may be related to long-lasting effect of the previous symbionts on the digestive efficiency of the paramecium host.

ACKNOWLEDGEMENTS

I wish to thank the US National Institutes of Health, the US National Science Foundation and the Max-Planck Gesellschaft for financial support during various phases of this work.

Note added in proof

Following submission of this manuscript, an article appeared (Brown, J. A. & Nielsen, P. J. (1974). Transfer of photosynthetically produced carbohydrate from endosymbiotic Chlorellae to *Paramecium bursaria*. *J. Protozool.*, **21**, 569–570.) in which it was demonstrated that symbiont-bearing *P. bursaria* incorporate $^{14}CO_2$ to a significantly greater extent than aposymbiotic animals. Large amounts of radioactive carbohydrates were identified in the host fraction of homogenates from which the algae had been removed by centrifugation, indicating that the symbiotic algae release photosynthate to the host *in situ*. Chromatographic analysis of these carbohydrates showed that maltose is the primary photosynthetic product released to the host. Since no radioactive carbohydrates could be detected in the algal fractions of homogenates, it must be inferred that the algae export essentially all of their photosynthate to the host under the experimental conditions used. Given the culture conditions employed by these investigators, it is probable that the paramecia had been maintained prior to experimentation in media low in bacterial flora, i.e. under conditions in which the symbionts would be making a large contribution to host growth.

REFERENCES

BEALE, G. H. & McPHAIL, S. (1967). Some additional results on the maintenance of kappa particles in *Paramecium aurelia* (Stock 51) after loss of the gene K. *Genet. Res., Camb.*, **9**, 369–373.

BOMFORD, R. (1965). Infection of alga-free *Paramecium bursaria* with strains of *Chlorella, Scenedesmus*, and a yeast. *J. Protozool.*, **12**, 221–224.

GIBSON, I. (1973). Transplantation of killer endosymbionts in *Paramecium*. *Nature, Lond.*, **241**, 127–129.

HIRSHON, J. (1969). The response of *Paramecium bursaria* to potential endo-cellular symbionts. *Biol. Bull. mar. biol. Lab. Woods Hole*, **136**, 33–42.

JONES, T. C. & HIRSCH, J. G. (1972). The interaction between *Toxoplasma gondii* and mammalian cells. II. The absence of lysosomal fusion with phagocytic vacuoles containing living parasites. *J. exp. Med.*, **136**, 1173–1194.

JONES, T. C., YEH, S. & HIRSCH, J. G. (1972). The interaction between *Toxoplasma gondii* and mammalian cells. I. Mechanism of entry and intracellular fate of the parasite. *J. exp. Med.*, **136**, 1157–1172.

KARAKASHIAN, M. W. & KARAKASHIAN, S. J. (1973). Intracellular digestion and symbiosis in *Paramecium bursaria*. *Expl Cell Res.* **81**, 111–119.

KARAKASHIAN, S. J. (1963). Growth of *Paramecium bursaria* as influenced by the presence of algal symbionts. *Physiol. Zool.*, **36**, 52–68.

(1970*a*). Morphological plasticity and the evolution of algal symbionts. *Annls N.Y. Acad. Sci.*, **175**, 474–487.

(1970*b*). Invertebrate symbioses with *Chlorella*. In *Biochemical Co-Evolution*, 29th Annual Biology Colloquium, 33–52 (ed. K. L. Chambers) Corvallis, Oregon: Oregon State University.

KARAKASHIAN, S. J. & KARAKASHIAN, M. W. (1965). Evolution and symbiosis in the genus *Chlorella* and related algae. *Evolution*, **19**, 368–377.

KARAKASHIAN, S. J., KARAKASHIAN, M. W. & RUDZINSKA, M. A. (1968). Electron microscopic observations on the symbiosis of *Paramecium bursaria* and its intracellular algae. *J. Protozool.*, **15**, 113–128.

MUSCATINE, L., KARAKASHIAN, S. J. & KARAKASHIAN, M. W. (1967). Soluble extracellular products of algae symbiotic with a ciliate, a sponge, and a mutant hydra. *Comp. Biochem. Physiol.*, **20**, 1–12.

OEHLER, R. (1922). Die Zellverbindung von *Paramecium bursaria* mit *Chlorella vulgaris* und anderen Algen. *Arb. aus dem Staatsinstitut für Experimentelle Therapie*, **15**, 3–19.

PADO, R. (1965). Mutual relation of protozoans and symbiotic algae in *Paramecium bursaria*. I. The influence of light on the growth of symbionts. *Folia biol., Praha*, **13**, 173–182.

(1967). Mutual relation of protozoans and symbiotic algae in *Paramecium bursaria*. II. Photosynthesis. *Acta Soc. Bot. Pol.*, **36**, 97–108.

PREER, J. R., PREER, L. B. & JURAND, A. (1974). Kappa and other endosymbionts in *Paramecium aurelia*. *Bact. Rev.*, **38**, 113–163.

PRINGSHEIM, E. (1928). Physiologische Untersuchungen an *Paramecium bursaria*. Ein Beitrag zur Symbioseforschung. *Archiv Protistenk.*, **64**, 289–418.

RICKETTS, T. R. (1971*a*). Periodicity of endocytosis in *Tetrahymena pyriformis*. *Protoplasma*, **73**, 387–396.

(1971*b*). Endocytosis in *Tetrahymena pyriformis*. Selectivity of uptake of particles and the adaptive increase in cellular acid phosphatase activity. *Expl Cell Res.*, **66**, 49–58.

SIEGEL, R. W. (1960). Hereditary endosymbiosis in *Paramecium bursaria*. *Expl Cell Res.*, **19**, 239–252.

SMITH, D., MUSCATINE, L. & LEWIS, D. (1969). Carbohydrate movement from autotrophs to heterotrophs in parasitic and mutualistic symbiosis. *Biol. Rev.*, **44**, 17–90.

TAYLOR, D. L. (1973). Algal symbionts of invertebrates. *A. Rev. Microbiol.*, **27**, 171–187.

VIVIER, E., PETITPREZ, A. & CHIVE, A. F. (1967). Observations ultrastructurales sur les Chlorellessymbiotes de *Paramecium bursaria*. *Protistologica*, **3**, 325–334.

WEIS, D. S. & BERRY, J. L. (1974) Correlation of mating behavior with species of endosymbiotic algae in *Paramecium bursaria*. *J. Protozool.*, **21**, 416.

EXPLANATION OF PLATES

PLATE 1

(a) Phase contrast micrograph of *Paramecium bursaria* with intracellular *Chlorella*. Cell fixed in osmium tetroxide vapor. × 900

(b) Electron micrograph showing intracellular algae in the region of the gullet and macronucleus (Karakashian, Karakashian & Rudzinska, 1968). × 4970

PLATE 2

(a) Three symbiont autospores still contained within the parental cell wall; acid phosphatase reaction product can be seen in adjacent cytoplasmic vesicles, but not in the perialgal vacuole. Gomori-stained cell (Karakashian & Karakashian, 1973). × 15 200

(b) Large amounts of acid phosphatase reaction product in digestion vacuole containing latex spheres and discarded cell wall fragments. Gomori-stained cell. × 15 400

(c) Reaction product in vacuole containing carmine particles. Gomori-stained cell (Karakashian & Karakashian, 1973). × 22 000

(d) Reaction product in food vacuole containing bacteria and a discarded cell wall fragment. Light spaces within algae and the food vacuole probably are the locations of dense material which fell out during preparation. Note old cell wall fragments clinging to two autospores within the perialgal vacuole on lower left. Gomori-stained cell. × 10 000

UPTAKE, RECOGNITION AND MAINTENANCE OF SYMBIOTIC *CHLORELLA* BY *HYDRA VIRIDIS*

By L. MUSCATINE, C. B. COOK,[*] R. L. PARDY,[†]
AND R. R. POOL

Department of Biology, University of California
at Los Angeles, California 90024, USA

INTRODUCTION

A wide range of eukaryotic cells harbor intracellular, self-reproducing symbionts clearly of foreign origin. Broadly categorized, these symbionts range from viruses and bacteria to complex eukaryotic cells. Falling within this last category are certain types of unicellular algae. These occur as endosymbionts in several hundred aquatic invertebrate genera representing five major phyla from arctic, temperate and tropical environments (McLaughlin & Zahl, 1966). Whereas many investigations on the nature of algae–invertebrate symbiosis have dealt with organismic physiology or ecological consequences (see reviews of Taylor, 1973a, b), few have probed experimentally into cellular interactions *per se*. This paper will outline some comparatively new lines of attack on the cell biology of algae–invertebrate symbiosis dealing generally with the mechanism of infection or invasion of metazoan cells by symbionts. The data will be limited to results of experiments on the symbiotic association of the fresh water hydra and unicellular algae (*Chlorella* sp.).

Historically, re-establishment of a symbiosis between hydra and unicellular algae was first reported by Goetsch (1924). His findings were confirmed by Park, Greenblatt, Mattern & Merril (1967) and extended by Pardy & Muscatine (1973) and will be discussed in detail in this paper. The re-establishment of a symbiosis with algae is not unique to hydra. *Anthopleura elegantissima*, a sea anemone, may be reinfected by repeated injection of its normal dinoflagellate symbionts (Trench, 1971). Provasoli, Yamasu & Manton (1968) have brought about the resynthesis of algal symbiosis in *Convoluta roscoffensis*, an acoel flatworm. Strains of *Para-*

---

[*] Present address: Department of Zoology, Ohio State University, Columbus, Ohio.

[†] Present address: Department of Developmental and Cell Biology, University of California at Irvine, California.

mecium bursaria and their symbiotic *Chlorella* have been separated, maintained independently and experimentally recombined (Karakashian, 1968).

Green hydra offer several advantages for the study of the interaction of metazoan cells and intracellular symbionts. The hydra can be reared in a defined medium (Muscatine & Lenhoff, 1965a). Under optimum conditions, the animals double in number every 1.5 days by asexual budding, giving rise to a large clone of animals of similar genetic, nutritional and developmental histories. The algae may be readily separated from the host cells and experimentally recombined with aposymbiotic hydra. The ease of recombination affords a rapid system for bioassay of features of the reinfection process. The algae serve as markers which may be easily located and counted by light and fluorescence microscopy and labeled with a variety of isotopic tracers. Finally, the morphology of the association is well known, as are many aspects of the physiology and developmental biology of both symbiotic and non-symbiotic hydras.

DESCRIPTION OF THE ASSOCIATION

Table 1 lists the hydras and algae used in our investigations. Since the systematics of green hydras is not yet firmly established (see footnote) those clones in our possession are given a strain designation relating to their geographic origin. The hydra used primarily in our experiments is *Hydra viridis* (Fla.). An individual *H. viridis* (Fla.) normally harbors approximately 1.5×10^5 symbiotic *Chlorella* sp. These algae are found only in digestive cells of the endoderm and never in any other differentiated cell types. Digestive cells of *H. viridis* contain varying numbers of algae depending on their location along the body tube. Those from the stomach and budding zone contain an average of 18 ± 2.6 algae per cell.* This value may fluctuate depending on the species of hydra, the feeding schedule, and the photoperiod under which the hydra are maintained (Pardy, 1974; Pardy & Muscatine, 1973). Light microscopy reveals that the algae are spherical or ellipsoidal, 5–6 μm in greatest dimension, and possess a single cup-shaped chloroplast. The algae are found normally at the base of digestive cells and thus lie adjacent to the mesolamella. Each alga is situated in an individual vacuole (Oschman, 1967; Plate 2b) distinct from the food vacuoles located at the tips of the digestive cells (Wood, 1959). Between the algae and the perialgal vacuolar membrane is an electron translucent 'space' (Plate 2b). We are alert to the possibility that this 'space' is a fixation artifact, sensitive to changing osmolarity. Such peri-

---

* Here, as elsewhere in this paper, measurements of algae per cell are given as means ± one standard deviation.

Table 1. *Summary of hydras and algae used by the authors in this study*

Organism	Status	Species[a]	Strain	Host	Source of stock organisms
Hydras	Symbiotic	*Hydra viridis*	Florida	—	1
		Hydra viridis	Carolina	—	2
		Hydra viridis	English	—	3
		Chlorohydra hadleyi	—	—	1
	Free-living	*Hydra attenuata*	—	—	1
Algae	Symbiotic	Not formally designated[b]	F	*H. viridis* (Fla.)	1
			C	*H. viridis* (Carolina)	2
			E	*H. viridis* (English)	3
			CH	*C. hadleyi*	1
			3H	*Paramecium bursaria* syngen 1	4
			34	*P. bursaria*, syngen 1	4
			NC64A	*P. bursaria*, syngen 1	4
			130C	*P. bursaria*, syngen 1	4
			42A	*P. bursaria*, syngen unknown	4
			838	*Spongilla sp.*	5
	Free-living	*Chlorella vulgaris*	397	—	5

1. Department of Cell and Developmental Biology, University of California, Irvine, California, USA.
2. Carolina Biological Supply Co., North Carolina, USA.
3. Vicinity of Cambridge, England, courtesy of Dr S. A. Wainwright.
4. Dr S. J. Karakashian, State University of New York, Old Westbury, New York, USA.
5. Indiana University Culture Collection of Algae.

[a] The taxonomy of green hydras is in a state of confusion. The use of the binomial *Hydra viridis* L. 1767 superceded *Hydra viridissima* Pallas 1766, and persisted as a result of Linnaeus's authority until Schulze erected the binomial *Chlorohydra viridissima* Pallas (Schulze, 1917). Several hydra workers have since argued both for and against the use of this binomial and the matter is still under debate (see discussion by Muscatine, 1974).

[b] Brandt (1881) coined the genus *Zoochlorella* to describe symbionts in hydra. The use of the genus has since fallen into vernacular usage (e.g. 'zoochlorellae'). Whereas most workers agree that free-living and symbiotic *Chlorella* are clearly related, no precise classification of the symbionts has yet been rendered.

symbiont spaces are not unique to this association, having been observed, for example, in *P. bursaria* and in other types of cells harboring endoparasitic bacteria (Draper & Rees, 1970; Friis, 1972). Details of algal fine structure are given by Oschman (1967). He also describes differences between algae from different strains of hydra and interprets these as either indicative of different symbiotic species or simply adaptive changes resulting from a symbiotic habit with different strains of hydra. (See discussion of 'morphological plasticity' of symbiotic *Chlorella* by Karakashian, 1970.)

METHODS USED IN REINFECTION STUDIES

Details of methods used by Park *et al.* (1967) and Pardy & Muscatine (1973), are briefly summarized here.

Harvesting symbiotic algae

Algae from *H. viridis* have not yet been cultured *in vitro* and must be obtained by isolation from the host. Gentle homogenization of about 400 one-day starved hydra followed by a series of low speed centrifugations and washings yields approximately 6×10^7 algal cells together with some nematocysts and other animal tissue contaminants. Other strains of algae (see Table 1) are maintained on sterile agar slants of modified Loefer's Medium (Karakashian, 1963). When needed for experiments, algae are inoculated into sterile liquid medium, incubated at room temperature in the light, and harvested several days later during log-phase growth.

Aposymbiotic H. viridis

Aposymbiotic *H. viridis* have been obtained in the past by treatment with 0.5 % (v/v) glycerol (Whitney, 1907; Muscatine & Lenhoff, 1965*b*). Curiously, this technique has no effect on any of the strains now in our possession. Our aposymbiotic *H. viridis* (Fla.) hydra were cloned from algae-free eggs and have been maintained algae-free for several years.

Injection of algae into H. viridis

Algae are introduced into the coelenteron of aposymbiotic *H. viridis* (Fla.) by microinjection. The cells are loaded into a glass micropipette prepared from capillary tubing drawn out to a fine tip. The pipette is attached to an oil-filled micrometer syringe. The pipette tip is inserted into the hydra's mouth and by slight pressure on the syringe enough algae are delivered to fill the coelenteron, ensuring a saturating dose.

Measuring uptake rate

In previous experiments (Pardy & Muscatine, 1973) we prelabeled isolated *H. viridis* (Fla.) algae *in vitro* with ^{14}C as $H^{14}CO_3^-$, and then measured the appearance of ^{14}C in hydra tissues at various intervals after injection. Recently, we have measured uptake rates by a sensitive fluorometric method developed in our laboratory by Dr C. D'Elia specifically for assay of algae in hydra. Both methods yield data on uptake kinetics of similar character. Numbers of algae taken up by individual digestive cells are determined from maceration preparations of injected hydra (see below); total algae taken up are computed from hemocytometer counts of algae from standard aliquots of homogenized hydra.

Cytology and ultrastructure

Hydra are macerated with a solution of acetic acid : glycerol : water (1:1:13, v/v/v) (David, 1973). Amongst the array of individual cell types obtained, the digestive cells, easily identified by shape and presence of algae, are examined by light and fluorescence microscopy (BG 12 excitation and 530 K barrier filters for chlorophyll fluorescence). These preparations may also be stained with aqueous toluidine blue and permanently mounted. The algae are readily identified by their staining characteristics and their numbers easily determined.

Generally, for electron microscopy whole hydra are fixed in 1% glutaraldehyde at room temperature at intervals from zero time to 300 minutes after injection, postfixed in 1% osmium tetroxide in 0.075 M phosphate buffer, and embedded in Spurr embedding medium (Spurr, 1969). Other methods will be outlined in the relevant sections below.

REINFECTION OF APOSYMBIOTIC *H. VIRIDIS* (FLA.)

It is convenient to consider reinfection of aposymbiotic hydra as occurring in a series of phases. Briefly, these are (1) the *contact phase*, during which contact is established between the symbiont and the host cell; (2) the *engulfment phase* (endocytosis), during which the algae become surrounded by a membrane-bound vacuole of host origin; (3) a *recognition phase*, in which potential symbionts are identified as such; (4) an *intracellular migration phase*, in which algae move from the site of uptake to the base of the cell; and finally (5) a *repopulation phase*, marked by the restoration of the normal number of algae per host cell.

The contact phase

Plate 1*a* shows a section of a hydra fixed within 10 minutes after injection of algae. Two algal cells are in the coelenteron. One of these is in close apposition to the hydra cell membrane. The host membrane is marked in this region and elsewhere in the section by microvilli, indentifiable from longitudinal and cross-section profiles. Where the alga appears to just touch the host membrane, the algal cell wall and host membrane run parallel (arrow). Between them is an electron-translucent gap about 40 nm wide. In this region, the host membrane exhibits a distinct increase in electron opacity (Plate 1a, Plate 1b). Increased electron opacity at contact zones between cells has been observed by other investigators in studies of cell adhesion and endocytosis (Bensch, Gordon & Miller, 1964; Sheffield, 1970). Based on their observations and interpretations, we conjecture that these areas of increased electron opacity are not only sites of contact but also sites of attachment of the algae to the host cell membrane. Attachment is known to precede engulfment in endocytosis by mammalian macrophages (Rabinovitch, 1967, 1968) and it is reasonable to assume that the endocytosis of algae by hydra digestive cells involves both processes. Much remains to be learned of the nature of the presumed attachment of algae to digestive cells, and whether these attachment entities are permanent or labile features of the cell surfaces.

The engulfment phase

With size as a criterion, engulfment of algae by digestive cells is, by definition, a phagocytic process (Casley-Smith, 1969; Gordon, 1973; Simson & Spicer, 1973). Engulfment begins with the development of microvilli in the vicinity of an algal cell (Plate 1a) followed by the enclosure of algae by long microvilli or possibly by a veil of cytoplasm (cf. Tizard & Holmes, 1974) as seen in Plate 1c. Free and anastomosing microvilli have been observed in other hydra species (Gauthier, 1963; Lentz, 1966) and in other hydroids (Lunger, 1963) in connection with endocytosis. Slautterback (1967) has described discoidal vesicles (0.2–1.0 μm diameter, 80 nm in thickness) in the apical cytoplasm of digestive cells of *Hydra oligactis*. These pre-formed vesicles possess a highly ordered coat of pegs and globules on their luminal surfaces. They function in the selective uptake of materials such as ferritin and particulate glycogen from the coelenteron. Slautterback also described their role in the phagocytosis of 1–2 μm 'heterogeneous dense bodies' which result from the initial extracellular digestion of food substrates such as *Artemia* larvae. Plates 1c and 2a show numerous discoidal coated vesicles in the cytoplasm of *H. viridis*, but as yet we have not observed them in association with phagocytosis of algae. Plate 2a shows the disposition of several algae at the tip of a digestive cell about 10 minutes after injection. The engulfment process is complete and the sequestration of the algae into individual vacuoles is evident.

The rate of uptake of algae from the coelenteron into digestive cells is shown in Fig. 1 (upper curve). Uptake is initially very rapid. About half the total number of algae incorporated by the digestive cells is taken up in the first few minutes after injection of algae. Uptake continues more slowly thereafter for about an hour and then appears to cease, even

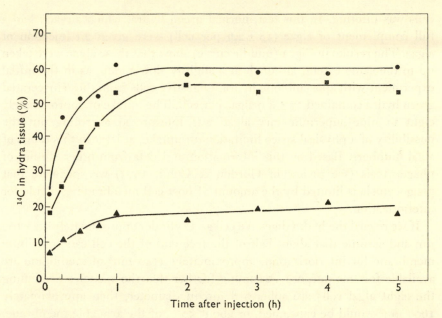

Fig. 1. Percent of radioactivity in *H. viridis* (Fla.) tissue plotted against time after injection of ¹⁴C-labeled algae. (●) Aposymbiotic hydra injected with freshly isolated *H. viridis* (Fla.) algae or (▲) NC64A algae cultured in Loefer's Medium; (■) normal green *H. viridis* (Fla.) injected with *H. viridis* (Fla.) algae (data of Pardy & Muscatine, 1973).

though algae may still be abundant in the coelenteron. Rapid initial uptake of algae is consistent with observations on uptake of algae by *Paramecium bursaria* (Hirshon, 1969) and with phagocytosis rates of several different mammalian cell types in which phagocytosed material is taken up within seconds to minutes (Christiansen & Marshall, 1965; Korn & Wiseman, 1962; Ryser, Caulfield & Aub, 1962; Zucker-Franklin & Hirsch, 1964). Uptake rates vary with temperature. Experiments of Dr C. D'Elia in our laboratory show that low temperature (5 °C) results in complete inhibition of engulfment of algae.

H. viridis normally regurgitates particulate material not taken up by digestive cells about five hours after feeding. Similarly, excess algae are also regurgitated from the coelenteron about five hours after injection. In a sample of digestive cells from the stomach and budding zone of hydra macerated at this time, Pardy & Muscatine (1973) counted an average of 8 ± 2 algae per cell. Since normal green hydra digestive cells contain an average of 18 ± 2.6 algae per cell, Pardy & Muscatine concluded that only limited numbers of algae can be initially phagocytosed by aposymbiotic *H. viridis* digestive cells. Since the supply of algae was presumed not to be a limiting factor, the possibility was tested that space within the digestive cells was limiting. In this test, normal green hydra, which already had a full complement of algae (18 ± 2.6 per cell) were given an injection of algae. The results in Fig. 1 (middle curve) show that these algae were taken up to the same extent, although at a slightly slower rate, as in the initial experiment with aposymbiotic hydra. Digestive cells from these macerated green hydra contained 27 ± 4.9 algae per cell. The uptake of approximately eight to nine supernumerary algae was interpreted as ruling out the possibility of a physical space limitation on uptake, at least in this range of algal numbers. Based on this information and data from other studies of phagocytosis (see review of Gordon & Cohn, 1973) we speculate that phagocytosis is limited by the amount of host cell membrane available for interiorization.

If we regard the hydra digestive cell as a cylinder approximately 20×100 μm and assume that about half of the free end of the cell can contribute membrane for interiorization, approximately 3140 μm^2 of membrane are available for this purpose. Assuming further that the vacuoles containing the eight algal cells are spheres 8 μm in diameter, then approximately 1600 μm^2 would be consumed, or about 50 % of the available membrane. This estimate agrees with that of Tsan & Berlin (1971) and Werb & Cohn (1972) who showed that mouse peritoneal and alveolar macrophages and rabbit leukocytes interiorize as much as 50 % of the initial plasma membrane surface area in response to a saturating phagocytic load of polystyrene latex particles. The consensus regarding macrophages is that 50 % is the maximum percentage of membrane which can be interiorized before the cell is incapable of further endocytosis (Simson & Spicer, 1973). Preliminary data in our laboratory suggest that full capacity for uptake of algae by digestive cells occurs no sooner than two hours and no later than 24 hours after the initial phagocytosis is completed.

The recognition phase

Two types of observations suggest that after uptake of symbiotic algae by *H. viridis* there are receptor–effector events which can be conveniently referred to as the recognition phase of reinfection. The first type of observation comes from experiments in which free-living *Chlorella vulgaris* and symbiotic algae are tested for ability to infect and maintain a symbiosis with aposymbiotic *H. viridis* and non-symbiotic *Hydra attenuata*. The results in Table 2 show clearly that only the combination of symbiotic algae and symbiotic hydra leads to a stable, persistent infection. Therefore, successful establishment of the symbiosis is a function of very specific mutual recognition between the algae and the hydra (see also Karakashian, 1968, and Hirshon, 1969). The second type of observation comes from

Table 2. *Ability of symbiotic and free-living algae to form a stable symbiosis with symbiotic and free-living hydra*

	Hydra	
Algae	H. viridis (Fla.)	H. attenuata
H. viridis (Fla.)	+	−
C. vulgaris (free-living)	−	−

+ , Successful infection; − , unsuccessful infection.

experiments on the injection of particles other than symbiotic algae into aposymbiotic *H. viridis*. Without exception, food particles, carmine particles, latex spheres, free-living *C. vulgaris* and heat-inactivated symbiotic algae, as well as living symbiotic algae, are all engulfed by digestive cells. We infer from this that phagocytosis is initially non-specific. That is, the digestive cells are initially not discriminating about the particles they take up. However, soon after phagocytosis we observe a fundamental dichotomy in the disposition of the phagocytosed particles and the symbiotic algae. Non-living particles and free-living *C. vulgaris* collect into large apical phagosomes and ultimately undergo exocytosis. Living symbiotic algae remain sequestered in individual vacuoles (Plate 2*a*), suggesting that some form of recognition of symbiotic algae *vis-à-vis* other particulate entities takes place after formation of phagosomes. Current models of cellular recognition are based generally on the inter-action of complementary receptor molecules at apposed cell surfaces (Smith & Good, 1969; Roth, 1973). The nature of the receptors presumed to be involved in specific recognition of symbiotic algae has not yet been investigated. However, it is reasonable to tentatively rule out such physio-logical parameters as oxygen production by the algae as a primary re-cognition event leading to specific sequestration of symbiotic algae, since free-living *C. vulgaris* produce oxygen in the light but are still rejected by *H. viridis*. Similarly, the selective release of maltose, a unique property of symbiotic *Chlorella* isolated from *Paramecium bursaria* and *H. viridis* (Smith, Muscatine & Lewis, 1969) may also be tentatively ruled out as the sole chemical cue leading to sequestration since all but two of these strains of symbiotic *Chlorella* are rejected by *H. viridis* (Table 3). The basis of cellular recognition in the system described here and in general remains as an extremely fertile area for investigation.

The intracellular migration phase

One of the striking features of the infection of hydra digestive cells by symbiotic algae is that, after engulfment, the algae migrate from the site of uptake to a position adjacent to the base of the digestive cell (Pardy & Muscatine, 1973). As noted earlier, this is the position of the algae in the normal green hydra digestive cells. The size and opacity of living hydra precludes direct study of migration. We have therefore studied this movement by observing (Plate 3) and measuring (Fig. 2) the positions of algae within cells from hydra freshly macerated at intervals after the injection of algae. Since the morphology of digestive cells is typically polarized, with a distal rounded end bearing one or two flagella projecting

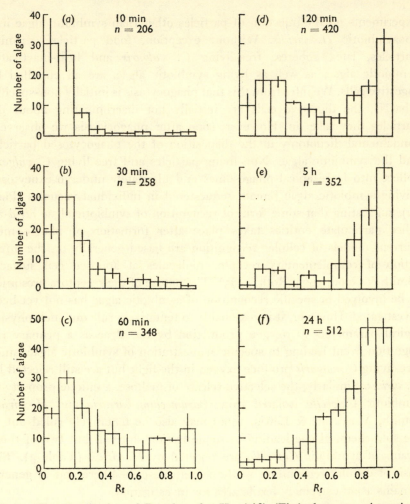

Fig. 2. The distribution of R_t values for *H. viridis* (Fla.) algae at various times after injection into aposymbiotic *H. viridis* (Fla.) hydra. Each graph represents values for all algae in 24 digestive cells from each of three hydra; histogram bars represent the mean for one hydra. Vertical lines represent mean ± one standard error of the mean: n = total number of algae measured.

into the coelenteron and a flattened end lying next to the mesoglea, a longitudinal axis may be defined and used to describe the relative position of an algal cell within a digestive cell. For this purpose we use the term 'R_f' which is the ratio of the distance of an algal cell from the coelenteron margin of the digestive cell to the total length of the digestive cell.

Figure 2 shows the frequency distributions of R_f values of algae in digestive cells from aposymbiotic hydra macerated at intervals after injection. Initially, algae accumulate at the tips of the cells for 1–2 hours after injection (Fig. 2a–d); by this time the uptake process is completed (see Fig. 1). During this period algae gradually appear at the mesoglea end of the digestive cells. Within five hours, the intracellular distribution of the algal cells is similar to that seen 24 hours after injection except for a small population of algae near the distal tip of the cell (R_f values 0.1 to 0.3, Fig. 2e). It is not clear whether these residual algae migrate to the base of the cell during the interval before the next sample or whether they are lost from digestive cells. A slight decline in fluorescence of injected hydra after 24 hours, compared to fluorescence measurements after two to five hours when uptake is complete suggests that either a small percentage of cells taken up are lost or that there is a change in the chlorophyll content of the algae (C. D'Elia, unpublished).

Figure 3 shows that the mean distance between an algal cell and the digestive cell tip increases at intervals after injection, as determined from maceration preparations. Within 24 hours after injection, a given algal cell is located at an average distance of 45 μm from the coelenteron edge of a digestive cell. This average distance is governed by the size of the

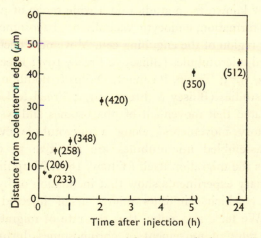

Fig. 3. The average distance between phagocytosed algae and the coelenteron tips of their host digestive cells as a function of time after injection. Vertical lines as in Fig. 2. Sample size in parentheses.

digestive cells; in larger cells, algae may migrate as far as 100 μm. The actual migration distances may be somewhat greater than these since our maceration technique produces some shrinkage of hydra cells (David, 1972). While the precise rate of algal movement can only be estimated from direct observations of the migration process, we know that algal cells are found near the base of digestive cells within 10 minutes after injection (Fig. 2a). In one such instance an algal cell was located 60 μm from the tip of the digestive cell. If we assume that this algal cell was taken up at the tip of the cell, rather than somewhere along the free lateral border, an average rate of migration of 5–6 μm per minute is indicated. In making this estimate we also assume that the algae are moving constantly throughout the interval. Obviously, higher absolute rates would prevail if movement were saltatory. Stages in the migration sequence are shown in Plate 3.

While we have studied quantitatively only the migration of Florida algae after injection into *H. viridis* (Fla.) hydra, algae isolated from the English green hydra and NC64A algae in culture also migrate within *H. viridis* (Fla.) digestive cells. Since all of the infective strains which we have examined exhibit migration and since non-infective* algae do not exhibit migration, we infer that the process is necessary to establish a stable symbiosis.

The mechanism responsible for the migration of infective algae appears to be a transport system intrinsic to the hydra digestive cells. *Chlorella* spp. are non-motile algae and do not possess locomotory organelles. In addition, the transport system must act on the perialgal vacuolar membrane. It is now known from studies on a wide range of phagocytic cells that, soon after formation, endocytic vacuoles or phagosomes move toward the perinuclear region of the engulfing cell. Movement occurs along paths close to oriented microtubules (Bhisey & Freed, 1971; Freed & Lebowitz, 1970; Margulis, 1973; Tiffon, Rasmont, DeVos & Bouillon, 1973; Tilney, 1971). Recent studies (Bhisey & Freed, 1971; Freed & Lebowitz, 1970) have demonstrated that movement of phagosomes and lysosomes occurs in 'long saltatory movements' along a microtubule cytoarchitectural system. The assembled microtubules are sensitive to colchicine and vinblastine as is the migration itself (Tilney, 1971; Margulis, 1973).

Our preliminary experiments show that incubation of hydra in culture medium with colchicine inhibits migration of injected algae (Cook, unpublished). We have also noted that the rate of migration of algae is similar to the rates of movement of chromosomes during mitosis and

* By non-infective algae, we mean not capable of establishing a permanent symbiosis. Non-infective algae are, however, still taken up by digestive cells.

axonal flow which are known to be dependent on oriented arrays of micro-tubules (Rehbun, 1972; Tilney, 1971). This suggests that microtubules may be involved in the algal migration process. Although we have observed oriented microtubules in injected *H. viridis* digestive cells (Cook, un-published), we have not yet observed a cytoarchitectural framework, such as described by Bhisey & Freed (1971), and Freed & Lebowitz (1970), in close association with migrating algae.

It is still unclear how this intracellular transport system is activated, how it operates, and how it discriminates between infective and non-infective algae. The result of migration is the curious polarized distribution of algae within digestive cells in normal green hydra. This distribution might serve to isolate the algae from host cell processes, such as digestive attack by lysosomes, which pose a threat to successful maintenance (see Karakashian & Karakashian, 1973).

The repopulation phase

After migrating to the base of digestive cells, the algae undergo asexual reproduction *in situ*, increasing in number until the host cell is repopulated. Typically, each mother cell divides to yield four autospores or daughter cells. In previous studies Park *et al.* (1967) used visual comparison with normal green hydra to estimate the time of repopulation of aposymbiotic *Chlorohydra hadleyi* infected with symbiotic algae. These authors reported full repopulation within 20 days after injection. Pardy & Muscatine (1973) used measurements of ^{14}C fixation to determine that functional repopula-tion was reached about 18 days after injection. The results of a more detailed analysis of repopulation are given in Fig. 4. In this experiment algae from *H. viridis* were injected into aposymbiotic individuals, some of which were fed daily and growing at maximum intrinsic logarithmic rates, and others starved during and after injection. The stomach and budding zone of each polyp was excised and macerated at 48 hour intervals and the numbers of algae per cell ascertained in fixed and stained cell preparations. Figure 4 shows that in fed hydra the numbers of algae per cell reached control levels about 10–11 days after injection. This rate of repopulation is faster than that cited above in the previous investigations but the sets of data were obtained by different assays. The former used whole hydra; the latter used only the stomach zone digestive cells. These cells are the predominant ones which initially engulf injected algae and would therefore be expected to repopulate relatively rapidly. In starved hydra the numbers of algae per cell reached control levels after 6–7 days so that there is a greater number of algae per cell than that in fed hydra. We interpret this as indicating that in starved hydra the algae grow faster than the host cells.

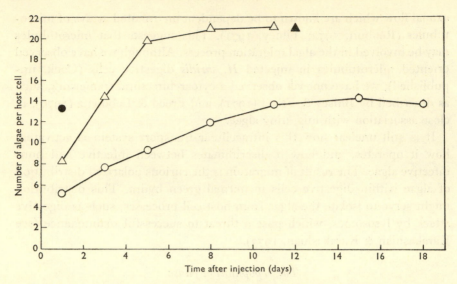

Fig. 4. Rate of repopulation of aposymbiotic *H. viridis* (Fla.) hydra after injection of *H. viridis* (Fla.) algae. Number of algae per digestive cell in (○) hydra fed daily during repopulation; (●) normal green control hydra, fed daily; (△) hydra starved during repopulation; (▲) normal green control hydra, 12-day starved. Each point represents the mean number of algae from 20 digestive cells from the stomach and budding region (Zone 2) of four injected hydra. (Buds excluded from samples.)

The possibility that starved digestive cells take up more algae than fed cells has not yet been rigorously tested.

While injection of algae into *H. viridis* results in uptake of algae by at least 90 % of the digestive cells (Table 5), we can conceive of situations in which only a small fraction of digestive cells in aposymbiotic *H. viridis* might take up algae. How do the rest of the digestive cells obtain symbionts? There appear to be three possibilities. First, the algae might move from cell to cell by transit across adjacent host cell membranes. Though not ruled out, this appears unlikely since bits of green hydra tissue grafted to aposymbiotic individuals maintain their mosaic appearance (Burnett & Garofalo, 1960). Second, fully populated digestive cells might produce and expel excess algae into the coelenteron which are then recaptured by other digestive cells. This possibility has not yet been tested. Third, and most likely, weakly populated individual hydra may not actually become repopulated. Rather, algae-laden cells dividing in the budding zone may produce buds with a greater number of infected cells, giving rise by this process to fully populated progeny after several generations.

OTHER INFECTIVE AND NON-INFECTIVE
TYPES OF ALGAE

As part of a comparative investigation of infectivity we have tested other symbiotic algae for their ability to establish a symbiosis with aposymbiotic *H. viridis*. Table 3 shows that symbionts from *H. viridis* (English) hydra

Table 3. *Strains of symbiotic algae infective (+) and non-infective (–) in* H. viridis (*Fla.*)

Algal strain	Condition	Fate in aposymbiotic *H. viridis* (Fla.)
3H		–
34		–
130C	All	–
42A	living	–
NC64A		+
838		–
H. viridis (English)		+
H. viridis (Fla.)	Heat-inactivated 60 °C, 5 min	–

(Data of Pardy & Muscatine, 1973, and R. Pool, unpublished)

can infect and persist in *H. viridis* (Fla.) hydra. We have designated this heterologous combination E/F. We have not yet tested the infectivity of algae from other hydra strains. A more interesting finding, perhaps, is that only one strain of algae from *P. bursaria*, strain NC64A, has proved infective in *H. viridis* (Fla.) hydra (designated NC/F). The rate of uptake of NC64A, previously grown in pure culture, is slower than that of normal symbionts (Fig. 1, lower curve) and the numbers taken up are substantially fewer. These algae are approximately the same size as the normal symbionts so that limited uptake cannot be explained simply in terms of the amount of membrane available for formation of vacuoles. Both the English and NC64A algae migrate in *H. viridis* cells and both E/F and NC/F hydra now flourish in our laboratory. The growth rates of NC/F fed daily, are virtually identical to those of normal *H. viridis* (Fla.) populations (Pardy & Muscatine, 1973).

The non-infective algae shown in Table 3 fall into two categories; normally competent algae which have been rendered non-infective by heating to 60 °C for 10 minutes, and living but incompetent algae. Among the latter are five strains of symbiotic algae from *P. bursaria* and the algae from a freshwater sponge. We have already referred to the non-infective status of free-living *C. vulgaris* (Table 2). Unlike infective algae, the non-infective cells are found in apical vacuoles. They do not migrate to the base of the digestive cells (Plate 4; compare Fig. 5 and Fig. 2e), and they rarely persist within the digestive cells beyond 1–2 days. The fate of non-infective algae is of considerable interest, though not yet completely understood. At present exocytosis seems to be the end result. There is

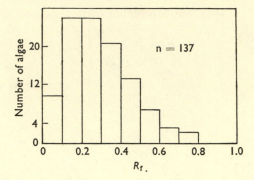

Fig. 5. The distribution of R_f values for *Chlorella vulgaris* (strain 397) five hours after injection into aposymbiotic *H. viridis* (Fla.) hydra, showing the failure of the free-living algae to migrate to the base of the digestive cells. Parameters same as Fig. 2 except S.E. not given.

little direct evidence to support the interpretation that such algae are digested by hydra. In a study of hydra fine structure, Oschman (1966) described 'phagolysosomes' fusing with the perialgal vacuoles in *Chlorohydra viridissima* but admitted that further study would be required to determine if the algae (in this case the *normal* competent symbionts) were undergoing digestion. In recent studies on *P. bursaria*, Karakashian & Karakashian (1973) reported that acid phosphatase activity, as revealed by electron microscope cytochemistry, appears in vacuoles containing latex spheres, carmine particles, and heat-killed algae, but not in vacuoles surrounding established symbionts, implying that the former sustain digestive attack while the latter resist or repress it. Further, when symbionts occur together in vacuoles with heat-killed algae, digestion of the latter is delayed (see Karakashian, this volume). Whereas there is no conclusive evidence yet for digestion of infective or non-infective algae in *H. viridis*, Mr T. Hohman in our laboratory has observed that *H. viridis* symbionts can pass rapidly through the gut of *Daphnia magna*, a natural herbivore, and emerge with no detectable change in morphology at the light microscope level and only minimum loss of photosynthetic function as determined by fixation of $^{14}CO_2$. In contrast, free-living *C. vulgaris* pass slowly through the gut and emerge with readily detectable morphological alterations and substantial loss of photosynthetic function. If the symbiotic algae are actually digestion-resistant, the possibility should be explored that they possess an acid hydrolase-resistant cell wall constituent such as the polycarotenoid sporopollenin, found in the cell wall in some strains of *Chlorella* (Atkinson, Gunning & John, 1972).

Specificity of infection of H. viridis

Goetsch (1924) carried out one of the first investigations on specificity of infection of hydra by algae. He reported that *Chlorohydra viridissima* could be successfully reinfected not only by its own algae but also by the free-living alga *Oocystis* sp. and by algae from the bivalve mollusc *Anodonta* sp. In mixed infections of free-living and symbiotic algae, the latter ultimately became the surviving species. Surprisingly, Goetsch claimed to have infected with symbiotic algae a few individuals of *Hydra attenuata* and *Hydra vulgaris* but not *Hydra circumcincta* or *Pelmatohydra oligactis*: all are non-symbiotic hydra. Goetsch concluded that there were degrees of infectivity between various hydra and algae. Evaluation of Goetsch's findings is difficult due to the uncertainty of his species identifications, the paucity of quantitative data and the limited nature of his cytological observations. Recently Park *et al.* (1967) have confirmed and clarified at least some of Goetsch's findings. Whereas Goetsch infected hydra by feeding them *Daphnia* with a bit of green hydra tissue or algae wedged beneath the carapace, Park *et al.* used microinjection of isolated algae. Table 4 summarizes this work and shows that, of the algae injected into aposymbiotic *Chlorohydra hadleyi*, only the host's native symbionts yielded rapid and stable infection. Trials with other types of algae, including symbionts from *Chlorohydra viridissima*, yielded limited infections of short duration. The cytology of heterologous infections differed from the normal infections in that algal cells occurred together in 'packets' rather than individually in vacuoles. These data, together with our own

Table 4. *Summary of reinfection trials by Park* et al., (*1967*). *Successful infections* (+), *unsuccessful* (−). (*Table from Muscatine, 1974*)

Potential symbionts	Fate in aposymbiotic *C. hadleyi*
C. hadleyi algae	+ Complete and rapid infection, morphology normal
Chlorohydra viridissima (Burnett strain) algae	± Slow and incomplete infection; algae in a 'few discrete and widely separated patches' 7 days after injection
Oocystis sp.	± Limited transient infection algae in 'packets'
Chorella sp.	−
Euglena sp.	−
Euglena sp. chloroplasts	−

observations, permit formulation of the salient features of the apparently
high degree of specificity involved in the establishment of an hereditary
symbiosis of algae and hydra. These features are: rapid initial uptake of
algae after injection, sequestration of the algae into individual vacuoles,
migration of algae to the base of the digestive cells, persistence through
time in the hydra, passage to buds in asexual reproduction and transmission via eggs in sexual reproduction (see Brien & Reniers-Decoen,
1950). These features should serve equally as criteria for those who claim
successful culture of hydra algae.

Experimental conditions which result in enhanced uptake:
evidence from infectivity data and serology

Data in Fig. 1 show that algae from hydra are taken up by aposymbiotic
H. viridis to a greater extent than algae cultures *in vitro*. An immediate
question raised by this finding is whether the determinants in this experiment are a function of the particular algal strain or a function of the
environment that the particular alga occupies prior to injection. Experiments carried out in our laboratory show that uptake of at least one
strain of cultured algae can be increased by altering its environment prior
to injection. In these experiments, aposymbiotic *H. viridis* (Fla.) were
first injected with cultured NC64A. Populations of these hydra were then
raised for three months. At that point, fresh aposymbiotic *H. viridis* hydra
were injected with either cultured NC64A algae (NC→F) or the re-
isolated NC/F algae (NC/F→F). After 24 hours the hydra were macerated
and the digestive cells examined for the presence of algae and their
numbers per host cell determined. The results are shown in Table 5. It
is evident that the NC/F→F algae show a dramatic increase in numbers
taken up per host cell and that a greater percentage of host cells have

Table 5. *The effect of residence in hydra on subsequent infectivity of NC64A*

		Fate in aposymbiotic *H. viridis* (Fla.)	
Condition	Designation	Number of algae per host cell	% Host cells infected
NC64A cultured *in vitro*; injected into *H. viridis* (Fla.)	NC→F	2.91 ± 2.21	8.5
NC64A algae grown for 3 months in hydra, re-isolated and injected into aposymbiotic *H. viridis* (Fla.)	NC/F→F	5.01 ± 3.04	80.1
Florida algae in Florida hydra	F/F	18.0 ± 2.6	>90

acquired the symbionts, when compared to the control (NC → F) injections.

These changes in uptake of algae in response to culture conditions can be correlated with changes in the antigenic properties of NC64A and other infective algae. These antigenic changes have been detected using the serological technique of microcomplement fixation. The method has been well-reviewed (Levine, 1967), and its use in serological comparison of materials ranging from macromolecules to whole cells is now routine. Full details will be published elsewhere and we will give only a summary of results here. We have examined the capacity of algae isolated from three geographical strains of *H. viridis* (Fla., Carolina, English) and algae from *C. hadleyi* to fix complement. Note that these experiments deal only with algae and not hydra. All experiments were performed using rabbit antibodies made against algae isolated from Florida hydra. The results (Fig. 6) are expressed as the percentage of complement fixed by the test antigen versus the concentration of antigen added to the reaction mixture. Proper control experiments were performed to eliminate the possibility that the observed serological reactions were in response to animal tissue contaminants rather than the algae. The complement fixation reactions typically show some peak fixation for each algal antigen at a specific antigen dilution. The peak represents the concentration of antigen that reacts with some set concentration of antibody to yield optimal complement fixation. The location of the test peak, *vis-à-vis* the reference peak, relates to differences between test and reference antigens (see legend, Fig. 6). In our experiments, the reference peak is the result of Florida algae reacting against Florida antiserum.

Figure 6a shows the patterns of complement fixation by various algae taken from their normal 'culture' medium. In the case of hydra algae, we consider the host as the 'culture medium'. Carolina and *C. hadleyi* algae show peaks at the same antigen dilution but complement fixation maxima below that of the Florida algae reference antigen. This indicates that the two algal strains are antigenically similar to the Fla. strain for a majority of antigenic determinants but some determinants may be modified or present in lower numbers as compared to the Florida algae. This is especially the case for the *C. hadleyi* algae. The English algae exhibit a reactivity different from the other algae. The peak fixation is laterally displaced from the others, indicating a dissimilarity in antigenic determinants compared to Florida algae. The peak also extends over a range of antigen dilutions. This may be due to anticomplementarity, the non-specific fixation of complement exhibited by some antigens when present in high concentrations. NC64A algae, originally isolated from *P. bursaria*, but long since cultured in Loefer's medium, showed no detectable

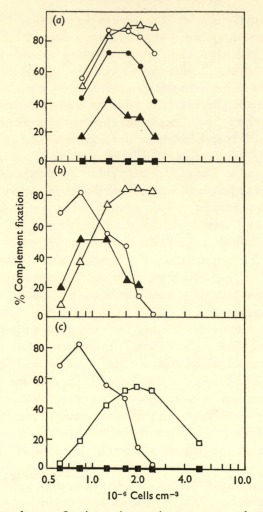

Fig. 6. Microcomplement fixation using antiserum prepared against *H. viridis* (Fla.) algae to assay (*a*) algae from *H. viridis* (Fla.) (○), *H. viridis* (English) (△), *H. viridis* (Carolina) (●), *C. hadleyi* (▲), and NC64A algae grown in Loefer's Medium (■). Antiserum dilution 1 : 15 000. (*b*) Algae from *H. viridis* (Fla.) (○), English algae in English hydra (△), and English algae grown in Florida hydra (▲). Antiserum dilution 1 : 13 000. (*c*) Florida algae from Florida hydra (○), NC64A algae grown in Florida hydra (□), and NC64A algae grown in Loefer's medium (■). Antiserum dilution 1 : 13 000. A vertical shift in fixation with respect to the reference peak indicates a quantitative difference in antigenic determinants. A horizontal shift from the reference peak indicates a change in the antigen quality.

complement fixation. Therefore, these NC64A algae possess no Florida-type antigenic determinants as characterized by this technique.

Figure 6*b*, *c* show the patterns of complement fixation of algae after three months growth in heterologous combinations with Florida hydra.

In Fig. 6*b* it can be seen that as a result of residence in Florida hydra, English algae (E/F) exhibit antigenic properties more closely approaching those of the Florida algae (F/F), compared to control English algae (E/E). This is indicated by the closeness of the peaks. The lower fixation maximum for E/F indicates slight changes in antigenic determinant concentration with respect to the F/F reference maximum. These changes in antigenicity are not necessarily correlated with changes in uptake capacity since the English algae, both before and after residence in Florida hydra, show uptake characteristics fully comparable with Florida algae (Pool, unpublished).

Whereas NC64A algae cultured *in vitro* show no Florida-type antigenic determinants, Fig. 6*c* indicates that after residence in Florida hydra for three months, NC/F algae now take on detectable antigenic properties similar to the Florida algae. Some qualitative difference is indicated by the lateral shift in peak fixation. In this case, changes in uptake characteristics of NC/F algae (i.e., NC/F→F) can be correlated with changes in reactivity to Florida algae antiserum.

Although we interpret these as phenotypic changes resulting from residence in hydra, our experimental protocol does not exclude the possibility that after injection of NC64A algae into hydra, selection of variants gives rise to a new or dominant line of algal cells in the hydra.

We have not yet investigated the nature of the cell moieties responsible for the observed antigenic character of the algae. We assume that some features of the algal cell surface are implicated. Pfeiffer *et al.* (1971) have shown that the antigenicity of rat glial cells changes with mode of culture of these cells. The authors attribute these changes to cell surface alterations.

SOME ASPECTS OF THE STABILITY OF ALGAL POPULATIONS IN *H. VIRIDIS*

A crucial feature of the maintenance of algae in an endocellular symbiosis is the establishment of a symbiont growth rate harmonious with that of the host. Where this does not occur, the symbionts may potentially overgrow and kill the host or be diluted out by failure to keep pace with host cell multiplication. In this section we shall describe briefly the pattern of distribution of algae in specific body regions of *H. viridis*, and how this pattern may be affected by illumination, darkness and feeding regimes.

Hydra grown in constant light with daily feeding maintain a steady-state ratio of algae per digestive cell. In random samples from stomach and budding regions this number ranges from 10 ± 3.7 in *C. hadleyi*, 17 ± 6.7 in *C. viridissima* (Burnett strain) (Park *et al.*, 1967), to 18 ± 2.6 in *H.*

viridis (Fla.) (Pardy & Muscatine, 1973). In *H. viridis* (Fla.), and probably the other strains as well, the algae are not distributed uniformly throughout the hydra. Figure 7 shows that they are most abundant in the digestive cells of the stomach and budding region (Zone 2; 18 ± 2.6) with fewer in the hypostome and tentacle region (Zone 1; 12.0 ± 3.2) and stalk and basal disk (Zone 3; 12 ± 2.9) (Pardy & Muscatine, 1973). Changes in illumination, darkness and feeding regimes can alter the steady-state ratio as described in Table 6 and the distribution pattern (Fig. 7). Table 6 shows that,

Table 6. *Effect of maintenance conditions on the number of algae per digestive cell in* H. viridis *(Fla.). Hydra were maintained at 20 °C. Constant illumination was 2000 lx; diurnal illumination was 10 h light (300 lx), 14 h dark. Data expressed as mean number of algae per cell ± standard deviation of the mean. n = 100 digestive cells from Zone 2 of five hydra*

	Number of algae/cell	
Condition	Day 2	Day 4
Constant light, starvation	21.0 ± 5.7	21.8 ± 4.6
Constant light, fed every 24 h	20.0 ± 4.7	20.8 ± 4.1
Diurnal light, starvation	20.4 ± 1.0	20.0 ± 1.7
Diurnal light, fed every 24 h	18.0 ± 2.6	18.3 ± 1.9
Constant dark, starvation	17.0 ± 1.4	11.0 ± 3.0
Constant dark, fed every 24 h	12.3 ± 3.8	6.8 ± 2.3

Data of Pardy & Muscatine (1973).

depending on maintenance conditions, the number of algae per digestive cell ranges from 7–22 after four days of treatment. The mean ratios fall generally into two groups; (*a*) about 18–22 algae per cell in hydra which were fed or starved but received some illumination (diurnal or constant), and (*b*) about 7–17 algae per cell in hydra which were fed or starved but received no illumination. Thus illumination and nutrition appear to be important variables governing numbers of algae per cell.

Lowest numbers of algae per cell are seen in hydra maintained in darkness with daily feeding. It might be expected that hydra which are fed daily and growing at maximum rates would eventually outgrow their symbionts and become aposymbiotic. On the contrary, as described by Pardy (1974), populations of hydra maintained in darkness for six months with daily feeding still appear pale green. Examination of digestive cells from these hydra, prepared by maceration, reveals the presence of several pale green algae per cell. Moreover, when returned to the light, these dark-grown hydra always become greener. Thus, a low level of algal infection is maintained even in the dark. It must be stated that animals maintained in the dark were actually exposed to 2–3 minutes of room light twice daily during feeding and routine maintenance. This periodic exposure to light plus the assumption of heterotrophic nutrition of the algae under dark conditions might serve to maintain a low level population of the symbionts. Cook (1972) has demonstrated a transfer of labeled material from hydra to algae. In his experiments hydra were kept in both light and dark, and were fed ^{14}C-labeled *Artemia* nauplii. Assay of the symbiotic algae 48 hours later showed that the symbionts in darkness had acquired 25–34 % of the total radioactivity. While the nature of the translocated

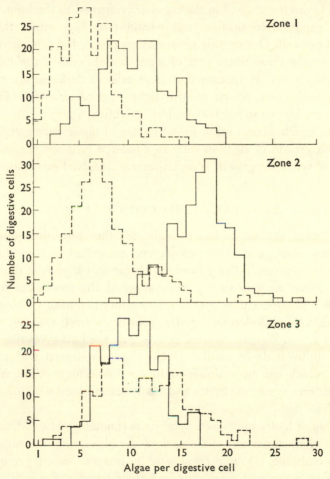

Fig. 7. Histograms showing the distribution of algae in hydra maintained in constant light and fed daily (solid line) and in continuous darkness and fed daily (dashed line). Individual hydra were cut transversely into three pieces; hypostome and tentacles (Zone 1) stomach and budding region (Zone 2), and the stalk and pedal disk (Zone 3). Pieces from each zone were macerated separately and the algae in a large number of digestive cells from each zone were counted (data of Pardy, 1974).

substance(s) is not known, Cook's work clearly demonstrates the existence of a flow of organic material from the hydra, or hydra's food, to the algae.

Figure 7 also shows that growth of hydra in darkness affects the distribution of algae in the various zones. Zones 1 and 2 have fewer algae per cell (averaging 6 ± 2.7 and 7 ± 3.2, respectively) and Zone 3 remains unchanged from the condition seen in hydra grown in continuous light (11–12 algae per cell). The reason for this change in distribution pattern is still not clear.

When hydra maintained in darkness are returned to the light, the algae undergo rapid multiplication and within two days attain the normal numbers per cell. During this repopulation phase, the growth rate of the algae is nearly three times that of algae maintained in the light under normal conditions. It appears that algal cells in darkness are poised for division, and upon return to the light multiply rapidly. Free-living *Chlorella* are known to behave similarly (Griffith, 1961).

Factors which permit rapid growth rates of algae during repopulation yet inhibit growth of algae in fully repopulated cells are not understood. Control of symbiont growth is an important unsolved problem.

SUMMARY AND CONCLUSIONS

The foregoing discussion has brought together published information, preliminary data, and some conjecture regarding the mechanism of infection of metazoan cells by symbiotic algae and aspects of their population dynamics, as seen in the association of the freshwater hydra and symbiotic *Chlorella*. Under specified laboratory conditions, about 18 algae dwell within the endodermal digestive cells of the host; each hydra harboring about 1.5×10^5 algae. Asexual budding by the host gives rise to a large clone of individuals of similar genetic, developmental and nutritional histories which can be maintained in the laboratory. Algae are readily separated from symbiotic hydra and easily recombined with aposymbiotic host hydra.

Infection of hydra by its own symbionts (homologous infection) has the following characteristics. (*a*) *Contact* of the algae with the digestive cell plasma membrane; (*b*) rapid *engulfment* by phagocytosis; (*c*) *recognition* of the algae as competent symbionts and their *sequestration* into individual perialgal vacuoles; (*d*) intracellular *migration* of the algae to the base of digestive cells soon after engulfment; and finally, (*e*) an attenuated *repopulation* of the aposymbiotic hydra.

Non-infective algae, such as free-living *Chlorella vulgaris* or heat-inactivated symbiotic algae, as well as carmine particles and latex spheres, are engulfed but sequestered together in large apical vacuoles. These do not migrate and are rejected from the digestive cells within a day or two after engulfment.

Comparative studies on infectivity show that there is a high degree of specificity involved in establishing a stable symbiosis with hydra. A few heterologous infections have been obtained. In these cases, the algae are initially only marginally infective. After maintenance in hydra for several months, these algae exhibit enhanced uptake when re-isolated and injected in aposymbiotic hydra. Serological studies using microcomplement fixation show that residence of these heterologous algae in hydra results in an algal flora with altered surface antigenicity, and, in one case, increased infective competence.

The steady-state ratio of algae per host cell and the regional distribution of algae in hydra can be perturbed by environmental factors such as illumination and feeding regimes of the host.

ACKNOWLEDGEMENTS

The authors are grateful for the excellent technical assistance of Miss Bibbi Wolowske, Mrs Katherine Bezy and Mr Herman Kabe in various phases of these investigations. We thank Dr F. S. Sjöstrand for his generous support in all phases of the electron microscopy, Dr H. Herschman for advice on serology and Dr C. D'Elia for a critique of the manuscript. We acknowledge support of research grants from the National Science Foundation (NSF GB 11940) and National Institutes of Health (NIH AI200070).

REFERENCES

ATKINSON, A. W., GUNNING, B. E. S. & JOHN, P. C. L. (1972). Sporopollenin in the cell wall of *Chlorella* and other algae. Ultrastructure, chemistry, and incorporation of ^{14}C acetate, studied in synchronous cultures. *Planta*, **107**, 1–32.

BENSCH, D., GORDON, G. & MILLER, L. (1964). The fate of DNA-containing particles phagocytized by mammalian cells. *J. Cell Biol.*, **21**, 105–114.

BHISEY, A. N. & FREED, J. J. (1971). Altered movement of endosomes in colchicine-treated cultured macrophages. *Expl Cell Res.*, **64**, 430–438.

BRANDT, K. (1881). Ueber das Zusammenleben von Thieren und Algen. *Sber. naturf. Ges. Berl.*, No. **9**, 140–146.

BRIEN, P. & RENIERS-DECOEN, M. (1950). Étude d'*Hydra viridis* (Linnaeus) (la blastogenèse, la spermatogenèse, l'ovogenèse). *Annls Soc. r. zool. belg.*, **81**, 33–110.

BURNETT, A. & GAROFALO, M. (1960). Growth pattern in the green hydra, *Chlorohydra viridissima*. *Science, Wash.*, **131**, 160.

CASLEY-SMITH, J. R. (1969). Endocytosis: The different energy requirements for the uptake of particles by small and large vesicles into peritoneal macrophages. *J. Microsc.*, **90**, 15–30.

CHRISTIANSEN, R. G. & MARSHALL, J. M. (1965). A study of phagocytosis in the ameba *Chaos chaos*. *J. Cell Biol.*, **25**, 443–457.

COOK, C. B. (1972). Benefit to symbiotic zoochlorellae from feeding by green hydra. *Biol. Bull mar. biol. Lab. Woods Hole.*, **142**, 236–242.

DAVID, C. N. (1973). A quantitative method for maceration of hydra tissue. *Wilhelm Roux Arch. EntwMech. Org.*, **171**, 259–268.

DRAPER, P. & REES, R. J. W. (1970). Electron transparent zone of mycobacterium may be a defense mechanism. *Nature, Lond.*, **228**, 860.

FREED, J. & LEBOWITZ, M. (1970). The association of a class of saltatory movements with microtubules in cultured cells. *J. Cell Biol.*, **45**, 334–354.

FRIIS, R. R. (1972). Interaction of L cells and *Chlamydia psittaci*: Entry of the parasite and host responses to its development. *J. Bact.*, **110**, 706–721.

GAUTHIER, G. (1963). Cytological studies on the gastroderm of *Hydra*. *J. exp. Zool.*, **152**, 13–39.

GOETSCH, W. (1924). Die Symbiose der Süsswasser-hydroiden und ihre künstliche Beeinflussung. *Z. Morph. Ökol. Tiere*, **1**, 660–731.

GORDON, A. H. (1973). The role of lysosomes in protein catabolism. In *Lysosomes in Biology and Pathology*, Vol. **3**, 89–137 (ed. J. T. Dingle). Amsterdam: North-Holland.

GORDON, S. & COHN, Z. A. (1973). The macrophage. *Int. Rev. Cytol.*, **36**, 171–214.

GRIFFITH, D. J. (1961). Light induced cell division in *Chlorella vulgaris* Beijerinck (Emerson Strain). *Ann. Bot.*, **25**, 85–93.

HIRSHON, J. B. (1969). The response of *Paramecium bursaria* to potential endocellular symbionts. *Biol. Bull mar. biol. Lab. Woods Hole*, **136**, 33–42.

KARAKASHIAN, M. W. & KARAKASHIAN, S. J. (1973). Intracellular digestion and symbiosis in *Paramecium bursaria*. *Expl Cell Res.*, **81**, 111–119.

KARAKASHIAN, S. J. (1963). Growth of *Paramecium bursaria* as influenced by the presence of algal symbionts. *Physiol. Zool.*, **36**, 52–67.

(1968). Invertebrate symbioses with *Chlorella*. In *Biochemical Co-evolution*, Biology Colloquium no. **29**, 33–52 (ed. K. L. Chambers). Corvallis: Oregon State University.

(1970). Morphological plasticity and the evolution of algal symbionts. *Annls N.Y. Acad. Sci.*, **175**, 474–487.

KORN, E. D. & WISEMAN, R. A. (1962). Phagocytosis of latex beads by *Acanthamoeba*. II. Electron microscope studies of the initial events. *J. Cell Biol.*, **34**, 219–227.

LENTZ, T. L. (1966). *The Cell Biology of Hydra*. Amsterdam: North-Holland.

LEVINE, L. (1967). Micro-complement fixation. In *Handbook of Experimental Immunochemistry*, 707–719 (ed. D. M. Weir). Oxford: Blackwell's.

LUNGER, P. (1963). Fine-structural aspects of digestion in a colonial hydroid. *J. ultrastruct. Res.*, **9**, 362–380.

MARGULIS, L. (1973). Colchicine-sensitive microtubules. *Int. Rev. Cytol.*, **34**, 333–361.

MCLAUGHLIN, J. J. A. & ZAHL, P. (1966). Endozoic algae. In *Symbiosis*, vol. **1**, 257–297 (ed. S. M. Henry). New York: Academic Press.

MUSCATINE, L. (1974). Endosymbiosis of cnidarians and algae. In *Coelenterate Biology: Reviews and New Perspectives*, 359–395 (eds. L. Muscatine and H. M. Lenhoff). New York: Academic Press.

MUSCATINE, L. & LENHOFF, H. M. (1965*a*). Symbiosis of hydra and algae. I.

PLATE I

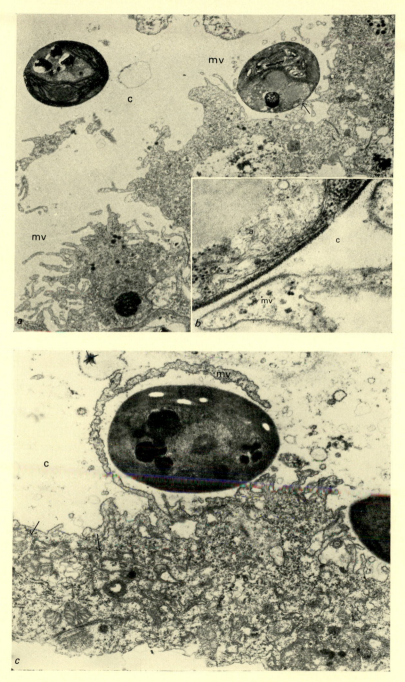

For explanation see p. 202

PLATE 2

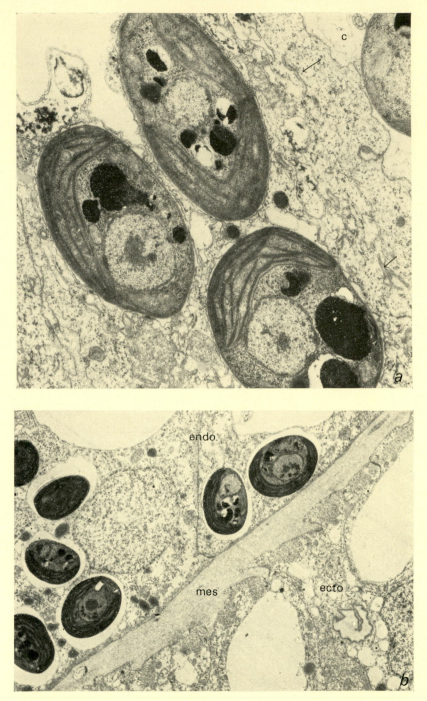

For explanation see p. 202

PLATE 3

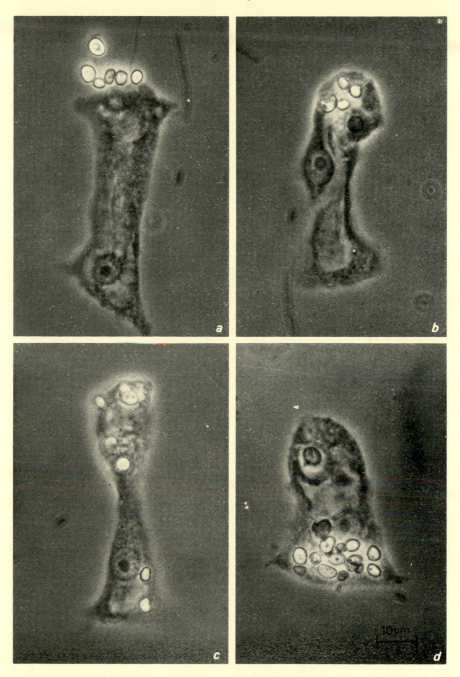

For explanation see p. 203

PLATE 4

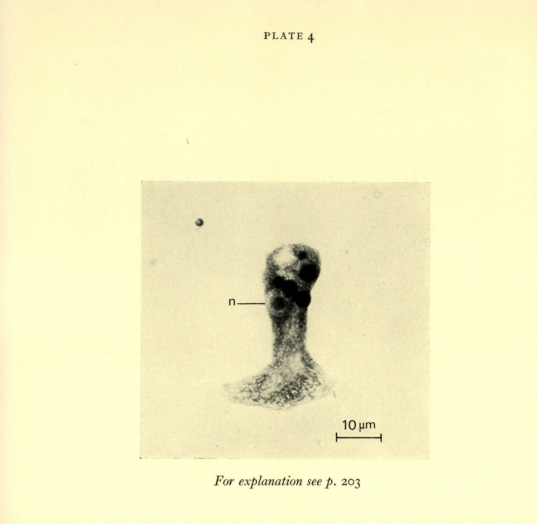

For explanation see p. 203

Effects of some environmental cations on growth of symbiotic and aposymbiotic hydra. *Biol. Bull. mar. biol. Lab. Woods Hole*, **128**, 415–424.

(1965*b*). Symbiosis of hydra and algae. II. Effects of limited food and starvation on growth of symbiotic and aposymbiotic hydra. *Biol. Bull. mar. biol. Lab. Woods Hole*, **129**, 316–328.

OSCHMAN, J. L. (1966). Apparent digestion of algae and nematocysts in the gastrodermal phagocytes of *Chlorohydra viridissima*. *Am. Zool.*, **6**, 320.

(1967). Structure and reproduction of symbiotic algae by *Hydra viridis*. *J. Phycol.*, **3**, 221–228.

PARDY, R. L. (1974). Some factors affecting the growth and distribution of the algal endosymbionts of *Hydra viridis*. *Biol. Bull. mar. biol. Lab. Woods Hole*, **147**, 105–118.

PARDY, R. L. & MUSCATINE, L. (1973). Recognition of symbiotic algae by *Hydra viridis*. A quantitative study of the uptake of living algae by aposymbiotic *H. viridis*. *Biol. Bull. mar. biol. Lab. Woods Hole*, **145**, 565–579.

PARK, H., GREENBLATT, C. L., MATTERN, C. F. T. & MERRIL, C. R. (1967). Some relationships between *Chlorohydra*, its symbionts and some other chlorophyllous forms. *J. exp. Zool.*, **164**, 141–162.

PFEIFFER, S. E., HERSCHMAN, H. R., LIGHTBODY, J. E., SATO, G. & LEVINE, L. (1971). Modification of cell surface antigenicity as a function of culture conditions. *J. Cell Physiol.*, **78**, 145–152.

PROVASOLI, L., YAMASU, T. & MANTON, I. (1968). Experiments on the resynthesis of symbiosis in *Convoluta roscoffensis* with different flagellate cultures. *J. mar. biol. Ass. UK*, **48**, 465–479.

RABINOVITCH, M. (1967). The dissociation of the attachment and ingestion phases of phagocytosis by macrophages. *Expl Cell Res.*, **46**, 19–28.

(1968). Phagocytosis. The engulfment stage. *Semin. Hematol.*, **5**, 134.

REHBUN, L. I. (1972). Polarized intracellular transport: saltatory movements and cytoplasmic streaming. *Int. Rev. Cytol.*, **32**, 93–137.

ROTH, S. (1973). A molecular model for cell interactions. *Q. Rev. Biol.*, **48**, 541–563.

RYSER, H., CAULFIELD, J. B. & AUB, J. C. (1962). Studies on protein uptake by isolated tumor cells. I. Electron microscope evidence of ferritin uptake by Ehrlich Ascites tumor cells. *J. Cell Biol.*, **14**, 255–268.

SCHULZE, P. (1917). Neue Beiträge zu einer Monographie der Gattung *Hydra*. *Arch. Biontol.*, **4**, 39–119.

SHEFFIELD, J. (1970). Studies on aggregation of embryonic cells: Initial cell adhesions and the formation of intercellular junctions. *J. Morph.*, **132**, 245–264.

SIMSON, J. V. & SPICER, S. S. (1973). Activities of specific cell constituents in phagocytosis (endocytosis). *Int. Rev. exp. Pathol.*, **12**, 79–118.

SLAUTTERBACK, D. B. (1967). Coated vesicles in absorptive cells of *Hydra*. *J. Cell Sci.*, **2**, 563–572.

SMITH, D. C., MUSCATINE, L. & LEWIS, D. H. (1969). Carbohydrate movement from autotrophs to heterotrophs in parasitic and mutualistic symbiosis. *Biol. Rev.*, **44**, 17–90.

SMITH, R. T. & GOOD, R. A. (1969). *Cellular Recognition*, 328 pp. New York: Appleton-Century-Crofts.

SPURR, A. R. (1969). A low-viscosity epoxy resin embedding medium for electron microscopy. *J. ultrast. Res.*, **26**, 41–43.

TAYLOR, D. L. (1973*a*). Algal symbionts of invertebrates. *A. Rev. Microbiol.*, **27**, 171–187.

(1973*b*). The cellular interactions of algal-invertebrate symbiosis. *Adv. mar. Biol.*, **11**, 1–56.

TIFFON, Y., RASMONT, R., DEVOS, L. & BOUILLON, J. (1973). Digestion in lower metazoa. In *Lysomes in Biology and Pathology*, Vol. **3**, 49–68 (ed. J. T. Dingle). Amsterdam: North-Holland.

TILNEY, L. G. (1971). Origin and continuity of microtubules. In *Origin and Continuity of Cell Organelles*, Vol. **2**, 222–260 (eds. J. Reinert and H. Ursprung). New York: Springer-Verlag.

TIZARD, I. R. & HOLMES, W. L. (1974). Phagocytosis of sheep erythrocytes by macrophages. Some observations under the scanning electron microscope. *J. reticuloendothel. Soc.*, **15**, 132–138.

TRENCH, R. K. (1971). The physiology and biochemistry of zooxanthellae. symbiotic with marine coelenterates. III. The effect of homogenates of host tissues on the excretion of photosynthetic products *in vitro* by zooxanthellae from two marine coelenterates. *Proc. R. Soc. Lond. B*, **177**, 251–264.

TSAN, M. F. & BERLIN, R. D. (1971). Effect of phagocytosis on membrane transport of nonelectrolytes. *J. exp. Med.*, **134**, 1016–1035.

WERB, Z. & COHN, Z. A. (1972). Plasma membrane synthesis in the macrophage following phagocytosis of polystyrene latex particles. *J. biol. Chem.*, **247**, 2439–2446.

WHITNEY, D. D. (1907). Artificial removal of the green bodies of *Hydra viridis*. *Biol. Bull. mar. biol. Lab. Woods Hole*, **13**, 291–299.

WOOD, R. L. (1959). Intercellular attachment in the epithelium of *Hydra* as revealed by electron microscopy. *J. biophys. biochem. Cytol.*, **6**, 343–352.

ZUCKER-FRANKLIN, D. & HIRSCH, J. G. (1964). Electron microscope studies on the degranulation of rabbit peritoneal leukocytes during phagocytosis. *J. exp. Med.*, **120**, 569–576.

EXPLANATION OF PLATES

Plates 1–2a show sections of aposymbiotic *H. viridis* (Fla.) hydra injected with hydra symbionts. Plate 2b shows a section of a normal green *H. viridis* (Fla.) hydra.

PLATE 1

(a) Low-power section through tip of digestive cell fixed 10 minutes after injection of E/F algae. An algal cell (top left) is free in the coelenteron (c) and another is adjacent to digestive cell membrane. × 6000

(b) Higher magnification of section indicated by arrow in (a). Algal cell (a) at upper left with cell wall visible as heavy black band. Parallel to wall is the microvillus (mv) cell membrane showing increased electron opacity. × 90 000

(c) Section through tip of digestive cell fixed 10 minutes after injection of Florida algae showing a microvillus (mv) or possibly a section through a veil of cytoplasm engulfing the algal cell. Discoidal coated vesicles are abundant (at arrows) in the digestive cell cytoplasm; c, coelenteron. × 18 200

PLATE 2

(a) Section through a digestive cell fixed 10 minutes after injection of Florida algae. Three center algae have just been engulfed and are already sequestered into individual vacuoles. Another alga is still in the coelenteron (c) and adjacent to the digestive cell membrane (upper right). Discoidal coated vesicles at arrows. × 26 000 (*b*) Cross-section through body wall showing endoderm (endo), with algae in vacuoles at base of cells, mesolamella (mes) and ectoderm (ecto). × 7000

PLATE 3

Phase contrast photomicrographs of digestive cells macerated from aposymbiotic *H. viridis* (Fla.) at various times after injection of Florida algae. Phase optics show algae as bright discs. Digestive cells are oriented with distal absorptive end up: (*a*) 30 s after injection showing algae close to tip of cell; (*b*) 10 min, algae are within the digestive cell as a result of phagocytosis; (*c*) 30 min, algae migrating to the base of the cell; (*d*) 5 h, algae now in normal basal location (from Pardy & Muscatine, 1973).

PLATE 4

Brightfield photomicrograph of a digestive cell macerated from an aposymbiotic *H. viridis* (Fla.) hydra 10 hours after injection of *C. vulgaris*. Cell fixed and stained with aqueous toluidine blue (0.25 %), oriented with distal absorptive end up. Algae appear as black discs; n = nucleus.

SYMBIOSIS IN
CONVOLUTA ROSCOFFENSIS

BY P. M. HOLLIGAN* AND G. W. GOODAY†

* Marine Biological Association, Plymouth PL1 2PB
† Department of Biochemistry, University of Aberdeen,
Aberdeen AB9 1AS

CONVOLUTA ROSCOFFENSIS: A PLANT-ANIMAL

Convoluta roscoffensis Graff is a free-living marine flatworm, 2–4 mm in length, that occurs in large colonies on beaches of coastal sand in the Channel Islands and along the coast of western France. It is a member of the simplest group of ciliated Platyhelminthes, the acoel turbellarians, in which the gut is formed from a syncytium of endodermal cells. In the adult animal numerous cells of an algal symbiont, *Platymonas convolutae* Parke et Manton (Class Prasinophyceae), are present between the epidermal and subepidermal tissues and give the whole organism a deep green colour. Although many other examples of algal–metazoan symbioses are known (Droop, 1963), *C. roscoffensis* is unique in that it ceases to ingest particulate food once the symbiont has been acquired and the symbiotic relationship established. The species may be grown through many generations in a seawater medium containing nitrate, phosphate, minerals and vitamins in the presence of light and carbon dioxide (Dorey, 1965; Provasoli, Yamasu & Manton, 1968). Thus the energy and organic substances required by the whole organism for growth, movement and reproduction can be totally supplied by algal metabolism. Any suggestion that bacteria might be an integral part of the symbiosis, possibly by providing micronutrients, appears to be ruled out by the successful axenic culture of *C. roscoffensis* (L. Provasoli, cited by Taylor, 1973).

Convoluta roscoffensis has attracted the attention of biologists since the latter half of the last century and, in a delightful monograph, Keeble (1910) summarized much of the early work on the species and the related *Convoluta convoluta* (Abildgaard) (=*C. paradoxa* Örsted). Geddes (1879, 1882) had shown that the green pigment has the properties of chlorophyll and that, when illuminated, the animals accumulate starch within the green 'cells' and evolve oxygen. However, he concluded that the green 'cells' were not algae but chloroplasts, and these were later considered by Haberlandt (1891) to be transmitted to successive generations of worms as colourless leucoplasts. Keeble & Gamble provided the first accurate and detailed account of the histology of the organism and the cytological relationship between alga and animal (Gamble & Keeble, 1904; Keeble & Gamble, 1907).

The behaviour of *C. roscoffensis* has been described by Keeble (1910) and Bohn & Drzewina (1928). The worms live in the mid-tidal zone and, during daylight hours, migrate upwards just after the tide recedes to lie on the surface of the beach. They are then exposed to the light for a few hours before burrowing back into the substratum as the tide rises again. Bohn (1903) and Keeble (1910) found that the animals will continue to move up and down through sand in time with the tides when they are kept in the laboratory under illumination and that this 'memory' of the tidal rhythm may last for several days. *C. roscoffensis* is positively phototactic, turning immediately towards a source of light and accumulating on the white half of an illuminated black and white background (Keeble & Gamble, 1907). Also the species shows a strong positive geotaxis and readily disappears into the beach when disturbed (Fraenkel, 1929).

On the beach the worms become aggregated in drainage channels and in shallow pools around stones. Within minutes of appearing on the surface, the colonies, each of which consists of a single layer of closely spaced worms, are enveloped in a mass of translucent mucus that may become buoyed up by trapped bubbles of oxygen (Geddes, 1879). This extracellular gel is released from epidermal gland cells and although its functions are not known several possible roles may be suggested: (*a*) It holds the animals together in large masses without any noticeable restriction on their movement, and prevents the loss of individuals that would otherwise be swept down the beach with the flow of drainage water. Also when the rising tide reaches the *Convoluta* colonies, the worms that are still on the surface are protected by the mucus layer to which sand carried on the wave fronts adheres. (*b*) Fraenkel (1961) suggested that the mucus may increase the rate of photosynthesis by raising the animals above the substratum so that they are exposed to light reflected from the sand beneath as well as to direct light from above. (*c*) The mucus could have adsorption or ion-exchange properties that help to control the chemical environment of the animals and to protect them from any rapid changes in the composition and salinity of the surrounding water. (*d*) The mucus may prevent desiccation of the colonies, especially on hot summer days. (*e*) Finally, as has been proposed for Acoels in general, the mucus may aid locomotion and adhesion of the worms (Oschman, 1967). The animals leave a slime trail when moving over a flat surface, but it is not known whether this is similar to the mucus seen on the beach.

In the laboratory, mucus is produced in visible amounts only when large numbers of *C. roscoffensis* are exposed to sunlight (Fraenkel, 1961) and therefore seems to be an indirect product of photosynthesis equivalent to the mucus of chloroplast-containing molluscs (Trench, Trench & Muscatine, 1972). Worms that have been kept at lower light intensities produce fluffy clumps of mucus-like material when agitated by swirling and, although this may or may not be the same as the substance produced on the beach, it does become labelled after exposure to radioactive bicarbonate (Muscatine, Boyle & Smith, 1974). Little is known about the chemical composition of the mucus, but it is completely dispersed when shaken in sea water. Histochemical staining of the mucus gland cells of *C. convoluta* indicates that they contain acidic mucopolysaccharides with carboxyl and sulphate groups (Pedersen, 1964, 1965).

As described by Gamble & Keeble (1904) and Keeble & Gamble (1907), the eggs of *Convoluta roscoffensis* are released at the onset of the spring tides and after a few days hatch out as active, ciliated worms which feed on diatoms and other algae. Although this food is digested, the animals grow and mature only after infection by *Platymonas convolutae*. The swimming cells of the symbiont are attracted by and settle on the egg cases (Keeble & Gamble, 1907), and the latter, when placed in a suspension of the alga, become coated by settled cells within five minutes (Gooday, 1975). However, this phenomenon is not species specific as a similar response is shown by *P. tetrathele*. Also on *Convoluta* egg cases collected in the Channel Islands we have found a variety of organisms, including diatoms, amoebae and a *Labyrinthula* sp. Nevertheless, the chances of the young worms ingesting the natural symbiont must be enhanced by this chemotactic response of the algal cells towards the egg cases, as well as the habit of *P. convolutae* in remaining non-motile within parental thecae for a long time after mitotic division (Gooday, 1970).

Oschman (1966) described the early stages of the incorporation of cells of the symbiont into the animal tissues as seen under the electron microscope. The flagella and then the algal theca (which determines the shape of free-living cells) are soon lost and the alga assumes an interdigitated form between animal cells. The theca, which by analogy to *P. tetrathele* is chiefly of pectin-like material with galactose and galacturonic acid as major constituents (Gooday, 1971a), is probably degraded by enzymatic lysis. Oschman has also observed algal cells within vacuoles in digestive parenchyma cells, and it is possible that symbionts in the epidermal tissues have to pass through an intracellular phase. Indeed, Dorey (1965) described the algal cells as being enclosed by a layer of animal cytoplasm within extensions of the digestive parenchyma.

The process of 'greening' has been investigated in detail by Provasoli *et al.* (1968) who showed that an exposure of one hour to the most effective clone of *P. convolutae* was sufficient for all of a batch of aposymbiotic worms to develop the symbiotic condition within five days. After four weeks these animals had increased in length up to six times, whereas the uninfected controls had not grown at all. In another experiment two white worms became green after each had ingested just two algal cells. Also other species of related algae, particularly *Prasinocladus* spp., were shown to form a symbiotic relationship with *C. roscoffensis*. However, these were generally less efficient than the natural symbiont in that longer exposure times were required for 100 % 'greening' and the infected worm grew more slowly despite equivalent rates of photosynthesis (Nozawa, Taylor & Provasoli, 1972). The most intriguing result of the experiments was the discovery of a 'pecking order' amongst algal species able to act as symbionts which results in the replacement of an unnatural symbiont when the worms are presented with *P. convolutae* or a more 'competent' unnatural symbiont. Temporary chimaeral conditions were observed under the electron microscope, with cells of different algal species lying adjacent to one another in the animal tissue.

Although the majority of algal cells in the animal appear viable and capable of division, digestion of the cells can take place. This phenomenon has been reported for old animals in natural populations (Keeble & Gamble, 1907) and for worms in culture (Dorey, personal communication). Also the algae may be ejected by being squeezed out between the epidermal cells of the animal.

Aposymbiotic worms can be obtained from laboratory cultures of *C. roscoffensis* by slitting open the egg cases, washing the eggs and allowing them to hatch in sterile media (Provasoli *et al.*, 1968). Also, adult animals extrude their algae after carbon dioxide has been bubbled through the medium for a few minutes (Bohn & Drzewina, 1928; Boyle & Smith, 1975), and *P. convolutae* grows readily in an artificial or enriched seawater medium (Gooday, 1970). Thus for physiological studies on *C. roscoffensis*, the following tissues may be compared experimentally: symbiotic adult worms, aposymbiotic juvenile worms, adult worms freed of algae, algae released from adult worms by careful homogenisation or by treatment with carbon dioxide, and finally algae grown in pure culture.

UPTAKE AND ASSIMILATION OF NUTRIENTS

Relatively little attention has been given to the nutrition of *Convoluta roscoffensis* in its natural habitat, due mainly to the ease with which the species can be cultured for laboratory experiments. Natural populations of worms, as they move from within the beach substratum to the surface and back again, are frequently exposed to marked and rapid changes in the temperature, salinity and composition of the surrounding water, as well as in illumination. For example, on hot summer days at low tide, the temperature of shallow pools containing *C. roscoffensis* exceeds that of sea water by as much as 15 °C, and in the winter the animals have been seen lying on snow (Dorey, personal communication). Heavy rain must cause considerable dilution of salts and other constituents in the drainage water. Even on days of average weather conditions changes occur in the environment of the worms (Table 1) that could have a significant effect on the assimilation of external nutrients.

Table 1. *Two examples of the variation in environmental conditions which affect colonies of* Convoluta roscoffensis *at low tide on Herm, Channel Islands*

Date	Period of observations	Air temp.	Sea temp.	Beach drainage water Temp. (°C)	pH	Salinity ($^0/_{00}$)	Incident light intensity[a] (mW cm^{-2})
9 April 1974	12.30– 18.30	13.1– 14.2	11.0	11.5– 14.7	8.0– 7.8	35.8– 34.4	16–25
18 June 1974	09.30– 14.30	15.1– 17.3	14.7	14.8– 20.9	8.0– 6.4	35.0– 34.2	12–42

[a] Determined with a Lambda L1–192S Quantum Sensor for the 400–700 nm wavelength range.

The levels of dissolved nutrients in the interstitial and drainage waters of beaches where *C. roscoffensis* occurs have not previously been investigated. Small metazoans, diatoms (Guérin, 1960) and probably bacteria are the most abundant associated organisms. From experiments with columns of sand, McIntyre, Munro & Steele (1970) concluded that in a sandy beach ecosystem the meiofaunal population feeds largely on bacteria, and the bacteria assimilate organic substances in solution and release nitrate.

Carbon dioxide

Little is known about the effects of changes in the concentration of dissolved carbon dioxide and in light intensity on the rate of carbon fixation by adult *Convoluta roscoffensis* under field conditions.

Carbon dioxide fixation in many species of marine pelagic phytoplankton is markedly inhibited by prolonged exposure to high light intensities (Hellebust, 1970), and data given by Ryther (1956) for a *Platymonas* sp. show that the rate of net photosynthesis under illumination equivalent to full sunlight was reduced to 5–10 % of that at light saturation. However, the photosynthetic efficiency of certain species of intertidal benthic marine diatoms decreases only by about 10 % after 1 h in sunlight (Taylor, W. R., 1964), suggesting that algae normally exposed to bright light are adapted to withstand the potentially harmful effects of direct solar radiation. The same appears to be true for symbiotic *Platymonas convolutae* as observations on wild populations of *C. roscoffensis* indicate that mucus production and oxygen evolution (seen as bubbles trapped in mucus) are maximal in bright light, and also the animals show no evasive response to sunshine. However, when worms that have been kept under low light intensities in the laboratory for several weeks are placed in the sun, the rate at which they fix carbon dioxide falls by about 40 % within 3 h (Holligan & Gooday, unpublished results). It is possible that the mucus layers produced on the beach, or even the epidermal pigments of the animals, have photoprotective functions; absorption spectra for samples of mucus, much diluted in sea water, show extinction maxima of 0.05 at 261 and 268 nm and 0.04 at 327 nm. The absorbing substances diffuse through a dialysis membrane.

At the normal pH and alkalinity of sea water, carbon dioxide is present mainly as bicarbonate and the total level in solution is in the range 2.0 to 2.2 mmol l^{-1} (Harvey, 1955). However, pH values as low as 6.4 for beach drainage water (Table 1) and 5.5 for the tissue fluids of *C. roscoffensis* (Taylor, 1971*a*) have been recorded. At these values dissolved carbon dioxide will be mainly in the form of carbonic acid rather than bicarbonate, and will be reduced in total amount by as much as one order of magnitude (Whitfield, 1974). The rate of photosynthesis by *C. roscoffensis* at pH 5.5 is approximately twice that at pH 7.8 under a light intensity equivalent to 20 % sunlight (Boyle & Smith, 1975), but these measurements were made with laboratory cultures that had been kept under very low illumination. The re-assimilation of carbon dioxide derived from animal respiration by algal photosynthesis has not yet been investigated.

Adult *C. roscoffensis* and freshly isolated cells of *P. convolutae* assimilate carbon dioxide in the dark, the major products being organic and amino acids (Muscatine *et al.*, 1974). Although the observed rates of dark fixation are less than 5 % of those in the light, this process could provide a significant proportion of the total carbon dioxide assimilated by the worms in their natural habitat where they are in darkness for an average of 19 hours each day. As non-symbiotic turbellarians also utilise carbon dioxide (Hammen & Lum, 1962), the animal as well as the algal cells of *C. roscoffensis* may have the capacity for dark fixation.

Products of photosynthesis

Mannitol and starch are the major products of photosynthesis in symbiotic *Platymonas convolutae* (Muscatine *et al.*, 1974) as in the free-living alga (Gooday, 1970; Phillips, 1970) and in other species of the Prasinophyceae (Craigie *et al.*, 1966, Craigie, McLachlan, Ackman & Tocher, 1967). The starch, which is produced as grains in the chloroplast and as a shell around the pyrenoid (Parke & Manton, 1967), is similar to that found in the Chlorophyceae (Phillips, 1970). Both compounds are also important chemical constituents of *C. roscoffensis*; mannitol (determined as the trimethylsilyl derivative by gas–liquid chromatography) comprising about 1 % and starch (determined as glucose following degradation of the water-insoluble glucan by amyloglucosidase) about 4.5 % of the total dry weight of worms collected on the beach. The presence of a major pool of mannitol in the algae may perhaps be compared with the central role of polyols suggested by Smith (this volume) for lichens, in providing a 'physiological buffer' to environmental stresses such as osmotic shock. However, due to the variability between samples, no significant changes in the absolute levels of each have been detected in the animals during their natural light/dark cycle, and it is only from laboratory experiments with $^{14}CO_2$ that information about their metabolic role has been acquired.

Data on the incorporation of radioactive carbon into various fractions of *C. roscoffensis* after incubation in sea water containing $NaH^{14}CO_3$ for 4 h are presented in Table 2. Of the three samples of worms kept in the light, one was taken for analysis immediately after the ^{14}C-feeding period, and the others after a further 18 h in the light or dark in non-labelled sea water.

After 4 h in the light, about 32 % of the soluble radioactivity in the worms was present as mannitol, a value similar to that reported by Muscatine *et al.* (1974), and a further 10 % as amino and organic acids. Subsequently the proportion of label in the hexitol declined, especially in the light, whereas that in the amino and organic acid fractions fell only slightly in the light and increased in the dark. Radioactivity in the starch component was stable in the light but was reduced gradually in the dark, indicating that starch is accumulated as long as photosynthesis continues but is degraded slowly when *C. roscoffensis* is placed in the dark. Observations made in Brittany by Keeble & Gamble (1907) support this conclusion. By staining animals with iodine, they described how the starch grains gradually disappear over a period of two weeks in darkness but became detectable again within minutes of illumination. Also they estimated the starch content to be higher during periods of the spring tides when low water is in the middle of the day, and suggested that this was due to

Table 2. [^{14}C]Carbon dioxide fixation by Convoluta roscoffensis[a]

| | Treatment | | | |
| | % Total radioactivity incorporated by worms[c] | | | |
Fraction	4 h light	4 h light + 18 h light	4 h light + 18 h dark	4 h dark
Chloroform–methanol–water extract[b]				
Lipids and pigments	13.3	15.7	19.2	13.9
Mannitol	12.3	2.5	6.5	}38.6
Organic and amino acids	3.9	3.1	5.0	
Water extract	8.3	15.8	16.2	14.3
Starch (amyloglucosidase hydrolsate)	57.4	56.1	42.8	20.5
Insoluble residue	4.8	6.8	10.3	12.7
Total ^{14}C incorporated (dpm $\times 10^{-3}$)	1018	977	836	26

[a] The conditions of the experiment were as follows: for each treatment 100 adult worms were incubated in filtered sea water plus 10 μCi NaH$^{14}CO_3$ at 15 °C and exposed to light (4.5 mW cm^{-2}) as indicated. The worms were then washed twice in sterile sea water before being killed by immersion in the chloroform–methanol–water (5–12–3 v/v) extractant.

[b] Partitioned to give chloroform and aqueous methanol phases. The latter was then concentrated and resolved into neutral, acidic and basic fractions by ion exchange chromatography. >90% of the radioactivity in the neutral fraction was recovered as mannitol.

[c] Radioactivity in each fraction was assayed by scintillation counting in a toluene–Triton X100 (2–1 v/v) fluor.

the worms being exposed to higher light intensities than during neap tides when low water is in the early morning and late evening. In *P. tetrathele* starch is also utilised only slowly in the dark (Gooday, 1971b) and appears to act chiefly as a substrate for the synthesis of thecal polysaccharides, rather than for respiration.

The reduction in radioactivity in starch and mannitol after the incubation period with $^{14}CO_2$ is accompanied by continued incorporation into the lipid, and water-soluble and water-insoluble fractions. Only the water-soluble component has been further analysed. After concentration by evaporation and the addition of two volumes of ethanol, this yields a white precipitate which can be partially degraded to glucose by treatment with amyloglucosidase. Radioactivity in the glucose is barely detectable after 4 h but forms a major proportion of the label in the water soluble fraction after 18 h in the light and dark. This water soluble α-glucan is therefore probably mainly glycogen from animal cells.

In laboratory experiments with *C. roscoffensis*, Muscatine *et al.* (1974) showed that animal mucus and eggs also become radioactive following the photosynthetic assimilation of $^{14}CO_2$. That the mucus holding the animals together on the beach is also an indirect product of photosynthesis has been suggested by our field experiments in which two samples each of about 1000 worms were washed in several changes of filtered, fresh sea water and incubated at 20 °C in Universal bottles containing 10 cm³ sea water, one in sunlight and one in the dark. When examined after 2 h, only the worms in the sunlight were surrounded by mucus and only their seawater medium gave a white gelatinous precipitate when acetone was dripped into the bottle. Presence or absence of mucus at the beginning of the light period has little or no effect on subsequent photosynthesis as washed worms and worms transferred to bottles with their mucus intact have shown the same rate and pattern of $^{14}CO_2$ fixation.

Nitrogenous compounds

On a beach in the Channel Islands levels of both nitrate and dissolved organic nitrogen in water flowing over colonies of *Convoluta roscoffensis* at low tide are high compared with those in local sea water (Tables 3 and 4). Rates of uptake could not be determined due to difficulties in making quantitative measurements of both the flow of water and the density of worms. However, samples of drainage water from above and below a

Table 3. *Nitrogen content of samples of sea and beach drainage water collected from a site for* Convoluta roscoffensis *on Herm, Channel Islands, 9 April 1974*

		Nitrogen ($\mu g\ l^{-1}$)[a]			
Water type	Time	NH₃	NO₃	Organic	Total
Sea	09.30 (high tide)	6.0	113.6	22.0	141.6
Sea	12.30	8.8	246.5	24.4	279.7
Beach[b]	12.30	6.3	279.3	209.0	494.6
Beach[b]	15.30 (low tide)	32.4	253.0	362.8	648.2
Beach[b]	18.30	8.1	192.1	90.5	290.7
Sea	18.30	7.4	113.3	82.1	202.8
Change in concentration during flow across colony[c]	15.30 (low tide)	−24.2	−35.8	−259.0	−319.0

[a] NH₃ determined by the method of Solorzano (1969). NO₃ determined by the method of Wood, Armstrong & Richards (1967). Total nitrogen determined by the method of Armstrong & Tibbitts (1968). Organic nitrogen estimated by subtraction of (NH₃ + NO₃) from total nitrogen.

[b] Samples collected from above colonies of *Convoluta*.

[c] Determined by comparison of sample collected above *Convoluta* colony with one taken at the same time from below colony.

Table 4. *Nitrogen and phosphorus content of samples of sea and beach drainage water collected from a site for Convoluta roscoffensis on Herm, Channel Islands, 18 June 1974*

Water type	Time	Nitrogen (µg l^{-1})[a]				Total phosphorus (µg l^{-1})[a]
		NH$_3$	NO$_3$	Organic	Total	
Sea	09.30	24.9	12.0	224.1	261.0	19.8
Beach[b] (mean and range of 11 samples taken at ½ h intervals)	09.30–14.30	22.0 (13.2 to 50.3)	273.8 (255.5 to 282.7)	276.2 (238.8 to 308.4)	571.9 (539.1 to 603.6)	36.1 (31.0 to 54.3)
Sea	14.30	34.6	46.1	197.3	278.0	19.2
Mean change in concentration during flow across colony[c] (range given for 6 estimates, made at 1 h intervals)	09.30–14.30	+12.3 (−33.1 to +108.8)	−72.0 (−2.9 to −154.7)	−94.3 (−48.7 to −190.3)	−154.0 (−70.4 to −254.4)	−14.1 (−5.2 to −34.0)

[a] Nitrogen determined by methods given in Table 3.
[b] Samples collected from above *Convoluta* colonies. No significant trends were observed during the low tide period.
[c] Determined by comparison of samples collected above *Convoluta* colonies with ones taken at the same time from below colony.
[d] Determined by the method of Murphy & Riley (1962) as modified by Armstrong & Tibbitts (1968).

colony were analysed to establish whether either form of nitrogen is utilised by the worms. Although it was not possible to ensure that the samples were derived from water of the same origin and initial chemical composition, the preliminary results (Table 3) indicated that both nitrate and organic nitrogen are taken up and that, as no accumulation of nitrogen could be detected in the water and mucus immediately surrounding the worms, the differences represent net assimilation. The second series of observations which were made at hourly intervals confirmed that the animals were utilising, in approximately equal amounts, nitrate and organic nitrogen (Table 4). Determinations of free amino acids in samples of the same drainage water, using the method of Troll & Cannan (1953) after desalting with a chelating ion exchange resin (Webb & Wood, 1967), showed that amino nitrogen at concentrations of 15 to 25 μmol l^{-1} accounted for more than 90 % of the organic nitrogen fraction. The only amino acids that could be positively identified by thin layer chromatography were glycine and serine. By contrast, levels of ammonium nitrogen were low, and the data, though variable, suggest that, when *C. roscoffensis* is exposed to the light, ammonium is not assimilated and may even be excreted.

The net nitrogen requirement of *Convoluta roscoffensis* cultured in media containing nitrate as the sole source of nitrogen must be met by the synthesis of amino acids in algal cells since nitrate reductase is an enzyme known only from plants and microbes. However, adult flatworms can absorb amino acids from concentrations as low as 1 μmol l^{-1} (Read, 1968), and aposymbiotic worms, symbiotic worms and isolated *P. convolutae* metabolise alanine (Boyle & Smith, 1975). Furthermore, nitrogen-deficient cells of other *Platymonas* spp. utilise a number of amino acids at concentrations of 1 μmol l^{-1} or less (North & Stephens, 1971; Wheeler, North & Stephens, 1974). Although the assimilation of organic substrates by marine invertebrates does not necessarily imply a net influx (Johannes, Coward & Webb, 1969), the evidence from field and laboratory investigations indicates that *C. roscoffensis* can and does utilise amino acids as well as nitrate. The most probable source of amino acids in the beach ecosystem is excretion by the meiofauna (see Johannes & Webb, 1970).

The uptake of both nitrate and organic nitrogen by the worms *in situ* on the beach does present a paradox, as nitrate reductase is usually repressed by the presence of reduced compounds of nitrogen such as ammonia and amino acids. Thus the occurrence of this enzyme in wild populations of *Convoluta roscoffensis* exposed to light on the beach surface (Gooday & Holligan, unpublished results) suggests that the algal cells are utilising nitrate but that the amino acids are unavailable to them, perhaps because the animal epidermal cells are retaining this organic nitrogen.

However, this need not exclude the possibility that amino acids could be used by the algal cells at some time, as certain marine phytoplankton show a diurnal periodicity in nitrate reductase activity (Collos & Lewin, 1974), and so reduced forms of nitrogen, perhaps excreted from the animal cells, could be utilised if the nitrate reductase activity declined during the long dark period that the worms experience each day.

Other nutrients

The concentration of total phosphorus decreases in the beach drainage water during flow across a colony (Table 4), and it is probable that phosphate, other inorganic ions and various micronutrients, are assimilated by *C. roscoffensis* in its natural habitat. Although the worms are able to incorporate a range of organic compounds (Meyer & Meyer, 1972; Taylor, 1974) there is no evidence that such substrates are ever present in sufficiently high concentrations to represent a significant source of nutrients.

BIOCHEMICAL RELATIONSHIP BETWEEN THE SYMBIONTS

Fitness for symbiosis of plant and animal

The acquisition and accommodation of algal symbionts by acoels appears to depend on the ingestion of living cells directly into a central mass of endodermal cells and the ability of the algae to resist digestion. Three totally unrelated algae are known to form symbiotic associations with these animals: the prasinophycean alga *Platymonas convolutae* with *Convoluta roscoffensis*; the diatom, *Licmophora* sp., with *C. convoluta* (Ax & Apelt, 1965; Apelt, 1969); and the dinoflagellate, *Amphidinium klebsii* Kofoid *et* Swezy, with *Amphiscolops langerhasi* (Graff) (Taylor, 1971*b*). Also, as described earlier, several other species of the Prasinophyceae are capable of establishing a successful symbiosis with *C. roscoffensis* (Provasoli *et al.*, 1968; Nozawa *et al.*, 1972).

Gooday (1970) investigated some aspects of the physiology of *P. convolutae* and reported that, in comparison with its free-living relative, *P. tetrathele*, it is poorly permeable to a range of carbohydrates and other compounds. It was suggested that this property is of importance to the symbiotic relationship in providing a barrier between, for example, the mannitol of the alga and the glucose of the animal, so allowing each symbiont to control its own internal pools of metabolites. As suggested by Droop (1963) poor permeability may also be an advantage to an alga in an environment as rich in possible toxic substances as the animal tissues.

Cytological relationships between symbionts

There was little understanding of the details of the cytological relationship between alga and animal in *C. roscoffensis* before the advent of electron microscopy, and even the cellular constitution of the acoels themselves has been a matter of controversy (Dorey, 1965). Gamble & Keeble (1904) and Keeble & Gamble (1907) described the presence of a chloroplast, nucleus and pyrenoid in the symbiotic alga but, contrary to modern reports, stated that in the early stages the infecting cell is colourless and, in later stages, tends to lose its nucleus and is eventually digested. The following description of *P. convolutae* in the animal, as compared to the free-living organism, is based on electron microscope descriptions of *C. roscoffensis* (Dorey, personal communication; Oschman & Gray, 1965; Oschman, 1966; Parke & Manton, 1967; Provasoli *et al.*, 1968) and of the very similar *Convoluta psammophila* Bekl. (Sarfatti & Bedini, 1965).

(*a*) The theca is lost after ingestion. The resultant protoplast becomes irregular in outline, with finger-like projections which contain lobes of chloroplast lying between epidermal, muscle and other subepidermal cells of the animal. This produces a large area of contact between each algal cell and the surrounding animal cells. The algae are also characteristically arranged so that the anterior ends, with the flagellar bases and the nuclei, point inwards and the ramifications of the chloroplasts face the outside of the animal. As division in culture produces daughter cells pointing in opposite directions, this orderly array of algal cells would seem to require re-arrangement of the cells following division within the worms.

(*b*) The structures associated with locomotion and phototaxis – the four flagella, their flagellar pit, and the eyespot which is normally towards the anterior end of one of the plastid lobes – are also lost, although the four flagellar roots are still present.

(*c*) The Golgi bodies are smaller and appear quiescent. In free-living *P. convolutae* and *P. tetrathele* the Golgi bodies are very active during cell division, when they are packed with particles that are released to form the extracellular theca (Manton & Parke, 1965; Parke & Manton, 1967; Gooday, 1971*a*). The theca, which is probably composed chiefly of polysaccharide (Gooday, 1971*a*; Manton, Oates & Gooday, 1973), is not formed following division of the algal cells in the animal and there is no sign of any release of thecal precursors to the outside of the cell via the Golgi bodies.

(*d*) The mitochondria have fewer and smaller cristae and are less electron-dense. This suggests that the algae may have a lowered capacity for oxidative electron transfer when in the animal tissue. By contrast, the adjacent animal mitochondria have many cristae.

Transfer of metabolites between symbionts

The identification of metabolites that are transferred between symbionts must be based on three major criteria: (*a*) synthesis by the donor; (*b*) release from the donor; and (*c*) assimilation by the recipient.

By analogy with other symbiotic systems and as there is no evidence to the contrary, we presuppose that the transfer of nutrients between alga and animal in *C. roscoffensis* involves soluble substances of low molecular weight and that assimilation of macromolecules plays no major role in the nutrition of either symbiont. However, there is no reason to assume that, as in certain other symbioses (Smith, Muscatine & Lewis, 1969), the transfer of nutrients is essentially uni-directional and confined to one or just a few metabolites. The animal cells of *C. roscoffensis* are completely dependent on the algal symbionts and probably utilise a whole range of algal products.

Radioactive carbon, generally in the form of $NaH^{14}CO_3$ as a substrate for carbon fixation in the light, is widely employed in studies on symbiotic organisms as a means of identifying and quantifying small amounts of organic substances that are transferred from autotrophic to heterotrophic cells. In such experiments the possible transfer of minor photosynthetic products from autotroph to heterotroph and of substances from heterotroph to autotroph is easily underestimated or overlooked, even though these may form an important part of the net interchange of metabolites. This is due to the relatively slow rate at which ^{14}C becomes incorporated into cell constituents other than the major products of photosynthesis. Although valuable data concerning synthesis and assimilation come from studies of the symbionts in isolation, investigations of the release of metabolites must be based on experiments with the intact associations since marked changes in the control and nature of excretion from cells, and in particular from those of the autotrophic partner, may occur following both the establishment and the disruption of the symbiosis. The 'inhibition' technique, which was originally developed for work on lichens (Drew & Smith, 1967; Smith, 1973, this volume) can be used for isolating metabolites that are released in symbiotic associations. However, care must be taken to distinguish the movement of substances between plant and animal cells from that between animal cells.

Muscatine *et al.* (1974) investigated the movement of photosynthetically fixed carbon from alga to animal in *C. roscoffensis*. Inhibition experiments, in which whole worms were incubated in 1 % solutions of a wide range of organic compounds in the presence of $NaH^{14}CO_3$ in the light, showed that alanine was the most effective substance in stimulating the release of radioactivity into the medium, followed by glycine, serine, pyruvic acid and the amides, glutamine and asparagine. The labelled compounds found in the alanine inhibition medium were alanine and, in lesser amounts, glycine, serine and glutamine. These were also the only excretory products detected from carefully crushed worms, algae isolated from homogenised worms, and algae ejected from worms that had been treated with carbon dioxide (Boyle & Smith, 1975). However, on account of the delay between the incorporation of ^{14}C into alanine during photosynthesis and the appearance of radioactive alanine in the inhibition medium, Muscatine et al. (1974) concluded that this amino acid is transferred between animal cells rather than from alga to animal. Although they estimated from inhibition experiments that only about 8% of the total fixed carbon is moved from alga to animal, the proportion of photosynthate remaining in worms that are induced to eject all their algae by treatment with carbon dioxide indicates that as much as 50% of the products of photosynthesis are incorporated into animal tissues (Boyle & Smith, 1975). Despite the fact that the inhibition experiments appear not to provide a quantitative measurement of the movement of carbon from alga to animal, Boyle & Smith considered that amino acids are the major form in which carbon is transferred, and as a result of observations on the metabolism of glutamic acid and glutamine suggested that the latter is the most important individual compound involved.

Taylor (1974) stated that the principal metabolites excreted by symbiotic cells of *P. convolutae* include mannitol, fructose and glucose, as well as lactic acid. However, mannitol despite being a major product of photosynthesis is excreted into the medium of pure cultures of algae in very small (0.06% of the total carbon fixed in 12 h) (Taylor, 1971a) or undetectable amounts (Gooday, 1970) and the presence of hexoses in *P. convolutae* has never been demonstrated. Also data from the inhibition experiments carried out by Muscatine et al. (1974) provide no evidence for the release of any of these carbohydrates in adult symbiotic worms. Although aposymbiotic worms appear able to metabolise mannitol (Taylor, 1974), assimilation by adults has been reported as negligible (Muscatine et al., 1974).

By contrast, lactic acid, a characteristic product of anaerobic glycolysis, is synthesised and released into the medium by both *P. convolutae* (Gooday, 1970) and *C. roscoffensis* (Boyle & Smith, 1975). In a typical experiment with a 7 day old culture of *P. convolutae* (grown at 20 °C and 5000 lx under a 16 h light + 8 h dark regime), a concentration of 2.2 mmol l^{-1} lactate was found in the medium and cells. It is found in large amounts as an excretory product of the worms only when they are incubated in the dark or in the presence of DCMU, an inhibitor of photosynthesis. However, when the incubation vials are gassed with oxygen, the release of lactic acid into the medium ceases almost completely. Boyle & Smith considered therefore that this compound is a product of anaerobic metabolism in animal cells and support for their view comes from the ability of many non-symbiotic species of flatworms to excrete lactic acid (Read, 1968). As an end product of fermentation and a poor source of energy, lactic acid seems unlikely to be an important mobile substance in symbiotic associations.

The lipid metabolism of *C. roscoffensis* has been investigated by Meyer & Meyer (1972). The animal cells lack the ability to synthesise fatty acids and sterols, and appear to depend on the alga for supplies of these compounds. Aposymbiotic worms are unable to incorporate [14C]acetate into their sterols, and the sterols of young and adult animals are primarily phytosterols identical to those in *P. convolutae*. The animal fatty acids are either structurally identical or homologous to algal fatty acids. The labelling patterns following assimilation of [14C]acetate are consistent with the inability of animal cells to synthesise fatty acids *de novo* but indicate that they can carry out chain elongation and desaturation of algal fatty acids. In the dark the incorporation of [14C]acetate into the fatty acids of adult worms is much reduced, and Meyer & Meyer suggested that this reflects the relatively inactive metabolism of algal cells in the absence of light.

Uric acid is produced in animals as a degradation product of purines that are synthesised to remove excess nitrogen or arise from the turnover of nucleotides. In most animals it is further catabolised by uricase and successive enzymes to allantoin, allantoic acid, urea and ammonia. Florkin & Duchateau (1943), in a comparative study of the phylogenetic distribution of uricase, reported that the platyhelminthes do not have this enzyme and excrete uric acid. The acoels have no flame excretory system and accumulate insoluble urates. Young aposymbiotic *C. roscoffensis* become packed with small (< 1 μm long) crystals of uric acid which in polarised light are strongly birefringent and appear as twinkling silver particles. These have been characterised by degradation with uricase: the worms after a brief wash in acetone, were incubated in a drop of 0.05 mol l^{-1} Tris-HCl buffer, pH 8.3, containing 100 μg cm^{-3} uricase (Sigma Type II from bovine kidney) at 25 °C on a microscope slide with free access to air. After 5 min the crystals had completely disappeared. Controls with boiled enzyme or with the addition or the uricase inhibitor, 1×10^{-3} mol l^{-1} 2, 6, 8, trichloropurine, showed no dissolution of the crystals.

Keeble & Gamble (1907) suggested that the utilisation of the animal's uric acid by the algal cells could be an important factor in the nitrogen economy of *C. roscoffensis*. *P. convolutae* contains uricase and will grow with uric acid as a sole source of nitrogen, although not as fast as *P. tetrathele* (Gooday, 1970). Thus transfer of uric acid from animal to alga, with its subsequent breakdown to ammonia and re-assimilation of the nitrogen, almost certainly occurs. However, the importance of this process in the symbiotic organism is difficult to assess. Uric acid crystals can still be seen in healthy adult worms, and although they are much less abundant than in the aposymbiotic worms, uric acid may not be produced in such large quantities after infection by the algal symbiont. Boyle & Smith (1975) detected uricase in adult worms at a similar activity to that found by Gooday (1970) in pure cultures of *P. convolutae*: 3 nmol urate oxidised min^{-1} mg^{-1} extracted protein, compared with 3.5 and 10 nmol min^{-1} mg^{-1} algal protein for algae grown with nitrate and uric acid as sole nitrogen sources respectively. Boyle & Smith (1975) could not detect uric acid in adult worms by a standard assay procedure, even after incubation of worms in the presence of 2, 6, 8-trichloropurine (a treatment which caused death in about three hours). However, they did show that uric acid accumulated in animals incubated with the photosynthetic inhibitor, DCMU, for two weeks, and suggested that this results from a lack of energy for algal growth and protein metabolism.

P. convolutae also grows on ammonia as a sole nitrogen source (Gooday, 1970) and presumably readily utilises ammonia released by the degradation of uric acid. Ammonia is excreted by many flatworms (Simmons, 1970) and if produced by *C. roscoffensis* may also be transferred from animal to algal cells.

CONCLUSIONS

The experimental data concerning the synthesis, release and re-assimilation of metabolites that may be transferred between animal and plant cells in *C. roscoffensis* are summarised in Table 5. Based on the three criteria of synthesis and release by donor and utilisation by recipient, the major mobile substances are considered to be amino acids, amides, fatty acids and sterols from alga to animal, and uric acid from animal to alga. It is probable that further research will provide additions to this list which, although perhaps not of equal importance in terms of carbon and nitrogen fluxes, may be as vital for the mutual survival of the two symbionts. The available information on the assimilation of external nutrients by *C. roscoffensis* and the metabolism of substances transferred between the symbionts is illustrated diagrammatically in Fig. 1. For convenience, processes that are thought to occur mainly or entirely in the dark are shown separately. These schemes are clearly incomplete but draw attention

Table 5. *Some probable major metabolic interrelationships in* Convoluta roscoffensis

Metabolite	Synthesised by		Metabolised[a] by		Released[b] by alga	Probability[c] of net interflow	References
	Alga	Animal	Alga	Animal			
Mannitol	+	−	(+)	(+)	−	Nil	Gooday (1970), Taylor (1974), Muscatine *et al.* (1974), Boyle & Smith (1975)
Glucose	+	+	(+)	+	−	Nil	
Lactate	+	+	+	+	+	Low	Gooday (1970), Taylor (1971a, 1974), Boyle & Smith (1975)
Fatty acids	+	−	?	+	?	High	Meyer & Meyer (1972)
Sterols	+	−	?	+	?	High	
Amino acids	+	+	+	+	+	High	Read (1968), Gooday (1970), Muscatine *et al.* (1974), Boyle & Smith (1975)
Ammonia	+	+	+	−	−	?[d]	
Urate	+	+	+		−	High[d]	

[a] Utilisation shown by uptake experiments or inferred from metabolic patterns. The algae are impermeable to exogenous glucose and mannitol, and the animals may or may not utilise mannitol.

[b] Release from cultured or freshly isolated algae, or shown by 'inhibition' experiments.

[c] See text for rationale.

[d] Animal to alga, remainder are alga to animal.

+, Yes; −, no; ?, no evidence.

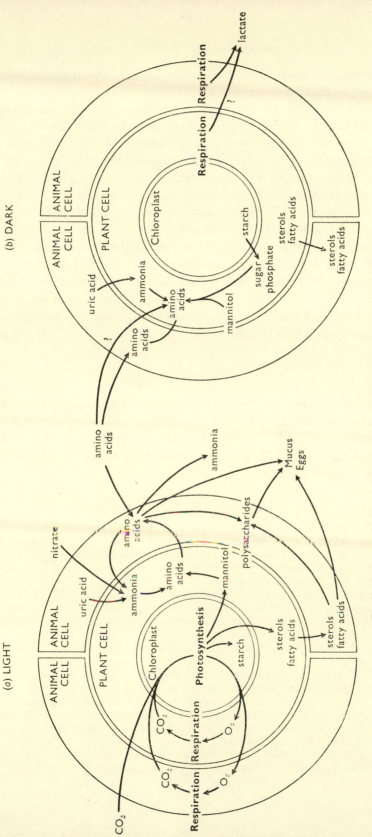

Fig. 1. Some probable major metabolic pathways in *Convoluta roscoffensis*.

to aspects of the biochemical relationship between the algal and animal cells which require further experimental investigation.

We have been primarily interested in the nutrition of *C. roscoffensis* in its natural habitat. During the summer months the beach drainage water is rich in nitrogen, phosphorus and probably other nutrients, and the worms appear to grow and reproduce rapidly. However, there are no comparable data for the winter months. At that time of year the rate of photosynthetic carbon fixation is probably lower due to less favourable light intensities and temperatures and the levels of dissolved nutrients in the beach ecosystem are likely to be much reduced. Survival of *C. roscoffensis* under such conditions would depend on the efficient retention of nutrients, such as nitrogen and phosphorus, by tight re-cycling between the symbionts. This could be achieved through an integration of intermediary metabolism that results from the transfer of specific metabolites from one symbiont to another in sufficient quantities to allow the continued growth and metabolism of both types of cells.

ACKNOWLEDGEMENTS

We thank R. N. Head for carrying out the seawater analyses, those who sent us reprints and unpublished information, and especially Professor D. C. Smith and Dr A. E. Dorey for valuable discussions. G. W. G. is grateful to the Carnegie Trust for research expenses to support the visit to Herm in April 1974.

REFERENCES

APELT, G. (1969). Die Symbiose zwischen dem acoelen Turbellar *Convoluta convoluta* und Diatomeen der Gattung *Licmophora*. *Mar. Biol.*, **3**, 165–187.

ARMSTRONG, F. A. J. & TIBBITTS, S. (1968). Photochemical combustion of organic matter in sea water for nitrogen, phosphorus and carbon determination. *J. mar. biol. Ass. UK*, **48**, 142–152.

AX, P. & APELT, G. (1965). Die 'Zooxanthellen' von *Convoluta convoluta* (Turbellaria Acoela) entstehen aus Diatomeen. *Naturwissenschaften*, **52**, 444–446.

BOHN, G. (1903). Sur les mouvements oscillatoires des *Convoluta roscoffensis*. *C. r. Acad. Sci.*, **137**, 576–578.

BOHN, G. & DRZEWINA, A. (1928). Les '*Convoluta*'. *Ann. Sci. nat.* (*Zool.*), **10**, 299–398.

BOYLE, J. E. & SMITH, D. C. (1975). Biochemical interactions between the symbionts of *Convoluta roscoffensis*. *Proc. R. Soc. Lond.* B **189**, 121–135.

COLLOS, Y. & LEWIN, J. (1974). Blooms of surf-zone diatoms along the coast of the Olympic peninsula, Washington. IV. Nitrate reductase activity in natural populations and laboratory cultures of *Chaetoceros armatum* and *Asterionella socialis*. *Mar. Biol.*, **25**, 213–221.

CRAIGIE, J. S., McLACHLAN, J., ACKMAN, R. G. & TOCHER, C. S. (1967). Photosynthesis in algae. III. Distribution of soluble carbohydrates and dimethyl-βpropiothetin in marine unicellular Chlorophyceae and Prasinophyceae. *Can. J. Bot.*, **45**, 1327–1334.

CRAIGIE, J. S., McLACHLAN, J., MAJAK, W., ACKMAN, R. G. & TOCHER, C. S. (1966). Photosynthesis in algae. II. Green algae with special reference to *Dunaliella* spp. and *Tetraselmis* spp. *Can. J. Bot.*, **44**, 1247–1254.

DOREY, A. E. (1965). The organization and replacement of the epidermis in acoelous turbellarians. *Q. Jl microscop. Sci.*, **106**, 147–172.

DREW, E. A. & SMITH, D. C. (1967). Studies in the physiology of lichens. VIII. Movement of glucose from alga to fungus during photosynthesis in the thallus of *Peltigera polydactyla*. *New Phytol.*, **66**, 389–400.

DROOP, M. R. (1963). Algae and invertebrates in symbiosis. *Symp. Soc. gen. Microbiol.*, **13**, 171–199.

FLORKIN, M. & DUCHATEAU, G. (1943). Les formes du système enzymatique de l'uricolyse et l'évolution du catabolisme purique chez les animaux. *Arch. int. Physiol.*, **53**, 267–307.

FRAENKEL, G. (1929). Über die Geotaxis von *Convoluta roscoffensis*. *Z. vergl. Physiol.*, **10**, 237–247.

—— (1961). Quelques observations sur le comportement de *Convoluta roscoffensis*. *Cah. Biol. mar.*, **2**, 155–160.

GAMBLE, F. W. & KEEBLE, F. (1904). The binomics of *Convoluta roscoffensis*, with special reference to its green cells. *Q. Jl. microscop. Sci.*, **47**, 363–431.

GEDDES, P. (1879). Observations on the physiology and histology of *Convoluta Schultzii*. *Proc. R. Soc. Lond.* B, **28**, 449–457.

—— (1882). On the nature and functions of the 'Yellow cells' of Radiolarians and Coelenterates. *Proc. R. Soc. Edin.*, **11**, 377–396.

GOODAY, G. W. (1970). A physiological comparison of the symbiotic alga *Platymonas convolutae* and its free living relatives. *J. mar. biol. Ass. UK*, **50**, 199–208.

—— (1971a). A biochemical and autoradiographic study of the role of the Golgi bodies in thecal formation in *Platymonas tetrathele*. *J. exp. Bot.*, **22**, 959–971.

—— (1971b). Control by light of starch degradation and cell-wall biosynthesis in *Platymonas tetrathele*. *Biochem. J.*, **123**, 3P.

—— (1975). Chemotaxis and chemotropism in fungi and algae. In *Primitive sensory and communication systems*, (ed. M. J. Carlile). London: Academic Press (in press).

GUÉRIN, M. (1960). Observations écologiques sur *Convoluta roscoffensis* Graf. *Cah. Biol. mar.*, **1**, 205–220.

HABERLANDT, G. (1891). Bau und Bedeutung der Chlorophyll-Zellen von *Convoluta roscoffensis*. In *Die Organisation der Turbellaria acoela*, 75–90 (ed. L. von Graff). Leipzig: Engelmann.

HAMMEN, C. S. & LUM, S. C. (1962). Carbon dioxide fixation in marine invertebrates. III. The main pathway in flatworms. *J. biol. Chem.*, **237**, 2419–2422.

HARVEY, H. W. (1955). *The Chemistry and Fertility of Sea Waters*. London: Cambridge University Press.

HELLEBUST, J. A. (1970). Light: plants. In *Marine Ecology*, vol. **1**, part 1, 125–158 (ed. O. Kinne). London: Wiley-Interscience.

JOHANNES, R. E., COWARD, S. J. & WEBB, K. L. (1969). Are dissolved amino acids an energy source for marine invertebrates? *Comp. Biochem. Physiol.*, **29**, 283–288.

JOHANNES, R. E. & WEBB, K. L. (1970). Release of dissolved organic compounds by marine and freshwater invertebrates. In *Symposium on organic matter in natural waters*, 257–273 (ed. D. W. Hood). Alaska: University of Alaska Press.

KEEBLE, F. (1910). *Plant–Animals: A Study in Symbiosis.* Cambridge: University Press.

KEEBLE, F. & GAMBLE, F. W. (1907). The origin and nature of the green cells of *Convoluta roscoffensis. Q. Jl microscop. Sci.*, **51**, 167–219.

MANTON, I., OATES, K. & GOODAY, G. W. (1973). Further observations on the chemical composition of thecae of *Platymonas tetrathele* West (Prasinophyceae) by means of the X-ray microanalyser electron microscope (EMMA). *J. exp. Bot.*, **24**, 223–229.

MANTON, I. & PARKE, M. (1965). Observations on the fine structure of two species of *Platymonas* with special reference to flagellar scales and the mode of origin of the theca. *J. mar. biol. Ass. UK*, **45**, 525–536.

McINTYRE, A. D., MUNRO, A. L. S. & STEELE, J. H. (1970). Energy flow in a sand ecosystem. In *Marine Food Chains*, 19–31 (ed. J. H. Steele). Edinburgh: Oliver & Boyd.

MEYER, F. & MEYER, H. (1972). Loss of fatty acid biosynthesis in flatworms. In *Comparative Biochemistry of Parasites*, 383–393 (ed. H. Van den Bossche). New York: Academic Press.

MURPHY, J. & RILEY, J. P. (1962). A modified single solution method for the determination of phosphate in natural waters. *Analytica Chim. Acta*, **27**, 31–36.

MUSCATINE, L., BOYLE, J. E. & SMITH, D. C. (1974). Symbiosis of the acoel flatworm *Convoluta roscoffensis* with the alga *Platymonas convolutae. Proc. R. Soc. Lond.* B, **187**, 221–234.

NORTH, B. B. & STEPHENS, G. C. (1971). Uptake and assimilation of amino acids by *Platymonas.* II. Increased uptake in nitrogen deficient cells. *Biol. Bull. mar. biol. Lab. Woods Hole*, **140**, 242–254.

NOZAWA, K., TAYLOR, D. L. & PROVASOLI, L. (1972). Respiration and photosynthesis in *Convoluta roscoffensis* Graff infected with various symbionts. *Biol. Bull. mar. biol. Lab. Woods Hole*, **143**, 420–430.

OSCHMAN, J. L. (1966). Development of the symbiosis of *Convoluta roscoffensis* Graff and *Platymonas* sp. *J. Phycol.*, **2**, 105–111.

(1967). Microtubules in the subepidermal glands of *Convoluta roscoffensis* (Acoela, Turbellaria). *Trans. Am. microscop. Soc.*, **86**, 159–162.

OSCHMAN, J. L. & GRAY, P. (1965). A study of the fine structure of *Convoluta roscoffensis* and its endosymbiotic algae. *Trans. Am. microscop. Soc.*, **84**, 368–375.

PARKE, M. & MANTON, I. (1967). The specific identity of the algal symbiont of *Convoluta roscoffensis. J. mar. biol. Ass. UK*, **47**, 445–464.

PEDERSEN, K. J. (1964). The cellular organization of *Convoluta convoluta*, an acoel turbellarian: A cytological, histochemical and fine structural study. *Z. Zellforsch. mikrosk. Anat.*, **64**, 655–687.

(1965). Cytological and cytochemical observations on the mucous gland cells of an acoel turbellarian, *Convoluta convoluta. Ann. N.Y. Acad. Sci.*, **118**, 930–965.

PHILLIPS, J. C. (1970). The reserve carbohydrates of the unicellular algae *Chlamydomonas moewusii, Polydriella helvetica* and *Platymonas convolutae.* Ph.D. Thesis, University of Newcastle upon Tyne.

PROVASOLI, L., YAMASU, T. & MANTON, I. (1968). Experiments on the resynthesis of symbiosis in *Convoluta roscoffensis* with different flagellate cultures. *J. mar. biol. Ass. UK*, **48**, 465–479.

READ, C. P. (1968). Intermediary metabolism of flatworms. In *Chemical Zoology*, vol. II, 327–357 (eds. M. Florkin and B. T. Scheer). New York: Academic Press.

RYTHER, J. H. (1956). Photosynthesis in the ocean as a function of light intensity. *Limnol. Oceanogr.*, **1**, 61–70.

SARFATTI, G. & BEDINI, C. (1965). The symbiont alga of the flatworm *Convoluta psammophila* Bekl. observed at the electron microscope. *Caryologia*, **18**, 207–223.

SIMMONS, J. E. (1970). Nitrogen metabolism in Platyhelminthes. In *Comparative Biochemistry of Nitrogen Metabolism*, vol. **1**, 67–90 (ed. J. W. Campbell). New York: Academic Press.

SMITH, D. C. (1973). *Symbiosis of algae with invertebrates*. Oxford Biology Reader, no. **43**. London: Oxford University Press.

SMITH, D. C., MUSCATINE, L. & LEWIS, D. (1969). Carbohydrate movement from autotrophs to heterotrophs in parasitic and mutualistic symbiosis. *Biol. Rev.*, **44**, 17–90.

SOLORZANO, L. (1969). Determination of ammonia in natural waters by the phenol-hypochlorite method. *Limnol. Oceanogr.*, **14**, 799–801.

TAYLOR, D. L. (1971*a*). Patterns of carbon translocation in algal-invertebrate symbiosis. *Proc. Int. Seaweed Symp.*, 590–597. Tokyo: University of Tokyo Press.

— (1971*b*). On the symbiosis between *Amphidinium klebsii* (Dinophyceae) and *Amphiscolops langerhansi* (Turbellaria: Acoela). *J. mar. biol. Ass. UK*, **51**, 301–313.

— (1973). The cellular interactions of algal–invertebrate symbiosis. *Adv. mar. Biol.*, **11**, 1–56.

— (1974). Nutrition of algal–invertebrate symbiosis I. Utilization of soluble organic nutrients by symbiont-free hosts. *Proc. R. Soc. Lond.* B, **186**, 357–368.

TAYLOR, W. R. (1964). Light and photosynthesis in intertidal benthic diatoms. *Helgolander wiss. Meers.*, **10**, 29–37.

TRENCH, M. E., TRENCH, R. K. & MUSCATINE, L. (1972). Utilization of photosynthetic products of symbiotic chloroplasts in mucus synthesis by *Placobranchus ianthobapsus* (Gould), Opisthobranchia, Saccoglossa. *Comp. Biochem. Physiol.*, **37**, 113–117.

TROLL, W. & CANNAN, R. K. (1953). A modified photometric method for the analysis of amino and imino acids. *J. biol. Chem.*, **200**, 803–811.

WEBB, K. L. & WOOD, L. (1967). Improved techniques for analysis of free amino acids in sea water. In '*Automation in Analytical Chemistry*'. Technicon Symposium 1966, vol. **1**, 440–444. White Plains, New York: Mediad Inc.

WHEELER, P. A., NORTH, P. B. & STEPHENS, G. C. (1974). Amino acid uptake by marine phytoplankters. *Limnol. Oceanogr.*, **19**, 249–259.

WHITFIELD, M. (1974). The ion-association model and the buffer capacity of the carbon dioxide system in sea water at 25 °C and 1 atmosphere total pressure. *Limnol. Oceanogr.*, **19**, 235–248.

WOOD, E. D., ARMSTRONG, F. A. J. & RICHARDS, F. A. (1967). Determination of nitrate in sea water by cadmium copper iodine reduction to nitrite. *J. mar. biol. Ass. UK*, **47**, 23–31.

OF 'LEAVES THAT CRAWL': FUNCTIONAL CHLOROPLASTS IN ANIMAL CELLS

By R. K. TRENCH

Biology Department, Osborn Memorial Laboratories,
Yale University, New Haven, Connecticut 06520, USA

> Should it be asked what it is that constitutes the difference between animal and vegetable life; what it is that lays the line that separates those two great kingdoms from each other, it would be difficult, perhaps we should find it impossible to return an answer.
>
> Oliver Goldsmith, 1825

INTRODUCTION

When visiting Dr Rachael Leech's Laboratory in York in 1971, I took a few live specimens of *Elysia viridis* along with me. Another visitor, knowing of Dr Leech's interest in plant leaves and photosynthesis, peered into an aquarium which contained the slugs crawling around with outspread parapodia and exclaimed, 'Oh look, leaves that crawl!' This expression sums up rather well, I think, the reason for much of the present interest in the phenomenon of 'chloroplast symbiosis'.

Since Kawaguti & Yamasu (1965) first described the occurrence of chloroplasts from the green siphonaceous seaweed *Codium fragile* in the cells of the digestive diverticulum of the gastropod mollusc *Elysia atroviridis*, investigation into the biology of this novel kind of autotroph–heterotroph symbiosis has gone on apace. It was soon recognized that this phenomenon could shed light on several different types of biological problems, notably in the field of chloroplast biology. Slugs with symbiotic chloroplasts provided a unique system for the study of interactions between a cell and a foreign organelle as well as a new method whereby the physiology of metabolism in sacoglossan molluscs could be investigated (Trench *et al.*, 1973*a*). When we consider that within a decade of the original report, as many as five review articles have appeared dealing with this subject (Smith, Muscatine & Lewis, 1969; Taylor, 1970; Greene, 1974; Muscatine & Greene, 1973; Smith, 1974), the significance of the phenomenon is made even more apparent. Moreover because plant chloroplasts can function in foreign environments (see Nass, 1969; Giles & Sarafis, 1971; Ridley & Leech, 1970; Taylor, 1970), it was proposed that

[229]

this property is a function of their biochemical and genetic autonomy. The phenomenon of 'chloroplast symbiosis' has been used by several authors as an example in support of the concept of chloroplast evolution through endosymbiosis (see Schnepf & Brown, 1971 and Greene, 1974).

Throughout this paper I shall use the term 'chloroplast symbiosis' in the broad sense of 'symbiosis' as defined by de Bary (1879), which has been modified into the concept of 'plasmid symbiosis' (Lederberg, 1952; Ris, 1961; Sagan, 1967). Although I shall use these concepts as a framework, I do not believe that the natural occurrence of functional plant chloroplasts in animal cells has any true phylogenetic significance (see Echlin, 1970), though the existence of such a phenomenon does demonstrate, as other intracellular associations do, that the potential for genetically distinct organisms to establish long-lasting, intracellular associations is widespread, and may have originated very early in evolutionary history. I prefer to place the emphasis here on the co-existence of distinct genomes within a common cellular environment without considering the related evolutionary problems. Thus viewed, 'chloroplast symbiosis' can be discussed within the same context as the other types of associations under consideration at this symposium.

'Chloroplast symbiosis' is defined as the phenomenon where, under natural circumstances an animal acquires intracellularly and retains undamaged plant chloroplasts, free from other associated plant organelles and cytoplasm. Such chloroplasts show sustained active photosynthesis, and photosynthetic products become available to and are utilized by the animal host. By 'sustained active photosynthesis' I mean much higher rates of carbon dioxide fixation in the light than in the dark for periods exceeding one week, readily detectable synthesis of primary photosynthetic products, and net oxygen evolution under adequate conditions of illumination (cf. Hinde & Smith, 1974).

In considering the phenomenon of chloroplast symbiosis, I shall discuss the three major themes of current investigation. Firstly, I shall discuss the data which establish the existence of functional chloroplasts in animal cells. Secondly, I shall treat the associations in the more classical sense of plant–animal symbiosis. In these studies emphasis has been placed on the translocation of photosynthate from the photosynthetic partner to the animal host, and subsequent utilization of such substances by the host. Thirdly, I shall discuss the significance of the phenomenon in the context of plastid function outside the plant cell, i.e. chloroplast biochemical autonomy (Bell, 1970).

OCCURRENCE AND MORPHOLOGICAL ASPECTS OF CHLOROPLAST SYMBIOSIS

It has been known for some time that sacoglossan gastropods show strict preferences for certain algae as food (see review by Kay, 1968). In a survey of 38 sacoglossan species, Greene (1970a) found that 56% showed a preference for seaweeds belonging to the Order Siphonales. Such data should not be interpreted as implying that all these molluscs which feed on siphonaceous seaweeds establish a symbiosis with the chloroplasts from their food plants. For example, although *Berthelinia chloris* and *Oxynoe panamensis* feed on *Caulerpa*, and *Placida dendritica* feeds on *Codium* or *Bryopsis*, none of these animals form a symbiosis with the chloroplasts from the seaweeds. In addition, several sacoglossans feed on non-siphonaceous algae. For example, *Hermaeina smithi* feeds on *Chaetomorpha* (Order Cladophorales), *Limapontia depressa*, *L. capitata* and *Alderia modesta* feed on *Vaucheria* (Order Xanthophyta), but no symbiosis is established (see Greene, 1970a; Greene & Muscatine, 1972; Muscatine & Greene, 1973; Hinde & Smith, 1974). The one possible exception to this is *Hermea bifida* which feeds on the red seaweed *Griffithsia flosculosa* and may retain functional plastids (Taylor, 1975).

The potential for establishing a symbiosis may well be a function of both animal and seaweed. Elysioid sacoglossans (e.g. *L. depressa* and *L. capitata*) do not form symbioses with chloroplasts from their food plant *Vaucheria*, which is not a member of the Siphonales. By contrast, all the more well-known examples of chloroplast symbiosis involve elysioid sacoglossans (e.g. *Elysia viridis*, *Tridachia crispata*, *Tridachiella diomedea* and *Placobranchus ianthobapsus*) and chloroplasts from Siphonales (see Taylor, 1968; Trench, Greene & Bystrom, 1969; Muscatine & Greene, 1973). With the possible exception of *H. bifida*, no eolidiform sacoglossans (e.g. *P. dendritica*) (Plate 1c) form symbioses with plant chloroplasts, even when such plastids are derived from siphonaceous seaweeds.

Chloroplast symbiosis is a much more restricted phenomenon than was initially recognized (see Taylor, 1967; Greene, 1970a) and appears to be demonstrated principally by elysioid sacoglossans (Plate 1a, b, d), and chloroplasts from the Siphonales.

MORPHOLOGY OF ANIMAL–CHLOROPLAST ASSOCIATIONS

Seaweeds belonging to the order Siphonales are characterized by being syncytial (or coenocytic) with very few cross walls (non-septate) (Dawson,

1966). Thus the nuclei, plastids and other organelles occur in the cytoplasm which surrounds a central vacuole. Chloroplasts are usually spindle-shaped to oval in outline (Plate 2), and, particularly amongst the tropical forms, may be of two types: photosynthetic plastids and amylogenic plastids (Feldman, 1946).

The digestive tract of elysioid sacoglossans has been described in detail by Fretter (1940) and Yonge & Nicholas (1940). Basically, the alimentary system consists of an oral tube, buccal mass and radula (the latter comprised of a single row of dagger-shaped, serrated teeth), esophagus, stomach, intestine and anus. The highly branched tubular system comprising the digestive diverticulum arises from the stomach, and ramifies throughout the organisms (see Taylor, 1968; Trench et al., 1969; Greene, 1970a). The digestive system of the eolidiform sacoglossans is essentially similar, except that the tubules of the digestive diverticulum extend into the cerata (see Muscatine & Greene, 1973).

There are two significant features of the digestive system of the sacoglossa which relate to their potential for establishing symbioses with chloroplasts. Firstly, the feeding apparatus is adapted for puncturing plant cells and sucking out the fluid contents (Yonge & Nicholas, 1940; Fretter, 1940). I have actually observed this process in the coral reef-dwelling species *Elysia cauzescops* (Trench, unpublished). Secondly, the cells which line the tubules of the digestive diverticulum are of two types, lime cells of unknown function and phagocytic digestive cells (see Fig. 1, and Plates 3, 4 and 5). These digestive cells sequester the chloroplasts. In establishing a symbiosis with chloroplasts, it must be assumed that, at least initially, the process of intracellular digestion is not initiated. The chloroplasts are therefore taken into the animals' cells in an intact state.

In parallel with the adaptations mentioned in the animals, it is possible that the peculiar 'robust' nature of chloroplasts from siphonaceous algae, as indicated by their behaviour *in vitro*, might enhance their survival in the animals' cells (Trench, Boyle & Smith, 1973a; Giles & Sarafis, 1974).

The mechanism whereby the other plant cell organelles which are undoubtedly sucked up along with the chloroplasts, are separated and eliminated is unknown. Histology of the stomach of *Tridachia crispata* and *E. viridis* shows a densely ciliated and highly folded epithelium (Trench, unpublished), but no sorting mechanism analogous to that of the stomach of lamellibranchs (Morton, 1958) could be demonstrated. Some observations (Trench, unpublished) on *E. cauzescops* bear on this problem. *E. cauzescops* feeds on *Caulerpa sertularioides* which has been shown to have both photosynthetic and amylogenic plastids. When animals were fixed for electron microscopy shortly after feeding, examination of the digestive

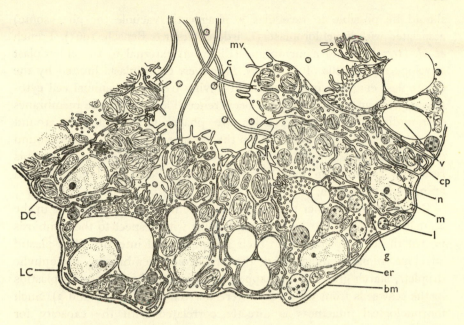

Fig. 1. Diagrammatic representation of a digestive gland tubule shown in transverse section. DC, digestive cell; LC, lime cell; mv, microvilli; c, cilia; v, vacuole; cp, chloroplast; n, nucleus; m, mitochondrion; l, lime spherule; g, golgi; er, endoplasmic reticulum; bm, basement membrane. (From Taylor, 1968, with permission of the Marine Biological Association, UK.)

cells showed only photosynthetic plastids to be present. It must be assumed therefore, that the amylogenic plastids were selected against. It is not known where this selection occurs.

The assumption has been made that the chloroplasts become engulfed by the animals' digestive cells through phagocytosis, but direct evidence of this is still lacking. Despite exhaustive examination with the electron microscope, Trench *et al.* (1973*b*) never observed chloroplasts in the process of phagocytosis, and the lumina of the tubules of the digestive diverticulum never contained recognizable plant cell organelles (see Plate 4).

The morphology of the intracellular relationship between the chloroplasts and the host cell cytoplasm was recognized as important very early in studies of chloroplast symbiosis (see Kawaguti & Yamasu, 1965; Taylor, 1968; Trench *et al.*, 1969). It is important to establish whether the chloroplasts reside in animal cell vacuoles or lie free in the cytoplasm. Unfortunately, this apparently simple distinction cannot even now be firmly made. The problem resides in the accurate recognition of intracellular membranes. If the chloroplasts are taken into the animal cells by phagocytosis, and there is really no other conceivable way they could enter, it should be possible to recognize a phagocytic vacuole (or phagosome) associated with the chloroplasts (Gordon, Miller & Bensch, 1965). Trench *et al.* (1973*b*) found a membrane (Plate 6*a*) external to the chloroplast envelope in the cells of *E. viridis*. However, chloroplasts limited by the chloroplast envelope membranes only, lying free in the animal cell cytoplasm, were an equally common occurrence (Plate 6*b*). If the membranes surrounding the healthy plastids are phagosome membranes around recently engulfed chloroplasts, then the absence of the membranes around other chloroplasts probably implies that the phagosome membrane is resorbed, resulting in more intimate contact between chloroplast and animal cell cytoplasm.

Regardless of whether the chloroplasts are in vacuoles or lie free in the animal cell cytoplasm, the important point with respect to the symbiosis is that the chloroplasts retain their morphological integrity. The plastid envelope remains intact, and the thylakoid membranes, osmophylic droplets, starch grains etc. all appear identical to those found in plastids in the seaweeds from which they were derived (cf. Plates 2 and 3). Such morphological intactness is directly correlated with the capacity for photosynthetic function, since chloroplasts with ruptured envelope membranes do not fix carbon dioxide (Walker, 1967).

PERSISTENCE OF CHLOROPLAST FUNCTION AND HERITABILITY

The capacity for sustained chloroplast function within the digestive cells of sacoglossans is one of the most important criteria in determining whether a symbiosis exists. Several assays can be used to determine chloroplast function. These include light-induced carbon dioxide fixation, oxygen evolution, pigment composition and structural integrity.

Of the sacoglossans studied, several show sustained chloroplast function. Based on electron microscope autoradiography, Taylor (1968) initially concluded that the chloroplasts in *E. viridis* were capable of only 'limited photosynthesis for a period of less than 24 hr. following ingestion'. Subsequent studies (Hinde & Smith, 1973) have shown however, that the chloroplasts in *E. viridis* may continue to function for more than two months. Similarly, Trench (1969) and Trench *et al.* (1969) found that chloroplasts in *T. crispata* and *T. diomedea* could maintain function for more than six weeks. Schmidt and Lyman (unpublished) detected ribulose diphosphate carboxylase activity in extracts of *T. crispata* for up to 30 days, while Greene (1970*c*) showed that the plastids in *E. hedgpethi* and *P. ianthobapsus* fixed carbon photosynthetically for 8 and 27 days respectively. Bear in mind however, that Testerman (unpublished) has detected net oxygen production in *P. ianthobapsus* after eight weeks 'starvation'. Perhaps the two photosystems in chloroplasts in *P. ianthobapsus* decay at different rates (cf. Leech, 1972). In the case of *E. hedgpethi*, loss of photosynthetic function accompanied loss of photosynthetic pigments, but such was not the case in *P. ianthobapsus*, where the photosynthetic pigments remained quantitatively constant, but the rate of carbon dioxide fixation per unit chlorophyll declined.

By contrast, in other animals photosynthetic function is either non-existent or is very rapidly lost. Greene & Muscatine (1972), following the patterns of $^{14}CO_2$ incorporation, found evidence for rapid loss of photosynthetic function in *P. dendritica*. Similarly, Hinde & Smith (1974) found non-existent function in 'dark' forms of *L. depressa*, *L. capitata* and *A. modesta*. In another study of *P. dendritica* which feeds on *Codium intertextum* in Bermuda, Trench (unpublished) found negligible quantitative or qualitative differences between light and dark $^{14}CO_2$ fixation. This was correlated with the morphological disruption of the chloroplast envelope membranes in the animal cells (see Plate 7).

Pigment analyses have also been useful in the determination of the functional integrity of symbiotic chloroplasts. Accumulation of chlorophyll degradation products in starved slugs has been used as a criterion of plastid destruction (digestion?) by the host animals. Taylor (1968) found increased divergence from the normal chromatographic pattern in *E. viridis* following starvation for one week or more, while Greene (1970a) found a similar situation in *Hermaeina* starved for 24 hours. It is not clear from any study so far conducted by what mechanism pigment degradation is effected. Taylor (1968) suggested degradation by animal enzymes, but also stated that the animals' digestive cells lacked the enzyme systems to destroy plastid membranes. How enzymes would enter intact chloroplasts to destroy the pigments remains unresolved. The correlation of the presence of chlorophyll degradation products and envelope-free plastids in *P. dendritica* suggests strongly that this slug does possess enzymes which digest chloroplast membranes (Trench, unpublished). Chlorophyll degradation is therefore probably effected after envelope destruction (see also Muscatine & Greene, 1973).

The question can readily be asked: In those animals that retain functional chloroplasts for extended periods, are the plastids ever destroyed? This problem is more difficult to solve unequivocally than may at first be apparent. Since the animal cells are digestive in function, it might be logical to assume .plastid digestion. Hence Kawaguti & Yamasu (1965) interpreted the moribund appearance of some chloroplasts in the cells of

E. atroviridis as evidence for digestion by the animals, but no concomitant biochemical data demonstrating hydrolytic activity was presented. Taylor (1968) concluded that *E. viridis* did not possess the enzyme systems necessary to digest chloroplast membranes. However, Trench *et al.* (1973*b*), in examining the ultrastructure of chloroplasts in *E. viridis* after 30 days starvation in the dark, found an abundance of 'degenerate' plastids in possible autophagic vacuoles in the animal cells (Plate 8). This observation revived the possibility of digestion, suggesting that the process of plastid degradation might be controlled through metabolite repression, whereby only non-functional plastids would be destroyed (see also Lewis, 1973).

In an attempt to gain further insight into this problem, Trench and Kanungo (unpublished) studied by electron microscope histochemistry, the distribution of phosphomonoesterase II (acid phosphatase) in the digestive cells of *T. crispata* which had been maintained in the light away from a source of plastid replenishment, for 12 weeks. Such animals were obviously paler green than when freshly collected. Acid phosphatase could readily be demonstrated in the digestive cells, and evidence for the presence of the enzyme associated with defunct plastids was demonstrable. Phosphatase activity was never found associated with intact plastids. Thus, at least in *T. crispata*, enzymes that could bring about intracellular digestion are present in the animal cells, but they only appear to function in chloroplast degradation after the plastids become defunct.

Symbiotic chloroplasts are not inherited, but are acquired *de novo* with each new generation (Greene, 1968, 1970*a*; Trench *et al.*, 1969). The eggs or developing veligers show no evidence of chlorophyll when analysed chemically or by ultraviolet microscopy (M. E. Trench & R. K. Trench, unpublished). Although the possibility of inheritance through proplastids has been suggested (Trench *et al.*, 1969), electron micrographs of freshly laid eggs or developing veligers never revealed recognizable proplastids (Trench, unpublished). Hence there appears to be no genetic continuity involved in this kind of 'symbiosis', in contrast to that demonstrated in some associations involving whole algae (see Muscatine, 1973, and Taylor, 1973, for recent reviews), where the symbionts are passed on from one generation to the other via the egg.

BIOCHEMICAL ASPECTS OF THE ASSOCIATION

Photosynthetic pigments

The photosynthetic pigments of chloroplasts in the cells of marine slugs are readily extracted by standard procedures (Strain, 1958). Analyses of

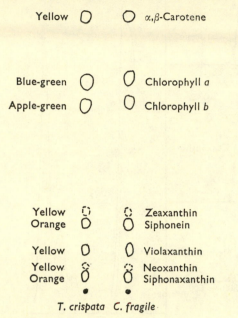

Yellow ○ ○ α,β-Carotene

Blue-green ○ ○ Chlorophyll *a*
Apple-green ○ ○ Chlorophyll *b*

Yellow ○ ○ Zeaxanthin
Orange ○ ○ Siphonein

Yellow ○ ○ Violaxanthin
Yellow ○ ○ Neoxanthin
Orange ○ ○ Siphonaxanthin

T. crispata *C. fragile*

Fig. 2. (*a*) Diagram comparing the pattern of separation of photosynthetic pigments from *T. crispata* and *C. fragile*. For details of methods etc. see Trench et al., 1969, 1970, 1973*b*.

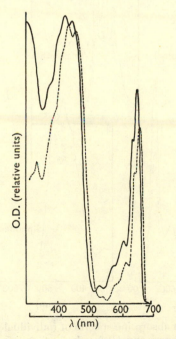

Fig. 2. (*b*) Comparison of absorbtion spectra of total photosynthetic pigments from *T. crispata* (solid line) and *C. fragile* (broken line) in diethyl ether.

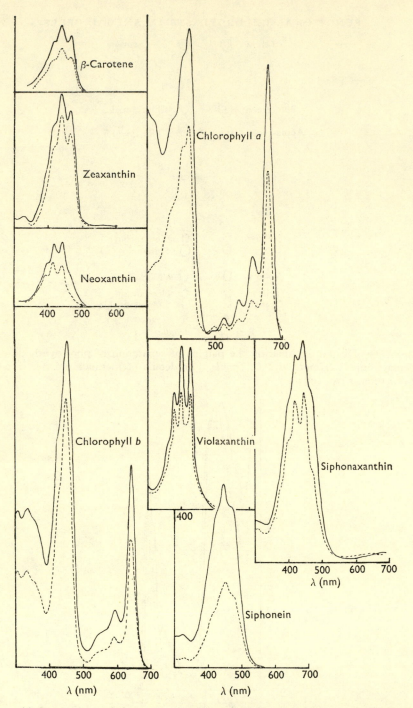

Fig. 2. (c) Comparison of absorption spectra of individual photosynthetic pigments from *T. crispata* (solid lines) and *C. fragile* (broken lines) after initial separation by thin-layer chromatography.

such extracts by thin layer chromatography have shown that the chloroplast pigments are indicative of the algal group from which the plastids are derived. This has proved to be a powerful tool in instances where the actual alga on which the animals feed is unknown, for example, *Tridachia crispata*, *Tridachiella diomedea* and *Placobranchus ianthobapsus* (see Trench *et al.*, 1969; Muscatine & Greene, 1973). In these instances, the presence of the xanthophyll pigments siphonaxanthin and siphonein in animal extracts demonstrates the siphonalean origin of the plastids (Strain, 1958).

Most of the slugs that retain functional chloroplasts appear to obtain them from algae belonging to the Siphonales. Comparison of photosynthetic pigments extracted from *Codium fragile* or *Caulerpa racemosa* with those from *T. crispata*, *T. diomedea*, *P. ianthobapsus*, *Elysia viridis* and *E. hedgpethi* shows a great deal of similarity following spectrophotometric analysis of total pigments or of individual pigments after separation by thin layer chromatography (Fig. 2*a*, *b*, *c*; see also Trench *et al.*, 1969; Trench & Smith, 1970; Trench *et al.*, 1973*b*; Taylor, 1968; Greene, 1970*a*). Similarly, in those instances where the alga involved was not a member of the Siphonales, pigment analysis proved an important tool in determining the origin of the chloroplasts. For example, Taylor (1974) found that *Hermea bifida* had photosynthetic pigments characteristic of red algae (including phycocyanin and phycoerythrin), and that the chloroplasts were derived from *Griffithsia flosculosa*, the food plant. Similarly, Greene (1970*a*) found that pigments from *Hermaeina smithi* were similar to those from *Chaetomorpha aerea*.

Oxygen evolution

Net oxygen production is the classical method used to indicate photosynthesis, and in the case of symbioses involving animals with photosynthetic symbionts, oxygen production in the light has been one of the major criteria used to indicate the presence of a functional photosynthetic unit.

Support for the notion of functional chloroplasts in *E. atroviridis* comes from the work of Kawaguti & Yamasu (1965) who found that the slugs produced more oxygen than they consumed in the light. Similarly, Trench (1969) and Trench *et al.* (1969) found net oxygen production in *Tridachia crispata* and reported similar data for *P. ianthobapsus* (from Testerman, unpublished). Haxo (unpublished) has found net oxygen production in *Tridachiella diomedea*, and in addition, found that the production of oxygen was inhibited by DCMU (4×10^{-5} mol l^{-1}). (See also Vandermeulen, Davis & Muscatine, 1972.) In *T. crispata*, oxygen evolution was inhibited by DCMU at a concentration of 5×10^{-6} mol l^{-1} (Trench, unpublished).

In a systematic analysis of oxygen evolution in *T. crispata*, Taylor (1971) found a photosynthetic saturation light intensity value of 5000 μW cm^{-2}, a compensation value around 1000 μW cm^{-2} and photosynthesis–respiration ratios of 1.6–2.4 at 2750 μW cm^{-2}, like intact algae.

From the data presented above, it is obvious that the symbiotic chloroplasts retain their functional integrity with respect to the Hill reactions.

Rates of carbon dioxide fixation

In any demonstration of photosynthesis, light mediated carbon dioxide fixation is the most important function to be established. Great care has to be taken to distinguish between light and dark fixation, both in terms of absolute rates of fixation and of the products of CO_2 fixation.

Initially, photosynthetic $^{14}CO_2$ fixation in symbiotic chloroplasts was demonstrated *in situ* using techniques of light and electron microscope autoradiography (Plate 9). In these instances, specimens incubated in the dark served as controls for dark CO_2 fixation, and in all cases dark fixation was found to be negligible compared to light fixation (Taylor, 1968; Trench, 1969; Trench *et al.*, 1969). However, because autoradiography only shows that radioactivity which remains after extraction by solvents used in tissue dehydration etc., it is possible that the pathways of carbon were different and that disproportionate amounts of ^{14}C went into soluble metabolites during dark fixation.

From more direct analyses, it appears that different associations have different levels of dark fixation. Greene & Muscatine (1972) found high dark fixation values in *E. hedgpethi* and *P. ianthobapsus* (10–16 % of light fixation), while Trench *et al.* (1973*b*) found values of 1–3 % for *E. viridis*.

In other examples of slugs suspected of having symbiotic chloroplasts, comparison of light and dark fixation has shown that the chloroplasts are not photosynthetically functional. Greene (1970*a*) found similar values for light and dark fixation in *Hermaeina smithi* and *Oxynoe antillarum* and Hinde & Smith (1974) found a similar situation in *Limapontia capitata*, *Alderia modesta* and 'dark forms' of *L. depressa*; the 'light forms' of *L. depressa* gave values of dark fixation varying from 11 to 59 % of light fixation. These animals, like *P. dendritica*, probably do not retain functional plastids for any extended period.

Very few studies have been conducted wherein the rates of carbon dioxide fixation by chloroplasts in their natural environment the plant cell, have been compared with rates in the animal cells or in artificial media. Trench *et al.* (1973*a*, *b*) found that chloroplasts from *C. fragile* in *E.*

viridis fixed carbon at 55 to 67 % of the rate in intact *Codium* (see Table 1). Isolated *Codium* chloroplasts or chloroplasts in homogenates of *E. viridis* fixed carbon dioxide at higher rates than in the intact systems (cf. Bidwell, Levin & Shephard, 1969, 1970).

Table 1. *Comparison of rates of CO_2 fixation in chloroplasts from* Codium fragile *in their natural environment, the plant cell, in an artificial medium, and in cells of* E. viridis. (*Data from Trench* et al., *1973a, b, by permission of The Royal Society of London*)

	Temperature (°C)	CO_2 fixation rate (μmol CO_2 mg^{-1} chlorophyll h^{-1})	
		Light	Dark
Intact *C. fragile*	18	11.9–12.0	—
Intact *E. viridis*	18	7.9	—
Isolated *Codium* plastids	18	18.0–23.0	—
Intact *C. fragile*	12	6.4	0.14
Intact *E. viridis*	12	—	0.13
Isolated *Codium* plastids	12	8.2	0.16

Patterns of $^{14}CO_2$ fixation

Trench, R. K., Trench, M. E. & Muscatine (1972) compared the distribution of photosynthetically fixed ^{14}C in *T. crispata*, *T. diomedea* and the siphonaceous seaweed *C. sertularioides* after 6 h photosynthesis in H^{14}CO$^-_3$. Biochemical fractionation of the tissues showed that compared with plants, the animals incorporated less ^{14}C into lipids, polysaccharide and proteins, but more into intermediary metabolites. Amongst the intermediary metabolites the animals showed most incorporation into galactose and glucose, while the plant extracts showed glucose and sucrose. Label incorporated into animal polysaccharide indicated mucopolysaccharide synthesis, while that in the plants showed the synthesis of plant cell wall polysaccharides.

In another study, Greene & Muscatine (1972) compared incorporation of ^{14}C into the ethanol soluble fractions from *E. hedgpethi*, *P. dendritica* and *P. ianthobapsus* with those from *C. fragile* after 2.5 h photosynthesis, and confirmed the differences between plant and animal patterns.

Although the comparison of photosynthetic products in animal and plant tissues has been very useful, it must be emphasized that this technique does not demonstrate chloroplast products *per se*. *A priori* it has to be recognized that chloroplasts in their natural cell environment may function in biochemically distinct ways from plastids in a foreign environment. Therefore differences, and some similarities, are to be expected. Similarly, a distinction has to be made between chloroplast products *sensu strictu* and host metabolism of products released by the chloroplasts. In order to understand more clearly which products are made and retained by the chloroplasts and those that are released to and metabolized by the animal hosts or the plant cell cytoplasm, comparative studies involving intact plants, isolated chloroplasts and chloroplasts in intact and homogenized animals were necessary.

Trench *et al.* (1973*a*, *b*) compared the photosynthetic products of intact *C. fragile* with those of chloroplasts isolated from *Codium*, those of intact animals, and those of homogenates (chloroplast enriched preparations) of animals (see Table 2). In initial comparisons between intact plants and

Table 2. *Comparison of the characteristics of photosynthetic* ^{14}C *fixation by* Codium *chloroplasts in different environments*

		Intact	Crushed	Isolated
	Codium	*Elysia*	*Elysia*	chloroplasts
Net rate of carbon fixation at 18 °C and 21 500 lx illumination (μmol C mg^{-1} chlorophyll h^{-1})	12	8	20	18 to 23
^{14}C incorporated during photosynthesis for 1 h at 21 500 lx and 18 °C {sucrose	+ +	–	–	–
glucose	+	Trace	+ +	Trace
galactose	–	+ +	–	–
glycolic acid	+	+	+	+ +
carotene+ xanthophyll	+	+		
chlorophyll	+	–		
galactolipid	+	–		
% Fixed ^{14}C released from chloroplasts	?	36	40	2

Data from Trench *et al.* (1973*b*): by permission of the Royal Society of London.

+, Readily detected; + +, abundant; –, not detectable.

isolated plastids, it became immediately apparent that the pattern of ^{14}C fixation was different. Although most of the primary photosynthetic products could be detected in extracts of both isolated plastids and intact plants, the radiochromatographic patterns obtained from isolated plastids were conspicuous in the absence of [^{14}C]sucrose and [^{14}C]glucose (cf. Greene, 1970*b*), even though the plastids maintained their photosynthetic function for very extended periods (Trench *et al.*, 1973*a*). The re-addition of the plant cytoplasmic fraction to the chloroplast suspension did promote glucose synthesis and the synthesis of numerous oligosaccharides, but little or no sucrose incorporated ^{14}C (see Plate 10*a*, *b*, *c*). It is therefore reasonable to conclude that *Codium* chloroplasts, away from the plant cytoplasm, do not synthesize sucrose (see Walker, 1967; cf. Shephard, Levin & Bidwell, 1968). This could account for the absence of sucrose in animal extracts (Plate 11), but the possibility that sucrose is rapidly metabolized by the animals after release by the chloroplasts cannot be ruled out.

In short-term photosynthesis experiments with *E. viridis*, $^{14}CO_2$ fixation (i.e. incorporation into sugar phosphates) could be detected within 2 min, and [^{14}C]glucose was identifiable within 5 min. 'Pulse-chase' experiments with intact *E. viridis* (Trench *et al.*, 1973*b*) and *T. crispata* (Trench, unpublished) indicated that glucose was the first sugar synthesized by the chloroplasts, but after 20 to 30 min galactose became the most labelled sugar in the extracts (see Table 3). When homogenates of *E. viridis* or

Table 3. *Relative incorporation of fixed ^{14}C into glucose and galactose during the first 60 min of photosynthesis by* E. viridis

Time (min) from first exposure of animals to $NaH^{14}CO_3$	$\dfrac{^{14}C \text{ in galactose}}{^{14}C \text{ in glucose}}$
5	0.40
10	0.60
15	0.75
20	0.90
30	4.00
45	45.00
60	5000–10 000

Animals were incubated in sea water containing $NaH^{14}CO_3$ at 19 °C and 26 900 lx. At intervals, animals were killed and extracted in methanol, and extracts run on one-dimensional paper chromatograms, scanned, and ^{14}C in glucose and galactose calculated by measuring the areas under the peaks. (Data from Trench *et al.*, 1973*b*; by permission of the Royal Society of London.)

T. crispata were incubated in $H^{14}CO^-_3$ in the light, the pattern of ^{14}C extracted from the tissues showed [^{14}C]glucose and no galactose (see Plate 12).

These data have led to the conclusion that chloroplasts in *E. viridis* and *T. crispata* may function in almost the same manner as they do in the parent plant cytoplasm with respect to the production of primary photosynthate, and that sucrose synthesis is the function of the plant cytoplasm just as galactose synthesis is the function of the cytoplasm of the animal cells. In any case, the precursor of sucrose or galactose must be some metabolite, either glucose or some derivative thereof, released to the cytoplasm by the chloroplasts. Some possible mechanisms for release of photosynthetic products by chloroplasts have been discussed by Heldt & Saur (1971), Smith (1974) and Heber (1974).

Utilization of chloroplast photosynthate by animals

The movement of photosynthetically fixed carbon from photosynthetic endosymbionts to non-photosynthetic hosts is of widespread occurrence

(Smith *et al.*, 1969). Translocation of photosynthate from symbiotic chloroplasts, and utilization of these products by molluscan hosts was first clearly demonstrated in *T. crispata* and *T. diomedea* (Trench, 1969; Trench *et al.*, 1969). It is significant that, within the limits of the auto-radiographic techniques used, ^{14}C could be detected in plastid free animal tissues after one hour incubation in the light. Such tissues include the renopericardium, cephalic ganglia, intestine and pedal gland (see Trench *et al.*, 1969). Of course these data threw no light on what substances moved from chloroplasts to animal cells.

In an attempt to obtain more direct evidence on the nature of the compounds moving from the chloroplasts, Trench *et al.* (1973*b*) used chloroplast-enriched preparations of *E. viridis* incubated in $H^{14}CO^-_3$ in the light. On chromatographic analysis of the media, after centrifugation of particulate matter, [^{14}C]glucose was the major substance detected. The one other compound was glycolic acid (see Plate 13). When similar preparations of *T. crispata* were tested, [^{14}C]glucose, [^{14}C]glycolic acid and [^{14}C]alanine were also detected (Trench, unpublished).

When these data, and those obtained in 'pulse-chase' experiments with intact animals are viewed together, it seems likely that glucose might be one of the major substances moving from chloroplasts to animal cells.

The observation that much of the carbon released by the symbiotic chloroplasts became associated with the pedal gland suggested that animals utilized chloroplast photosynthetic products in mucus synthesis. Trench *et al.* (1972), following the initial demonstration of photosynthetic-ally fixed ^{14}C in glucose and galactose moieties of mucus secreted by *P. ianthobapsus* (Trench *et al.*, 1970), used the 'pulse-chase' technique to follow ^{14}C from the chloroplasts to secreted mucus in *T. crispata*. Animals were 'pulse-labelled' for 1 h and 'chased' for a further 10 h period in the light. Analysis of 'free' [^{14}C]glucose and galactose, and determination of the specific activity of the secreted mucus, suggested that galactose and glucose were incorporated into the mucopolysaccharide. After a 1 h 'pulse', galactose was the most heavily labelled free sugar, but soon declined with a concomitant increase in glucose, which in turn coincided with the maximum specific activity detected in the mucus.

Since Trench *et al.* (1973*b*) subsequently found that in *E. viridis* glucose was first to be labelled, followed by galactose, a reinvestigation of *T. crispata* was conducted using shorter sampling times during the initial 'pulse'. Figure 3 shows that glucose initially incorporated most ^{14}C, and after 30 min photosynthesis, was much more abundant than galactose. However, the radioactivity in glucose declined thereafter with a con-comitant increase in galactose. Continuation of the experiment yielded

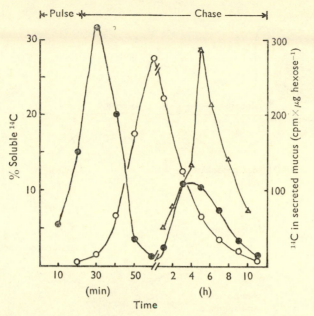

Fig. 3. Relationship in *T. crispata* between [^{14}C]glucose, [^{14}C]galactose and [^{14}C]mucopolysaccharide synthesis after 20 min 'pulse-labelling' in NaH^{14}CO$_3$ in the light. ●, glucose; ○, galactose; △, secreted mucus.

the same results as those reported by Trench *et al.* (1972). It is therefore proposed that glucose moves from the chloroplasts to the animal tissues where it is converted to galactose. The site of conversion is unknown, but results of 'inhibition' studies (Trench *et al.*, 1974) strongly suggest that it may be some distance away from the chloroplasts. Galactose is then transported, probably via the haemocoel, to sites of utilization in the pedal gland, where some is converted to glucose, glucosamine, galactosamine or incorporated directly as galactose into the mucopolysaccharide (see Fig. 4 and Trench, 1973).

Based on two independent estimates, Trench *et al.* (1973*b*) suggested that the chloroplasts in *E. viridis* contributed 36–40 % of the carbon fixed photosynthetically to the animals. In *T. crispata* a value of 50 % was suggested (Trench *et al.*, 1972), while Greene (1970*b*) found an average value of 21 % in *P. ianthobapsus*. These values are within the range of those found in symbioses involving whole algae (see Smith *et al.*, 1969; Trench, 1971*a*, *b*), but the different ways that the host may utilize metabolites contributed by the chloroplasts is not clearly understood. For example, although the incorporation of photosynthetically fixed carbon from chloroplasts into animal tissues has been documented, no information on what proportion of such substrates is used in respiratory metabolism is

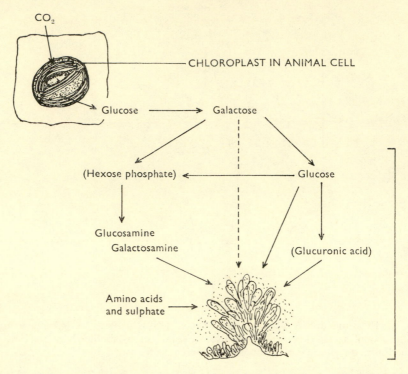

Fig. 4. Schematic representation of a hypothetical pathway of carbohydrate moving between symbiotic chloroplasts and the pedal mucus gland of *T. crispata*.

available. From data on total carbon fixed (Trench *et al.*, 1973*b*), percentage translocation to animal tissues (Trench *et al.*, 1973*b*) and photosynthesis: respiration ratios (Taylor, 1971), it is apparent that the chloroplasts can satisfy much of the respiratory carbon requirements of the animals. Apparently the animals utilize carbon substrates for the synthesis of substances with high turnover rates (see Trench, 1973). An analogous situation exists in tridacnid clams which possess endozoic dinoflagellate symbionts (Goreau, Goreau & Yonge, 1973). Hinde & Smith (unpublished) have found evidence suggesting rapid respiratory metabolism of exogenously supplied [14C]glucose in *E. viridis*, as well as limited utilization of exogenously supplied [14C]glycolate. Such experiments were conducted in the dark, in order to circumvent the possibility that animals may oxidize substrates to carbon dioxide which the plastids can subsequently fix in the light, resulting in the production of a mixture of heterotrophic and photosynthetic intermediates (e.g. see Trench *et al.*, 1973*b*). It is clear that these slugs possess the capacity for utilization of dissolved organic compounds (see Trench *et al.*, 1974; Trench, 1974).

From the data presented above, it can be readily discerned that slugs with symbiotic chloroplasts really fit conventional definitions of 'plants'. The animals possess a photoautotrophic mode of nutrition, and utilize primary photosynthate. However, there is no mechanism for plastid inheritance. Kirk & Tilney-Bassett (1967) states 'it is the possession of plastids which, at the cellular level, more than anything else distinguishes the Plant Kingdom from the Animal Kingdom'. It is not the *possession* of plastids, but the capacity for the *inheritance* of plastids that really needs to be emphasized as the criterion for separation of the two Kingdoms, if such a separation is really necessary.

SYMBIOTIC CHLOROPLASTS AND CHLOROPLAST BIOCHEMICAL AUTONOMY

Since chloroplasts in the digestive cells of elysioid sacoglossans remain functional for very extended periods, it is possible to use these symbiotic chloroplast systems for the study of chloroplast biochemical autonomy. To be useful, such plastids should be able to carry out normal physiological functions just as they do in the plant cells (Bell, 1970). It is reasonable to assume that within the animal cells, although the plastids may receive small molecular weight metabolites, e.g. amino acids from the host cell, they are essentially independent with respect to macromolecular synthesis. It is unlikely that the animal cell genome would code information which directs chloroplast macromolecular biosynthesis. Therefore, any synthesis of macromolecules by symbiotic chloroplasts would have to be directed by the plastid genome.

A voluminous literature is available on the co-operation between nuclear and cytoplasmic genes in organelle development and maintenance (for reviews, see for example Kirk & Tilney-Bassett, 1967; Surzycki, Goodenough, Levine & Armstrong, 1970; Schiff & Epstein, 1968; Tewari & Wildman, 1970; Sager, 1972; Gibbs, 1971; Boardman, Linnane & Smillie, 1971; Boulter, Ellis & Yarwood, 1971). Many of these studies are based either on the use of isolated chloroplasts, or chloroplasts in the plant cytoplasm functioning under the influence of inhibitors. Some really elegant contributions have also been made taking advantage of plastid and nuclear mutations in algae and higher plants.

Symbiotic chloroplasts offer yet another opportunity for the study of plastid biochemistry. In the preceding section I have shown that chloroplasts in slug cells can function photosynthetically for extended periods. To what extent are they capable of autonomous macromolecular synthesis? Specifically, do symbiotic chloroplasts synthesize pigments, lipids, proteins and nucleic acids?

Synthesis of photosynthetic pigments

Trench & Smith (1970) studied pigment synthesis by chloroplasts in *E. viridis*, *T. crispata* and *C. fragile*. They found that the chloroplasts in the slugs did not incorporate photosynthetically fixed ^{14}C into either chlorophyll *a* or chlorophyll *b*, but did incorporate ^{14}C into carotene and xanthophyll pigments. By contrast, in *C. fragile* all the photosynthetic pigments incorporated ^{14}C. It was concluded that chlorophyll synthesis did not occur in symbiotic chloroplasts since the enzymes necessary for many of the reactions in chlorophyll biosynthesis are coded for in the plant nucleus and synthesized on plant cytoplasmic ribosomes.

In a later study, Trench *et al.* (1973*b*) repeated their initial experiments, and again found no chlorophyll synthesis by symbiotic chloroplasts (Fig. 5). Since it was possible that the symbiotic chloroplasts needed more

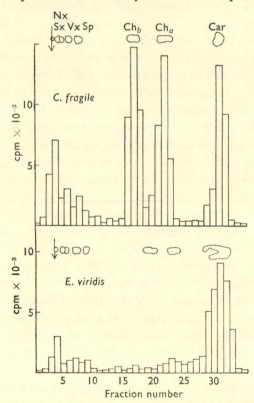

Fig. 5. Distribution of radioactivity on one-dimensional thin-layer chromatograms of extracts of photosynthetic pigments of *E. viridis* and *C. fragile* after 10 h photosynthesis in NaH^{14}CO$_3$. Arrows indicate origin; Car, α,β-carotene; Ch$_a$, chlorophyll *a*; Ch$_b$, chlorophyll *b*; Sp, siphonein; Vx, violaxanthin; Nx, neoxanthin; Sx, siphonaxanthin. (From Trench *et al.*, 1973*b*, by permission of the Royal Society of London.)

PLATE I

For explanation see p. 263

(Facing p. 248)

PLATE 2

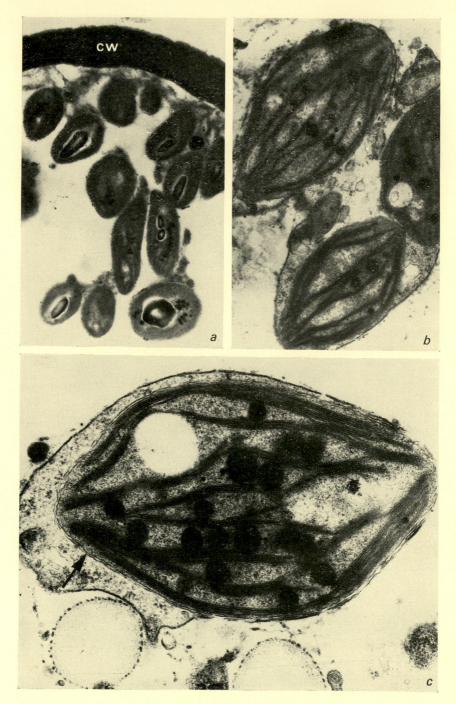

For explanation see p. 264

PLATE 3

For explanation see p. 264

PLATE 4

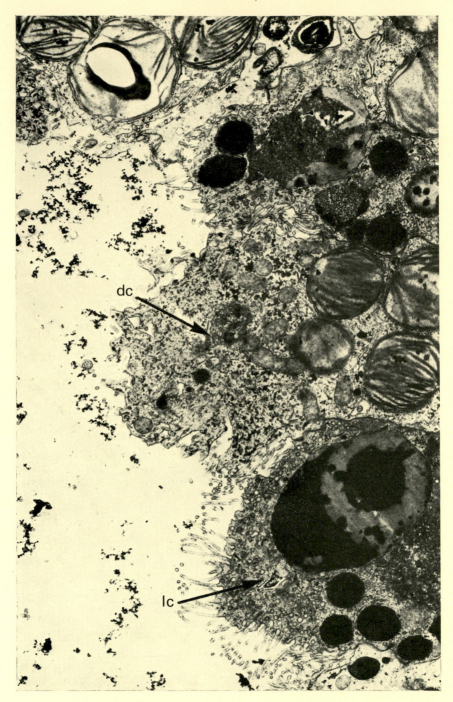

For explanation see p. 264

PLATE 5

For explanation see p. 264

PLATE 6

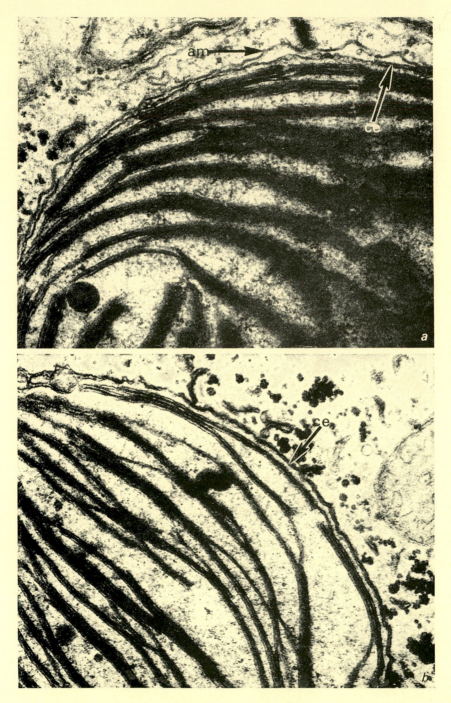

For explanation see p. 264

PLATE 7

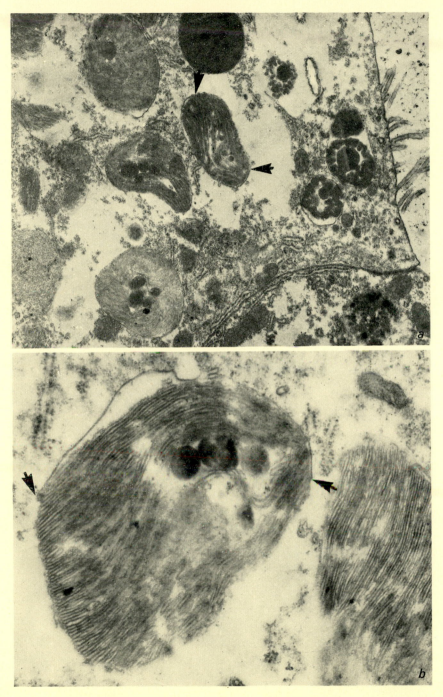

For explanation see p. 264

PLATE 8

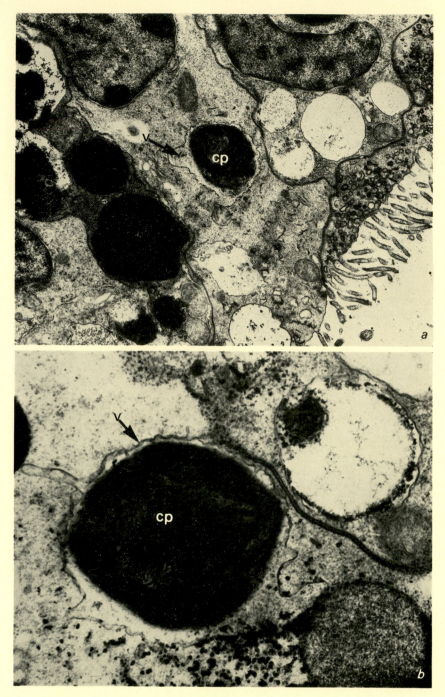

For explanation see p. 264

PLATE 9

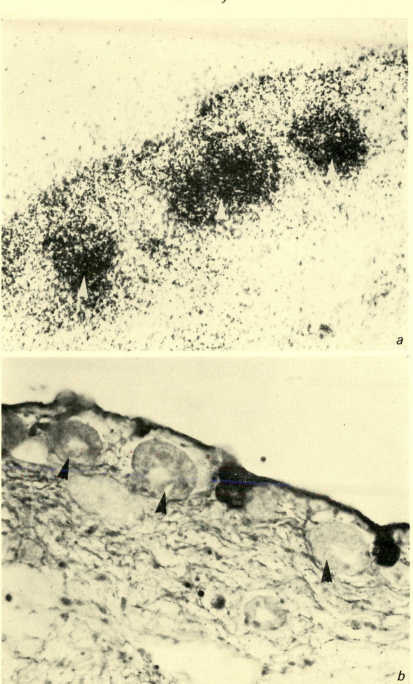

For explanation see p. 265

PLATE 10 A AND B

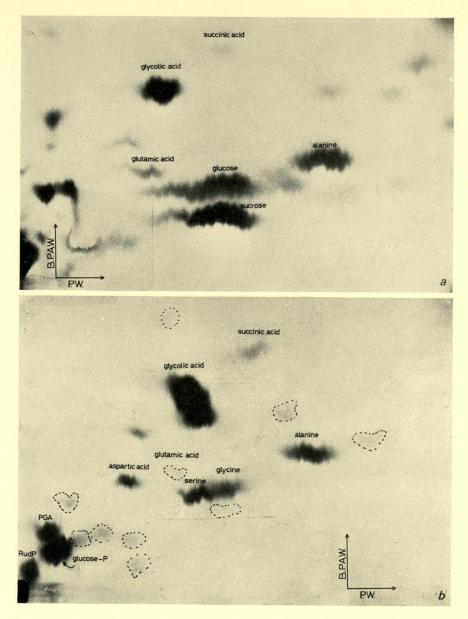

For explanation see p. 265

PLATE 10 C

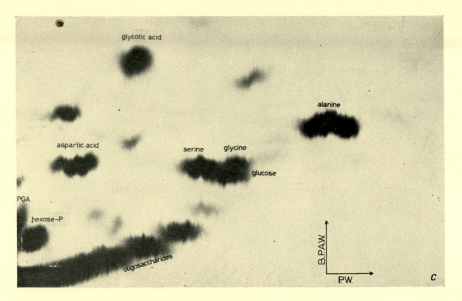

For explanation see p. 265

PLATE 11

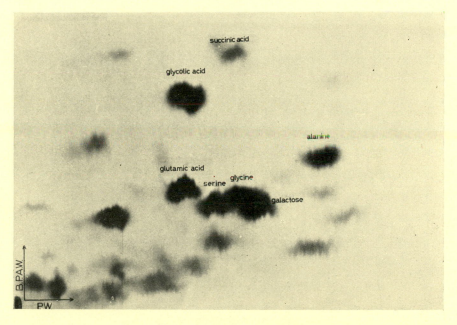

For explanation see p. 265

PLATE 12

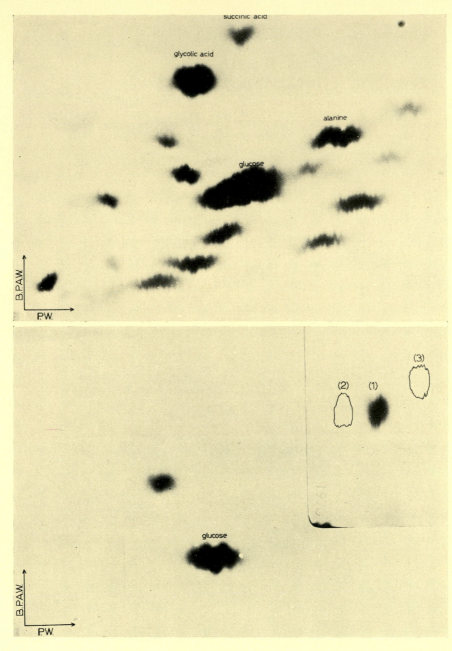

For explanation see p. 265

complex preformed substrates (see Trench & Smith, 1970; Rebeiz & Castelfranco, 1970), [^{14}C]glycine, [^{14}C]succinic acid and [^{14}C]δ-amino levulinic acid, precursors of chlorophyll, were supplied exogenously. Although uptake and utilization of these substrates could readily be demonstrated, no incorporation into chlorophyll was found.

Taking a slightly different approach, Trench *et al.* (1973*b*) reasoned that, if the absence of enzymes from the plant cell cytoplasm rendered symbiotic chloroplasts incapable of synthesizing chlorophyll, then similar results to those obtained in studies of chloroplasts in slug cells could be produced in the plant through the use of inhibitors of cytoplasmic protein synthesis. Therefore growing pieces of *C. fragile* were treated with 5-fluorodeoxyuridine (FUdR), cycloheximide and chloramphenicol, and the pattern of incorporation of ^{14}C into pigments compared with an untreated control plant. Figure 6 shows that although both cycloheximide and chloramphenicol inhibited incorporation of ^{14}C into chlorophyll, cycloheximide had a more marked effect. These inhibitors had only a limited inhibitory effect on carotenoid synthesis. FUdR by contrast, inhibited the synthesis of all pigments.

Since the inhibitors produced in *C. fragile* results similar to those obtained in *E. viridis* and *T. crispata*, it might be reasonable to conclude that the chloroplasts in the animal cells do not synthesize chlorophyll because of the absence of enzyme systems originating from the plant cytoplasm. By the same logic, I am forced to the conclusion that symbiotic chloroplasts are autonomous with respect to carotene and xanthophyll synthesis. This contrasts with higher plant chloroplasts (see Kirk & Tilney-Bassett, 1967).

It could be argued that chlorophyll synthesis is demonstrated only by growing chloroplasts (see Boasson & Laetsch, 1969), and that what our data show is that chloroplasts in slug cells cease to grow soon after they are taken up. The results of experiments described in the following sections tend to support this alternative explanation.

Lipid synthesis by symbiotic chloroplasts

Among the plant lipids, the glycolipids are known to be principally associated with chloroplast membrane systems (see Benson, 1966; Weir & Benson, 1966; Kreutz, 1966, 1970). A study of lipid synthesis by symbiotic chloroplasts should therefore yield pertinent information on plastid autonomy in lipid biosynthesis as well as growth. The extent to which chloroplasts are autonomous with respect to lipid and fatty acid biosynthesis is unresolved and the literature is relatively confused (see for example Erwin, 1968; Webster & Chang, 1969; Matson, Fei & Chang,

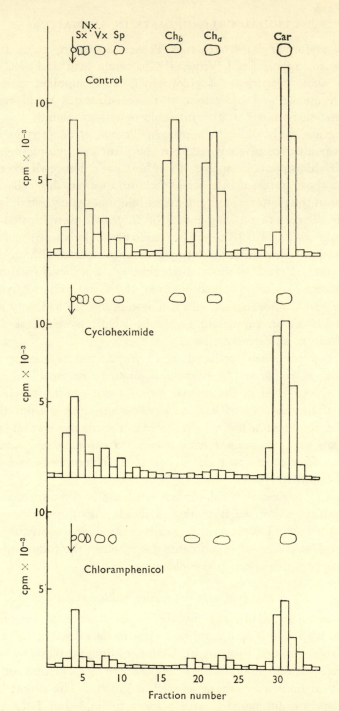

Fig. 6. Distribution of radioactivity on one-dimensional thin-layer chromatograms of extracts of photosynthetic pigments of *C. fragile* treated with the antibiotics cycloheximide and chloramphenicol before incubation in $NaH^{14}CO_3$ in the light. Abbreviations as in Fig. 5. (From Trench *et al.*, 1973*b*, by permission of the Royal Society of London.)

1969; Delo, Ernst-Fonberg & Block, 1971; Ernst-Fonberg & Block, 1971).

Trench *et al.* (1973*b*) compared the patterns of incorporation of photosynthetic ^{14}C into lipids in *C. fragile* and *E. viridis*. In *C. fragile* they found evidence (Fig. 7) for high levels of synthesis of monogalactosyl, digalactosyl and sulphoquinovosyl diglycerides whereas phosphatidyl glycerol, phosphatidyl choline and phosphatidyl ethanolamine incorporated lower quantities of ^{14}C. The pattern of incorporation in *E. viridis* was distinct from that of *C. fragile* in that no label was found in the glycolipids, though the phospholipids did incorporate ^{14}C, particularly phosphatidyl glycerol and phosphatidyl choline (see also Trench *et al.*, 1973*b*).

Since the phospholipids in animal extracts could just as well have been derived from animal tissues which synthesize lipids using substrates derived from chloroplast photosynthesis (see Trench *et al.*, 1969), the precise origin of the ^{14}C-labelled phospholipids in animal extracts is unclear. Therefore the possibility that symbiotic chloroplasts do synthesize phospholipids cannot be ruled out. Some studies (e.g. Webster & Chang, 1969) have shown synthesis of glycolipids by isolated spinach chloroplasts, but such chloroplasts were provided with nucleotide sugars.

Protein synthesis

Trench & Gooday (1973) studied protein synthesis by symbiotic chloroplasts in *E. viridis* using electron microscope autoradiography. They found that chloroplasts in slug cells incorporated exogenously supplied [^{3}H]leucine. This was possible because the slugs are able to take organic molecules up from solution (see also Trench *et al.*, 1974). Such incorporation of [^{3}H]leucine by symbiotic chloroplasts was insensitive to cycloheximide, but was later found to be chloramphenicol sensitive (Trench, unpublished).

There are four possible ways that the protein synthesis observed in symbiotic chloroplasts could have occurred. These are: protein is synthesized (*a*) on slug ribosomes, under the direction of slug mRNA or chloroplast mRNA, and transferred to the chloroplasts; (*b*) on chloroplast ribosomes using slug mRNA; (*c*) on chloroplast ribosomes using 'long-lived' mRNA from the plant nucleus and (*d*) on chloroplast ribosomes using chloroplast mRNA.

Alternatives (*a*) and (*b*) were ruled out by Trench & Gooday (1973) because no evidence for transfer of protein from animal cells to chloroplasts was obtained and because alternative (*b*) would imply the coding of information directing plastid protein synthesis by slug genome. They could not distinguish between alternatives (*c*) and (*d*). Although I cannot even now make this distinction, inhibition of protein synthesis in sym-

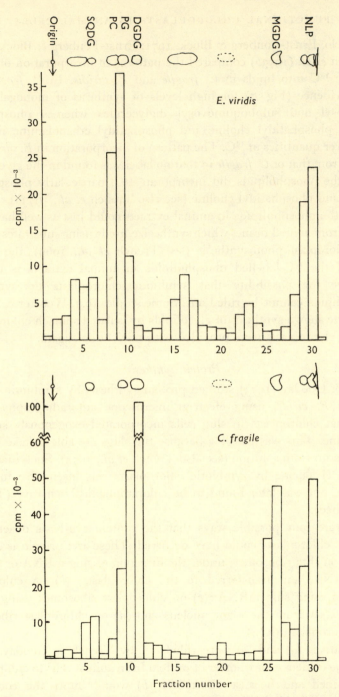

Fig. 7. Distribution of radioactivity in one-dimensional thin-layer chromatograms of ^{14}C-labelled lipids extracted from *E. viridis* and *C. fragile* after 10 h incubation in the light in $NaH^{14}CO_3$. NLP, neutral lipids and pigments; MGDG, monogalactosyl diglyceride; DGDG, digalactosyl diglyceride; PG, phosphatidyl glycerol; PC, phosphatidyl choline; SQDG, sulphoquinovosyl diglyceride.

biotic chloroplasts by chloramphenicol does demonstrate that protein is synthesized on chloroplast ribosomes, but the source of mRNA directing such synthesis is unknown.

Although the evidence from electron microscope autoradiography has demonstrated protein synthesis by symbiotic chloroplasts, the nature of the protein synthesized, i.e. whether soluble or structural proteins, remains unresolved. Some attempts have been made to study the synthesis of soluble and structural chloroplast proteins by symbiotic chloroplasts (Trench & Ohlhorst, unpublished), and I shall now summarize these results.

Studies were first initiated to determine whether symbiotic chloroplasts might be able to synthesize the CO_2 fixing enzyme ribulose-diphosphate carboxylase (RudP-carboxylase). This enzyme was selected because (a) it is the major soluble protein in higher plant chloroplasts (Wildman & Bonner, 1947; Lyttleton & Ts'o, 1958; Heber, Pon & Heber, 1963) and (b) its activity in symbiotic chloroplasts is retained for more than 30 days

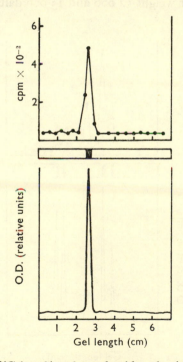

Fig. 8. Distribution of ^{14}C in 7% polyacrylamide gel with labelled RudP-carboxylase from *C. fragile*. Enzyme was isolated from *C. fragile* by the method of Goldthwaite & Bogorad (1971). Disc electrophoresis was conducted as described by Maizel (1971). *Upper part*: scan of ^{14}C in gel; *center part*: diagram of stained gel; *lower part*: densitometer tracing of stained gel.

(Schmidt and Lyman, unpublished). It was recognized at the outset however, that RudP-carboxylase from higher plants is a protein composed of two non-identical subunits (see for example Rutner & Lane, 1967; Kawashima, 1969), and that the larger subunit is synthesized on chloroplast ribosomes while the smaller is synthesized on plant cytoplasmic ribosomes (see Kawashima, 1970; Criddle, Dan, Kleinkopf & Huffaker, 1970). If the same is true for chloroplasts in seaweeds, then there should be no synthesis of RudP-carboxylase in symbiotic chloroplasts unless the small subunit can be assembled on slug ribosomes. However, Blair & Ellis (1972) have reported synthesis of the large subunit of RudP-carboxylase by isolated pea chloroplasts, and this encouraged us to investigate whether symbiotic chloroplasts would do likewise.

Our studies of RudP-carboxylase isolated from chloroplasts of *C. fragile*, using polyacrylamide gel disc electrophoresis, shows that the enzyme from *Codium* is similar in many respects to RudP-carboxylase isolated from spinach or pea chloroplasts. It can be resolved into two subunits of molecular weight 52 000 and 13 000 daltons respectively, after

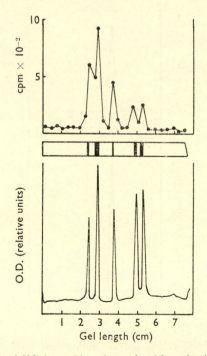

Fig. 9. Distribution of ^{14}C in 10% polyacrylamide gel with labelled chloroplast membrane proteins from *C. fragile*. Membrane proteins were prepared by slight modifications of the method of Hoober (1972), and analysed by SDS-disc electrophoresis.

treatment with dodecyl sodium sulphate (SDS). However, on natural gels, RudP-carboxylase from *Codium* always migrated slightly faster than enzyme from spinach or pea. Thus it is possible that the charge distribution on the *Codium* enzyme is different from that on enzyme from the higher plants.

When *C. fragile* is incubated with $^{14}CO_2$ in the light, incorporation of ^{14}C into RudP-carboxylase can readily be demonstrated (see Fig. 8). However, after numerous attempts, it has not been possible to demonstrate the synthesis of this enzyme by symbiotic chloroplasts in *T. crispata* or *E. viridis* using either [^{3}H]leucine or $^{14}CO_2$ as substrates. This is remarkable in view of the sustained enzyme activity found by Schmidt and Lyman, and the well-documented capacity for sustained photosynthetic CO_2 fixation (see Trench *et al.*, 1969; Hinde & Smith, 1973). By contrast, Kawashima (1970) could show incorporation of ^{14}C into RudP-carboxylase after 15 min photosynthesis in $^{14}CO_2$ by tobacco leaves, indicating either marked de-novo synthesis or rapid turnover of the molecule. On the basis of existing evidence, I am forced to conclude that RudP-carboxylase in symbiotic chloroplasts is very stable.

Comparative studies on the synthesis of chloroplast structural proteins have also suggested high stability of these proteins in symbiotic chloroplasts. When *Codium* plants are incubated in $^{14}CO_2$ and lamellar proteins from the chloroplasts are analysed after SDS treatment, it is possible to demonstrate incorporation of ^{14}C into several polypeptides (see Fig. 9). However, it has not been possible to demonstrate a similar phenomenon in symbiotic chloroplasts in *T. crispata* (Fig. 10).

When taken together, the evidence on pigment, lipid and protein synthesis suggests very strongly that symbiotic chloroplasts do not grow while they reside in animal cells. It is possible that such chloroplasts might be autonomous with respect to the synthesis of some plastid components. However, the stability of symbiotic chloroplasts may frustrate attempts at obtaining direct evidence for synthesis or turnover of macromolecules.

Nucleic acid synthesis

To my knowledge, no information on the synthesis of RNA or DNA by symbiotic chloroplasts is available. Greene (personal communication) attempted to demonstrate DNA synthesis by plastids in *P. ianthobapsus* using [^{3}H]thymidine and light microscope autoradiography, but his results were equivocal. Greene also used ^{32}P and CsCl density gradient centrifugation of DNA from *P. ianthobapsus*, but could not resolve labelled chloroplast-specific DNA because of heavy bacterial contamination. Trench (unpublished) also investigated the possibility of DNA

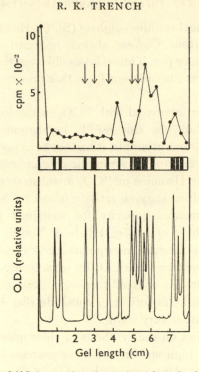

Fig. 10. Distribution of ^{14}C in 10% polyacrylamide gel with membrane proteins from chloroplasts from *T. crispata*. Note incorporation into non-plastid polypeptides. Electrophoretic conditions as in Fig. 9 above. Arrows indicate recognized chloroplast derived membrane polypeptides. Other peaks probably represent contamination from animal tissue.

synthesis by chloroplasts in *E. viridis* using electron microscope autoradiography but, like Taylor (personal communication), found no evidence for DNA synthesis.

The apparent lack of DNA synthesis, when viewed together with the absence of protein synthesis, suggests that neither growth nor replication occurs in symbiotic chloroplasts. Such findings are similar to those reported by Nass (1969) for chloroplasts in mouse fibroblast cells, but are different from those of Giles & Sarafis (1971), Ridley & Leech (1970) and Steffensen & Sheridan (1965), who found evidence for chloroplast division with either pea chloroplasts or plastids of siphonalean algae in Honda or Erd-Schreiber media or hen's eggs. The problem of whether chloroplasts can *divide and grow* autonomously is therefore open to further study, but I believe that the probability to achieve true chloroplast culture *in vitro* is probably very slim because of the major dependence of plastids on the nuclear–cytoplasmic synthetic processes.

CONCLUSIONS

It is now quite clear that chloroplasts from seaweeds belonging to the order Siphonales can remain photosynthetically active for extended periods while in the cells of elysioid sacoglossan molluscs. These symbiotic associations have been used with some success to study the biology and biochemistry of interactions between cells and foreign organelles. In addition, they have provided a convenient naturally occurring system with which to study the problem of chloroplast autonomy in a non-plant environment. Our knowledge of both features of the association, that is, of the nature of the symbiosis itself, as well as the extent of plastid autonomy, is as yet quite limited. Both features present interesting problems for continued investigation.

Within the context of symbiosis, it is now well-documented that chloroplasts are able to contribute photosynthetic carbon to their animal hosts. The quantity of carbon contributed is very likely enough to satisfy the animals' respiratory requirements. However, animals with symbiotic chloroplasts are not necessarily photoautotrophic with respect to their total nutrition. There is virtually no information available on the nutrition of slugs with symbiotic chloroplasts other than the data on carbon translocation and utilization. Nonetheless, the study of carbon utilization has provided us with much new information on carbon metabolism in sacoglossans, and has led to a fairly detailed analysis of mucopolysaccharide synthesis by sacoglossan opisthobranchs. In addition, the study of chloroplast symbiosis has demonstrated, unambiguously, ways in which animals are able to utilize the photosynthetic contributions of their endosymbionts for the synthesis of substances with high metabolic turnover, a phenomenon demonstrated by other symbiotic associations involving marine invertebrates and endozoic algae (see for example, Trench *et al.*, 1969; Trench, 1973; Goreau *et al.*, 1973).

The mode of establishment of chloroplast symbiosis is not yet clearly understood. It is quite apparent that the digestive processes in the host's cells must be suppressed in some manner so that the enzymes of the herbivorous animal are not activated and the organelles remain intact. Certainly the chloroplasts become established within the cells of the gut and remain functional with respect to carbon fixation although the metabolic pathways of these plastids are probably altered by some host factor. Gallop (1974) has found that a soluble 'factor' in homogenates of *Elysia* tissues causes release of photosynthetically fixed ^{14}C from prelabelled chloroplasts isolated from *Codium*. Such release was both quantitatively and qualitatively distinct from patterns observed with chloroplasts in a mineral medium.

The restricted distribution of these associations (i.e. elysioid saco-glossans and chloroplasts from Siphonales) suggests that the establishment of functional chloroplasts in the digestive cells of herbivorous gastropods probably evolved relatively recently, and very likely involved 'pre-adaptations' in both plant and animals. Such adaptation in the chloroplasts appears to be their remarkable robustness. In the animals, intracellular digestion appears to be activated only when the chloroplasts become non-functional. It is possible that the continued movement of metabolites from functional plastids may effect a system of metabolite repression which prohibits the activation of host intracellular digestive mechanisms (see also Lewis, 1973). Some indirect evidence now tends to support this inter-pretation (Trench and Kanungo, unpublished). This hypothesis is testable, and should be investigated further.

The most interesting aspect of chloroplast symbiosis is its potential for studies on plastid biochemical autonomy. Unfortunately only a limited number of studies have been undertaken with this approach in mind. Available ultrastructural and biochemical evidence (e.g. inability to synthesize sucrose) tends to support the conclusion that chloroplasts in slug cells are free of any plant cytoplasm or organelles. If this is accurate, then symbiotic chloroplasts represent a system of chloroplasts functioning outside their normal environment, the plant cell. Such experimental systems have been a major quest of plant physiologists and biochemists, and it is probably significant that amongst those of the artificial chloroplast 'cultures' that have achieved some success, plastids from Siphonalean algae have been involved (Steffensen & Sheridan, 1965; Giles & Sarafis, 1971). Chloroplast symbiosis therefore provides us with a system amenable to the experimental analysis of chloroplast biochemical autonomy. For example, using the symbiotic chloroplast system, it has been possible to provide supporting evidence for the hypothesis that sucrose synthesis, a predominant feature of intermediary metabolism of intact plants, is a property of the plant cytoplasm (cf. Walker, 1967; Bidwell et al., 1970), although chloroplasts can synthesize several amino acids, organic acids and sugar phosphates independently.

As regards the synthesis of more complex plastid components by symbiotic chloroplasts, most of the evidence has been negative. Out-standing among the chloroplast components which are, apparently, not synthesized by symbiotic chloroplasts are glycolipids, sulpholipids, RudP-carboxylase, membrane proteins and nucleic acids. These observations may imply that the code directing the synthesis of all these macromolecules resides in the plant cell nucleus (see for example Surzycki, 1969; Arm-strong, Surzycki, Moll & Levine, 1971). Of course in our attempts to demonstrate the synthesis of these substances in symbiotic chloroplasts, no preformed precursors were supplied, except in the case of chlorophyll synthesis and nucleic acid synthesis. Certainly, if the chloroplasts were autonomous, they should have been able to synthesize proteins etc. from carbon dioxide. That symbiotic chloroplasts appear not to be able to accomplish this suggests that the system represents an unusual situation, wherein photosynthetic function is maintained but macromolecular turn-over is held in 'suspended animation'. To my knowledge there are no other biological systems which show a parallel phenomenon. The stability and persistence of symbiotic chloroplasts are fascinating, but this very stability may frustrate attempts at analysis of macromolecular synthesis. Perhaps the use of invertebrate tissue culture techniques, wherein isolated slug digestive cells could be made to phagocitose chloroplasts *in vitro*, could prove a fruitful system for controlled experimental analysis.

This work was supported by grants from the Science Research Council (UK), and the US National Science Foundation (GB-34375).

REFERENCES

ARMSTRONG, J. J., SURZYCKI, S. J., MOLL, B. & LEVINE, R. P. (1971). Genetic transcription and translation specifying chloroplast components in *Chlamydomonas reinhardii*. *Biochemistry, N.Y.*, **10**, 692–701.

DE BARY, A. (1879). *Die Erscheinung der Symbiose*. Strassburg: Trübner.

BELL, P. R. (1970). Are plastids autonomous? *Symp. Soc. exp. Biol.*, **24**, 109–127.

BENSON, A. A. (1966). On the orientation of lipids in chloroplast and cell membranes. *J. Am. Oil Chem. Soc.*, **43**, 265–270.

BIDWELL, R. G. S., LEVIN, W. B. & SHEPHARD, D. C. (1969). Photosynthesis, photorespiration and respiration of chloroplasts from *Acetabularia mediterrania*. *Pl. Physiol.*, **44**, 946–954.

—— (1970). Intermediates of photosynthesis in *Acetabularia mediterrania* chloroplasts. *Pl. Physiol.*, **45**, 70–75.

BLAIR, G. E. & ELLIS, R. J. (1972). Light-driven synthesis of the large subunit of Fraction I protein of intact chloroplasts. *Biochem. J.*, **127**, 42.

BOARDMAN, N. K., LINNANE, A. W. & SMILLIE, R. M. (1971). *Autonomy and Biogenesis of Mitochondria and Chloroplasts*. Amsterdam: North-Holland.

BOASSON, R. & LAETSCH, W. M. (1969). Chloroplast replication and growth in tobacco. *Science, Wash.*, **166**, 749–751.

BOULTER, D., ELLIS, R. J. & YARWOOD, A. (1971). Biochemistry of protein synthesis in plants. *Biol. Rev.*, **47**, 113–175.

CRIDDLE, R. S., DAN, B., KLEINKOPF, G. E. & HUFFAKER, R. C. (1970). Differential synthesis of ribulose diphosphate carboxylase subunits. *Biochem. biophys. Res. Commun.*, **41**, 621–627.

DAWSON, E. Y. (1966). *Marine Botany*. New York, Chicago, San Francisco, Toronto and London: Holt, Rinehart and Winston, Inc.

DELO, J., ERNST-FONBERG, M. L. & BLOCK, K. (1971). Fatty acid synthetases from *Euglena gracilis*. *Arch. Biochem. Biophys.*, **143**, 384–391.

ECHLIN, P. (1970). The photosynthetic apparatus in prokaryotes and eukaryotes. *Symp. Soc. gen. Microbiol.*, **20**, 221–248.

ERNST-FONBERG, M. L. & BLOCK, K. (1971). A chloroplast-associated fatty acid synthetase system in *Euglena*. *Arch. Biochem. Biophys.*, **143**, 392–400.

ERWIN, J. A. (1968). Lipid metabolism. In *The Biology of Euglena*, vol. II, 133–147 (ed. D. E. Buetow). London and New York: Academic Press.

FELDMAN, J. (1946). Sur l'hétéroplastie de certaines Siphonales et leur classification. *C. r. hebd. Séanc. Acad. Sci. Paris*, **222**, 752–753.

FRETTER, V. (1940). On the structure of the gut of the sacoglossan nudibranchs. *Proc. zool. Soc. Lond.*, **110**, 185–189.

GALLOP, A. (1974). Evidence for the presence of a 'factor' in *Elysia viridis* which stimulates photosynthate release from its symbiotic chloroplasts. *New Phytol.* **73**, 1111–1117.

GIBBS, M. (1971). *Structure and Function of Chloroplasts*. Berlin and New York: Springer-Verlag.

GILES, K. L. & SARAFIS, V. (1971). On the survival and reproduction of chloroplasts outside the cell. *Cytobios*, **4**, 61–74.

GILES, K. L. & SARAFIS, J. (1974). Implications of rigescent integuments as a new structural feature of some algal chloroplasts. *Nature, Lond.*, **248**, 512–513.

GOLDSMITH, O. (1825). *A History of the Earth and Animated Nature*. New York: Thomas Kinnersley.

GOLDTHWAITE, J. J. & BOGORAD, L. (1971). A one step method for the isolation and determination of leaf ribulose-1,5-diphosphate carboxylase. *Analyt. Biochem.*, **41**, 57–66.

GORDON, G. B., MILLER, L. R. & BENSCH, K. G. (1965). Studies on the intracellular digestive process in mammalian tissue culture cells. *J. Cell Biol.*, **25**, 41–55.

GOREAU, T. F., GOREAU, N. I. & YONGE, C. M. (1973). On the utilization of photosynthetic products from zooxanthellae and of a dissolved amino acid in *Tridacna maxima* f. *elongata* (Roeding). *J. Zool.*, **169**, 417–454.

GREENE, R. W. (1968). The egg masses and veligers of southern California sacoglossan opisthobranchs. *Veliger*, **11**, 100–104.

(1970a). Symbiosis in sacoglossan opisthobranchs: symbiosis with algal chloroplasts. *Malacologia*, **10**, 357–368.

(1970b). Symbiosis in sacoglossan opisthobranchs: translocation of photosynthetic products from chloroplast to host tissue. *Malacologia*, **10**, 369–380.

(1970c). Symbiosis in sacoglossan opisthobranchs: functional capacity of symbiotic chloroplasts. *Mar. Biol.*, **7**, 138–142.

(1974). Sacoglossans and their chloroplast endosymbionts. In *Symbiosis in the Sea*, 21–27 (ed. W. B. Vernberg). University of South Carolina Press.

GREENE, R. W. & MUSCATINE, L. (1972). Symbiosis in sacoglossan opisthobranchs: photosynthetic products of animal–chloroplast associations. *Mar. Biol.*, **14**, 253–259.

HEBER, U. (1974). Metabolic exchange between chloroplasts and cytoplasm. *A. Rev. Pl. Physiol.*, **25**, 393–421.

HEBER, U., PON, N. G. & HEBER, M. (1963). Localization of carboxydismutase and triosephosphate dehydrogenases in chloroplasts. *Pl. Physiol., Lancaster*, **38**, 355–360.

HELDT, H. W. & SAUR, F. (1971). The inner membrane of the chloroplast envelope as the site of specific metabolite transport. *Biochem. biophys. Acta*, **234**, 83–91.

HINDE, R. & SMITH, D. C. (1973). Persistence of functional chloroplasts in *Elysia viridis* (Opisthobranchia, Sacoglossa). *Nature, Lond.*, **239**, 30–31.

(1974). 'Chloroplast symbiosis' and the extent to which it occurs in Sacoglossa. *Biol. J. Linn. Soc.* **6**, 349–356.

HOOBER, J. K. (1972). A major polypeptide of chloroplast membranes of *Chlamydomonas reinhardii*. Evidence for synthesis in the cytoplasm as a soluble component. *J. Cell Biol.*, **52**, 84–96.

KAWAGUTI, S. & YAMASU, T. (1965). Electron microscopy on the symbiosis between an elysioid gastropod and chloroplasts from a green alga. *Biol. J. Okayama Univ.*, **11**, 57–65.

KAWASHIMA, N. (1969). Comparative studies on fraction I protein from spinach and tobacco leaves. *Pl. Cell Physiol., Tokyo*, **10**, 31–40.

—— (1970). Non-synchronous incorporation of $^{14}CO_2$ into amino acids of the two subunits of fraction I protein. *Biochem. biophys. Res. Commun.*, **38**, 119–124.

KAY, E. A. (1968). A review of the bivalved gastropods and a discussion of evolution within the Sacoglossa. *Symp. zool. Soc. Lond.*, **22**, 109–134.

KIRK, J. T. O. & TILNEY-BASSETT, R. A. E. (1967). *The Plastids.* London and San Francisco: W. H. Freeman and Co.

KREUTZ, W. (1966). The structure of the lamellar system of chloroplasts. In *Biochemistry of Chloroplasts*, vol. I, 83–88 (ed. T. W. Goodwin). London and New York: Academic Press.

—— (1970). X-ray structure research on the photosynthetic membranes. In *Advances in Botanical Research*, vol. **3**, 54–169 (ed. R. D. Preston). London and New York: Academic Press.

LEDERBERG, J. (1952). Cell genetics and hereditary symbiosis. *Physiol. Rev.*, **32**, 403–430.

LEECH, R. M. (1972). The behaviour of plastids in artificial environments. In *Biology and Radiobiology of Anucleate Systems*, II, *Plant Cells*, 27–49 (eds. S. Bonotlo, R. Goutier, R. Kirchmann and J. Maisin). London and New York: Academic Press.

LEWIS, D. H. (1973). Concepts in fungal nutrition and the origin of biotrophy. *Biol. Rev.*, **48**, 261–278.

LYTTLETON, J. W. & TS'O, P. O. P. (1958). The localization of fraction I protein of green leaves in the chloroplasts. *Arch. Biochem. Biophys.*, **73**, 120–126.

MAIZEL, J. V. (1971). Polyacrylamide gel electrophoresis of viral proteins. In *Methods in Virology*, 179–246 (eds. K. Maramoresch and H. Koprowski). London and New York: Academic Press.

MATSON, R. S., FEI, M. & CHANG, S. B. (1969). Comparative studies of biosynthesis of galactolipids in *Euglena gracilis* strain Z. *Pl. Physiol., Lancaster*, **45**, 531–532.

MORTON, J. E. (1958). *Molluscs.* London: Hutchinson and Co. Ltd.

MUSCATINE, L. (1973). Nutrition of corals. *In Biology and Geology of Coral Reefs*, vol. II, 77–115 (eds. O. A. Jones and R. Endean). London and New York: Academic Press.

MUSCATINE, L. & GREENE, R. W. (1973). Chloroplasts and algae as symbionts in molluscs. *Int. Rev. Cytol.*, **36**, 137–169.

NASS, M. M. K. (1969). Uptake of isolated chloroplasts by mammalian cells. *Science, Wash.*, **165**, 1128–1131.

REBEIZ, C. A. & CASTELFRANCO, P. A. (1970). Chlorophyll biosynthesis in a cell free system from higher plants. *Pl. Physiol.*, **47**, 33–37.

RIDLEY, S. M. & LEECH, R. M. (1970). Division of chloroplasts in an artificial environment. *Nature, Lond.*, **227**, 463–465.

RIS, H. (1961). Ultrastructure and molecular organization of genetic systems. *Can. J. Genet. Cytol.*, **3**, 95–120.

RUTNER, A. C. & LANE, M. D. (1967). Non-identical subunits of ribulose diphosphate carboxylase. *Biochem. biophys. Res. Commun.*, **28**, 531–537.

SAGAN, L. (1967). On the origin of mitosing cells. *J. theoret. Biol.*, **14**, 225–274.

SAGER, R. (1972). *Cytoplasmic Genes and Organelles.* London and New York: Academic Press.

SCHIFF, J. A. & EPSTEIN, H. T. (1968). The continuity of chloroplasts in *Euglena.* In *The Biology of Euglena,* vol. II, 285–333 (ed. D. E. Buetow). London and New York: Academic Press.

SCHNEPF, E. & BROWN, R. M. Jr (1971). On relationships between endosymbiosis and the origin of plastids and mitochondria. In *Origin and Continuity of Cell Organelles,* 299–322 (eds. J. Reinert and H. Ursprung). New York, Heidelberg and Berlin: Springer-Verlag.

SHEPHARD, D. C., LEVIN, W. B. & BIDWELL, R. G. S. (1968). Normal photosynthesis by isolated chloroplasts. *Biochem. biophys. Res. Commun.,* **32,** 413–420.

SMITH, D. C. (1974). Transport from symbiotic algae and symbiotic chloroplasts to host cells. *Symp. Soc. exp. Biol.,* **8,** 473–508.

SMITH, D. C., MUSCATINE, L. & LEWIS, D. H. (1969). Carbohydrate movement from autotrophs to heterotrophs in parasitic and mutualistic symbiosis. *Biol. Rev.,* **44,** 17–90.

STEFFENSEN, D. M. & SHERIDAN, W. E. (1965). Incorporation of ^{3}H-thymidine into chloroplast DNA of marine algae. *J. Cell Biol.,* **25,** 619–626.

STRAIN, H. H. (1958). Chloroplast pigments and chromatographic analysis. *32nd Annual Priestley Lectures,* 1–180. Pittsburgh: The Pennsylvania State University.

SURZYCKI, S. J. (1969). Genetic functions of the chloroplast of *Chlamydomonas reinhardii:* Effect of rifampin on chloroplast DNA-dependent RNA polymerase. *Proc. natn. Acad. Sci. USA,* **63,** 1327–1334.

SURZYCKI, S. J., GOODENOUGH, U., LEVINE, P. R. & ARMSTRONG, J. J. (1970). Nuclear and chloroplast control of chloroplast structure and function in *Chlamydomonas reinhardi. Symp. Soc. exp. Biol.,* **24,** 13–35.

TAYLOR, D. L. (1967). The occurrence and significance of endosymbiotic chloroplasts in the digestive glands of herbivorous opisthobranchs. *J. Phycol.,* **3,** 234–235.

(1968): Chloroplasts as symbiotic organelles in the digestive gland of *Elysia viridis* (Gastropoda, Opisthobranchia). *J. mar. biol. Ass. UK,* **48,** 1–15.

(1970). Chloroplasts as symbiotic organelles. *Int. Rev. Cytol.,* **27,** 29–64.

(1971). Photosynthesis of symbiotic chloroplasts in *Tridachia crispata* Bergh. *Comp. Biochem. Physiol.,* **38A,** 233–236.

(1973). Cellular interactions of algae-invertebrate symbiosis. *Adv. mar. Biol.,* **11,** 1–56.

(1975). Symbiosis between the chloroplasts of *Griffithsia flosculosa* (Rhodophyta) and *Hermea bifida* (Gastropoda, Opisthobranchia). *Pubb. Staz. Zoologica Napoli* (in press).

TEWARI, K. K. & WILDMAN, S. G. (1970). Information content in the chloroplast DNA. *Symp. soc. exp. Biol.,* **24,** 147–179.

TRENCH, M. E., TRENCH, R. K. & MUSCATINE, L. (1970). Utilization of photosynthetic products of symbiotic chloroplasts in mucus synthesis by *Placobranchus ianthobapsus* Gould, Opisthobranchia, Sacoglossa. *Comp. Biochem. Physiol.,* **37,** 113–117.

TRENCH, R. K. (1969). Chloroplasts as endosymbiotic organelles in *Tridachia crispata* (Bergh), Sacoglossa, Opisthobranchia. *Nature, Lond.,* **222,** 1071–1072.

(1971a). The physiology and biochemistry of zooxanthellae symbiotic with marine coelenterates. I. The assimilation of photosynthetic products of zooxanthellae by two marine coelenterates. *Proc. R. Soc. Lond. B,* **177,** 225–236.

(1971*b*). The physiology and biochemistry of zooxanthellae symbiotic with marine coelenterates. II. Liberation of fixed ^{14}C by zooxanthellae *in vitro*. *Proc. Soc. Lond. B*, **177**, 237–250.

(1973). Further studies on the mucopolysaccharide secreted by the pedal gland of the marine slug *Tridachia crispata* (Opisthobranchia, Sacoglossa). *Bull. mar. Sci.*, **23**, 299–312.

(1974). Nutritional potentials in *Zoanthus sociatus* (Coelenterata, Anthozoa). *Helgolander wiss. Meeresunters.*, **26**, 174–216.

TRENCH, R. K., BOYLE, J. E. & SMITH, D. C. (1973*a*). The association between chloroplasts of *Codium fragile* and the mollusc *Elysia viridis*. I. Characteristics of isolated *Codium* chloroplasts. *Proc. R. Soc., Lond. B*, **184**, 51–61.

(1973*b*). The association between chloroplasts of *Codium fragile* and the mollusc *Elysia viridis*. II. Chloroplast ultrastructure and photosynthetic carbon fixation in *E. viridis*. *Proc. R. Soc. Lond. B*, **184**, 63–81.

(1974). The association between chloroplasts of *Codium fragile* and the mollusc *Elysia viridis*. III. Movement of photosynthetically fixed ^{14}C in tissues of intact living *E. viridis* and in *Tridachia crispata*. *Proc. R. Soc. Lond. B*, **185**, 453–464.

TRENCH, R. K. & GOODAY, G. W. (1973). Incorporation of ^{3}H-leucine into protein by animal tissues and by endosymbiotic chloroplasts in *Elysia viridis* Montagu. *Comp. Biochem. Physiol.*, **44A**, 321–330.

TRENCH, R. K., GREENE, R. W. & BYSTROM, B. G. (1969). Chloroplasts as functional organelles in animal tissues. *J. Cell Biol.*, **42**, 404–417.

TRENCH, R. K. & SMITH, D. C. (1970). Synthesis of pigment in symbiotic chloroplasts. *Nature, Lond.*, **227**, 196–197.

TRENCH, R. K., TRENCH, M. E. & MUSCATINE, L. (1972). Symbiotic chloroplasts; their photosynthetic products and contribution to mucus synthesis in two marine slugs. *Biol. Bull. mar. biol. Lab. Woods Hole*, **142**, 335–349.

VANDERMEULEN, J. H., DAVIS, N. D. & MUSCATINE, L. (1972). The effect of inhibitors of photosynthesis on zooxanthellae in corals and other marine invertebrates. *Mar. Biol.*, **16**, 185–191.

WALKER, D. A. (1967). Photosynthetic activity of isolated pea chloroplasts. In *Biochemistry of Chloroplasts*, vol. II, 53–69 (ed. T. W. Goodwin). London and New York: Academic Press.

WEBSTER, D. E. & CHANG, S. B. (1969). Polygalactolipids in spinach chloroplasts. *Pl. Physiol.*, *Lancaster*, **44**, 1523–1527.

WEIR, T. E. & BENSON, A. A. (1966). The molecular nature of chloroplast membranes. In *Biochemistry of Chloroplasts*, 91–113 (ed. T. W. Goodwin). London and New York: Academic Press.

WILDMAN, S. G. & BONNER, J. (1947). The proteins of green leaves. I. Isolation, enzymatic properties and auxin content of spinach cytoplasmic proteins. *Arch. Biochem. Biophys.*, **14**, 381–413.

YONGE, C. M. & NICHOLAS, H. M. (1940). Structure and function of the gut and symbiosis with zooxanthellae in *Tridachia crispata* (Oerst.) Bgh. *Pap. Tortugas Lab.*, **32**, 287–301.

EXPLANATION OF PLATES

PLATE I

(*a*) Underwater photograph of *Tridachia crispata* Mörch. (Trench, 1973, by permission from *Bulletin of Marine Science* and Allen Press, Inc.) ×2.5

(*b*) *Elysia viridis* Montagu. ×5

(c) *Placida dendritica*, Alder and Hancock. ×6
(d) *Placobranchus ianthobapsus* Gould. (Photograph by R. W. Greene.) ×1

PLATE 2

(a) Electron micrograph of chloroplasts in *C. fragile*. Note prominent starch grains and oil droplets; cw, cell wall. ×7500
(b) Electron micrograph of chloroplasts in *C. fragile*. Note arrangement of thylakoid membranes and oil droplets. Two plant mitochondria can be seen. ×19 000
(c) Electron micrograph of a chloroplast in *C. fragile*. Arrow points to chloroplast envelope. (From Trench *et al.*, 1973a, by permission of the Royal Society of London.) ×50 000

PLATE 3

Chloroplasts from *C. fragile* in a digestive cell of *E. viridis*. (From Trench *et al.*, 1973b, by permission of the Royal Society of London.) ×10 000

PLATE 4

Electron micrograph of a portion of a tubule of the digestive diverticulum of *E. viridis* cut in the longitudinal plane. Note the presence of numerous microvilli projecting into the lumen. Several flagella can be seen in transverse section, as well as pinocytotic vesicles and animal cell mitochondria. lc, lime cell; dc, digestive cell. (From Trench *et al.*, 1973b: by permission of the Royal Society of London.) ×10 000

PLATE 5

Electron micrograph of a portion of the digestive tubule of *E. viridis* in transverse section. The lumen with microvilli and flagella projecting into it can be seen on the lower right. (From Trench *et al.*, 1973b, by permission of the Royal Society of London.) ×10 000

PLATE 6

(a) Electron micrograph of a portion of a chloroplast in a digestive cell of *E. viridis* showing the presence of an undulating membrane (am) external to the chloroplast envelope (ce). ×75 000
(b) Electron micrograph of a portion of a chloroplast in a digestive cell in *E. viridis* showing the chloroplast lying free in the animal cell cytoplasm bounded only by the chloroplast envelope (ce). An animal cell mitochondrion may be seen to the right. (From Trench *et al.*, 1973b, by permission of the Royal Society of London.) ×87 500

PLATE 7

(a) Electron micrograph of chloroplasts from *C. intertextum* in the digestive cells of freshly collected *P. dendritica*. Arrows indicate ruptured chloroplast envelope. ×14 000
(b) Electron micrograph of a chloroplast from *C. intertextum* in a digestive cell of *P. dendritica*. Arrows indicate ruptured chloroplast envelope. ×50 000

PLATE 8

(a) Electron micrograph of a portion of a digestive cell of *E. viridis* maintained in the dark for 30 days. Note the absence of intact chloroplasts and the presence of many membrane-bound vesicles or vacuoles (v). Arrow indicates one such vacuole containing a structure interpreted as the remnants of a chloroplast (cp). × 12 600

(b) Electron micrograph of a defunct plastid in a vacuole of a digestive cell of *E. viridis* maintained in the dark for 30 days. Abbreviations as in Plate 8a. (From Trench *et al.*, 1973b, by permission of the Royal Society of London.) × 37 500

PLATE 9

(a) Autoradiograph of a portion of the parapodium of *T. crispata* (unstained) showing the uptake of ¹⁴C by chloroplasts within the cells of the tubules of the digestive diverticulum after 300 min photosynthesis. Arrows point to the 'end bulbs' of the digestive tubules. × 350

(b) Autoradiograph of a portion of the parapodium of *T. crispata* treated as above but incubated in the dark for 300 min. (From Trench *et al.*, 1969, by permission of *The Journal of Cell Biology* and The Rockefeller University Press.) × 350

PLATE 10

(a) Autoradiograph of a two-dimensional paper chromatogram of 80 % ethanol extract of *C. fragile* after 1 h photosynthetic ¹⁴C fixation in sea water at 20 °C at 21 500 lx illumination.

(b) Autoradiograph of a two-dimensional paper chromatogram of products of photosynthetic ¹⁴C fixation by isolated *C. fragile* chloroplasts incubated for 60 min under 21 500 lx illumination at 8 °C. Chloroplast suspension pipetted directly onto the origin of chromatogram and killed over boiling methanol.

(c) Autoradiograph of a two-dimensional paper chromatogram of a methanol extract of products of photosynthetic ¹⁴C fixation by isolated chloroplasts incubated in the supernatant fraction of centrifuged *Codium* homogenates. All other conditions as in Plate 10a, b above. (From Trench *et al.*, 1973a, by permission of the Royal Society of London.)

PLATE 11

Autoradiograph of a two-dimensional paper chromatogram of a hot-methanol extract of *E. viridis* after 60 min photosynthesis in NaH¹⁴CO₃. (From Trench *et al.*, 1973b, by permission of the Royal Society of London.)

PLATE 12

Autoradiograph of a two-dimensional paper chromatogram of a hot-methanol extract of the pellet after photosynthesis by 'chloroplast-enriched' preparations of *E. viridis*. (From Trench *et al.*, 1973b, by permission of the Royal Society of London.)

PLATE 13

Autoradiograph of a two-dimensional paper chromatogram of the supernatant medium after photosynthesis by a 'chloroplast-enriched' preparation from *E. viridis*. The insert (*upper right*) is a one-dimensional radiochromatogram of the major radioactive species found in the medium. (1), [¹⁴C]glucose from medium; (2), standard glucose; (3), standard galactose. (From Trench *et al.*, 1973b, by permission of the Royal Society of London.)

SYMBIOTIC DINOFLAGELLATES*

By D. L. TAYLOR

Rosenstiel School of Marine and Atmospheric Science,
University of Miami, 4600 Rickenbacker Causeway,
Miami, Florida 33149, USA

INTRODUCTION

Symbiosis involving unicellular algae and heterotrophic hosts is a common feature of many terrestrial and aquatic environments (Droop, 1963; Ahmadjian, 1967; Muscatine, 1973; Taylor, 1973a, b). The algal species concerned are generally recognized as members of the Cyanophyceae, Chlorophyceae or Dinophyceae. Dinophyceae are dominant as symbionts among marine invertebrate hosts (Taylor, 1973a, b, 1974a), and are probably the most ubiquitous algal symbionts to be found. Their existence constitutes a significant natural resource, both as an element of benthic primary productivity (notably in tropical seas), and as the principal driving force behind the specialized metabolic activities of their invertebrate hosts (e.g. hermatypic corals, gorgonians).

Dinoflagellates are known to be unique among the algae. Studies of their fine structure confirm that they are a clearly defined assemblage, recognized by the possession of characteristic nuclear and cytoplasmic organization (Dodge, 1971). Although incomplete, our knowledge of the nutrition of free-living forms suggests considerable diversity. Habits may range from strict autotrophy in photosynthetic forms, to obligate phagotrophy in non-photosynthetic species. The majority fall somewhere in between, exercising the more flexible nutritional options of auxotrophy and mixotrophy. These nutritional opportunists are perhaps the more successful species, although there is no sound experimental data to substantiate this view. With respect to symbiotic species, we know that their ultrastructure conforms with that observed in free-living forms (Taylor, 1968, 1969a, 1973a, b), but lack sufficient data to enable a similar conclusion with regard to their use of nutritional resources. One might surmise that the benefits of symbiosis would be more readily secured by symbionts capable of utilizing the maximum number of options presented by both the host and the external environment. As part of a general assessment of the

---

* Contribution from the Rosenstiel School of Marine and Atmospheric Science, University of Miami, 4600 Rickenbacker Causeway, Miami, Florida 33149.

cellular and metabolic interactions of the symbiotic state, it is essential that these be defined experimentally.

Two routes may be followed in seeking this data, (1) controlled laboratory experiments on symbionts cultured axenically, and (2) observations made directly on symbionts *in situ*. Other alternatives, such as parabiosis (see below), are developments of (1). In-vitro experimentation provides exceptional opportunities to study the organism's response to all potential conditions of existence, especially those of the culture vessel itself (e.g. volume, shape, nutrient gradients) (Levandowsky, 1972). They may clearly go beyond those conditions relative to a symbiotic state, and may possibly be less useful as a measure of the nutritional interactions with a host than studies *in situ* which have a higher degree of variability. While the latter can be extremely difficult to interpret, they are clearly a necessary complement to any series of experiments *in vitro*. Comparative studies focus the design of media and other requirements for culture, on conditions which exist *in situ*. In this fashion they serve to illuminate the factors governing symbiont nutrition and growth, host metabolism, and the mechanisms of regulation, which serve to maintain cellular equilibrium in the functional unit.

The discussion which follows is an attempt to analyze and bring together existing data on the growth and nutritional requirements of three symbiotic dinoflagellates that have been successfully isolated in axenic culture, *Gymnodinium microadriaticum* (Freudenthal), *Amphidinium chattonii* (Hovasse) and *A. klebsii* Kof. et Swezy. These observations can be related to a broader understanding of the symbiotic relationship which these algae form with marine invertebrates. A discussion of symbiont systematics has been omitted since this has received treatment in recent reviews (Taylor, 1973*b*, 1974*a*).

GROWTH AND NUTRITION *IN VIVO* AND *IN VITRO*

Growth kinetics and growth state

All three species under consideration have been successfully isolated using techniques originally described by McLaughlin & Zahl (1957, 1959, 1962). Routinely this has been done using ASP-8A, the medium developed specifically for this purpose (McLaughlin & Zahl, 1966). Experience has shown that there is apparently very little justification for the use of such a complex medium for the routine maintenance of cultures. Comparable growth rates have been obtained using ASP 2, ASP 6 (Provasoli, McLaughlin & Droop, 1957), ES-1 (Provasoli, 1968) and 'Erdschreiber'. The conditions which all of these media pose may, however, be far removed from those sought when attempting to duplicate the symbiotic milieu. It seems significant, however, that ASP-8A does yield a higher rate of success during periods of initial isolation. Its use at that time is recommended.

Using these media, growth rates have been analyzed under a variety of conditions of temperature and light. Optimal rates are obtained at 24–27 °C with illumination levels of 5000 μW cm^{-2}. Analysis of growth constant (Jitts, McAllister, Stephens & Strickland, 1964) gives values of 15×10^{-3} for *G. microadriaticum*, 20×10^{-3} for *A. chattonii* and 18×10^{-3} for *A. klebsii*. Inoculum size greatly influences the resulting pattern of growth, with lower cell numbers acting to extend the length of the lag phase. The phenomenon is known from studies of other algal species (Bunt, 1968), and may be indicative of a fundamental nutritional or physiological requirement.

The ability to infect the aposymbiotic larvae of the acoel, *Amphiscolops langerhansi* with experimental symbionts provides an opportunity to examine patterns of symbiont growth *in situ* (Taylor, 1971*a*). Due to problems of host acceptance, *G. microadriaticum* is not compatible with this system (Taylor, 1971*a*). However, useful data do exist for *A. chattonii* and *A. klebsii* (the natural symbiont). Growth of these symbionts *in situ* may be analyzed by either direct cell counts (useful during periods of initial growth, i.e. 7–10 days) or chlorophyll determinations (useful for extended studies). The former has the advantage that the same animal may be examined continuously. Growth curves for *A. chattonii* and *A. klebsii* based on cell counts following artificial infection have been determined. The growth constant for the maximum rate during logarithmic growth is 29×10^{-3} for *A. chattonii* and 31×10^{-3} for *A. klebsii*. These values exceed those obtained under optimum conditions of axenic culture. During the experimental period a true stationary phase is never reached, although some slowing of the maximum rate is seen (cf. Taylor, 1973*b*) (Fig. 1). Results of longer studies based on chlorophyll determinations confirm this. Determinations were made following the methods detailed by Strickland & Parsons (1968). These data show that the gradual slowing down of growth rate seen in cell count studies continues, until the host attains maximum size and an animal/algal equilibrium is established. Regulation of this equilibrium in other symbioses has been discussed previously (Taylor, 1969*b*). In the present instance, the growth state of both *A. chattonii* and *A. klebsii* in experimental hosts appears to be analogous with the stationary phase seen with these same species growing axenically in batch culture. There is no data to suggest, however, that the physiological state of these algae is similar in both systems.

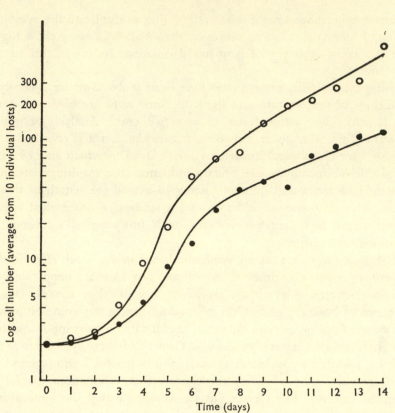

Fig. 1. Growth curve for *A. klebsii* (closed circles) and *A. chattonii* (open circles) based on cell counts *in situ*. See text for explanations.

This question is being pursued through the use of continuous cultures. A small volume chemostat has been constructed following the design used by Droop (1966). With this system, the mean physiological state of a symbiont population can be poised and stabilized, to provide uniform material for study. Successful cultures of *G. microadriaticum* have been established when either carbon dioxide or nitrogen are limited. Under these conditions, growth can be poised in continuous stationary phase. Analyses of materials excreted by symbionts in this phase suggest that the physiological condition of symbionts is nearly comparable to that encountered with cells growing *in situ* with a natural host. In continuous culture, *G. microadriaticum* will excrete 20–25 % of the total carbon fixed in photosynthesis. Chromatographic analysis shows that the principal components are glycerol, glucose and alanine, with measurable quantities of glutamate, aspartate, succinate and glycolate and several unidentified organic acids. Glycerol accounts for 85–90 % of the total excreted, glucose 5 % and alanine 1–2 %. These data compare favorably with published work on patterns of excretion *in situ* in this same species, although the absolute amounts encountered here are admittedly lower (Trench, 1971*a*,*b*). Studies of excreted photosynthate in batch culture are not nearly as favorable. Under these conditions, *G. microadriaticum* will excrete only 1–2 % of the total photosynthate, although the patterns of excretion are similar (Taylor, 1971*b*). It seems likely therefore, that the maintenance of a suitable growth state for symbionts *in situ* is a significant factor effecting the nutritional interactions which serve to sustain the symbiosis. It is clear also that comparative studies utilizing continuous culture systems (both chemostats and turbidostats) are necessary for further analysis of the symbiosis.

Growth in artificial systems

In-vitro growth responses of isolated dinoflagellate symbionts may also be examined experimentally with parabiotic systems. These are of special value since they permit metabolic exchanges between hosts and symbionts, while excluding physical contacts. In this way parabiosis forms a potentially useful bridge between the sometimes conflicting conditions of in-vitro culture and in-vivo studies involving cellular integration with a host.

Two systems have been used successfully in our laboratory. (1) Dialysis cells (No. 261 Technilab Instruments), employing permeable cellulose membranes to separate aposymbiotic hosts and axenic cultures of algal symbionts, and (2) Hollow-fibre devices (b/HFD-1 Beaker Dialyzer, Bio-Rad Laboratories). Both systems provide adequate separation and rapid exchange of metabolites. Hollow-fibre devices have been adopted for routine studies because they permit longer experiments and provide for rapid sampling of the host and algal population. A typical experimental set-up is shown in Fig. 2. Axenic symbiont cultures are placed in the smaller 'minibeaker' along with 10 cm³ sterile ASP-8A medium; bacteria-free, aposymbiotic hosts are placed in the larger beaker with 100 cm³ sterile ASP-8A medium. Closed circulation of sterile 'carrier' medium (ASP-8A) is maintained between the fibre bundles with a peristaltic pump. This circulation is held at a constant rate of 5 cm³ min^{-1} and, following equilibration, serves to maintain homogeneous conditions within the two beakers.

Growth rates for all three dinoflagellate symbionts have been determined using parabiosis with either single aposymbiotic *Aiptasia pallida*, or 100–200 aposymbiotic larval *Amphiscolops langerhansi* (see Taylor, 1974*b*). Temperature and light were held to previously stated values. With *A*.

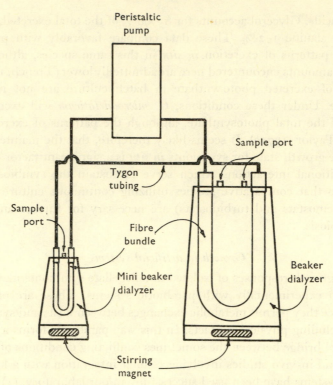

Fig. 2. Apparatus used for parabiosis with Hollow Fibre Devices.

pallida as the experimental host, *A. klebsii* exhibits a growth constant of 30×10^{-3}, *A. chattonii* a growth constant of 29×10^{-3} and *G. microadriaticum* a growth constant of 29×10^{-3}. Similarly, growth constants in the presence of *A. langerhansi* are 31×10^{-3} for *A. klebsii*, 29×10^{-3} for *A. chattonii* and 27×10^{-3} for *G. microadriaticum*. Growth constants for symbionts cultured in this system without an aposymbiotic host are 16×10^{-3} for *A. klebsii*, 18×10^{-3} for *A. chattonii* and 17×10^{-3} for *G. microadriaticum*. Growth constants in the absence of a host agree well with data presented earlier in reference to the growth of symbionts in batch culture with ASP-8A medium, and give confidence about an apparent stimulation of growth rate in the presence of a host. This may be attributed to a number of factors (i.e., gas or nutrient/waste exchange) that have not been adequately monitored in these early experiments. One would expect that these same factors are operable in intact associations found in nature. The separation of host and symbiont in parabiosis ought to make them amenable to direct analysis.

Nutritional requirements in vitro

The nutritional requirements of dinoflagellate symbionts are known only from growth studies in batch culture. As such, they may have but slight relevance to requirements experienced while growing in the host milieu. However, they do serve to place knowledge of these organisms within the context of known patterns of micro-algal nutrition. Much of this work has been carried out by McLaughlin & Zahl with the symbiont *G. micro-adriaticum* (1959, 1966). More recent studies employing *A. klebsii, A. chattonii* and *G. microadriaticum* have failed to expand this information significantly. Growth requirements in culture indicate that they are similar to other, free-living dinoflagellates in all respects. Maintenance in culture for several years clearly demonstrates that these symbiotic species can exist autotrophically in defined mineral media without organic supplements. This suggests that any postulated utilization of host-supplied organic nutrients would be purely facultative (see below). Like many free-living species, there is an apparent requirement for vitamins B_{12} and thiamine. Early studies suggested that this requirement for B_{12} was not absolute for *G. microadriaticum* (McLaughlin & Zahl, 1966), but that the vitamin did serve to stimulate growth in culture. More recent studies utilizing vitamin depleted cells show a requirement for B_{12} in all three symbiotic species. Addition of thiamine is stimulatory, particularly in the presence of B_{12}, while the addition of biotin has no observable effect on the growth rate.

In defined mineral media, requirements for nitrogen may be satisfied by the addition of a range of inorganic and organic sources (e.g. nitrate, ammonia, urea, uric acid, guanine, adenine, arginine, tryptophan, lysine). Phosphate requirements are readily satisfied by the addition of glycerophosphate, adenylic, guanylic or cytadylic acids.

The potential for heterotrophy and photoassimilation of organic substrates is poorly studied in the Dinophyceae. Available data suggest that, when it occurs, it is of limited significance in the overall nutritional picture (Provasoli & McLaughlin, 1963; Elbrächter, 1972). Symbiotic species are generally regarded as strict autotrophs (Droop, 1963; McLaughlin & Zahl, 1966), an unusual view if one considers that heterotrophy is a primary characteristic of species living in nutrient-rich environments such as the host milieu. As a potential nutritional option in symbioses with invertebrate hosts, it deserves careful experimental study. Preliminary screening of *A. klebsii, A. chattonii* and *G. microadriaticum* for possible heterotrophy and photo-heterotrophy, employing the radio-isotope techniques used by Parsons & Strickland (1962), indicates that acetate, lactate, succinate, pyruvate and glycerol are assimilated by all three species. Glucose is assimilated by *A. klebsii* and *A. chattonii* but not *G. microadriaticum*. Utilization of glucose has been studied using axenic cultures of *A. klebsii*. This species will incorporate [^{14}C]glucose (uniformly labelled) at rates up to 2.9 μmoles h^{-1} (mg N)$^{-1}$. Analysis of the pathways for utilization show that glucose contributes substantially to respired carbon dioxide. Radiochromatographic analysis indicates that comparatively little label is incorporated into stable cellular constituents (i.e. protein, lipid, nucleic acid). Available data are summarized in Table 1. Comparative studies, utilizing all three symbionts and a range of substrates, are needed before the potential heterotrophic abilities of these algae can be adequately assessed.

Table 1. *Distribution* (% *total* ^{14}C *recovered*) *of label incorporated in fractions of* G. microadriaticum *following 2 h exposure to* [^{14}C]glucose (*1 mmol l^{-1}*). *Fractions identified according to Roberts* et al. (*1955*)

Fraction	% ^{14}C recovered
Soluble	44.1
Lipids	11.8
Nucleic acid	2.9
Protein and insoluble matter	16.3
Respired CO_2	24.9

CONCLUSIONS

Detailed knowledge of microalgal physiology and biochemistry is confined principally to those organisms that come most readily into culture. These laboratory 'weeds' are nutritional opportunists in the extreme. They find adaptation to life in a culture vessel a facile matter and, as a result, have become the standard against which other species are evaluated. No single dinoflagellate species is known to the same extent as the ubiquitous *Chlorella vulgaris*, despite the fact that as a group they may be far more significant in terms of their environmental impact. The Dinophyceae are known to be cytologically unique, and, although this property alone is insufficient as proof, it may be a reflection of underlying biochemical and physiological uniqueness. As such, it deserves careful consideration. Parasitic and mutualistic symbiosis with other organisms is a widespread habit among dinoflagellates, and may be taken as a characteristic trait in certain genera (Chatton, 1920; Taylor, F. J. R., 1973; Taylor, 1973*a, b*). This in itself may have some fundamental biochemical or physiological basis, effecting cellular compatability with potential hosts, the quality and quantity of excreted metabolites or some underlying nutritional requirement of the alga. In view of present knowledge, the latter seems highly unlikely.

The discussion presented here is an attempt to focus attention on these questions by examining data from some rather fundamental and elementary experiments, relating to the patterns of (dinoflagellate) symbiont growth and nutritional requirements *in vitro* and *in vivo*. In batch culture, the three species examined here exhibit growth constants that are equivalent to those reported previously for other, free-living species (Jitts *et al.*, 1964). Association with a host in parabiotic experiments, or in studies involving artificial infection of an aposymbiotic host, clearly enhances growth rate with a corresponding increase in the growth constant. In parabiotic experiments, this may be attributable to a supplementing of the basic medium with nutrients or wastes released by the host. Correspondence between maximal growth rates *in situ* and in parabiosis is encouraging, and suggests that a carefully managed parabiosis could provide a fruitful avenue of study, directed at the basics of nutrient exchange between symbiotic partners.

Analysis of growth patterns *in situ*, and in batch culture, reveals useful information about the growth state of dinoflagellate symbionts living in host tissues. Continuous culture allows an extension of these observations by providing physiologically uniform cells poised in a known growth state. Assuming that these approximate to the condition of symbionts *in situ*, it should be possible to examine the way in which growth and nutritional states effect the quality and quantity of excreted metabolites in functional symbioses. Chromatographic analysis of chemostat-grown *G. microadriaticum* shows patterns of excretion that are similar to those reported from in-situ studies (Smith, Muscatine & Lewis, 1969; Trench, 1971*a*, *b*). The percentage of the total carbon fixed in photosynthesis that is excreted by *G. microadriaticum* is much lower than that found in experiments involving a host. Excretory rates of this same symbiont are known to be stimulated by a host tissue factor (Muscatine, 1967; Muscatine, Pool & Cernichiari, 1972). The present data suggest that an algal 'factor', dependent upon the growth state of the entire symbiont population, may also contribute to the selective release of carbon compounds *in situ*.

Growth conditions and the nutritional state of symbionts will undoubtedly have a critical effect upon the character of potential nutrients which are made available to the host. Careful study of these parameters should lead eventually to a more complete understanding of the way in which symbiont metabolism contributes to host growth and the production of specialized structures and materials (e.g. the photosynthetically dependent calcification of hermatypic corals, terpenoid biosynthesis in the Octocorallia).

ACKNOWLEDGEMENTS

Studies reported here were supported by grants GB 19790-1, GA-43533X from the National Science Foundation. Ms J. Fontana assisted with experimental studies.

REFERENCES

AHMADJIAN, V. (1967). *The Lichen Symbiosis*. London: Blaisdell.

BUNT, J. S. (1968). The influence of inoculum size on the initiation of exponential growth by a marine diatom. *Z. allg. Mikrobiol.*, **8**, 287–290.

CHATTON, E. (1920). Les Péridiniens parasites. Morphologie, reproduction, éthologie. *Arch. Zool. exp. Gen.*, **59**, 1–475.

DODGE, J. D. (1971). Fine structure of the Pyrrophyta. *Bot. Rev.*, **37**, 481–508.

DROOP, M. R. (1963). Algae and invertebrates in symbiosis. *Symp. Soc. gen. Microbiol.*, **13**, 171–199.

— (1966). Algae. In *Methods in Microbiology*, **38**, 269–313 (eds. J. R. Norris and D. W. Ribbons.) London: Academic Press.

ELBRÄCHTER, M. (1972). Begrenzte Heterotrophie bei *Amphidinium* (Dinoflagellate). *Kieler Meeresforsch.*, **28**, 84–91.

JITTS, H. R., McALLISTER, C. D., STEPHENS, K. & STRICKLAND, J. D. H. (1964). The cell division rates of some marine phytoplankters as a function of light and temperature. *J. Fish. Res. Bd. Canada*, **21**, 139–157.

LEVANDOWSKY, M. (1972). Ecological niches of sympatric phytoplankton species. *Am. Natur.*, **106**, 71–78.

McLAUGHLIN, J. J. A. & ZAHL, P. A. (1957). Studies in marine biology. II. *In vitro* culture of zooxanthellae. *Proc. Soc. expl Biol. Med.*, **95**, 115–121.

— (1959). Axenic zooxanthellae from various invertebrate hosts. *Ann. N.Y. Acad. Sci.*, **77**, 55–72.

— (1962). Axenic cultivation of the dinoflagellate symbiont from the coral *Cladocora*. *Arch. Mikrobiol.*, **42**, 40–41.

— (1966). Endozoic algae. In *Symbiosis*, **1**, 257–297 (ed. S. M. Henry). New York: Academic Press.

MUSCATINE, L. (1967). Glycerol excretion by symbiotic algae from corals and *Tridacna* and its control by the host. *Science, Wash.*, **156**, 576–579.

— (1973). Nutrition of corals. In *Biology and Geology of Coral Reefs*, **2**, 77–115 (eds. O. A. Jones and R. Endean). New York: Academic Press.

MUSCATINE, L., POOL, R. R. & CERNICHIARI, E. (1972). Some factors influencing selective release of soluble organic material by zooxanthellae from reef corals. *Mar. Biol.*, **13**, 298–308.

PARSONS, T. R. & STRICKLAND, J. D. H. (1962). On the production of particulate organic carbon by heterotrophic processes in seawater. *Deep Sea Res.*, **8**, 211–222.

PROVASOLI, L. (1968). Media and the prospects for the cultivation of marine algae. In *Cultures and Collections of Algae*, Proceedings of US–Japan Conference, Hakone 1966, 63–75 (eds. A. Watanabe and A. Hattori). Tokyo: Japan Society of Plant Physiologists.

PROVASOLI, L. & McLAUGHLIN, J. J. A. (1963). Limited heterotrophy of some photosynthetic dinoflagellates. In *Symposium on Marine Microbiology*, 105–113 (ed. C. H. Oppenheimer). Springfield, Illinois: Thomas.

PROVASOLI, L., McLAUGHLIN, J. J. A. & DROOP, M. R. (1957). The development of artificial media for marine algae. *Arch. Mikrobiol.*, **25**, 392–428.

ROBERTS, R. B., COWIE, D. B., ABELSON, P. H., BOLTON, E. T. & BRITTON, R. J. (1955). Studies of biosynthesis in *Escherichia coli*. *Publs Carnegie Instn*, **607**, 13–30.

SMITH, D., MUSCATINE, L. & LEWIS, D. H. (1969). Carbohydrate movement from autotrophs to heterotrophs in parasitic and mutualistic symbiosis, *Biol. Rev.*, **44**, 17–90.

STRICKLAND, J. D. H. & PARSONS, T. R. (1968). *A Practical Handbook of Seawater Analysis*. Bulletin **167**, 185–196, Fish. Res. Bd. Canada.

TAYLOR, D. L. (1968). *In situ* studies on the cytochemistry and ultra-structure of a symbiotic marine dinoflagellate. *J. mar. biol. Ass. UK*, **48**, 349–366.

—— (1969a). Identity of zooxanthellae isolated from some Pacific Tridacnidae. *J. Phycol.*, **5**, 336–340.

—— (1969b). On the regulation and maintenance of algal numbers in zooxanthellae-coelenterate symbiosis, with a note on the relationship in *Anemonia sulcata*. *J. mar. biol. Ass. UK*, **49**, 1057–1065.

—— (1971a). On the symbiosis between *Amphidinium klebsii* (Dinophyceae) and *Amphiscolops langerhansi* (Turbellaria: Acoela). *J. mar. biol. Ass. UK*, **51**, 301–313.

—— (1971b). Patterns of carbon translocation in algal–invertebrate symbiosis. In *7th International Seaweed Symposium, Sapporo*, 590–597 (ed. K. Nisizawa). Tokyo: University of Tokyo Press.

—— (1973a). Algal symbionts of invertebrates. *A. Rev. Microbiol.*, **27**, 171–187.

—— (1973b). The cellular interactions of algal–invertebrate symbiosis. *Adv. mar. Biol.*, **11**, 1–56.

—— (1974a). Symbiotic marine algae; taxonomy and biological fitness. In *Symbiosis in the Sea*, 245–262 (ed. W. B. Vernberg). Columbia: University of South Carolina Press.

—— (1974b). Nutrition of algal-invertebrate symbiosis. I. Utilization of soluble organic nutrients by symbiont-free hosts. *Proc. R. Soc. Lond. B*, **186**, 357–368.

TAYLOR, F. J. R. (1973). General features of dinoflagellate material collected by the 'Anton Bruun' during the International Indian Ocean Expedition. In *Ecological Studies, Analysis and Synthesis*, **3**, 155–169 (ed. B. Zeitschel). Berlin: Springer-Verlag.

TRENCH, R. K. (1971a). The physiology and biochemistry of zooxanthellae symbiotic with marine coelenterates. I. The assimilation of photosynthetic products of zooxanthellae by two marine coelenterates. *Proc. R. Soc. Lond. B*, **177**, 225–235.

—— (1971b). The physiology and biochemistry of zooxanthellae symbiotic with marine coelenterates. II. Liberation of fixed ^{14}C by zooxanthellae *in vitro*. *Proc. R. Soc. Lond. B*, **177**, 237–250.

THE EFFECT OF FUNGI (PARTICULARLY OBLIGATE PATHOGENS) ON THE PHYSIOLOGY OF HIGHER PLANTS

By J. G. MANNERS and A. MYERS

Department of Biology, University of Southampton,
Southampton SO9 5NH

Host–parasite relationships involving obligate pathogens of higher plants have much in common from a physiological point of view with those occurring in symbiotic associations involving green plants, but the net effect on the host is deleterious, not beneficial. Host–pathogen relationships involving non-obligate biotrophic pathogens (Wood, 1967) resemble in many respects those found in relationships with obligates; in both cases, food materials are abstracted from the living tissues of the host.

Some necrotrophic pathogens are very different from biotrophs in that they rapidly kill the entire host plant, and then live on the dead remains, e.g. many organisms causing blights or soft rots. Apart from the production of toxins which may be host-specific, this relationship is essentially saprophytic in nature, and is of little relevance to the subject of the present symposium. Other necrotrophic pathogens, however, have a much more local effect on the host, and exist for appreciable lengths of time on living host plants. The effects of such pathogens on the host may have much in common with those brought about by biotrophic pathogens, though the means by which they are brought about may be somewhat different.

EFFECTS OF PATHOGENIC ATTACK AT THE CELL LEVEL

Until recently the distinction between obligate pathogens, i.e. those which could not be grown in artificial culture, and non-obligate pathogens was closely correlated with the presence or absence of haustoria. Now that a number of rust fungi have been grown in axenic culture (Scott & Maclean, 1969), this relationship no longer holds good if the terms are used in a strict sense. However, the distinction between haustorial and non-haustorial fungal pathogens appears to be of considerable physiological importance, even though it can no longer be correlated with the obligate or non-obligate nature of the pathogen, and even though there is little or no direct evidence that the haustorium plays any special physiological role not played by ordinary vegetative hyphae (Bushnell, 1972). In non-haustorial (traditionally non-obligate) pathogens, the organism may usually be readily grown in culture, and its nutritional requirements are often relatively simple. In such pathogens much of the interest lies in the means by which host resistance is overcome. This topic has recently been fully reviewed (Wood, Ballio & Graniti, 1971) and will not be further considered here. In haustorial pathogens, the nutritional requirements are presumably more complex: there is some evidence for this from the rusts, which require certain amino acids for satisfactory growth in axenic culture (Coffey & Shaw, 1972; Kuhl, Maclean, Scott & Williams, 1971) and from *Tieghemiomyces parasiticus*, a haustorial parasite of *Cokeromyces*, which also has such requirements (Binder & Barnett, 1971). In other groups of haustorial pathogens, which apparently have still not been grown in axenic culture, for example the Erysiphales, the requirements may well be more complex, and related to biochemical deficiencies of a rather fundamental nature. In the case of the viruses, the obligate nature of the pathogen is associated with major deficiencies in the protein synthesizing machinery of the pathogen, which it makes good by utilising that of the host. It therefore seemed worthwhile to investigate the behaviour of nucleic acids in powdery mildew infections.

Nucleic acids of powdery mildews and their hosts

In a preliminary investigation, Plumb, Manners & Myers (1968) noted the high RNA content of the conidia of *Erysiphe graminis*, but were unable to detect the presence of DNA in the spores. In a recent re-examination carried out in Southampton by Mr J. Woolston, values for RNA per spore in the range of 40–70 pg were confirmed for conidia of *E. graminis* and *E. cichoracearum* (cucumber mildew), and polyacrylamide gel electrophoresis revealed that this RNA is attributable to normal, mostly ribosomal, components. By use of fluorescence assay for DNA (Kissane & Robins, 1958), Woolston was able to avoid interference from a brown pigment which had previously prevented the estimation of DNA in *E. graminis* by the diphenylamine method, and values of DNA per spore of 0.63 pg and 1.5 pg were obtained for *E. graminis* and *E. cichoracearum* respectively; values which are, in fact, higher than those for other ascomycetes (Fincham & Day, 1971), and in no way suggest an explanation for the apparent inability of the Erysiphales to grow in the absence of an appropriate host plant.

In the course of measuring nucleic acid levels in wheat, Plumb *et al.* (1968) made the observation that the extractability of the host DNA in perchloric acid was enhanced in infected leaves, relative to that in un-infected material. This difference in behaviour appeared during the treatment of ethanol-extracted material in 10 % perchloric acid at 30 °C for 18 h, a process designed to remove RNA from the tissues, but which also results in considerable degradation of DNA to acid soluble products. Whilst this increased susceptibility to hydrolysis is not in itself a fundamental change, it can be assumed to be an indication of a significant change in the state of host DNA as a result of infection, and might be related to the means by which the physiology of host cells is influenced by the pathogen. Woolston has made a further examination of the conditions under which this change in properties of host DNA takes place and has made some preliminary observations concerning the biochemical basis of this effect.

Factors affecting the rate of extractability of host DNA in perchloric acid

Host plants were grown in a partially temperature-controlled glasshouse at *c.* 20 °C, with daylight supplemented by mercury-vapour lamps. To assess DNA extractability, harvested leaves were cut into 1 cm pieces, fixed in boiling ethanol and further extracted in ethanol for 3 h. After briefly washing in water (a test run using a cold 2 % perchloric acid treatment did not give significantly different results) and shredding the material into approximately 2 mm diameter pieces, the extraction of DNA was carried out in 10 % perchloric acid at 30 °C. At the end of the experiment, any residual DNA was removed by extracting the residue in 10 % perchloric acid at 60 °C for 30 min. DNA in the perchloric acid samples was estimated by diphenylamine assay (Giles & Myers, 1965) with a calibration curve produced using calf thymus DNA partially hydrolysed in 10 % perchloric acid.

Using this method, DNA extractabilities of mildew-infected and uninfected barley leaves (cv. Sultan) were measured (Fig. 1) and extractability differences, at their greatest between 8–16 h incubation, and of the same order as those reported by Plumb *et al.* (1968) were observed.

It seemed possible, however, that this effect of infection might be due to changes in host tissue permeability to either perchloric acid or the DNA degradation products, rather than a more direct effect on the nuclear materials. This was tested by blending the leaf material immediately after ethanol treatment, using a Silverson mixer–emulsifier which, as judged by microscopic examination, gave a very high percentage of cell wall breakage. Subsequent comparison of DNA extractability in 10 % perchloric acid at 30 °C for 14 h gave average values of 55 % and 97 % for uninfected and

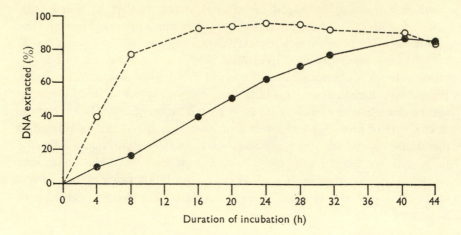

Fig. 1. DNA extracted from mildew-infected (○ - - - ○) and uninfected (●——●) third leaves of barley by 10 % perchloric acid at 30 °C. Each point is the mean of two determinations obtained from the extraction of twelve leaves.

mildew-infected barley leaves respectively, results very similar to those obtained from unbroken tissues.

In a further experiment, uninfected and mildew-infected cucumber eaves (cv. Long Green Ridge) were compared in terms of DNA extractability (Fig. 2): again, an infection-induced increase in extractability was found. In uninfected leaves, the first perchloric acid sample, taken about

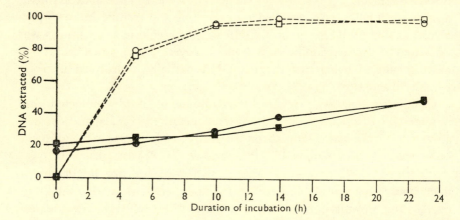

Fig. 2. DNA extracted from mildew-infected and uninfected fifth leaves of cucumber by 10 % perchloric acid at 30 °C with and without a pretreatment with 2 % perchloric acid at 2 °C for 18 h. Each point is the mean of two determinations obtained from the extraction of four leaves. ○ - - - ○, Infected washed; □ - - - □, infected unwashed; ●——●, uninfected washed; ■——■, uninfected unwashed.

5 min after the addition of the 10 % perchloric acid, contained approximately 20 % of the total leaf DNA. This situation was not altered by a preliminary 2 % perchloric acid wash at 2 °C, but the rapidly extracted fraction was absent from infected leaves. At present we can offer no explanation for this result.

The effect of mildew on host DNA extractability is not unique however. A comparison between barley leaves infected with mildew and leaves infected with barley yellow dwarf virus (Fig. 3) gave strikingly similar results in each case.

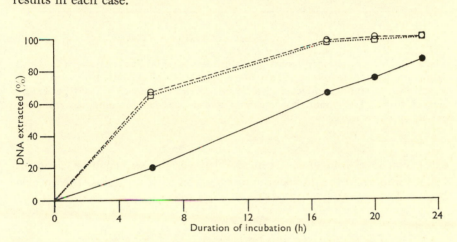

Fig. 3. DNA extracted from healthy (●—●) barley, yellow dwarf virus-infected (□ ... □) and mildew-infected (○ - - - ○) third leaves of barley by 10 % perchloric acid at 30 °C. Each point is a mean of two determinations obtained from the extraction of eight leaves.

A frequently observed effect of infection is the early onset and increased rate of leaf senescence, and since some of the more obvious symptoms of infection are also those of senescence it was felt necessary to investigate the possibility that the increase in DNA extractability resulting from mildew or virus infection could be a result of enhanced senescence induced by infection. The results of an experiment in which levels of DNA extractability of uninfected barley leaves were followed as the leaves aged (Fig. 4) show that there is an increase with age of leaf, so that at harvest 3 (28 days after sowing), at a time before the third leaf shows visible signs of senescence, the rate is approaching that of mildew infected young leaves.

These increases in DNA extractability may be either the result of changes brought about by senescence (whether natural or brought about by infection), or associated with fundamental changes that at the gene

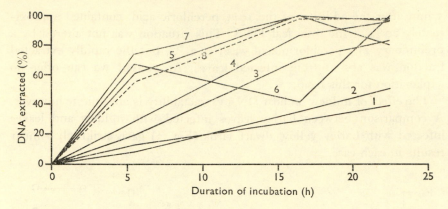

Fig. 4. DNA extracted from uninfected third leaves (solid line) of barley of various ages (samples 1–7 represent 17, 21, 28, 33, 37, 41, and 46 days after sowing respectively), and mildew-infected (dashed line) barley (sample 8) by 10 % perchloric acid at 30 °C. Each point is a mean of two determinations obtained from the extraction of six leaves.

expression level lead to senescence. Whatever the sequence of events, it seemed important to establish the biochemical basis for these observations of changes in DNA behaviour towards perchloric acid.

The biochemical basis of increased DNA extractability

The changes in DNA extractability could be due either to molecular changes in the DNA itself, of which a drastic reduction in molecular weight might be the most important, or to changes in the associated proteins, which together with DNA and a small proportion of RNA make up the chromatin. The second explanation seemed to be the more likely and the more accessible to investigation. Accordingly an experiment was carried out in which relative DNA extractability from uninfected and mildew-infected barley leaf material was observed at various stages during the DNA isolation process, in the hope that if the extractability difference was lost, the stage of the disappearance would point to the factors causing the difference in behaviour of intact material.

The technique for DNA extraction was based on that of Smith & Halvorson (1967) and samples were taken at each stage for evaluation of DNA extractability in perchloric acid. The leaves were first chopped and suspended in a solution containing 0.15 mol l⁻¹ NaCl and 0.1 mol l⁻¹ EDTA at pH 8.0, and broken by four passages through an X-press (A. B. Biox, Sweden). The resulting frozen blocks were thawed, made up to standard volume and sampled (Fig. 5, I). The suspension was filtered

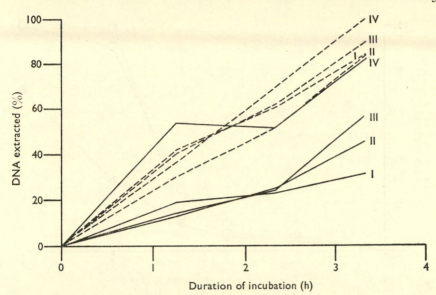

Fig. 5. DNA extracted from samples taken at various stages (see text) of the isolation of DNA from mildew-infected (dashed line) and uninfected (solid line) third leaves of barley by 10 % perchloric acid at 30 °C. Each point is a mean of two determinations.

through nylon mesh and the filtrate sampled (II). DNAase activity was destroyed by heating to 60 °C for 10 min (sample III) and the suspension taken up in a solution containing 2 % SDS and 1 mol l⁻¹ NaClO₄ (sample IV). The results (Fig. 5) show that the operations carried out between samplings III and IV are responsible for the loss of extractability differences between the preparations from infected leaves and those from uninfected leaves, and this suggests that the critical stage is the dissociation of proteins from the DNA.

In a further series of experiments, barley leaf chromatin was isolated by a method based on that of Huang & Bonner (1962) and shown to retain the DNA extractability differences when isolated from uninfected and infected leaves (Fig. 6), although the time required for the perchloric acid treatment to render the bulk of the DNA soluble was reduced compared to that for intact leaves. Analysis of the chromatin isolated from uninfected, mildew-infected, and senescent barley leaves indicated that whilst the DNA: histone ratio was about 1 : 1 in all cases, the proportion of non-histone chromatin protein was increased from approximately 21 % to around 52 % in infected or senescent material.

A preliminary qualitative analysis of histones and non-histone chromatin proteins by polyacrylamide gel electrophoresis was carried out. Of the

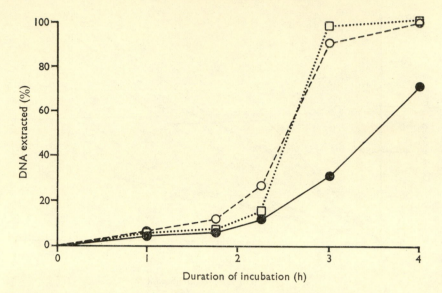

Fig. 6. DNA extracted from chromatin isolated from mildew-infected (○ - - - ○) and uninfected (●——●) third leaves of barley, and DNA made soluble from high molecular weight calf thymus DNA (□ ... □) by 10% perchloric acid at 30 °C.

seven major bands of histone in chromatin from chromatin protein of healthy, young leaves (Fig. 7a), band 2 (fraction I*b* of Bonner *et al.*, 1968) was missing from samples from infected material, and from those from senescent leaves; in senescent leaves there was also a marked reduction in the relative amounts of bands 3 and 4. With non-histone chromatin proteins (Fig. 7b), bands were again seen to disappear with both infection and senescence, so that in the infected sample only one band remained, in spite of the increase in the proportion of total non-histone protein.

Finally, DNA extractability was examined in normal and dehistonised chromatin from uninfected and mildew-infected barley. The results were not conclusive (Fig. 8) but suggest that removal of histones does not remove the extractability difference, and in fact may increase it, and therefore suggest that the difference is due to a change in non-histone chromatin proteins. So far it has not proved possible to carry out the reciprocal experiment involving removal of non-histone proteins whilst leaving the histones intact. Further experiments to test the behaviour of reassembled chromatin would also be of great interest.

Shaw and his colleagues (Bhattacharya, Naylor & Shaw, 1965; Bhattacharya & Shaw, 1968; Bhattacharya, Shaw & Naylor, 1968) have obtained evidence of changes in nuclear proteins following the infection of wheat by rusts: in particular, the proportion of histones decreased while non-

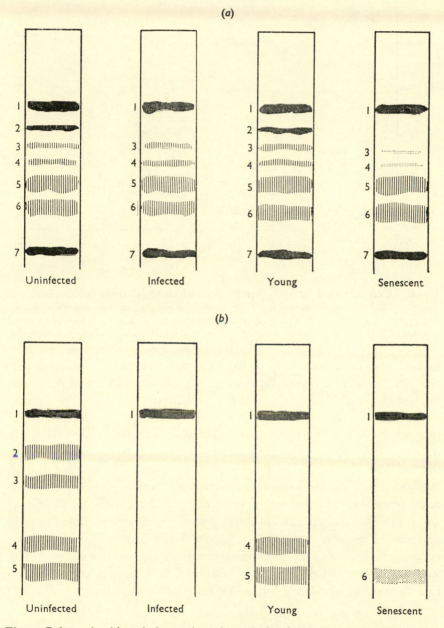

Fig. 7. Polyacrylamide gel electrophoretic analysis of proteins obtained from the chromatin of mildew-infected third leaves, uninfected young third leaves and senescent third leaves of barley. (*a*), histone proteins; (*b*), non-histone proteins.

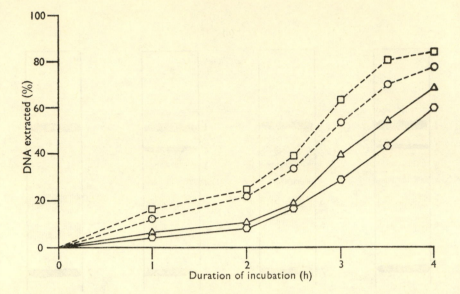

Fig. 8. DNA extracted by 10% perchloric acid at 30 °C from dehistonised and untreated chromatin obtained from mildew-infected and uninfected third leaves of barley. □ - - - □, Infected material dehistonised; ○ - - - ○, infected material normal; ⬡—⬡, uninfected material dehistonised; △—△, uninfected material normal.

histone protein increased. This, together with changes in nuclear RNA, was interpreted as evidence for the activation or inactivation of particular host genes, possibly being a primary event following infection. An elucidation of these matters and a fuller understanding of the relevance of phenomena resembling those observed during senescence (a resemblance noted by Uritani, 1971) must await further investigation, though it is of interest that the supply of kinetin or benzimidazole both retards leaf senescence and increases resistance to rusts and powdery mildews. Accelerated senescence is a frequent feature of attack by biotrophic pathogens and is probably an indication of the imperfect nature of the quasisymbiotic relationship between host and pathogen: for a biotrophic pathogen to obtain the maximum advantage from the relationship, the rate of senescence would need to be unaffected or even retarded, as in fact sometimes occurs, e.g. when green islands are formed (Wood, 1967).

EFFECT OF PATHOGENIC ATTACK ON TRANSLOCATION AND WATER RELATIONS

A major factor in the attack of many fungal pathogens is not a reduction in the amount of host assimilation, or a diversion to the pathogen of a major proportion of that assimilation, but involves instead a disturbance to the transport systems of the host. The mechanisms, in so far as they have been elucidated, seem to be quite diverse.

Biotrophic pathogens

Doodson, Manners & Myers (1965), working on *Puccinia striiformis* on wheat, and using $^{14}CO_2$ showed that when a single mature leaf was infected, translocation from an infected leaf, expressed as a percentage of assimilation, was greatly reduced by infection. Fourteen days after inoculation only negligible proportions (0.4 %) of assimilate were leaving the presentation (infected) leaf of infected plants, whereas in the control the proportion was 21 %. The pattern of distribution of labelled translocate was, however, relatively unaffected, except that the proportion going to the roots was much reduced. When the entire plant was infected (Siddiqui & Manners, 1971) assimilation was affected in the same way as before, but the percentage of assimilate leaving the presentation leaf was much less affected: 14 days after inoculation the percentage of assimilate translocated from infected leaves was nearly half that moving from healthy leaves. In these generally infected plants, not only was the proportion of assimilate moving to the roots reduced in infected plants, but that going to other (infected) leaves was increased. Livne & Daly (1966) found similar distortions of translocation patterns in legumes, where the diversion of assimilates in infected leaves was much more marked than in cereals.

It was thought at first (Doodson *et al.*, 1965) that the distortion of the translocation pattern in such cases could be explained by the pathogen acting as a sink. However this does not account for the massive accumulation of carbohydrates around lesions of some biotrophic pathogens (Crowdy & Manners, 1971). Gaunt & Manners (1971*a*) showed that in the early stages of the development of loose smut (caused by *Ustilago nuda* (Jens.) Rostr.) on wheat the reduction in host growth is far too great to be attributed to the very small amount of fungal mycelium present in the plant at that stage. Since rust infection in some hosts at least e.g. certain Leguminosae (Kiraly, El Hammady & Pozsar, 1967) was known to affect the concentration of certain growth substances in the host, an investigation of the possible implication of growth factors was initiated at Southampton, again using *U. nuda* on wheat. Preliminary results (Kinloch, personal communication) indicate that this fungus produces IAA and at least one related compound *in vitro*.

Another factor which would tend to distort translocation patterns is the conversion of host carbohydrates into substances not metabolised by the host, e.g. certain polyols. This is comparatively frequent in both symbiotic and parasitic associations (Smith, Muscatine & Lewis, 1969; Gaunt & Manners, 1971*b*) though how important quantitatively such processes are is much less certain.

Necrotrophic pathogens

It has already been mentioned that translocation into the roots, and hence root growth, is reduced by reduced translocation from the leaves in plants attacked by certain biotrophic foliar pathogens. The situation is somewhat different when a necrotrophic pathogen attacks the root system. Asher (1972*a*) has shown that infection by *Gaeumannomyces* (*Ophiobolus*) *graminis* reduces the dry weight, leaf area, tiller number, and water content of the shoot in both wheat and barley. The growth of an attacked root was severely retarded, but there was an increased production of adventitious roots, and the proportion of assimilates going to the roots was greater than that in healthy plants, especially in barley. Asher (1972*b*) showed that, following the exposure of the second youngest leaf to $^{14}CO_2$, assimilates were not translocated past *G. graminis* lesions; no accumulation of activity around the lesions was detected.

It seemed to us that the mechanism by which *G. graminis* disrupted translocation and water movement in invaded roots was worthy of further investigation. Accordingly Mr M. A. Kararah, collaborating with us, has investigated the anatomical and physiological effects of attack by *G. graminis* on wheat roots. The distribution of the fungus in the host root, and its effect on host anatomy were first investigated, using both light and electron microscopy. Samples for light microscopy were embedded, sectioned and stained with safranin and light green or AMB stain (Juniper, Gilchrist, Cox & Williams, 1970). For electron microscopy, material was fixed in 6 % acrolein, followed by 1 % osmium tetroxide, both in sodium cacodylate buffer, a fixation method found to be satisfactory for both host and pathogen, and then sectioned.

When the fungus invades the root tip, the sloughed root cap cells are first penetrated, then the thick outer wall of the protoderm is penetrated; a deposit of disorganised host material usually occurs at the point of penetration. The tissues within the tip are then invaded: the plasma membrane becomes electron dense and diffuse, and occasionally the cell walls become electron dense (light microscopy indicated that this is probably due to lignification) (Plate 1*a–d*). Later, the cell walls become completely dissolved, the plasma membrane becomes discontinuous, and only a few disorganised remnants of cell contents are visible (Plate 1*e*).

When older parts of the root are invaded, the epidermal wall is penetrated by dissolution at the penetration site (Plate 2*a*). Lignitubers are frequently found as the fungus invades successive cortical cells (Plate 2*b*), and modification of the cortical cell walls, as indicated by their becoming thickened and electron-dense, is frequent (Plate 3*c*). The fungus may penetrate the cell walls at any point, but sometimes makes use of pits (Plate 2*d*). The endodermal cells are then penetrated (Plate 2*e*) and the fungus then penetrates the thickened inner walls of the endodermal cells, sometimes via pits, but often by dissolving the wall itself, including those layers impregnated with suberin (Plate 3*a–c*). Following penetration of the endodermis, the pathogen spreads to the pericycle and phloem, and finally invades and disintegrates the xylem (Plate 3*d, e*).

Experiments were then carried out with the object of determining the effect of fungal invasion on translocation, and the way in which any such effect was correlated with the extent of mycelial invasion. Healthy and diseased plants were fed with $^{14}CO_2$, using leaf chambers similar to those employed by Quinlan & Sagar (1962). The diseased plants were inoculated either by placing a strip of agar bearing mycelium in contact with a root, to obtain a lesion in a specific position, or by growing the plants in silver sand (no. 21) containing the fungus to obtain general infection. After the required period of incubation, the plants were exposed for 3 h to $^{14}CO_2$, and left for a further 3 h for translocation to continue; each plant was then sampled. Samples for anatomy were taken (in the case of diseased plants, usually 0.5 cm of root at the centre point of the lesion), and the remainder of the root system was exposed to X-ray film (Kodirex) for 15–21 days. On the basis of the results of the autoradiography, the infected roots were classified into groups as follows:

1. Roots translocating normally (Plate 4*a–c*). Radioactivity is distributed throughout the whole root system, but is particularly concentrated in the apices. In such roots, the anatomical investigation revealed anything from no infection to heavy invasion of the cortex, endodermis and pericycle, but hardly ever penetration of the phloem.

2. Roots translocating little radioactivity past the lesion (Plate 4*d–f*). Here the fungus had invaded not only the cortex, but also the endodermis and pericycle, and around 75 % of the phloem – in some cases only one of the phloem bundles was still uninvaded.

3. Roots in which translocation past the lesion was completely inhibited (Plate 4*g–i*). In such roots, most of the phloem cells had been disintegrated by the fungus, and all the remainder had been invaded.

Thus, the conclusion may be drawn that the damaging effect of *G. graminis* on the phloem is essentially local: as long as some phloem elements are uninvaded at the point of maximum infection, translocation continues.

It is only when all the phloem elements have been invaded that translocation ceases. Clarkson, Drew, Ferguson and Sanderson (personal communication) have come to the same conclusion, from experiments where ^{86}Rb was injected into the plant at the base of the shoot.

One of the most obvious symptoms of plants attacked by *G. graminis* is that they become desiccated (Asher, 1972*b*). The movement of water itself is still under investigation, but Clarkson *et al.* (personal communication) have shown that active transport of such ions as potassium and calcium is inhibited within 48–72 hours of the completion of the invasion of the phloem. Whether water uptake is affected at this stage, or whether it is only interfered with when the xylem is completely disrupted has not yet been established.

It may be concluded that disruption of translocation, and probably also of water movement is caused by invasion of the vascular elements by *G. graminis*. As long as the pathogen is confined to the cortex, no appreciable damage appears to be done to the host plant, and in this connection it is of interest to note that *Phialophora radicicola*, a fungus related to *G. graminis* which is confined to the cortex, has no noticeable effect on the growth of the host (Deacon, 1973). It may be that the pathogenicity of *G. graminis* is due to an ability to penetrate the thick and heavily suberised cell walls of the endodermis of the wheat root, though the level of cellulolytic activity is probably also important (Pearson, 1974).

GENERAL CONCLUSIONS

The means by which pathogens damage their hosts can be grouped into a number of categories.

1. The production of toxins (Wood *et al.*, 1971) or disturbance of normal growth by interference of growth factor balance (Kiraly *et al.*, 1967).

2. Physical damage to the host tissues, e.g. rupture of cuticle by cereal rusts (the analysis of such damage is usually relatively simple).

3. Water relations may be affected through damage to roots, which has been discussed here, stems, as in the vascular wilts (Dimond, 1970) or leaves.

4. The host plant, or specific host plant organs, may be starved, due to the activities of the pathogen. Starvation may be brought about by one or more of the following mechanisms:

(*a*) The pathogen acting as a metabolic sink. This must, by definition, be operative in all biotrophic pathogens, but its importance in any given disease is not easy to assess.

(*b*) The accumulation of nutrients in the host tissues around the site of infection, e.g. the accumulation of starch around cereal rusts and mildews (Allen, 1942; Shaw & Samborski, 1956).

(*c*) The disorganisation of conducting elements, as in attack by *Gaueman-nomyces graminis* – already discussed.

(*d*) The diversion of food materials from particular organs, for example the diversion of nutrients from the root system in many shoot diseases, e.g. cereal rusts and smuts – already discussed.

(*e*) The induced formation of abnormal host tissue, e.g. galls.

(*f*) Increased host respiration rates: such increases have been demonstrated in many diseases (Wood, 1967) and are probably general.

(*g*) The induction of premature senescence in host plant organs. Rather little attention has been paid to this phenomenon, but the work on cereal mildew and certain other diseases described in this paper suggests that it may be one of the more important means by which biotrophic pathogens damage their hosts. The reduction in photosynthesis rate reported in many diseases caused by biotrophic pathogens, e.g. *Puccinia striiformis* (Doodson *et al.*, 1965), may perhaps best be interpreted as one of the components of premature senescence.

Any one investigator normally examines only one or two of these aspects, but it should be borne in mind that in any one disease several factors are likely to be operative, either simultaneously or sequentially. It should also be noted that the categories listed above are overlapping and not mutually exclusive.

At the biochemical and anatomical levels, the relationships in symbiotic and parasitic associations may be remarkably similar, involving in both cases such phenomena as the conversion of host carbohydrates into substances unavailable to the host, the alteration of translocation patterns and the penetration of host cell walls by the micro-organism. The essential difference is at the whole plant level, where the host is enabled to grow normally in the one case and prevented from growing normally in the other.

REFERENCES

ALLEN, P. J. (1942). Changes in the metabolism of wheat leaves induced by infection with powdery mildew. *Am. J. Bot.*, **29**, 425–435.

ASHER, M. J. C. (1972*a*). Effect of *Ophiobolus graminis* infection on the growth of wheat and barley. *Ann. appl. Biol.*, **70**, 215–223.

(1972*b*). Effect of *Ophiobolus graminis* infection on the assimilation and distribution of ^{14}C in wheat. *Ann. appl. Biol.*, **72**, 161–167.

BHATTACHARYA, P. K., NAYLOR, J. M. & SHAW, M. (1965). Nucleic acid and protein changes in wheat leaf nuclei during rust infection. *Science, Wash.*, **150**, 1605–1607.

BHATTACHARYA, P. K. & SHAW, M. (1968). The effect of rust infection on DNA, RNA and protein in nuclei of Khapli wheat leaves. *Can. J. Bot.*, **46**, 96–99.

BHATTACHARYA, P. K., SHAW, M. & NAYLOR, J. M. (1968). The physiology of host-parasite relations. XIX. Further observations on nucleoprotein changes in wheat leaf nuclei during rust infection. *Can. J. Bot.*, **46**, 11–16.

BINDER, F. L. & BARNETT, H. L. (1971). Nutrition and metabolism of the haustorial mycoparasite *Tieghemiomyces parasiticus* in axenic culture. *Phytopathology*, **61**, 885.

BONNER, J. *et al.* (1968). Isolation and characterization of chromosomal nucleoproteins. In *Methods in Enzymology*, vol. XII B, 3–65 (eds. L. Grossman and K. Moldave). New York: Academic Press.

BUSHNELL, W. R. (1972). Physiology of fungal haustoria. *A. Rev. Phytopathol.*, **10**, 151–176.

COFFEY, M. D. & SHAW, M. (1972). Nutritional studies with axenic cultures of the flax rust, *Melampsora lini. Physiol. Pl. Pathol.*, **2**, 37–46.

CROWDY, S. H. & MANNERS, J. G. (1971). Microbial disease and plant productivity. *Symp. Soc. gen. Microbiol.*, **21**, 103–123.

DEACON, J. W. (1973). *Phialophora radicicola* and *Gauemannomyces graminis* on roots of grasses and cereals. *Trans. Br. mycol. Soc.*, **61**, 471–485.

DIMOND, A. E. (1970). Biophysics and biochemistry of the vascular wilt syndrome. *A. Rev. Phytopathol.*, **8**, 301–322.

DOODSON, J. K., MANNERS, J. G. & MYERS, A. (1965). Some effects of yellow rust (*Puccinia striiformis*) on ^{14}carbon assimilation and translocation in wheat. *J. exp. Bot.*, **16**, 304–317.

FINCHAM, J. R. S. & DAY, P. R. (1971). *Fungal genetics*. Oxford: Blackwell.

GAUNT, R. E. & MANNERS, J. G. (1971*a*). Host–parasite relations in loose smut of wheat. I. The effect of infection on host growth. *Ann. Bot.*, **35**, 1131–1140.

(1971*b*). Host-parasite relations in loose smut of wheat. III. The utilization of ^{14}C-labelled assimilates. *Ann. Bot.*, **35**, 1151–1161.

GILES, K. W. & MYERS, A. (1965). An improved diphenylamine method for the estimation of deoxyribonucleic acid. *Nature, Lond.*, **206**, 93.

HUANG, R. C. & BONNER, J. (1962). Histone, a suppressor of chromosomal RNA synthesis. *Proc. natn. Acad. Sci. USA*, **48**, 1216–1222.

JUNIPER, B. E., GILCHRIST, A. J., COX, G. C. & WILLIAMS, P. R. (1970). *Techniques for Plant Electron Microscopy*. Oxford: Blackwell.

KIRALY, Z., EL HAMMADY, M. & POZSAR, B. I. (1967). Increased cytokinin activity of rust-infected bean and broad bean leaves. *Phytopathology*, **57**, 93–94.

KISSANE, J. M. & ROBINS, E. (1958). The fluorimetric measurement of deoxyribonucleic acid in animal tissues, with special reference to the central nervous system. *J. biol. Chem.*, **233**, 184–188.

KUHL, J. L., MACLEAN, D. J., SCOTT, K. J. & WILLIAMS, P. G. (1971). The axenic culture of *Puccinia* species from uredospores: experiments on nutrition and variation. *Can. J. Bot.*, **49**, 201–209.

LIVNE, A. & DALY, J. M. (1966). Translocation in healthy and rust-affected beans. *Phytopathology*, **56**, 170–175.

PEARSON, V. (1974). Virulence and cellulolytic enzyme activity of isolates of *Gauemannomyces graminis. Trans. Br. mycol. Soc.*, **63**, 199–202.

PLUMB, R., MANNERS, J. G. & MYERS, A. (1968). Behaviour of nucleic acids in mildew-infected wheat. *Trans. Br. mycol. Soc.*, **51**, 563–573.

QUINLAN, J. D. & SAGAR, G. R. (1962). An autoradiographic study of the movement of ^{14}C-labelled assimilates in the developing wheat plant. *Weed Res.*, **2**, 264–273.

Scott, K. J. & Maclean, D. J. (1969). Culturing of rust fungi. *A. Rev. Phytopathol.*, **7**, 123–146.

Shaw, M. & Samborski, D. J. (1956). The physiology of host-parasite relations: 1. The accumulation of radioactive substances at infections of facultative and obligate parasites including tobacco mosaic virus. *Can. J. Bot.*, **34**, 389–405.

Siddiqui, M. Q. & Manners, J. G. (1971). Some effects of general yellow rust (*Puccinia striiformis*) infection on ^{14}carbon assimilation, translocation and growth in a spring wheat. *J. exp. Bot.*, **22**, 792–799.

Smith, D. & Halvorson, H. O. (1967). The isolation of DNA from yeast. In *Methods in Enzymology*, vol. XII A, 538–541 (eds. L. Grossman and K. Moldave). New York: Academic Press.

Smith, D. C., Muscatine, L. & Lewis, D. H. (1969). Carbohydrate movement from autotrophs to heterotrophs in parasitic and mutualistic symbiosis. *Biol. Rev.*, **44**, 17–90.

Uritani, L. (1971). Protein changes in diseased plants. *A. Rev. Phytopathol.*, **9**, 211–234.

Wood, R. K. S. (1967). *Physiological Plant Pathology*. Oxford: Blackwell.

Wood, R. K. S., Ballio, A. & Graniti, A. (1971). *Phytotoxins in Plant Diseases*. New York: Academic Press.

EXPLANATION OF PLATES

Electron micrographs of wheat roots infected by *Gaeumannomyces graminis*

PLATE 1

Root tip. (*a*) Undifferentiated cells of healthy root tip. (*b*) Undifferentiated cells of diseased root tip, showing fungus hyphae (h) and disorganised cell contents. (*c*) Portion of a cell adjoining an invaded cell, showing electron-dense deposits in cell wall (CW) and disorganised nucleus (N). (*d*) Fungus hypha penetrating a cell wall. Note the accumulation of electron dense material in the cell being penetrated, around the point of penetration. (*e*) Late stage in dissolution of host tissues, with cell wall (CW) and cell contents almost completely dissolved, and cell membrane (M) fragmented.

In all parts, host organs are denoted by capital letters, fungus organs by lower case letters.

(*a*) ×5500; (*b*) ×8250; (*c*) ×25 000; (*d*) ×5400; (*e*) ×21 000.

PLATE 2

Mature root: cortex. (*a*) Wall of epidermal cell (EP) being penetrated by the fungus. (*b*) Fungus hyphae in cortical tissue, showing lignituber (LT) formation. (*c*) Electron dense deposits (D) on cell walls (CW) of cortical cells. (*d*) Fungus hypha passing from one cortical cell to another through a pit. (*e*) Penetration of endodermal cell (E) from cortex (CT).

(*a*) ×14 000; (*b*) ×3000; (*c*) ×9000; (*d*) ×9400; (*e*) ×30 000.

PLATE 3

Mature root: stele. (*a*) Penetration of inner, thickened wall of endodermal cell (E), and entry into pericycle (PE). (*b*) Penetration of inner wall of endodermal cell (E) through a pit and entry into pericycle (PE). (*c*) Late stage, showing complete dissolution of inner wall (IW) of endodermal cell (E). (*d*) Invasion of vascular

tissues by the fungus. E, endodermal cell; PE, pericycle cell; S, sieve tube; C, companion cell; X, xylem parenchyma cell; V, vessel. (*e*) Late stage in invasion of xylem, showing partial dissolution of cell walls and blocking of the lumen.
(*a*) × 7000; (*b*) × 6750; (*c*) × 12 200; (*d*) × 2650; (*c*) × 5000.

PLATE 4

Fungus invasion and translocation. (*a–c*) Control, uninfected plant exposed to $^{14}CO_2$.
(*a*) Primary root (A) and a typical secondary root (B). (*b*) Autoradiographs of roots A and B. (*c*) Trans-section of stele of root A. × 200.
(*d–e*) Lightly infected plant exposed to $^{14}CO_2$.
(*d*) Primary root (A) and a typical secondary root (B). The break towards the top root A denotes the infected site (removed for sectioning before autoradiography).
(*e*) Autoradiographs of roots A and B. (*f*) Trans-section of stele of root A at point of maximum infection. × 200.
(*g–i*) Heavily infected plant exposed to $^{14}CO_2$.
(*g*) Primary root (A) and a typical secondary root (B). The break towards the top of root A denotes the infected site (removed for sectioning before autoradiography).
(*h*) Autoradiographs of roots A and B. (*i*) Trans-section of stele of root A at point of maximum infection. × 200.

PLATE I

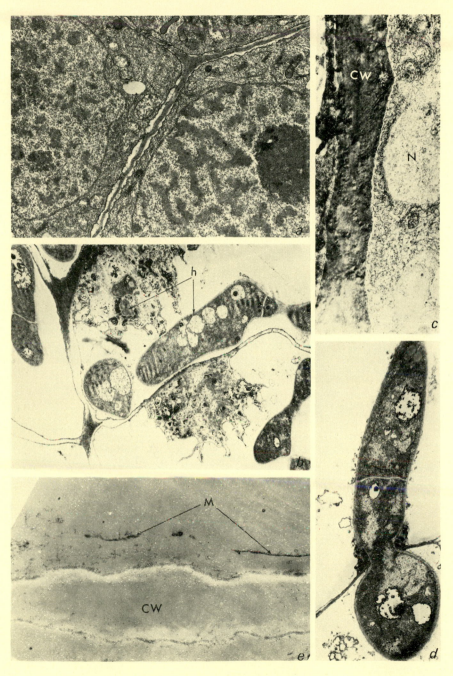

For explanation see p. 295

PLATE 2

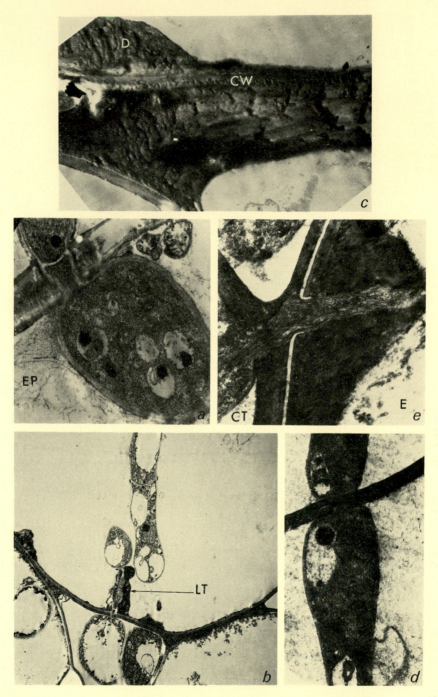

For explanation see p. 295

PLATE 3

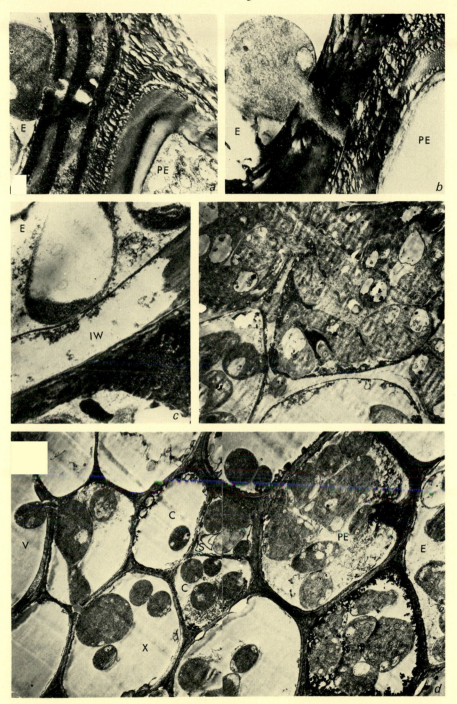

For explanation see p. 295

PLATE 4

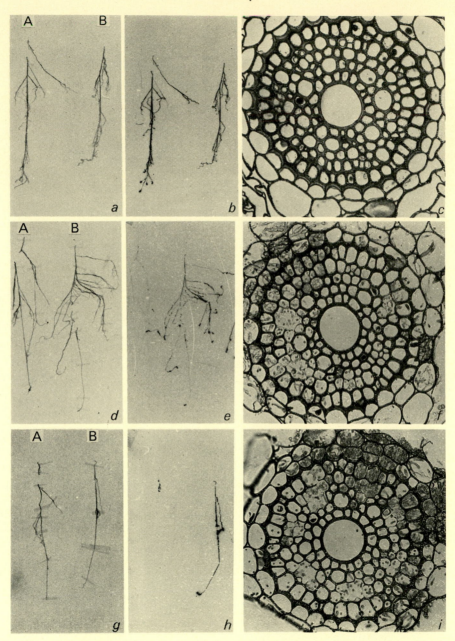

For explanation see p. 296

OBLIGATE PARASITES OF HIGHER PLANTS, PARTICULARLY RUST FUNGI

By M. D. COFFEY

Botany School and Electron Microscope Unit,
Trinity College, Dublin, Eire

INTRODUCTION

The study of obligate parasites of higher plants has never been more relevant to the survival of man than it is at the present time. The powdery and downy mildew fungi and particularly the rust fungi are still major causes of crop damage despite intensive control measures.

Their importance economically is paralleled by their significance as objects of keen academic interest. Their relationships with their host plants, the difficulty encountered in culturing them in the laboratory and the large number of physiologically different populations which exist within one species, are all indications of their complex nature. In recent years, a basic understanding of the biology of these pathogens and their associations with higher plants has begun to emerge. This information forms the basis of this review and it is my intention to provide not only a comprehensive account of the achievements but also to expose the obvious deficiencies in our present knowledge.

The rust diseases have attracted much attention and consequently my main emphasis will be on these pathogens and their associations. I will consider the germination process, the formation of infection structures and the growth of the fungal colony both in axenic culture and in the host plant. The host–parasite association will be treated in terms of those physiological events which benefit the growth of the parasite. Obviously, the specific mechanisms whereby host and parasite interact to determine the compatibility or incompatibility of the association are implicit in such a consideration. Finally, obligate parasitism in the rusts will be compared and contrasted with other types of fungus–higher plant association.

GROWTH AND DEVELOPMENT OF THE RUST SPORE

Spore germination

In an attempt to decipher some of the physiological interactions which constitute obligate parasitism, it is useful to consider the rust parasite in isolation. Uredospores which facilitate spread of the pathogen and which act as infective propagules when they come into contact with a new host are a good starting point for such physiological studies. In order for rust uredospores to germinate *en masse*, it is usually necessary to remove self-inhibitors by leaching with water. These inhibitors have recently been identified as cinnamic acid derivatives (Macko, Staples, Renwick & Pirone, 1972) and their function is presumably to prevent spore germination within the uredosorus prior to their dispersal by air currents.

In isolation from its host, the apparent inability of the rust uredospore to grow beyond the germ tube stage of development, even when provided with an exogenous nutrient supply has presented an exciting challenge to the physiologist. Staples and co-workers studied the uptake of ^{14}C-labelled acetate and leucine by germinating uredospores of three rust species and conidia of three imperfect fungi, all saprophytes (Staples, Syamananda, Kao & Block, 1962). In the rust uredospores, acetate assimilation remained at a low level compared with conidia and there was no net incorporation of label into protein. As much as 60 % of the label was incorporated into protein with *Neurospora* but no more than 9 % was incorporated with uredospores. With [^{14}C]leucine, incorporation into protein occurred at a steady linear rate in uredospores. More significantly, however, the rate of loss of leucine label from protein was at a similar level. In contrast, there was a substantial net protein synthesis in germinating conidia.

More recent work with bean rust uredospores, whose endogenous substrates had been ^{14}C-labelled, indicated that a 25 % increase in protein content occurred after $1\frac{1}{2}$ h germination (Trocha & Daly, 1970). After an initial increase in protein content, reaching a peak at 3 h germination, there was a gradual decline in protein levels. The fact that Staples and co-workers (1962) used a trichloracetic acid extraction while Trocha & Daly (1970) used a milder enzymatic hydrolysis probably accounts for the different results. A study has been made of the ability of ribosomes isolated *in vitro* to incorporate amino acids into protein at different stages in the germination process of the bean rust (Staples, 1968). A two-fold increase in amino acid incorporation occurred after 2 h germination but between 6 and 8 h the ability to incorporate amino acids into protein had dropped to 60 % of that of resting spores (Staples, 1968). Ribosomes from sunflower rust behaved in a similar manner, but ribosomes from uredospores of two races of wheat stem rust, and of flax rust remained fully active during germination and there was no obvious impairment of protein synthesising ability (Staples & Yaniv, 1973).

The conclusions to be reached from these studies are that uredospores fail to carry out appreciable synthesis of proteins. The poor utilisation of exogenously supplied substrates may be a reflection of this, though additional factors are probably involved. Amongst these, the presence of substantial amounts of endogenous lipid reserves in the uredospores, comprising up to 20 % of the spore weight, is significant (Tulloch & Ledingham, 1960).

Wheat stem rust uredospores have the capacity to carry out the β-oxidation of exogenously supplied [1 – ^{14}C]valerate (Reisener, McConnell & Ledingham, 1961). That the β-oxidation of endogenous fatty acids occurs in germinating wheat stem rust uredospores is indicated by the decline in their fatty acid content (Caltrider, Ramachandran & Gottlieb, 1963). Exogenously supplied [^{14}C]acetate is metabolised predominantly via the glyoxylate cycle (Staples, 1962) and the key enzymes of this cycle are present in uredospores (Frear & Johnson, 1961; Caltrider et al., 1963). Indeed, there is abundant evidence for the operation of the normal routes of intermediary metabolism in these fungi, including the tricarboxylic acid cycle and Ebden-Meherhoff and pentose phosphate pathways (Shaw, 1964). The importance of soluble carbohydrates as metabolites for the germinating wheat stem rust uredospore has been emphasised (Daly, Knoche & Wiese, 1967). A significant amount of soluble carbohydrates, representing between 4 and 9 % of the total spore carbon, was found in the germination medium. The soluble carbohydrates remaining within the uredospore comprised 3.4 % of the initial spore carbon and consisted mainly of trehalose, mannitol and arabitol. In the first $\frac{1}{2}$ hour of germination, approximately 30 % of these carbohydrates were metabolised, reaching over 50 % after 7 h (Daly et al., 1967).

The main function of the catabolism of lipids and carbohydrates during germination would appear to be in providing the starting materials for synthetic processes such as cell wall formation and the production of new cell membranes. A detailed study of the cell wall composition of germinating bean rust uredospores revealed the existence of glycoprotein (Trocha, Daly & Langenbach, 1974). Using mild hydrolytic procedures, it was established that there were covalent linkages between amino acids and carbohydrates. Further analysis of the uredospore and germ tube cell walls established that the carbohydrates were present as glucans and glucomannans while, in addition, the germ tube wall contained a chitin-like polymer (Trocha & Daly, 1974).

Phospholipid metabolism has been studied in some detail in germinating bean rust uredospores (Langenbach & Knoche, 1971). Levels of phospholipids dropped by 40 % in the first 20 min but, after a further 2 to 3 h, increased to some 80 % of levels initially present. These changes were mainly attributable to the catabolism and anabolism of phosphatidyl choline and phosphatidylethanolamine. It is likely that these phospholipids are involved in membrane formation during growth of the fungal germ tube.

The initial phases of germination are characterised by a catabolic phase involving the utilisation of endogenous substrates, both lipids and carbohydrates. An anabolic phase, involving synthesis of phospholipids and carbohydrates, follows and is related to the production of new cell membranes and development of a cell wall as the germ tube undergoes rapid elongation. A limited amount of protein synthesis takes place and it is probable that the major part of the protein is incorporated into the cell wall (Trocha & Daly, 1970).

Infection structure formation

If the spore germinates on the waxy cuticle surface of a leaf or stem epidermis, then this initial germination phase is followed by a precise sequence of morphological and cytological events (Maheshwari, Allen & Hildebrandt, 1967). The formation of specialised infection structures precedes the growth of the mycelial colony in the host plant. The factors which trigger these morphological events have been extensively studied (Dickinson, 1949a; Maheshwari, Allen & Hildebrandt, 1967).

When uredospores of the wheat leaf rust were germinated on collodion–wax membranes on which a thin layer of plant cell wall debris had been placed, Dickinson (1949a) observed the development of structures identical to the appressoria, infection pegs, substomatal vesicles and infection hyphae which occur in naturally infected leaves. Maheshwari, Allen & Hildebrandt (1967) reinvestigated the effects of the host epidermis and artificial membranes on the development of infection structures using several rust species. Isolated cuticles from respective host plants were capable of inducing infection structure formation, though the degree of effectiveness varied.

Crude ether extracts, hydrocarbon fractions obtained from the isolated cuticles, as well as high concentrations of mineral oil incorporated into collodion membranes generally elicited good responses. Of particular interest was the fact that the wheat stem rust and corn rust failed to develop infection structures on the oil–collodion membranes, though the corn rust did respond to membranes incorporating a hydrocarbon fraction. Maheshwari, Allen & Hildebrandt (1967) were able to obtain 95 % infection structure development of wheat stem rust uredospores on a dilute buffer solution, by incubating first at 20 °C for 2 h then at 30 °C for $1\frac{1}{2}$ h and finally at 20 °C again for 12–16 h.

A plant cuticle, not necessarily that of the host plant, is a determinant in triggering infection structure development in many rust species and a period of high temperature (30 °C) during germination is a requirement for one rust species, *P. graminis*, and perhaps others. In addition, a series of mitotic nuclear divisions coincides with infection structure formation (Dickinson, 1949*a*; Maheshwari, Hildebrandt & Allen, 1967; Wisdom & Shaw, 1974). Associated with these changes is the formation of a conspicuous nucleolus in the nuclei of the infection structures (Dunkle, Wergin & Allen, 1970; Ramakrishnan & Staples, 1970).

Dunkle, Maheshwari & Allen (1969) added metabolic inhibitors at sequential stages in order to gain more understanding of the role of nucleic acid and protein synthesis in the germination process and subsequent development of infection structures. Inhibitors of RNA synthesis, actinomycin D and 5-fluorouracil, prevented infection structure formation but did not interfere with the initial germination of the uredospore. The protein inhibitor cycloheximide inhibited completely both germination and infection structure formation.

The involvement of RNA synthesis in infection structure development of bean rust uredospores, is implied by the work of Ramakrishnan & Staples (1970). Total RNA did not change markedly, nor did the relative amounts of ribosomal and transfer RNA, but the amounts of a template RNA capable of stimulating amino acid incorporation into a cell-free system prepared from *E. coli*, increased temporarily between 4 to 8 h. This is the period of time just prior to appressorium formation. Incorporation of [^{3}H]uridine into RNA was three times as high in germlings forming infection structures as compared with those germlings forming straight germ tubes. A higher proportion of the label was found in the 4 to 19 S fraction in germlings with infection structures. Polyribosomes were extracted and treated with ribonuclease. No increase in the radioactivity profile of monosomes was detected, even though the activity in the polyribosome area declined. These results were taken to indicate that [^{3}H]-uridine had been mainly incorporated in messenger RNA although some label was found in ribosomal and transfer RNA.

Combined with the inhibitory effect of actinomycin D in bean rust (unpublished results cited in Ramakrishnan & Staples, 1970), wheat stem rust (Dunkle *et al.*, 1969) and crown rust (Naito, Lee & Tani, 1971), these biochemical results suggest that DNA-mediated RNA synthesis may initiate the series of metabolic events leading to infection structure formation. With the development of infection structures, the rust fungus establishes itself as a parasitic colony in the host plant.

GROWTH OF RUST FUNGI IN AXENIC CULTURE

In 1951, a paper appeared in the *Proceedings of the National Academy of Sciences*, by Hotson & Cutter, reporting the axenic culture of a rust, *Gymnosporangium juniperi-virginianae*, isolated from telial galls found on *Juniperus*. In a subsequent paper, Cutter (1959) described how, starting with nearly 14 000 gall slices, he obtained 358 callus cultures from which were eventually obtained seven rust isolates in axenic culture. Recently, *Cronartium fusiforme* has been grown in axenic culture, starting from branch galls of field-grown slash pine (Hollis, Schmidt & Kimbrough, 1972). Similarly, growth of white pine blister rust, *Cronartium ribicola* in tissue cultures of *Pinus monticola* (Harvey & Grasham, 1969) eventually led to the growth of this rust in axenic culture (Harvey, 1973).

Tissue culture techniques, while being a very useful method of attempting to culture the rusts, obviously do not represent the most direct approach. In 1966, Williams, Scott & Kuhl were successful in growing an isolate of the wheat stem rust (race 126-ANZ-6, 7) from aseptic uredospores placed directly on an agar medium containing sucrose, Czapek's minerals and 0.1 % Difco yeast extract. This remarkable achievement marks the beginning of a new era in obligate parasitism since nutritional and physiological studies of the rust can now be carried out apart from its host.

The direct seeding method has now been successfully used in allowing growth in axenic culture of uredospore isolates of wheat stem rust (Kuhl, Maclean, Scott & Williams, 1971; Bushnell & Stewart, 1971), flax rust (Turel, 1969a, b; Coffey & Shaw, 1972), sunflower rust (Coffey & Allen, 1973), wheat leaf rust (Raymundo & Young, 1974) and carnation rust (Jones, 1972).

The density of the uredospore inoculum used has an important bearing on the growth of rust fungi. In general, very high densities of spores are necessary to initiate axenic growth, especially in the case of the wheat stem and carnation rusts (Williams *et al.*, 1966; Bushnell & Stewart, 1971; Jones, 1972). This dependence can be overcome to some extent by the addition of defatted bovine serum albumin to the medium (Scott, 1968) and this effect is further enhanced by using gelatin suspensions of uredospores which are allowed to disperse as a very thin gel on the agar surface (Coffey, Bose & Shaw, 1969). Both the flax and sunflower rusts have been cultured using relatively low densities of uredospores (Turel, 1969a; Coffey *et al.*, 1970; Coffey & Allen, 1973). Germination is generally inhibited where high concentrations of spores are used, presumably due to the presence of endogenous inhibitors.

The requirement for a dense inoculum is most likely related to the presence of endogenous nutrients or, possibly, growth hormones which have a stimulatory effect on growth (Scott & Maclean, 1969). These unidentified substances leak out into the surrounding medium and are capable of stimulating growth from uredospores present at much lower densities (Kuhl *et al.*, 1971).

Nutrition of rusts in axenic culture

Optimal growth of the rust fungi normally occurs on media containing sucrose or glucose, mineral salts and an organic nitrogen source such as peptone, yeast extract or casamino acids (Kuhl *et al.*, 1971; Coffey & Allen, 1973). Usual levels of carbohydrates used are 3–4 %, though in the case of carnation rust, maximum dry weights of mycelium were achieved with 10 % sucrose (Jones, 1973).

The levels of organic nitrogen necessary for optimal growth can be quite critical in certain instances. Bushnell & Rajendren (1970) found that with the wheat stem rust, 0.5 % casamino acids plus 0.1 % peptone gave dry weight yields 1.4 times that on 0.1 % peptone alone. In unpublished experiments, we found that a strain of flax rust (race 1) could only be grown from uredospores when concentrations of Difco casamino acids exceeded 0.5 %. The highest frequency of growth was achieved with 2.0 % casamino acids though no growth occurred at higher concentrations (Coffey and Wall, unpublished). Jones (1973) found that subcultures of the carnation rust produced their maximum growth with 1 % peptone or yeast extract and no growth occurred at 0.1 % or 2 % levels.

The carbohydrates utilised by the wheat stem rust grown from uredospores in axenic culture (Table 1) included sucrose, glucose, fructose mannose and mannitol (Kuhl *et al.*, 1971). Likewise, growth of the sunflower rust from uredospores occurred with sucrose, glucose, mannose, mannitol, fructose and raffinose (Coffey & Allen, 1973). In experiments with two isolates of flax rust grown from uredospores (Table 1) sucrose, glucose, trehalose and mannitol supported growth of both, and in addition, raffinose, mannose, ribitol and sorbitol supported some growth of one isolate (Coffey & Allen, 1973). With colonies of an isolate of the flax rust which had been subcultured for over three years on a sucrose–peptone medium, the range of carbohydrates utilised was greater (Coffey & Shaw, 1972; Coffey & Allen, 1973).

Kuhl *et al.* (1971) were the first to show that the essential amino acid requirements for growth of the wheat stem rust from uredospores were aspartic acid and cysteine, cystine or glutathione but not methionine. Similarly, the flax rust grown from uredospores required glutamic acid and either cysteine or cystine, but not methionine (Coffey & Allen, 1973).

Table I. *Carbohydrates utilised by axenically grown rust fungi*[a]

	Isolates grown directly from uredospores				Subcultured isolates		
C source	Flax rust strain I	Flax rust race 79	Wheat stem rust race 126	Sunflower rust race I	Flax rust strain I	Wheat stem rust strain VIA	Carnation rust
Sucrose	+ +	+ +	+ +	+ +	+ +	+ +	+ +
Glucose	+ +	+ +	+ +	+ +	+ +	+ +	+ +
Fructose	o	o	+ +	+ +	+ +	+ +	+ +
Mannitol	+ +	+ +	+ +	+ +	+ +	+ +	+ +
Trehalose	+ +	+	+	o	+ +	+ +	+ +
Raffinose	+ +	+ +	+	+	+ +	+ +	+ +
Mannose	+ +	+ +	+	o	+ +	+ +	
Maltose	o	o		+	+ +	o	
Glycerol	o	o		o	+ +	+	
Ribitol	+	o			+ +	+	
Arabitol	o	o		o	+ +	o	
Sorbitol	+	o			o	+	o
Galactose	o	o		o	o		o
Starch	o	o		o			

+ +, Good growth; +, poor growth; o, no growth.

[a] Data taken from Kuhl et al., 1971; Coffey & Allen, 1973; Jones, 1973; Maclean, 1974.

Table 2. *Amino acids (other than those containing sulphur) and amides utilised by axenically grown rust fungi*[a]

	Isolates grown directly from uredospores			Subcultured isolates	
N source	Flax rust strain 1	Flax rust race 79	Sunflower rust race 1	Flax rust strain 1	Wheat stem rust strain VI
Alanine	+	+	+ + +	+ +	+ +
Glutamic acid	+ + +	+ + +	+ +	+ +	+ +
Glutamine				+ +	+ + +
Aspartic acid	+ +	o	+ +	+ +	+ +
Asparagine				+ +	+ +
Serine				+ +	+ +
Glycine				+ + +	+ +
Arginine				+ +	

+ +/+ + +, Good growth; +, poor growth; o, no growth
[a] Data taken from Howes & Scott, 1972; Coffey & Allen, 1973.

With subcultures of the flax rust, a range of amino acids (Table 2) with either cysteine or cystine supported growth (Coffey & Shaw, 1972; Coffey & Allen, 1973). The requirement for a sulphur amino acid was also established for the sunflower rust (Coffey & Allen, 1973). Either aspartic acid, glutamic acid or alanine in combination with cysteine allowed growth of this rust from uredospores (Table 2).

Metabolism of rusts in axenic culture

The unpredictability of rusts in culture, particularly those cultures grown directly from uredospores, has largely pre-empted many physiological studies which might otherwise have been undertaken. However, some studies have been made with subcultured material which has become more or less stabilised in culture.

The protein-synthesising capacity of the wheat stem rust has been investigated (Staples, Yaniv & Bushnell, 1972). Using a rust isolate which had been subcultured for four years, it was found that ribosomes isolated *in vitro* polymerised about three times as much phenylalanine as germinating uredospores of the same rust. Interestingly, the spore germlings of the wheat stem rust possessed ribosomes which maintained their ability to bind polyuridylic acid *in vitro* and incorporate leucine *in vivo*, whereas ribosomal activity in bean rust germlings declined markedly over the same 20 h period (Staples *et al.*, 1972). The bean rust has not yet been grown routinely in axenic culture, though small colonies of mycelium have been obtained after a prolonged lag phase of 12 weeks in one experiment (Coffey and Wall, unpublished).

Howes & Scott (1973) have studied sulphur metabolism in a sub-cultured line of the wheat stem rust which had a requirement for either cysteine, homocysteine or methionine. The mycelium synthesised large amounts of sulphur amino acids from [^{35}S]sulphide, but not [^{35}S]sulphate, indicating a block in inorganic sulphur metabolism. Over 70 % of the ^{35}S-labelled amino acids and peptides were released into the medium, mainly as cysteine and cysteinylglycine, and this loss probably accounts for the inability of sulphide to provide the nutritional requirements of the fungus.

Further metabolic studies of the rust fungi in axenic culture, particularly with respect to their apparent inability to retain sufficient quantities of nutrients, should not only enhance our understanding of their saprophytic capabilities but may well provide valuable clues as to the nature of their parasitic association with higher plants.

Variability of rusts in axenic culture

That some strains of the rust fungi can be grown, and others not, suggests a genetic component. However, this differing capacity may not entirely reflect true saprophytic abilities but rather the inability at present to define the nutritional requirements more exactly. Simply raising the levels of both carbohydrates and nitrogen compounds in the medium has improved the growth of some saprophytic rusts and has allowed new strains to be cultured.

In general, the dikaryotic uredospore germling gives rise to a mycelial colony which is predominantly binucleate, as determined by cytological techniques (Williams *et al.*, 1966; Coffey *et al.*, 1970). The wheat stem rust (race 126-ANZ-6,7) appears to possess a capacity to develop mycelial colonies which may be either binucleate or uninucleate (Maclean & Scott, 1970). In terms of their growth characteristics, the binucleate types formed within two weeks and growth was terminated either by staling or the formation of stromata, whereas the uninucleate types did not form for at least four weeks (often two to three months) and the colonies, remaining vegetative, were readily subcultured.

These uninucleate variants were found to be very stable in culture unlike the binucleate strains. A number of different uninucleate variants were isolated, differing in growth rates and other cultural characteristics (Maclean, 1974). The prevailing evidence now suggests that at least some of these variants are diploid (Williams & Hartley, 1971; Maclean, un-published) rather than haploid as was first thought (Maclean, Scott & Tommerup, 1971).

Infection structure formation and the dikaryon

Under natural conditions infection structure formation is a pre-requisite to the establishment of the dikaryotic parasitic colony (Dickinson, 1949a; Chakravarti, 1966). Williams (1971) also believes that in axenic growth of the wheat stem rust, infection structure formation precedes the establishment of dikaryotic mycelium. When uredospores were dispersed very thinly on a medium suitable for axenic culture, germling hyphae initiated from substomatal vesicles were dikaryotic, while those derived from straight germ tubes usually had only one nucleus. Unfortunately no attempt was made to obtain axenic cultures and consequently no correlation was obtained between the nuclear condition of the germling and the cultured mycelium.

Until direct evidence is produced that axenically grown dikaryotic mycelia are produced from groups of germlings in which at least a high percentage of infection structure formation occurs, Williams' findings must be treated with caution. Ideally, of course, the mycelium should be initiated from a single germling, but this does not appear possible at the present time.

Pathogenicity of rust fungi in axenic culture

The pathogenicity of axenically grown rusts has been demonstrated with wheat stem rust and flax rust (Williams, Scott, Kuhl & Maclean, 1967; Bushnell, 1968; Turel, 1971). In no instance has a re-isolated rust been known to change its race characteristics though differences in the abundance and type of spore formed have been noted (Turel, 1971; Maclean & Scott, 1974).

The methods used to obtain infection usually necessitate removal of the host epidermis so that the axenic mycelium or spores are in direct contact with the mesophyll tissue. In both wheat stem and flax rusts this method resulted in the formation of uredosori 10 to 12 days after inoculation (Williams et al., 1967; Bushnell, 1968; Bose & Shaw, 1974).

This inability to infect the host plant in the normal way probably reflects incomplete or abnormal spore development (Bose & Shaw, 1971; Maclean & Scott, 1974). Indeed, when the cultural conditions for sporulation were modified, flax rust uredospores were obtained which could infect the host in the normal way (Bose & Shaw, 1974).

Saprophytic cultures and parasitic colonies

One might well ask: are these saprophytic cultures similar to parasitic colonies in their physiology, nutrition and physico-chemical requirements?

After all the variability and low frequency of axenic growth encountered may, to a large extent, reflect the selection of a small number of spore germlings in a population, with a unique ability to grow as saprophytes. The long lag phase typical of many rusts in culture may be determined by the time necessary for mutations, reassortments of genetic information through nuclear migration and parasexuality, or depression of genetic information normally repressed in the parasitic colony. If the parasitic colony could be directly isolated from the host with a minimum of damage and found to grow on culture media fairly readily, many of these possibilities could be eliminated.

Lane & Shaw (1972), using a mixture of cell wall-hydrolysing enzymes, succeeded in isolating parasitic colonies of the flax rust from its host. All of these colonies grew when placed on a medium known to support growth of saprophytic colonies produced from germinating uredospores. These colonies sporulated and subsequently produced more vegetative mycelium which grew on transfer to fresh medium (Lane & Shaw, 1972). This technique should allow direct comparisons to be made of the physiology of saprophytic and parasitic colonies. It would seem to offer great potential in determining the role of the parasite in host–parasite relations.

GROWTH OF RUST FUNGI IN THEIR HOSTS

The structure of the haustorial apparatus

The uredospore germinates on the host surface and the germ tube elongates linearly. The germ tube adheres strongly to the surface of the cuticle and when it reaches a guard cell an appressorium is formed. The appressorium forms a slender infection peg which penetrates the stomatal aperture and from it develops the substomatal vesicle.

An infection hypha is produced and grows until it contacts a host cell where the hyphal tip may be cut off by a septum, so forming the haustorial mother cell (Heath & Heath, 1971; Littlefield, 1972). The wall of the haustorial mother cell becomes thickened where it contacts the host mesophyll cell wall (Plate 1a, b). The penetration peg, which later forms the haustorial neck, is continuous with the inner layer of the thickened mother cell wall (Coffey, Palevitz & Allen, 1972a; Littlefield & Bracker, 1972). About midway along the haustorial neck is a specialised region of the fungal cell wall (Plate 2b) which stains intensely with osmium tetroxide (Heath & Heath, 1971; Hardwick, Greenwood & Wood, 1971). This neck band or ring is formed during elongation of the penetration tube prior to formation of the haustorial body (Littlefield, 1972). The penetration peg elongates for several micrometres and upon cessation of growth expands to form the haustorial body (Littlefield, 1972). The haustorium expands until it reaches about 5 μm in diameter (Littlefield & Aronson, 1969). Initially, it is roughly spherical in shape (Plate 1a) but becomes asymmetrical and several lobes may develop from the body of the haustorium (Plate 1b). From estimating the frequency with which different developmental stages were observed, Littlefield (1972) postulates that the elongation of the penetration tube and initial expansion of the haustorial body are relatively rapid processes.

Electron micrographs of the haustorial neck (Plate 2b) occasionally show nuclei or mitochondria apparently migrating into the haustorial body (Heath & Heath, 1971; Littlefield & Bracker, 1972). The haustorial mother cell is frequently highly vacuolated in contrast to the dense cytoplasmic contents in the haustorium (Plate 1a, b). The fully developed haustorial body (Plate 1a, b) contains two nuclei, mitochondria with many plate-like cristae, occasional microtubules, and glycogen granules (Coffey et al., 1972a). There are many free ribosomes present, as well as an abundance of endoplasmic reticulum, mostly smooth but some associated with ribosomes (Plate 2c). In younger haustoria very few vacuoles are present, though large vacuoles appear and lipid bodies accumulate in older structures.

In a number of ultrastructural studies, the host plasmalemma has been clearly shown to be invaginated by the haustorium (Littlefield & Bracker, 1970; Heath & Heath, 1971; Coffey et al., 1972a). In most cases it is very closely applied to the fungal wall in the penetration region (Plate 2b) but where it surrounds the haustorial body an electron-lucent zone exists between it and the fungal cell wall (Plate 2c).

Using a stain specific for the plasmalemma, Littlefield & Bracker (1972) demonstrated that below the neck ring the invaginated host plasmalemma stained only faintly, whereas the remainder of the plasmalemma, as well as that of the fungus, stained very intensely. Using freeze-etching techniques, fractured faces of the host plasmalemma were shown to have a granular texture and were characterised by linear cleft-like depressions. In contrast, the invaginated host plasmalemma was smooth and the granules were absent (Littlefield & Bracker, 1972). It would appear that this extrahaustorial membrane (Bushnell, 1972) is quite different structurally, and therefore presumably functionally, from the remainder of the host plasmalemma.

The zone between the haustorial cell wall and the extrahaustorial membrane is a region about which we know very little. This extrahaustorial matrix (Bushnell, 1972) or sheath (Bracker & Littlefield, 1973) is usually electron-lucent but may contain fibrils or particles of various sizes (Plate 2c). Plasmolytic evidence from powdery mildews indicates that the matrix material is in osmotic equilibrium with the host cytoplasm and vacuole and is probably largely fluid in nature (Bushnell, 1972), but similar studies have not been carried out with rusts.

Host cell protoplasmic organisation

There are surprisingly few changes to organelle structure and organisation of the host cell cytoplasmic content during early stages of infection (Coffey et al., 1972b). Chloroplasts and mitochondria are frequently seen clustered around the haustorium (Plate 3). This arrangement of host organelles may simply result from the physical displacement of the peripheral cytoplasmic layer into the host cell vacuole by the haustorium. Nevertheless, the presence of many organelles around the haustorium is highly suggestive of, though it does not prove, a functional relationship (Plate 3). A proliferation of host organelles has even been suggested to occur (Bracker & Littlefield, 1973) but in the absence of quantitive data no good evaluation can be made. The host nucleus (Plate 1a) is frequently appressed to or even invaginated by the haustorium (Rice, 1927; Van Dyke & Hooker, 1969; Coffey et al., 1972a). Occasional cisternae of host endoplasmic reticulum are arranged approximately parallel to the extrahaustorial membrane (Bracker & Littlefield, 1973) and such cisternae are frequently seen in the region of the haustorial neck (Plate 2a).

The structure of the intercellular hyphae

The intercellular hyphae are septate and, like the haustorium, usually contain two nuclei with nucleoli (Plate 4a). The cytoplasm contains mitochondria, endoplasmic reticulum, ribosomes, glycogen particles and lipid bodies (Coffey et al., 1972a). The mitochondria usually contain less cristae than comparable organelles in the haustorium and their matrix is less dense.

The fine structure of axenically grown aerial mycelium is basically similar and cisternae of the endoplasmic reticulum are particularly evident (Coffey et al., 1972a). A simple septum is found in the rust fungi, the septal wall consisting of three layers which taper in thickness towards the centrally located pore (Plate 4c). The pore is plugged by a 'pulley-wheel' structure which is usually electron-lucent in appearance (Littlefield & Bracker, 1971; Coffey et al., 1972a). At the margins of the amorphous zone which surrounds the central pore are located single-membrane-bound organelles which contain crystals with a periodicity of 8 nm (Plate 4c). Coffey et al. (1972a) suggested that their morphological characteristics were very similar to microbodies, although histochemical tests showed that they lacked catalase, an enzyme found in most plant and animal microbodies.

The hyphal tip in the rust fungi has not been examined in any previous published work. As the structure and organisation of fungal hyphal tips may well reflect their growth capacity (Grove & Bracker, 1970) this would seem worthwhile. In preliminary studies conducted with flax rust axenic mycelium it has been established that in hyphal tips there are a small number of apical vesicles, and in the subapical region there is a zone of exclusion, containing a group of tubular cisternae which appear to give rise to more vesicles (Coffey, unpublished; Plate 4b). Although no vesicle has been seen fusing with the plasmalemma, its crenellated appearance is strongly suggestive of the addition of new membrane material (Plate 4b). In other septate fungi, a special structure of variable composition, known as the Spitzenkörper, is usually found just behind the tip of the hypha (Grove & Bracker, 1970). The apparent absence of this structure in the rust fungus may be correlated with the slow growth of this organism since it has been shown that when the Spitzenkörper is not present, growth ceases in other fungi (Grove & Bracker, 1970). The organisation of the tip, and particularly the small number of secretory vesicles present, is strongly reminiscent of the relatively slow growing germ tube tips of other fungi (Bracker, 1971).

A typical rust disease is marked by an early phase of five or six days in which the parasitic colony slowly becomes established both intra- and intercellularly within the host. This is followed by a sporulation phase which may last up to 14 days in which the infected tissue, and more particularly the tissue adjacent to the infection site, senesces.

PHYSIOLOGY OF THE HOST–PARASITE ASSOCIATION

Because of its small size and position within the host cell, the fungal haustorium does not easily lend itself to physiological studies. At present, we can only make deductions from studies using the light microscope, where resolving power is limited, or from biochemical techniques which, in general, do not distinguish between host and parasite.

Cytochemical studies

Using microspectrophotometric techniques, it has been shown that the host nuclei associated with the rust parasite undergo a number of changes (Bhattacharya, Naylor & Shaw, 1965; Bhattacharya et al., 1968). The DNA content of the host nucleus remained constant for up to about 12 days, but thereafter a decline occurred so that by 15 days 51 % of the DNA was lost. This loss was presumably associated with senescence and death of the parasitised cell. The RNA content of the nucleus, in contrast, increased by about 50 % in only two days and by six days had reached a peak, being about 90 % more than that of control nuclei (Bhattacharya *et al.*, 1965). The increases in RNA and protein in the host nucleoli were particularly striking with a threefold change in both after six days. Nucleolar volume increased dramatically from fourfold at two days to nearly eightfold after six days (Bhattacharya *et al.*, 1968). Similarly, using [^{3}H]cytidine, [^{3}H]uridine and [^{3}H]leucine, Bhattacharya & Shaw (1967) determined that at a six-day stage of infection there was a significant increase in nucleic acid and protein synthesis in both host and parasite.

Biochemical studies

Increases both in soluble nucleotide pools and in total RNA as well as changes in their base compositions are found in rust-infected tissues. These quantitative changes are first evident between six and eight days after the beginning of infection, at a time when the fungus occupies a considerable portion of the host tissue, the uredosorus is present and active sporulation has begun. There is a pronounced increase in incorporation of ^{32}P into nucleic acid fractions containing both ribosomal RNA and soluble RNA at the time of sporulation (Heitefuss & Bauer, 1969; Tani, Yoshikawa & Naito, 1973) and two- to threefold increases in ribosomal activity have been detected as early as two days after infection (Chakravorty & Shaw, 1971). Unfortunately, there is still no reliable data on the relative contributions of host and parasite to these changes in nucleic acid synthesis.

The diverse physiological changes which occur in the host–parasite association have been documented by Shaw (1963), Thrower (1965) and more recently, Scott (1972). The metabolism of starch is altered by the presence of the parasite. Initially, starch is depleted from a small area of the host surrounding the infection site and an unidentified β-amylase activator has been isolated from germinating spores (Schipper & Mirocha, 1969). Subsequently, starch accumulates in host chloroplasts in the infection site and reaches peak concentrations at the time when the uredosorus starts to develop, about five or six days after inoculation and later when sporulation has begun (Mirocha & Zaki, 1966). The parasite is able in some way to regulate the starch biosynthesis of the host and the changes induced reflect the metabolic needs of the parasite during growth and sporulation (MacDonald & Strobel, 1970). In fact, ultrastructural evidence indicates that starch biosynthesis is altered not only quantitatively

PLATE I

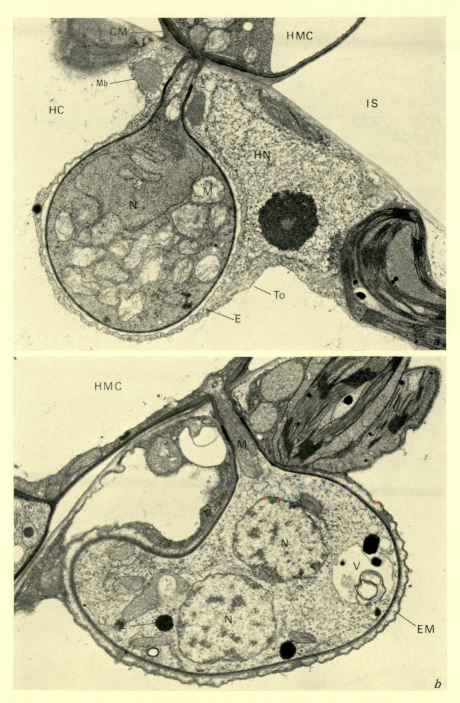

For explanation see p. 322

(Facing p. 312)

PLATE 2

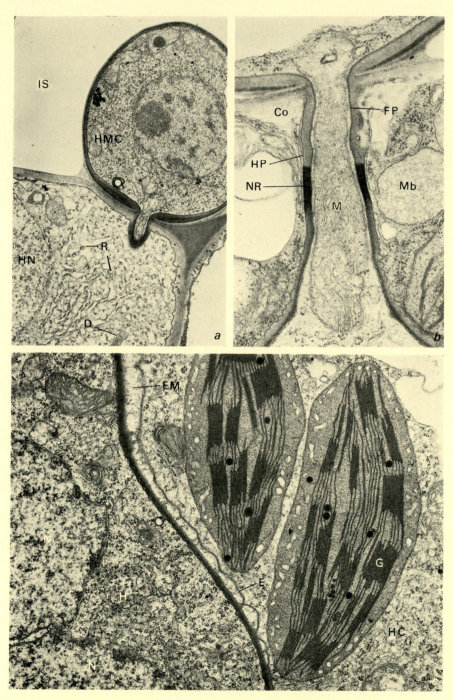

For explanation see p. 322

PLATE 3

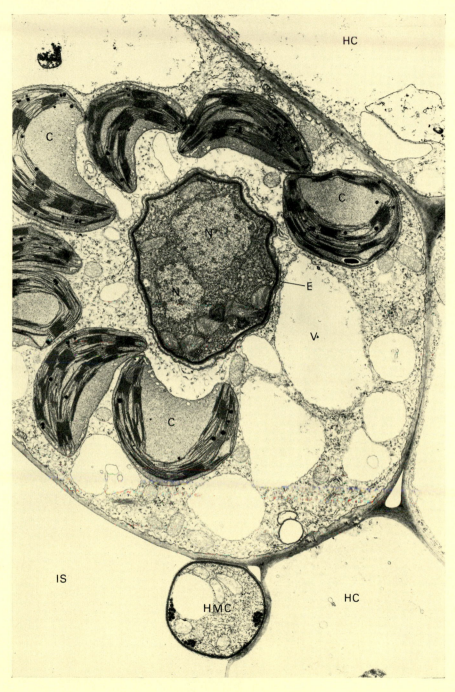

For explanation see p. 323

PLATE 4

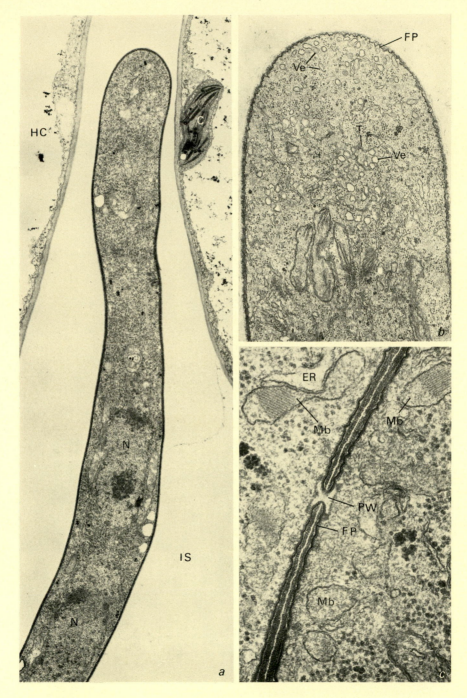

For explanation see p. 323

PLATE 5

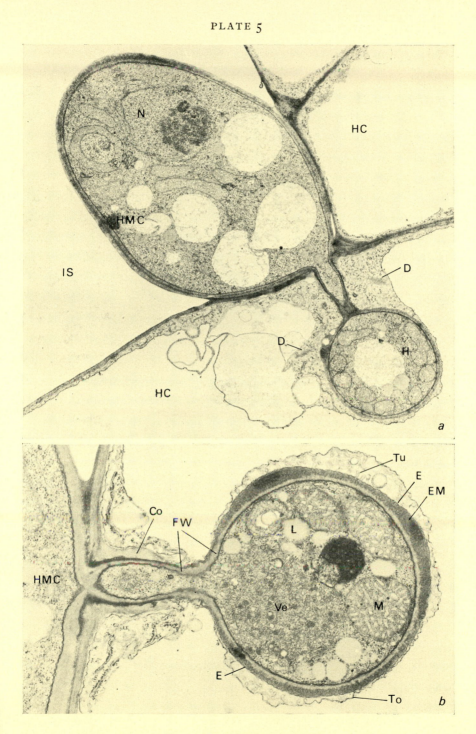

For explanation see p. 323

PLATE 6

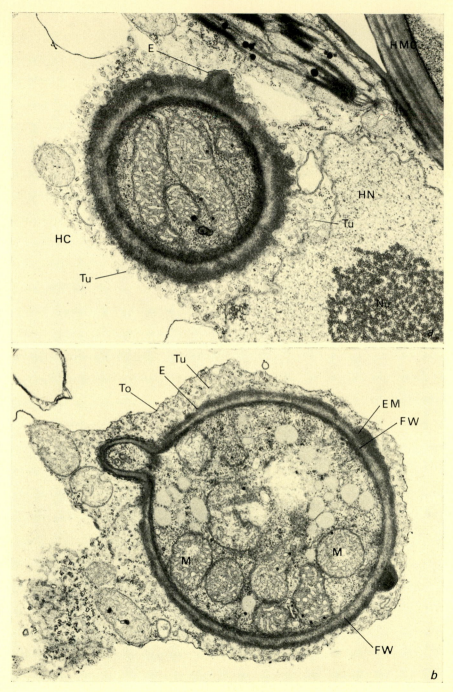

For explanation see p. 323

but also qualitatively, with the presence of clusters of glycogen-like granules in the chloroplasts during the phase of active sporulation (Coffey et al., 1972b).

During sporulation, accumulation of a wide range of inorganic and organic substances occurs and there is an increase in respiration which does not involve uncoupling of oxidative phosphorylation (Shaw, 1963; Thrower, 1965). An increased utilisation of the pentose phosphate pathway and the production of trehalose and sugar alcohols, particularly mannitol and arabitol, reflect the predominance of the metabolism of the parasite at this stage (Livne, 1964; Lewis & Smith, 1967; Mitchell & Roberts, 1973). This conversion of host sugars into trehalose and sugar alcohols which can only be metabolised by the parasite presumably helps to set up a concentration gradient which favours the growth and development of the parasite (Smith, Muscatine & Lewis, 1969). Growth hormones, including auxins, gibberellins and cytokinins, have been implicated in the increased metabolic activities that result from infection by the rust fungus, but unfortunately their site of production and precise mode of action have not been determined (Sequeira, 1973).

It is abundantly clear that the metabolic changes which occur in the infection site favour development of the pathogen. It is less clear whether the metabolism of host or parasite is principally involved. The work of Thatcher (1943) and more recently that of Hoppe & Heitefuss (1974) are perhaps relevant to this consideration. Their results indicate that there is a marked increase in the permeability of host membranes to inorganic ions, amino acids and sugars. Such an increase in permeability of host membranes would be expected to facilitate an interchange of metabolites between host and parasite. One important implication of an increased leakage from host cells is the possibility that intercellular hyphae take up nutrients directly; the host in effect being the living substrate for the parasite.

One major consideration remains and this concerns the specific mechanisms which determine the ability of the host and parasite to co-exist.

Compatibility and incompatibility

The classical work of Flor (1971) has established that a gene-for-gene relationship exists between the rust parasite and its host. The host has a series of major genes which determine resistance/susceptibility and the parasite a series of genes which govern virulence/avirulence. Moreover, for each gene conditioning a reaction in the host there is a complementary gene in the parasite which conditions pathogenicity (Flor, 1971).

In general, physiological and biochemical studies have yet to yield any definitive information as to the nature of specificity in the host–parasite association. However, cytological and ultrastructural investigations of incompatible associations have yielded useful information (Leath & Rowell, 1969; Littlefield & Aronson, 1969; Heath, 1972; Littlefield, 1973). In general, a parasite which invades a non-host elicits a response even before penetration of the first mesophyll cell is accomplished (Leath & Rowell, 1969; Heath, 1972). In some cases, highly incompatible reactions between hosts and their parasites follow a similar pattern, but more usually a haustorium is formed (Heath, 1972). Development of the haustorium is the same as in a compatible reaction for the first 48 h after inoculation, but then changes in the structural characteristics of the extra-haustorial membrane are followed by rapid necrosis of the host cell and eventually the haustorium (Heath, 1972). In intermediate-type reactions, the parasite establishes itself in an apparently normal fashion (see Plate 1*b*) but this is followed by a slow disorganisation of the host cells with the plasmalemma, tonoplast and the membranes of individual organelles losing their integrity prior to the development of uredosori (Heath, 1972).

Using isogenic lines of flax differing only in the presence of various single genes for incompatibility, it has been shown that even the highly incompatible reaction differs in the speed and extent to which cell necrosis occurs (Littlefield, 1973). It may be tentatively concluded that incompatibility is a specific event regulated by a particular gene in the host, and elicited by a corresponding avirulent gene in the parasite.

Conversely, it is tempting to suggest that susceptibility in the host and virulence in the pathogen may represent the absence of specific genes. If this is true, then the susceptibility state is achieved when the host and parasite do not interact in a specific fashion and when the host is able to provide the parasite with a suitable living substrate. That the rust fungi can produce infection structures, grow vegetatively and sporulate on a non-living medium suggests that their physical and nutritional requirements may be relatively non-specific.

SYMBIOSIS AND PARASITISM

The classical obligate parasites include the rust fungi, and powdery and downy mildews. The period of co-existence of these fungi with their hosts is generally much shorter than in mutualistic symbiotic associations such as the lichens and mycorrhizae. During this short period of co-existence, however, there is no reason to believe that the physiological mechanisms permitting this essentially balanced state are in any way less sophisticated than in more constant associations. The mechanisms which permit co-existence of two separate organisms are poorly understood. It is apparent, however, that on a more superficial level there are many similarities between different types of associations. In many cases, the formation of a haustorium is implicit in the establishment of the association, though this is by no means universal (e.g. ectoparasitic mycorrhizae and many lichens). Besides, it must not be assumed that all haustoria are structurally and functionally alike. Even amongst obligate parasites it is clear that the structure of the haustorial apparatus is dependent not only on the type of association, but also on the stage of development of the association (see Fig. 83 of Bracker & Littlefield, 1973).

One of the classical studies on fungal haustoria is that of Berlin & Bowen (1964) with the host–parasite interface of *Albugo candida* on *Raphanus sativus*. The capitate haustorium did not contain nuclei but was full of mitochondria and lomasomes. There was no evidence for a fungal cell wall around the portion of the haustorial neck proximal to the haustorial body. An amorphous electron-dense material was found between the haustorium and the extrahaustorial membrane (invaginated host plasmalemma). Glandular-like systems of tubules and connecting vesicles were found in the host cytoplasm surrounding the haustorium and some were found to be continuous with the extrahaustorial membrane. These tubules and associated vesicles were interpreted as evidence of a secretory process induced in the host by the parasite (Berlin & Bowen, 1964). As no subsequent study of any consequence has been carried out with this parasite, I decided to examine these earlier findings.

Cabbage cotyledons cv. Wheeler's Imperial were inoculated with *Albugo candida* spores. The material was examined under the electron microscope at a stage when small white pustules had appeared mainly on the undersurface of the cotyledons, some 12 days after inoculation. Numerous small spherical-shaped haustoria were observed and these were connected by a narrow neck to much larger intercellular hyphae (Plate 5a, b). Both the hyphae and haustoria were full of cytoplasm and organelles (Plates 5 and 6). Ribosomes were abundant in both haustoria and hyphae, and the former contained numerous mitochondria with tubular cristae (Plates 5 and 6). No nuclei were found in the haustoria, though they were seen in the intercellular hyphae (Plate 5a). No evidence was found for a discontinuity in the fungal wall in the haustorial neck (Plate 5b). The region between the extrahaustorial membrane and the thin haustorial wall was filled with an amorphous electron-dense material (Plate 5b). In the host cytoplasm surrounding the haustorium many tubules were seen and in some cases there was a clear continuity between these and the extrahaustorial membrane (Plates 5b and 6a). Occasionally large groups of secretory vesicles were seen in the body of the haustorium (Plate 5*b*). A collar of material which resembled that present in the host cell wall surrounded the haustorial neck at the site of penetration (Plate 5*a*, *b*).

Preliminary evidence indicates that the composition of the extra-haustorial matrix, the presence of many tubules in the host cytoplasm surrounding the haustorium and the absence of a nucleus are features which clearly distinguish it from the haustorial apparatus of the typical rust fungus (cf. with Plates 1 and 2). Although it would be unwise to speculate about the possible functions of the host tubules on the basis of this preliminary investigation, it is clear that the physiology of this host–parasite system is unlikely to be anything like that of rusts.

Supporting evidence for this idea comes from investigations on the presence of sugar alcohols in infected tissues. In a wide range of different host–parasite associations involving either Basidiomycetes or Ascomycetes, mannitol and either arabitol or erythritol were found as the predominant fungal carbohydrates. But these were not detected in associations involving Phycomycetes (Lewis, Webster and Thorpe, unpublished, cited in Smith *et al.*, 1969; Lewis, 1974; Thornton & Cooke, 1974).

All parasitic and mutualistic symbiotic associations have in common the fact that they have evolved successful mechanisms for co-existing in relative harmony. Each type of association seems to have evolved its own physiological mechanisms and a superficial study of these may lead us to believe that there are many common factors. However, the deeper we investigate these mechanisms, the more different these associations are likely to appear.

GENERAL SUMMARY AND CONCLUSIONS

Physiological studies have shown that the rust uredospore has a limited synthetic capacity associated with cell wall and phospholipid synthesis during germination. The waxy cuticle of the host plant acts as a non-specific trigger in initiating the morphogenesis of special infection structures and resulting in a series of mitotic divisions of the dikaryotic nuclei. The ability to culture many strains of the rust fungi axenically is beginning to yield valuable information about their nutrition and physiology. In future studies, a closer attention will have to be paid to the role of both infection structures and haustoria in the physiology of the parasite.

Axenic culture techniques may well prove invaluable in such studies and the work of Dickinson (1949*b*; 1971) on the induction of haustoria formation *in vitro* should be critically re-examined.

However useful the work with axenic cultures may prove to be, in the final analysis we have to return to the host–parasite association. Our present limited knowledge of this condition is to some extent a reflection of the resolving power of the techniques available to study it. The increasing sophistication of biochemical, histochemical and ultrastructural technology will have an important role to play in the study of obligate parasitism in the future. For many decades to come, we can expect that the major challenge will be the nature of specificity in obligate parasitism. In the end, though, progress in this field, as in any other, can be expected to come from those men and women of vision who deem it necessary to tackle the seemingly impossible.

ACKNOWLEDGEMENTS

I wish to express my gratitude to Mr Patrick Walsh for his excellent photographic work, and to Mrs Gillian Coffey for typing the manuscript and assisting in its editing.

REFERENCES

BERLIN, J. D. & BOWEN, C. C. (1964). The host–parasite interface of *Albugo candida* on *Raphanus sativus*. *Am. J. Bot.*, **51**, 445–452.

BHATTACHARYA, P. K., NAYLOR, J. M. & SHAW, M. (1965). Nucleic acid and protein changes in wheat leaf nuclei during rust infection. *Science, Wash.*, **150**, 1605–1607.

BHATTACHARYA, P. K. & SHAW, M. (1967). The physiology of host–parasite relations. XVIII. Distribution of tritium-labelled cytidine, uridine and leucine in wheat leaves infected with the stem rust fungus. *Can. J. Bot.*, **45**, 555–563.

BHATTACHARYA, P. K., SHAW, M. & NAYLOR, J. M. (1968). The physiology of host–parasite relations. XIX. Further observations on nucleoprotein changes in wheat leaf nuclei during rust infection. *Can. J. Bot.*, **46**, 11–16.

BOSE, A. & SHAW, M. (1971). Sporulation and pathogenicity of an Australian isolate of wheat stem rust grown *in vitro*. *Can. J. Bot.*, **49**, 1961–1964.

——— (1974). *In vitro* growth of wheat and flax rust fungi on complex and chemically defined media. *Can. J. Bot.*, **52**, 1183–1195.

BRACKER, C. E. (1971). Cytoplasmic vesicles in germinating spores of *Gilbertella persicaria*. *Protoplasma*, **72**, 381–397.

BRACKER, C. E. & LITTLEFIELD, L. J. (1973). Structural concepts of host-pathogen interfaces. In *Fungal pathogenicity and the plant's response*, section III, 159–318 (eds. R. J. W. Byrde and C. V. Cutting). London and New York: Academic Press.

BUSHNELL, W. R. (1968). *In vitro* development of an Australian isolate of *Puccinia* f. sp. *tritici*. *Phytopathology*, **58**, 526–527.

——— (1972). Physiology of fungal haustoria. *A. Rev. Phytopath.*, **5**, 343–374.

BUSHNELL, W. R. & RAJENDREN, R. B. (1970). Casein hydrolysates and peptones for artificial culture of *Puccinia graminis* f. sp. *tritici*. *Phytopathology*, **60**, 1287 (Abstr.).

BUSHNELL, W. R. & STEWART, D. M. (1971). Development of American isolates of *Puccinia graminis* f. sp. *tritici* on an artificial medium. *Phytopathology*, **61**, 376–379.

CALTRIDER, P. G., RAMACHANDRAN, S. & GOTTLIEB, D. (1963). Metabolism and function of glyoxylate enzymes in uredospores of rust fungi. *Phytopathology*, **53**, 86–92.

CHAKRAVARTI, B. P. (1966). Attempts to alter infection processes and aggressiveness of *Puccinia graminis* var. *tritici*. *Phytopathology*, **56**, 223–229.

CHAKRAVORTY, A. K. & SHAW, M. (1971). Changes in the transcription pattern of flax cotyledons after inoculation with flax rust. *Biochem. J.*, **123**, 551–557.

COFFEY, M. D. & ALLEN, P. J. (1973). Nutrition of *Melampsora lini* and *Puccinia helianthi*. *Trans. Br. mycol. Soc.*, **50**, 245–260.

COFFEY, M. D., BOSE, A. & SHAW, M. (1969). *In vitro* growth of gelatin suspensions of uredospores of *Puccinia graminis* f. sp. *tritici*. *Can. J. Bot.*, **47**, 1291–1293.

—— (1970). *In vitro* culture of the flax rust. *Melampsora lini*. *Can. J. Bot.*, **48**, 773–776.

COFFEY, M. D., PALEVITZ, B. A. & ALLEN, P. J. (1972*a*). The fine structure of two rust fungi, *Puccinia helianthi* and *Melampsora lini*. *Can. J. Bot.*, **50**, 231–240.

—— (1972*b*). Ultrastructural changes in rust-infected tissues of flax and sunflower. *Can. J. Bot.*, **50**, 1485–1492.

COFFEY, M. D. & SHAW, M. (1972). Nutritional studies with axenic cultures of the flax rust. *Melampsora lini*. *Physiol. Pl. Path.*, **2**, 37–46.

CUTTER, V. M. Jr (1959). Studies on the isolation and growth of plant rusts on host tissue cultures and upon synthetic media. 1. *Gymnosporangium*. *Mycologia*, **51**, 248–295.

DALY, J. M., KNOCHE, H. W. & WIESE, M. V. (1967). Carbohydrate and lipid metabolism during germination of uredospores of *Puccinia graminis tritici*. *Pl. Physiol.*, *Lancaster*, **42**, 1633–1642.

DICKINSON, S. (1949*a*). Studies in the physiology of obligate parasitism. II. The behaviour of the germ-tubes of certain rusts in contact with various membranes. *Ann. Bot.*, **13**, 219–236.

—— (1949*b*). Studies in the physiology of obligate parasitism. IV. The formation on membranes of haustoria by rust hyphae and powdery mildew germ-tubes. *Ann. Bot.*, **13**, 345–353.

—— (1971). Studies in the physiology of obligate parasitism. VIII. An analysis of fungal responses to thigmo-tropic stimuli. *Phytopath. Z.*, **70**, 62–70.

DUNKLE, L. D., MAHESHWARI, R. & ALLEN, P. J. (1969). Infection structures from rust urediospores: effect of RNA and protein synthesis inhibitors. *Science, Wash.*, **158**, 481–482.

DUNKLE, L. D., WERGIN, W. P. & ALLEN, P. J. (1970). Nucleoli in differentiated germ tubes of wheat rust uredospores. *Can. J. Bot.*, **48**, 1693–1695.

FLOR, H. H. (1971). Current status of the gene-for-gene concept. *A. Rev. Phytopath.*, **9**, 275–296.

FREAR, D. S. & JOHNSON, M. A. (1961). Enzymes of the glyoxylate cycle in germinating uredospores of *Melampsora lini* (Pers.) lev. *Biochem. biophys. Acta*, **47**, 419–421.

GROVE, S. N. & BRACKER, C. E. (1970). Protoplasmic organization of hyphal tips among fungi: vesicles and Spitzenkörper. *J. Bact.*, **104**, 989–1009.

HARDWICK, N. V., GREENWOOD, A. D. & WOOD, R. K. S. (1971). The fine structure of the haustorium of *Uromyces appendiculatus* in *Phaseolus vulgaris*. *Can. J. Bot.*, **49**, 383–390.

HARVEY, A. E. (1973). *In vitro* culture of *Cronartium ribicola*. *Second International Congress of Plant Pathology* Minneapolis, 5–12 September 1973 (Abstr).

HARVEY, A. E. & GRASHAM, J. L. (1969). Growth of the rust fungus *Cronartium ribicola* in tissue cultures of *Pinus monticola*. *Can. J. Bot.*, **47**, 663–666.

HEATH, M. C. (1972). Ultrastructure of host and nonhost reactions to cowpea rust. *Phytopathology*, **62**, 27–38.

HEATH, M. C. & HEATH, I. B. (1971). Ultrastructure of an immune and a susceptible reaction of cowpea leaves to rust infection. *Physiol. Pl. Path.*, **1**, 277–287.

HEITEFUSS, R. & BAUER, M. (1969). Der Einfluss von 5-Fluorouracil auf den Nukleinsäurestoffwechsel von *Phaseolus vulgaris* nach Infektion mit *Uromyces phaseoli*. *Phytopath. Z.*, **66**, 25–37.

HOLLIS, C. A., SCHMIDT, R. A. & KIMBROUGH, J. W. (1972). Axenic culture of *Cronartium fusiforme*. *Phytopathology*, **62**, 1417–1419.

HOPPE, H. H. & HEITEFUSS, R. (1974). Permeability and membrane lipid metabolism of *Phaseolus vulgaris* infected with *Uromyces phaseoli*. 1. Changes in the efflux of cell constituents. *Physiol. Pl. Path.*, **4**, 5–9.

HOTSON, H. H. & CUTTER, V. M. Jr (1951). The isolation and culture of *Gymnosporangium juniperi-virginianae* Schw. *Proc. natn. Acad. Sci. USA*, **37**, 400–403.

HOWES, N. K. & SCOTT, K. J. (1972). Sulphur nutrition of *Puccinia graminis* f. sp. *tritici* in axenic culture. *Can. J. Bot.*, **50**, 1165–1170.

(1973). Sulphur metabolism of *Puccinia graminis* f. sp. *tritici* in axenic culture. *J. gen. Microbiol.*, **76**, 345–354.

JONES, D. R. (1972). *In vitro* culture of carnation rust, *Uromyces dianthi*. *Trans. Br. mycol Soc.*, **58**, 29–36.

(1973). Growth and nutritional studies with axenic cultures of the carnation rust fungus, *Uromyces dianthi* (Pers.) Niessl. *Physiol. Pl. Path.*, **3**, 379–386.

KUHL, J. L., MACLEAN, D. J., SCOTT, K. J. & WILLIAMS, P. G. (1971). The axenic culture of *Puccinia* species from uredospores: experiments on nutrition and variation. *Can. J. Bot.*, **49**, 201–209.

LANE, D. W. & SHAW, M. (1972). Axenic culture of flax rust isolated from cotyledons by cell-wall digestion. *Can. J. Bot.*, **50**, 2601–2603.

LANGENBACH, R. J. & KNOCHE, H. W. (1971). Phospholipids in the uredospores of *Uromyces phaseoli*. II. Metabolism during germination. *Pl. Physiol., Lancaster*, **48**, 735–739.

LEATH, K. T. & ROWELL, J. B. (1969). Thickening of corn mesophyll cell walls in response to invasion by *Puccinia graminis*. *Phytopathology*, **59**, 1654–1656.

LEWIS, D. H. & SMITH, D. C. (1967). Sugar alcohols (polyols) in fungi and green plants. 1. Distribution, physiology and metabolism. *New Phytol.*, **66**, 143–184.

LEWIS, S. J. (1974). An investigation of host responses to a fungal pathogen: *Phytophthora infestans*. B.A.(Mod.) Thesis, University of Dublin, Trinity College.

LITTLEFIELD, L. J. (1972). Development of haustoria of *Melampsora lini*. *Can. J. Bot.*, **50**, 1701–1703.

(1973). Histological evidence for diverse mechanisms of resistance to flax rust, *Melampsora lini* (Ehrenb.) Lev. *Physiol. Pl. Path.*, **3**, 241–247.

LITTLEFIELD, L. J. & ARONSON, S. J. (1969). Histological studies of *Melampsora lini* resistance in flax. *Can. J. Bot.*, **47**, 1713–1717.

LITTLEFIELD, L. J. & BRACKER, C. E. (1970). Continuity of host plasma membrane around haustoria of *Melampsora lini*. *Mycologia*, **62**, 609–614.

(1971). Ultrastructure of septa in *Melampsora lini*. *Trans. Br. mycol. Soc.*, **56**, 181–188.

(1972). Ultrastructural specialization at the host–pathogen interface in rust-infected flax. *Protoplasma*, **74**, 271–305.

LIVNE, A. (1964). Photosynthesis in healthy and rust-infected plants. *Pl. Physiol.*, *Lancaster*, **39**, 614–621.

MACDONALD, P. W. & STROBEL, G. A. (1970). Adenosine diphosphate–glucose pyrophosphorylase control of starch accumulation in rust-infected wheat leaves. *Pl. Physiol.*, *Lancaster*, **46**, 126–135.

MACKO, V., STAPLES, R. C., RENWICK, J. A. A. & PIRONE, J. (1972). Germination self-inhibitors of rust uredospores. *Physiol. Pl. Path.*, **2**, 347–356.

MACLEAN, D. J. (1974). Cultural and nutritional studies on variant strains of the wheat stem rust fungus. *Trans. Br. mycol. Soc.*, **62**, 333–349.

MACLEAN, D. J. & SCOTT, K. J. (1970). Variant forms of saprophytic mycelium grown from uredospores of *Puccinia graminis* f. sp. *tritici*. *J. gen. Microbiol.*, **64**, 19–27.

(1974). Pathogenicity of variant strains of the wheat stem rust fungus isolated from axenic culture. *Can. J. Bot.*, **52**, 201–207.

MACLEAN, D. J., SCOTT, K. J. & TOMMERUP, I. C. (1971). A uninucleate wheat-infecting strain of the stem rust fungus isolated from axenic cultures. *J. gen. Microbiol.*, **65**, 339–342.

MAHESHWARI, R., ALLEN, P. J. & HILDEBRANDT, A. C. (1967). Physical and chemical factors controlling the development of infection structures from urediospore germ tubes of rust fungi. *Phytopathology*, **57**, 855–862.

MAHESHWARI, R., HILDEBRANDT, A. C. & ALLEN, P. J. (1967). The cytology of infection structure development in urediospore germ tubes of *Uromyces phaseoli* var. *typica* (Pers.) Wint. *Can. J. Bot.*, **45**, 447–450.

MIROCHA, C. J. & ZAKI, A. I. (1966). Fluctuation in amount of starch in host plants invaded by rust and mildew fungi. *Phytopathology*, **56**, 1220–1224.

MITCHELL, D. T. & ROBERTS, S. M. (1973). Carbohydrate composition of tissues infected by autoecious rusts. *Puccinia pelargonii-zonalis* and *P. malvacearum*. *Physiol. Pl. Path.*, **3**, 481–488.

NAITO, N., LEE, M-C. & TANI, T. (1971). Inhibition of germination and infection structure formation of *Puccinia coronata* uredospores by plant growth regulators and antimetabolites. *Tech. Bull. Fac. Agric. Kagawa Univ.*, **23**, 51–56.

RAMAKRISHNAN, L. & STAPLES, R. C. (1970). Changes in ribonucleic acids during uredospore differentiation. *Phytopathology*, **60**, 1087–1091.

RAYMUNDO, S. A. & YOUNG, H. C. Jr (1974) Improved methods for the axenic culture of *Puccinia recondita* f. sp. *triticia*. *Phytopathology*, **64**, 262–263.

REISENER, H., McCONNELL, W. B. & LEDINGHAM, G. A. (1961). Studies on the metabolism of valerate-1-C^{14} by uredospores of wheat stem rust. *Can. J. Biochem. Physiol.*, **39**, 1559–1566.

RICE, M. A. (1927). The haustoria of certain rusts and the relation between host and pathogen. *Bull. Torrey bot. Club*, **54**, 63–153.

SCHIPPER, A. L. Jr & MIROCHA, C. J. (1969). The mechanism of starch depletion in leaves of *Phaseolus vulgaris* infected with *Uromyces phaseoli*. *Phytopathology*, **59**, 1722–1727.

SCOTT, K. J. (1968). The initiation of saprophytic growth of *Puccinia graminis*. In *First International Congress of Plant Pathology London*, 14–26 July 1968 (Abstr.).

(1972). Obligate parasitism by phytopathogenic fungi. *Biol. Rev.*, **47**, 537–572.

SCOTT, K. J. & MACLEAN, D. J. (1969). Culturing of rust fungi. *A. Rev. Phytopath.*, **7**, 123–146.

SEQUEIRA, L. (1973). Hormone metabolism in diseased plants. *A. Rev. Pl. Physiol.*, **24**, 353–380.

SHAW, M. (1963). The physiology and host–parasite relations of the rust fungi. *A. Rev. Phytopath.*, **1**, 259–294.

(1964). The physiology of rust uredospores. *Phytopath. Z.*, **50**, 159–180.

SMITH, D., MUSCATINE, L. & LEWIS, D. (1969). Carbohydrate movement from autotrophs to heterotrophs in parasitic and mutualistic symbiosis. *Biol. Rev.*, **44**, 17–90.

STAPLES, R. C. (1962). Initial products of acetate utilization by bean rust uredospores. *Contr. Boyce Thompson Inst. Pl. Res.*, **21**, 487–498.

(1968). Protein synthesis by uredospores of the bean rust. *Neth. J. Pl. Path.*, **74** (suppl. 1), 25–36.

STAPLES, R. C., SYAMANANDA, R., KAO, V. & BLOCK, R. J. (1962). Comparative biochemistry of obligately parasitic and saprophytic fungi. II. Assimilation of C^{14}-labeled substrates by germinating spores. *Contrib. Boyce Thompson Inst. Pl. Res.*, **21**, 345–362.

STAPLES, R. C. & YANIV, Z. (1973). Spore germination and ribosomal activity in the rust fungi. II. Variable properties of ribosomes in the Uredinales. *Physiol. Pl. Path.*, **3**, 137–145.

STAPLES, R. C., YANIV, Z. & BUSHNELL, W. R. (1972). Spore germination and ribosomal activity in the rust fungi. 1. Comparison of a bean rust fungus and a culturable wheat rust fungus. *Physiol. Pl. Path.*, **2**, 27–35.

TANI, T., YOSHIKAWA, M. & NAITO, N. (1973). Effect of rust infection of oat leaves on cytoplasmic and chloroplast ribosomal ribonucleic acids. *Phytopathology*, **63**, 491–494.

THATCHER, F. S. (1943). Cellular changes in relation to rust resistance. *Can. J. Res.*, **21**, sect. *C*, 151–172.

THORNTON, J. D. & COOKE, R. C. (1974). Changes in respiration, chlorophyll content and soluble carbohydrates of detached cabbage cotyledons following infection with *Peronospora parasitica* (Pers ex Fr.) Fr. *Physiol. Pl. Path.*, **4**, 117–125.

THROWER, L. B. (1965). Host physiology and obligate fungal parasites. *Phytopath. Z.*, **52**, 319–334.

TROCHA, P. & DALY, J. M. (1970). Protein and ribonucleic acid synthesis during germination of uredospores. *Pl. Physiol.*, Lancaster, **46**, 520–526.

(1974). Cell walls of germinating uredospores. II. Carbohydrate polymers. *Pl. Physiol.*, Lancaster, **53**, 527–532.

TROCHA, P., DALY, J. M. & LANGENBACH, R. J. (1974). Cell walls of germinating uredospores. 1. Amino acid and carbohydrate constituents. *Pl. Physiol.*, Lancaster, **53**, 519–526.

TULLOCH, A. P. & LEDINGHAM, G. A. (1960). The component fatty acids of oils found in spores of plant rusts and other fungi. *Can. J. Microbiol.*, **46**, 425–434.

TUREL, F. L. M. (1969a). Saprophytic development of the flax rust *Melampsora lini*, race No. 3. *Can. J. Bot.*, **47**, 821–823.

(1969b). Low temperature requirement for saprophytic flax rust cultures. *Can. J. Bot.*, **47**, 1637–1638.

(1971). Pathogenicity and developmental differences of three saprophytically growing isolates of the flax rust fungus *Melampsora lini*, race 3. *Can. J. Bot.*, **49**, 1993–1997.

VAN DYKE, C. G. & HOOKER, A. L. (1969). Ultrastructure of host and parasite in interactions of *Zea mays* with *Puccinia sorghi*. *Phytopathology*, **59**, 1934–1946.

WILLIAMS, P. G. (1971). A new perspective of the axenic culture of *Puccinia graminis* f. sp. *tritici* from uredospores. *Phytopathology*, **61**, 994–1002.

WILLIAMS, P. G. & HARTLEY, M. J. (1971). Occurrence of diploid lines of *Puccinia graminis tritici* in axenic culture. *Nature New Biol.*, **229**, 181–182.

WILLIAMS, P. G., SCOTT, K. J. & KUHL, J. L. (1966). Vegetative growth of *Puccinia graminis* f. sp. *tritici in vitro. Phytopathology*, **56**, 1418–1419.

WILLIAMS, P. G., SCOTT, K. J., KUHL, J. L. & MACLEAN, D. J. (1967). Sporulation and pathogenicity of *Puccinia graminis* f. sp. *tritici* grown on an artificial medium. *Phytopathology*, **57**, 326–327.

WISDOM, C. J. & SHAW, M. (1974). Morphogenesis and nuclear behaviour in germinating uredospores of the wheat rust fungus. *17th Annual Meeting Canadian Federation of Biological Societies*. Hamilton, Ontario, 25–28 July 1974 (Abstr.), 58.

EXPLANATION OF PLATES

PLATE 1

(a) A median section through the haustorial apparatus of the flax rust, *Melampsora lini* (race 1), in a 5 day old infection of the susceptible host. The haustorial mother cell (HMC) is vacuolated in contrast to the haustorial body which is full of organelles, including nuclei (N) and mitochondria (M). The invaginated host plasmalemma (E) surrounds the haustorium and a thin band of host cytoplasm is seen between it and the tonoplast (To) of the host cell (HC). The host nucleus (HN) is closely appressed to the side of the haustorium. Host microbodies (Mb) are present adjacent to the neck ring. Note the dense amorphous material (CM) between the wall of the HMC and the host cell. IS, intercellular space. × 11 000

(b) A median section through the haustorial apparatus of flax rust (race 1) in a 5 day old incompatible host. The development of the haustorium is still relatively normal at this early stage. Small vacuoles (V) and several lipid bodies (L) are present. The extrahaustorial matrix (EM) contains some fibrillar material. Note the fungal mitochondrion (M) present in the haustorial neck. Other abbreviations as in Plate 1a. × 11 000

PLATE 2

(a) A near-median section through a penetration site of the flax rust in a susceptible host. An abundance of host rough endoplasmic reticulum and ribosomes (R) is commonly found in this region. Note the presence of the host nucleus (HN) and a dictyosome (D). HMC, haustorial mother cell; IS, intercellular space. × 13 000

(b) A higher magnification of Plate 1b. A collar (Co) of host material is present around the penetration site. No clear distinction between the haustorial neck wall and the host cell wall can be made at this site. Note the presence of the neck ring (NR) midway along the haustorial neck wall. The fungal plasmalemma (FP) is clearly continuous through the penetration site, but the continuity of the host plasmalemma (HP) appears to be interrupted by the presence of the collar. M, mitochondria; Mb, host microbodies. × 7000

(c) A section through the host–parasite interface of 5 day old rusted flax. The haustorium (H) contains two nuclei, endoplasmic reticulum and an abundance of ribosomes. The electron-lucent extrahaustorial matrix (EM) contains some fibrillar material. The extrahaustorial membrane (E) separates the matrix from the cytoplasm of the host cell (HC). Note the normal appearance of the host chloroplasts, especially their grana (G). N, nucleus. × 19 000

PLATE 3

A flax rust haustorium occupying a central position in a host mesophyll cell. The host cytoplasm has been displaced from its normal position at the periphery of the cell. Numerous chloroplasts (C) surround the haustorium and the usual central vacuole of the host has been subdivided into several smaller ones (V). A haustorial mother cell (HMC) occupies the intercellular space (IS) between two adjacent host cells (HC). N, nucleus. × 12 000

PLATE 4

(a) The apical and subapical regions of an intercellular hypha of the flax rust. The dikaryotic nuclei (N) occur some distance back from the tip of the hypha. HC, host cell; C, chloroplast; IS, intercellular space. × 8000

(b) The hyphal tip of axenically grown flax rust. A small number of vesicles (Ve) are present at the tip and the crenulated appearance of the fungal plasmalemma (FP) in this region suggests that fusion of vesicles has occurred. Behind the tip is a zone of exclusion where ribosomes are absent. Several tubular structures (T) appear to be giving rise to more vesicles. × 20 000

(c) The septal pore region of axenically grown flax mycelium. The pore is plugged by a 'pulley-wheel' structure (PW). The fungal plasmalemma (FP) is continuous around the septal wall on each side of the plugged pore. Microbodies (Mb) containing crystals are present in the pore region and one is associated with a cisterna of endoplasmic reticulum (ER). × 68 000

PLATE 5

(a) A near-median section through the haustorial mother cell (HMC) and haustorial apparatus of *Albugo candida* in cabbage cotyledons. The HMC contains a nucleus (N) but the spherical haustorium (H) is anucleate. HC, host cell; IS, intercellular space; D, dictyosome. × 8000

(b) A near-median section through the haustorium of *Albugo candida*. A large cluster of vesicles (Ve) are present as are several mitochondria (M) with many tubular cristae. Lipid bodies (L) are also seen. The penetration site is surrounded by a host collar (Co). The fungal wall (FW) is clearly continuous from the haustorial mother cell (HMC) through the neck region and surrounds the haustorial body. The extrahaustorial matrix (EM) is full of electron-dense material which appears to be intermingled with the outer layers of the fungal wall. Many tubules (Tu) are seen in the host cytoplasm and some are clearly continuous with the extrahaustorial membrane (E). × 17 000

PLATE 6

(a) A haustorial profile of *Albugo*. The thick wall of the haustorial mother cell (HMC) lies against the host cell wall. The extrahaustorial membrane (E) surrounds a region consisting of the extrahaustorial matrix and the haustorial wall. The host cell (HC) cytoplasm contains many tubules (Tu). The host nucleus (HN) has a large nucleolus (Nu) and is seen immediately adjacent to the haustorium. × 16 000

(b) A haustorial profile of *Albugo*. A thin layer of host cytoplasm bounded by the host tonoplast (To) and extrahaustorial membrane (E) contains many tubules (Tu). The extrahaustorial matrix (EM) is electron dense and merges with the lighter-staining haustorial wall (FW). The haustorial body contains mitochondria (M), lipid bodies (L) and many ribosomes. × 15 000

EFFECTS OF VESICULAR-ARBUSCULAR MYCORRHIZAS ON HIGHER PLANTS

By P. B. H. TINKER

Department of Plant Sciences, University of Leeds, Leeds LS2 9JT

INTRODUCTION

Lewis (1973) stated that the most meaningful classification of mycorrhizas is to group ectotrophic and vesicular-arbuscular mycorrhizas together, since these involve biotrophic fungi. His further classification of mycorrhizas as 'sheathing', 'vesicular-arbuscular', 'ericaceous', 'orchidaceous' or 'miscellaneous' seems sensible (but see Lewis (1975)) and in this paper I confine myself almost entirely to the vesicular-arbuscular grouping.

Endo-mycorrhizas have been known since last century, and the vesicular-arbuscular (v-a) type was described excellently, and in great detail. For over half a century they received rather little attention, and this was mainly directed towards the identification and classification of the endophyte. In the 1950s interest started to grow in the effects on the host, and during that period the endophyte was agreed to be in the genus *Endogone*. This early work has been well reviewed by Nicolson (1967), Gerdemann (1968) and Mosse (1973a). The existence of these excellent reviews allows me to deal quite briefly with the general background of the subject, and to concentrate more on certain questions of the mechanism of the growth-promoting effects.

MORPHOLOGY

Typical aspects of the symbiosis are shown in Plates 1 and 2 for grass and onion roots. Infection develops either from chlamydospores in the soil, or via mycelium originating from a previously infected root. The germ tube or invading hypha forms an appressorium on the root surface, and a hypha usually penetrates down between two cells. Cox & Sanders (1974) showed that the intercellular mycelium then ramifies rapidly, up to distances of some 0.5 cm from the entry point in onion root. They called the internal fungal structures arising from one single entry point an 'infection unit'. Shortly afterwards arbuscules develop: these are intracellular hyphae, which branch dichotomously and repeatedly until they end in a mass of fine hyphae, down to 0.2 μm in diameter (Plate 3). Vesicles containing lipid globules which are usually intercellular may form later still. The external mycelium develops over this period, from the original infecting hypha, and consists of wider main network hyphae of diameter 8–10 μm with shorter and narrower side branches of 2–3 μm which die off rapidly, and are then closed off by septa. Large spores develop on this mycelium (Plate 2). The form of the mycelium can vary widely, and may appear different in various host plants (see Gerdemann, 1974).

OCCURRENCE AND ECOLOGY

V–a mycorrhizas have been reported from all continents except Antarctica, and from an exceptionally wide range of hosts. As Gerdemann (1974) says: ' It is far easier to list the plant families in which they have not been found than to compile a list of plant families in which they are known to occur.' This list of families includes some forming ecto-mycorrhizas, or non-v-a endo-mycorrhizas, and a short list of families forming no or very few mycorrhizas, of which the most important are probably the Chenopodiaceae and the Cruciferae. Vesicular-arbuscular mycorrhizas are rare in aquatic plants, but some bog plants do develop the infection.

Reports continue to appear showing that species believed to be non-mycorrhizal may, in fact, become infected (e.g. Kruckelmann, 1975). Nevertheless, there are very marked differences in the ease with which different host species become infected, the extent to which this occurs, and in the degree to which any one host is infected by different endophytes (Hayman, 1975; Strzemska, 1974; Kruckelmann, 1975).

Apart from this difference in susceptibility, there is a quite remarkable lack of host specificity amongst different fungal strains in v-a mycorrhizas. Thus, Mosse (1973*a*) lists 20 host species which could be infected by the fungal strain or species called 'yellow vacuolate' (*Endogone mosseae* or *Glomus mosseae*). This does not, however, indicate that all hosts receive the same benefit from all potential endophytes, since there are very clear differences in yield improvement depending upon host, endophyte and environment.

The concept of specificity is not very precise, but might be greatly clarified by a mathematical model of infection spread. A spreading infection in growing roots is a highly dynamic system, and the result may depend upon the arbitrary date of harvest. It is possible to model this by assuming that the rate of formation of new infected root per unit volume of soil $\left(\dfrac{\mathrm{d}L_i}{\mathrm{d}t} \right)$ is proportional to a constant S, and to the amount of uninfected but receptive root $[=(nL_t - L_i)]$ and the amount of infected root (L_i), both per unit soil volume. If the plant roots are growing exponentially

$$L_t = Ae^{Rt}, \quad \text{and} \quad \frac{dL_i}{dt} = S\, L_i\, (nAe^{Rt} - L_i), \tag{1}$$

where L_t is total root length at time t, R is the relative growth rate of the host, and n and A are constants.

This approach has successfully modelled some sets of data (Fig. 1). S may thus be a definite specificity or susceptibility parameter, though the idea requires very much further testing.

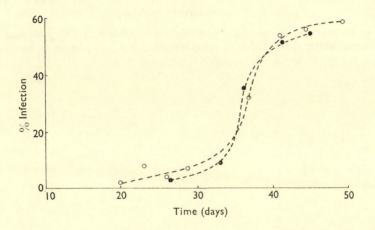

Fig. 1. Percentage of total root length of onions infected, with time, with bulbous reticulate endophyte. ○, Measured; ●, theoretical points calculated from solution to equation (1) $S = 1.25$, $n = 0.6$, $R = 0.8$ week^{-1}.

Environmental variables such as temperature, light, water regime and nutrient supply in the growth medium can affect the development of infection greatly (see Mosse, 1973a), but the full picture is by no means clear. The variability of natural environments makes comparisons between field results highly confusing, and studies in controlled environmental conditions are essential. Despite these difficulties certain points are by now fairly clear.

(a) Factors affecting host photosynthesis directly
Light shading has little effect (Redhead, 1975) but when the host is heavily shaded, spore production is reduced. The effect of different light intensities on infection varied drastically between soils (Hayman, 1974).

Low temperatures greatly reduce spore formation and infection (Fig. 2)

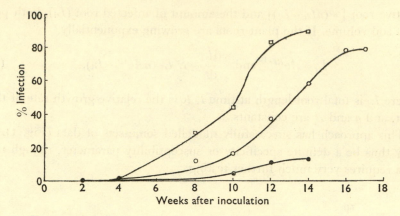

Fig. 2. Percentage infection of onion roots with *Endogone calospora* endophyte, with time, under three different temperature regimes (redrawn from Furlan & Fortin, 1973). ●, 11/16 °C; ○, 16/21 °C; □, 21/26 °C.

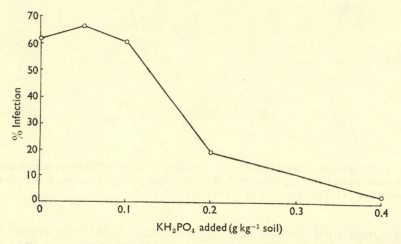

Fig. 3. Percentage infection of onion roots with yellow vacuolate endophyte, at 8 weeks old, in relation to phosphate in soil (after Sanders & Tinker, 1973).

(Furlan & Fortin, 1973). As might be expected, cutting off plant tops reduced spore numbers (Baylis, 1969).

(b) Factors affecting soil conditions

Mosse (1972) found that 'yellow vacuolate' strain (*Glomus mosseae*) did not establish in four soils below pH 4.6, but did so when they were limed to pH 5.6. The percentage infection and spore numbers are usually reduced by phosphorus and nitrogen fertilizers (Hayman, 1970; Hayman, 1975; Kruckelmann, 1975), though the effect is by no means consistent. Additional supplies of phosphorus generally diminished infection in pot culture (Fig. 3, Sanders & Tinker, 1973), and in the field (Hayman, 1975) with differences between crops (Table 1). The spore numbers were surprisingly constant, and indicate either lack of correlation of percentage infection and spore production rate, or a quite remarkable longevity of spores in this soil.

Table 1. *Mycorrhizal development in field plots given phosphorus fertilizers (from Hayman, 1975)*

Total P applied over 12 years kg/ha	Crop	Length of root with v-a infection (%)	Total v-a spores per 100 g soil
0	Potatoes	20	133
	Barley	21	163
165	Potatoes	25	119
	Barley	8	171
330 (annual application)	Potatoes	8	129
	Barley	5	273
660	Potatoes	3	131
	Barley	2	159

The situation is somewhat complex, since it is uncertain whether soil conditions exert their influence directly, by preventing or slowing external hyphal growth, or indirectly by altering the composition or metabolism of the host plant. pH or water potential effects, for example, might arise directly from an inhibition of hyphal growth in soil. However, Sanders (1975) applied phosphorus to plant leaves and showed that it is the concentration of this element inside the plant which reduces the infection. At present the only possible generalization seems to be that increased supply of mineral nutrients may tend to reduce infection, whereas increased photosynthesis may tend to increase it, but there are many apparent exceptions.

ENDOPHYTE SPECIES OR STRAINS

Following the work of Peyronel and of Mosse (see Mosse, 1973a), it has become accepted that the v-a endophytes are in the genus *Endogone*. A number of authors have described a variety of 'types' or 'strains', it being often uncertain whether they should be regarded as separate species. However, Gerdemann & Trappe (1974) have recently revised the taxonomy of these fungi, and have classified them as being in the genera *Glomus*, *Acaulospora*, *Gigaspora*, and *Sclerocystis*. None of them has yet been grown in axenic culture. Their classification depends entirely upon differences in the appearance of spores (Plate 2) since the mycelia of the various types are virtually indistinguishable.

EFFECTS ON HOST PLANT

Growth increase of host

Size and origin of effect

The observation of the effect of v-a mycorrhizas on their host plants has probably been delayed by the very ubiquity of the infection. The effects could only be noted with certainty when identifiable species of endophyte were obtained, by infecting host plants in sterilized soil with single spores (see Hayman & Mosse, 1971). This provided a reliable source of inoculum, and allowed comparison with uninfected plants in similar sterilized soil. A common practice at present is to sterilize with γ-radiation and to inoculate with a mixture of spores and mycelium, wet-sieved from soil.

Experiments of this type rapidly produced evidence that the endophyte could greatly improve growth in several host plants, and that a growth increase could almost always be reproduced by improved supplies of phosphorus to the plant (Table 2) (Hayman & Mosse, 1971). Many similar results are listed in the reviews mentioned earlier.

Table 2. *Dry weight of tops of 10 week old onions grown in ten different soils: left as a control, with the addition of 0.4 g $CaH_2PO_4 \cdot H_2O$ per kg of soil, and after inoculation with yellow vacuolate endophyte (after Hayman & Mosse, 1971)*

Soil no.	Dry weight of shoot (g)		
	Control	+ P	Inoculated
7	0.05	0.33	0.21
8	0.02	0.06	0.15
9	0.03	0.57	0.28
10	0.02	0.28	0.25
11	0.01	0.05	0.06
12	0.08	0.37	0.31
15	0.01	0.18	0.05
16	0.13	0.30	0.17
17	0.02	0.35	0.17
19	0.01	0.06	0.01

The majority of such results refer to plants growing in soil, but some have employed sand culture, with various phosphorus-containing materials of low solubility such as apatite (Murdoch, Jakobs & Gerdemann, 1967; Ross & Gilliam, 1973). The general characteristic of any growth medium in which mycorrhizal infection causes a yield response is that the concentration of some nutrient is present in extremely low concentration in the aqueous phase, but that there is a much larger 'buffer store' of the nutrient in some solid form. No clear case has yet been reported in which infection with a v-a endophyte has produced significantly greater growth than can be produced by better nutrition, and currently it is believed that this is the sole cause of growth response. Possible non-nutritional effects are not considered here.

A great majority of such experiments have pointed to phosphorus as the nutrient in question, though other elements can also be implicated. Gilmore (1971) obtained evidence that zinc nutrition could be improved by mycorrhiza formation. Increased uptakes of sulphur (Gray & Gerdemann, 1973) and strontium (Jackson *et al.* 1973) have been reported, but without yield response, and Powell (1975) has claimed that yield and potassium nutrition of *Griselinia littoralis* were improved by mycorrhizal infection. Evidence of improved nitrogen nutrition has, so far, been conspicuously absent, which is interesting in relation to the work on this subject in the ectomycorrhizas. However, Stribley & Read (1974) have found evidence of increased uptake of soil nitrogen by ericaceous mycorrhizas.

A most interesting suggestion (Safir, Boyer & Gerdemann, 1971) was that water supply from soil could be improved by mycorrhizas, though later work (Safir *et al.*, 1972) has indicated that changes in phosphorus nutrition may have been the primary effect. This point is of such enormous potential importance that it certainly deserves further investigation. The theory of water transport in soils shows close analogies with that of phosphate movement, and improved water uptake by mycorrhizas in dry soil is an attractive possibility. The present position is, therefore, that whereas other elements may occasionally be involved, phosphorus is by far the most important nutrient concerned in responses to mycorrhiza formation.

Mechanism of nutritional effect

Increased phosphorus uptake per plant could most probably arise from:

(i) Morphological changes in the plant;
(ii) increase in the ability of the plant's root surface to absorb phosphorus;
(iii) provision of additional or more efficient absorbing surface in fungal hyphae, with subsequent transfer to the host;
(iv) ability of the mycorrhizal root or hyphae to utilize sources of phosphorus not available to the uninfected root;
(v) longer duration of mycorrhiza than uninfected root as absorbing organ (see Bowen & Theodorou, 1967).

Of these, (iv) has been suggested in relation to ecto-mycorrhizas (see Bowen, 1973), and appears attractive since there is evidence (Bartlett &

Lewis, 1973) that ecto-mycorrhizal hyphae possess surface phosphatases. The ability of endo-mycorrhizal plants to use poorly soluble forms of phosphorus such as apatite and rock phosphorus also seemed to indicate a special process of some kind. This possibility has been directly tested with an isotopic dilution method, by labelling soil with ^{32}P, and comparing the specific activity of plants with and without mycorrhizas grown on it (Sanders & Tinker, 1971; Hayman & Mosse, 1972a, b; Tinker, 1975). The results (Fig. 4) show that both mycorrhizal and non-mycorrhizal plants

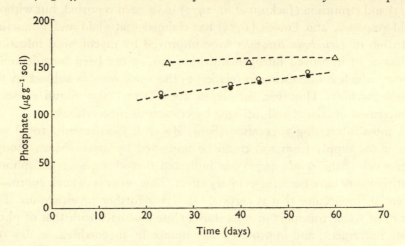

Fig. 4. Size of labile-phosphate pool in equilibrium with the soil solution, at different times: ○ as measured by specific activity of phosphorus in mycorrhizal onions; ● as measured by specific activity of phosphorus in non-mycorrhizal onions; △ as measured by extraction of phosphate with anion-exchange resin.

are obtaining their phosphate from the same source. In the same experiment, the specific activity of phosphate in the soil solution was also measured. The conclusion was that both mycorrhizal and non-mycorrhizal plants were drawing their phosphate from the soil solution, or adsorbed phosphate in equilibrium with it. Benians & Barber (1972) have reported a slight, but not statistically significant, difference in the specific activity in infected and uninfected barley. Further work on this point seems advisable, on a wider range of soils and conditions, but for the present the conclusion must be that mycorrhizal roots absorb directly from the soil solution in the same way as non-mycorrhizal roots.

Any further examination of the response mechanism must use some means of defining the rate or quantity of the response. This has usually been the overall yield increase of a plant, or the increase in its phosphorus uptake or percentage, at some arbitrary harvest date. In general a yield response is accompanied by an increase in the percentage of phosphorus and it seems that there is no difference in the efficiency of its utilization in the plants (see Bowen, 1973).

However, a far better understanding of the system is possible by measuring the uptake rate per unit amount of root. The form of the diffusion equation for cylindrical geometry (Nye & Tinker, 1969) suggests that the best measure is the uptake per unit length, or inflow (*I*). Sanders & Tinker (1971, 1973) (Table 3), found a four-fold increase in *mean I* for

Table 3. *Inflows and fluxes of phosphorus during uptake by onions infected with yellow vacuolate endophyte (after Sanders & Tinker, 1973)*

	Root infected (%)	Inflows: Mean for myc. onions (mol cm^{-1} s^{-1} × 10^{-14})	Inflows: Mean for non-myc. onions (mol cm^{-1} s^{-1} × 10^{-14})	Inflows: Via hyphae[a] (mol cm^{-1} s^{-1} × 10^{-14})	Calculated flux in hyphae[b] (mol cm^{-2} s^{-1} × 10^{-8})
Expt A					
44–45 days	50	13	4.2	17.6	3.8
Expt B					
39–54 days	45	11.5	3.2	18.5	

[a] Calculated from difference between mycorrhizal and non-mycorrhizal inflows, divided by fraction of root infected.

[b] Based on six main entry hyphae (not actual entry points) per cm of infected root.

phosphorus when onion roots became mycorrhizal. Explanations based solely on plant morphology (i) can thus be dismissed.

Calculations of the diffusive flow of phosphate ions to a single cylindrical root in this soil (Sanders & Tinker, 1971; Tinker, 1975) indicated that the *I* value into non-mycorrhizal roots (*c.* 4×10^{-14} mol cm^{-1} s^{-1}) was about the maximum rate possible, hence the additional phosphate inflow into mycorrhizal roots of about 18×10^{-14} mol cm^{-1} s^{-1} must have entered via the external hyphae (Table 3). This result also eliminates possibility (v), and there is in any event good evidence that roots may absorb and translocate phosphorus for a long period (Russell & Sanderson, 1967).

There is now additional evidence that the theory of direct hyphal uptake is correct. Hattingh, Gray & Gerdemann (1973) showed greatly increased uptake of ^{32}P, placed at a distance of 2.7 cm from the root surface, when roots were mycorrhizal. In an experiment of Pearson & Tinker (1975) the distance of transport could not be determined precisely, but must have been of the order of 1–2 cm.

I therefore regard hypothesis (iii) as correct, whilst emphasizing the fact that it is not so much the extra absorbing surface which matters, as its distribution at an appreciable distance from the root. From the data of Sanders & Tinker (1973), the total hyphal surface area was only some 50 % of the total root surface, which is not a large increase in relation to the greater uptake. A strongly absorbing root will be surrounded by a depleted zone in which the soil solution concentration of phosphorus will be greatly reduced. The value of the external hyphae is that they extend into soil beyond this depleted shell, which for phosphorus may be up to about 1 mm thick. There is, of course, no precise boundary to the depleted zone, but it can be considered to be around $\sqrt{(Dt)}$ cm thick, where D is the diffusion coefficient of the ion in question, and t is the time since uptake began (Baldwin, Tinker & Nye, 1971). D is usually of the order 10^{-8}–10^{-9} cm^2 s^{-1}. Bieleski (1973) made a theoretical calculation that a root with 80 cm hyphae per cm root could absorb at 60 times the rate of the root alone. The basis of the calculation is not clear, and according to Sanders & Tinker (1973) the figure was about 6.

A mycorrhizal root is therefore regarded as a highly efficient diffusion sink in soil, due to the widely ramified hyphae. Baylis (1970, 1975) has stressed the analogy with root hairs, and the latter appear to confer similar benefits. The relative advantage of root hairs and hyphae have been discussed by Tinker (1975). Short root hairs are inherently rather inefficient, since the individual root hairs compete in uptake, in that the separate depletion shells which surround each one overlap greatly. Hyphae are of the order of several centimetres long, hence they may absorb more from undepleted soil well away from the root.

This assumes that there is undepleted soil further away. If the root density is very great, as found by Sparling & Tinker (1975) in several grass species, where the absorbing surfaces (roots or root hairs) are much less than 1 mm distant on average, this argument becomes less convincing, and the situation will then depend upon the dimensions of the individual depletion zones.

Baylis (1970, 1975) has also noted the relation between growth responses due to endo-mycorrhizas and the root development of the host plant: if roots are fibrous and long and have root hairs there is little benefit from infection. In our terminology, this gives low I values, and normal diffusion may be capable of bringing these quantities up to a single root and its associated hairs. This interpretation suggests which nutrients may be absorbed more rapidly by mycorrhizal roots and which hosts may benefit. It is necessary that the rate of uptake per unit of *active* uninfected root length required shall be such that normal diffusion or mass flow cannot supply it readily; this depends upon the relative growth rate of the plant, the fraction of the root which absorbs, the existence of root hairs and the necessary minimum percentage of the nutrient in the tissues (Nye & Tinker, 1969), the diffusion coefficient of the ion, and upon the soil solution concentration (see Tinker, 1975).

Hyphal length and properties

The above arguments direct attention towards the external hyphae and their length, distribution and properties, and stress the importance of transport in the growth-promoting effect. This is in contrast to work on ecto-mycorrhizas, where the main interest has been directed towards the sheath. Hyphae extend at least 1–2 cm into the soil (Hattingh *et al.*, 1973; Sanders, personal communication) and it seems likely that hyphal networks many centimetres in extent may form linked to several hosts. However, we have been unable to detect transfer of ^{32}P between infected onions growing in the same pot of soil in preliminary experiments.

The determination of hyphal length is difficult. Sanders & Tinker (1973) sieved out and weighed mycelium from sandy soils without much organic debris. The mean radius was easily measured, and by assuming a mean specific gravity of 1, they obtained a length of 80 cm external hyphae per cm of infected root length in onions. No distinction between living and dead hyphae was possible in this work, and further attention to this point seems necessary. They also found a surprisingly uniform amount of hyphae per unit length of root (Fig. 5), and later work with other endophytes has supported this. This result therefore does not conflict with the idea that total hyphal length is important, but also suggests that an important factor may be length of infected root. However, Mosse (1972) did not claim a good relationship between growth and percentage infection

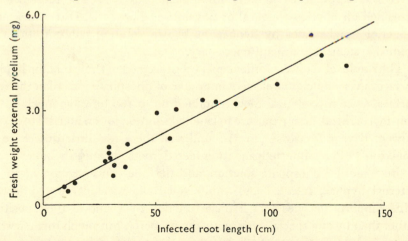

Fig. 5. Relationship between fresh weight of external hyphae and length of infected root in onions, with yellow vacuolate endophyte.

of onion roots in tests of six fungal strains (Table 4), though poor strains always produced both low infection and low growth responses.

We know little of the uptake properties of v-a hyphae, but comparison of work with excised ecto-mycorrhizas of beech and other fungi is interesting. Thus Jennings (1964) found an uptake rate with ecto-mycorrhizas of about 6×10^{-6} mol g^{-1} day^{-1} in 5×10^{-5} mol l^{-1} phosphate, and Lyon & Lucas (1969) one of 10^{-5} mol g^{-1} day^{-1} with *Rhizopus*. From the results of Sanders & Tinker (1973), the calculated uptake rate of mycorrhizal hyphae on onions is about 5×10^{-4} mol g^{-1} day^{-1}, which implies that they are highly effective. It is, however, very difficult to assess Jennings's and other data since the total hyphal surface in ecto-mycorrhizas and the real solution concentration maintained below the gross surface of the mycorrhizas, are not known. It is clear from work on a series of soils in our laboratory that the hyphae of v-a mycorrhizas, with onion as host, can absorb at a rapid rate from soil solutions of about 10^{-6} mol l^{-1}, but that the response to infection, in absolute terms, may be small on soils which have an extremely low soil solution concentration of phosphate, possibly between 10^{-7} and 10^{-8} mol l^{-1}. This may be due to the impossibility of causing a useful flux of phosphate ions with such extremely small concentration gradients, or to a physiological inability to absorb at useful rates from extremely dilute solutions. At 10^{-8} mol l^{-1}, the inflow, to a zero concentration sink of radius 10^{-3} cm in a homogenous soil, would be only of the order 10^{-16}–10^{-17} mol cm^{-1} s^{-1}, whereas the mean inflow into unit length of hyphae suggested by the calculations of Sanders & Tinker (1973) was 2×10^{-15} mol cm^{-1} s^{-1}, when a good response was found. Low mycorrhizal responses in very low-phosphorus soils can therefore be explained on physico-chemical or physiological grounds. Direct measurements of uptake rates by hyphae, at known solution concentrations in defined systems, are urgently necessary.

The results of Gray & Gerdemann (1969) seemed to show that the hyphal surface is very much more active in uptake of phosphate ions than the root surface, since mycorrhizal roots contained up to 160 times as much ^{32}P as non-mycorrhizal, after exposure to labelled solution for 90 hours. However, Bowen, Bevege & Mosse (1975) found that there was little difference in uptake of ^{32}P by infected and uninfected roots during 16 hours from 5×10^{-6} mol l^{-1} phosphate solution, and that the concentration of ^{32}P in external hyphae, infected and uninfected roots was not very different. This supports the idea that hyphae function by virtue of their position rather than by any special surface uptake property, but much further work is required. In particular, it will be interesting to determine the uptake rate of unit amount of mycelium attached to different hosts.

Translocation of phosphate in hyphae

This interpretation of the mycorrhizal response mechanism focuses interest on the translocation mechanisms of the hyphae, since this could be rate limiting. Work in this field originally concentrated on the *velocity* of movement, or the speed at which a detectable front of radio isotopes travel after uptake by a mycelium. In our situation this question is largely irrelevant; what is required is the steady-state flux of phosphate ions or the flow rate per hyphae. I am only aware of two other attempts to measure flux in fungi; Cowan, Lewis & Thain (1972) found a value of around 10^{-9} mol cm^{-2} s^{-1} for potassium in *Phycomyces blakesleeanus* mycelium, and Jennings, Thornton, Galpin & Coggins (1974) deduced a flux of 1.27×10^{-8} mol cm^{-2} s^{-1} for carbohydrate (hexose) in *Serpula lacrymans*. Sanders & Tinker (1973) (Table 3) computed a flux of around 3.8×10^{-8} mol cm^{-2} s^{-1} from the mean uptake of phosphate into their plants, and the mean number and size of the entry points (see Plate 1) per cm of onion root. Pearson & Tinker (1975) attempted to substantiate this by a completely different method. Clover plants were grown in sterile conditions in soil–agar mixtures in one half of divided Petri dishes, and infected with *Glomus mossaeae*. When hyphae had grown over the dividing barrier into agar in the other half, the latter was flooded with radio-labelled phosphate solution, chosen to give a final mean concentration in the agar of 8×10^{-6} mol l^{-1} phosphate. This method gave a flux of up to 10^{-9} mol cm^{-2} s^{-1} in hyphae at some distance from the host root. The difference was accepted as reasonable in view, not only of the fact that the flow is concentrated at the entry points, but also of the different environmental conditions and hosts. Any acceptable translocation mechanism must thus be able to explain fluxes of the order 10^{-8}–10^{-9} mol cm^{-2} s^{-1}. Simple diffusion is quite unacceptable, since the gradient needed to drive a 10^{-9} mol cm^{-2} s^{-1} flux of phosphate in water is about $0 \cdot 1$ mol l^{-1} cm^{-1}. Cowan *et al.* (1972) discussed this point in detail, and calculated a phosphorus flux of 6×10^{-11} mol cm^{-2} s^{-1} from results of Lyon & Lucas (1969). They are probably justified in their conclusion that this last value could be accounted for by diffusion, but it seems most unlikely that a flux of 10^{-9} mol cm^{-2} s^{-1} could be. Cowan *et al.* (1972) concluded that they could explain the results in their system by diffusion and electrical field strength, but I feel that this is doubtful, for reasons connected with their techniques for measuring the ionic mobility of potassium. Jennings *et al.* (1974) stated three possible transfer mechanisms for material (other than nuclei) in fungal hyphae. These are (i) diffusion, (ii) protoplasmic streaming, which may be mono- or bidirectional, and (iii) bulk flow, which is generated by pressure differences within the hyphae, and involves a bulk monodirectional movement of all the hyphal contents. It may well be difficult to distinguish clearly between (ii) and (iii).

In the v–a mycorrhizas, all the mycelium is in the root or in the soil, and direct transpiration from the mycelium must be minimal. The host will, however, transpire, and there is the possibility that lowered water potential in the host cell causes net movement of hyphal contents towards the host. There is evidence (Melin & Nillson, 1958) that transpiration accelerates transfer of ^{32}P from ectomycorrhizal sheath to host in pine. Since there are no septa in the v–a mycelium, viscous drag should not be too severe (see Isaac, 1964; Jennings *et al.*, 1974). There may thus be a net movement inside hyphae towards a transpiring host, but it seems unlikely that it can explain the whole flux, since it certainly cannot account for the movement of carbon in the opposite direction, away from the host. Pearson & Tinker's (1975) experiments were mainly conducted under low transpiration conditions but the largest flux value was found when one plant was left exposed to the atmosphere for a period.

The remaining explanation is protoplasmic streaming (cyclosis) (Burnett, 1968). Sanders (personal communication) observed rapid bidirectional streaming in germ tubes of spores, at speeds of the order of a few centimetres per hour, which agrees with most of the published work (Wilcoxon & Sudia, 1968).

We have determined the phosphorus content of dry extra-radicular hyphae grown in sand by chemical analysis, and obtained a figure of 0.4 % P. Using the fresh:dry weight ratio of the hyphae and the relative size of fungal wall and hyphal lumen and assuming that none of the phosphorus is in the hyphal wall, this gives around 0.3 % in the cytoplasm/vacuolar mixture contained in the hyphae. A mean bulk flow speed of the hyphal contents of the order of 1 cm h^{-1} would then give a flux of 3×10^{-8} mol cm^{-2} s^{-1}, which is about the same as the maximum calculated value. Continuous, monodirectional bulk flow is not credible however, and protoplasmic streaming alone seems insufficient, since it is necessary to account for bidirectional movement, reversal of direction or restriction of the flow to part of the lumen of the hyphae. It may be that flow is more rapid, say up to 10 cm h^{-1}, but there is clearly not a very great margin between the theoretical calculations and the maximum inferred value. It appears most likely that both bulk flow and protoplasmic streaming occur, perhaps simultaneously.

We have recently made observations which suggest an elaboration of this idea. Histological work on hyphae in the roots detected small granules of diameter 0.1–0.2 μm showing metachromasy when stained with toluidine blue (Cox, Sanders, Tinker & Wild, 1975), which is a well-known property of polyphosphates (Harold, 1966) (Plate 3). Similar granules have also been identified in electron micrographs of arbuscules (Plate 3); they are particularly frequent when the infection is in a young and vigorous state. The granules have always been detected inside vacuoles, and the metachromatic reaction can sometimes be seen dispersed throughout the whole vacuole. This suggests that much of the phosphate in v–a endophytes is segregated as polyphosphate in the small vacuoles, sometimes forming granules.

Much further work on this topic is needed, but tentatively it appears possible that the transport mechanism for phosphate is cyclosis, plus bulk flow, with loading and unloading of polyphosphate into vacuoles as the method of varying the phosphorus concentration of the streaming protoplasm. A characteristic of polyphosphate granules in bacteria, yeasts and fungi is their rapid increase or decrease in amount, depending upon the external circumstances (Harold, 1966) and this characteristic would be ideal for the mechanism we propose.

Transfer to host

It has usually been assumed that the transfer of phosphorus from endophyte to host occurred in arbuscules (Plate 3). Early workers assumed that the arbuscules were 'digested' by the host (Harley, 1969). The development and degeneration of the arbuscule have been described in considerable detail, at the ultrastructural level, by Cox & Sanders (1974) and others (see Plate 3). During senescence, the arbuscule branches gradually collapse, starting with the finest. These collapsed fungal walls may in certain circumstances be surrounded by an encasement layer (Cox & Sanders, 1974). The hyphal contents are lost during collapse, but it is at least as credible that they are transferred back into the intercellular hyphae, as that they leak out into the cell. None of the published electron micrographs show incontrovertible evidence of the latter process, but it may well be difficult to detect. Degenerating arbuscules are often accompanied by lipid globules, in the host cell (see Mosse, 1973a), which might be derived from the arbuscule.

It is possible to make a rough calculation of the possible inflow rate which could be generated by such a 'digestion' process, using the figures for the phosphorus content of external hyphae, and the estimated mean life of arbuscules of about 7–11 days in onions (Bevege, Bowen & Skinner, 1975). We have made measurements on onion roots which support this figure. When an infected plant enters a state of roughly constant percentage infection, the relative amount of 'young' and 'old' arbuscules is related to the duration of the 'young' (i.e. active and uncollapsed) stage (Cox & Tinker, unpublished). From this approach, we estimate a 'young' life of

5 days in onions infected with *Glomas mosseae* and grown under standard conditions in a growth chamber. The total volume fraction of arbuscule in the cortex (Cox & Tinker, unpublished) (Fig. 6) is about 1 %, hence in a root of 0.020 cm radius with 95 % cortex, the maximum transfer rate can be calculated to be about 10^{-15} mol cm^{-1} s^{-1}. The measured mean 'hyphal inflow' into infected roots of some 2×10^{-13} mol cm^{-1} s^{-1} (Table 3) can thus hardly be accounted for. This rather crude calculation cannot prove the case one way or the other, but it attempts to introduce some quantitative arguments into this discussion.

Alternatively, it is tempting to regard arbuscules as organs in which direct transfer of phosphate and other materials may occur, across the fungal and host plasmalemma. The mechanism for this is as yet unknown, but Woolhouse (1975) has suggested permeability effects based on changes in the host cell membrane. However, there are reasons to doubt whether arbuscules are necessarily so predominant a site of transfer as is commonly assumed. Firstly, direct measurements of total surface area from electron micrographs (Fig. 6) (Cox & Tinker, unpublished) have shown that the total area of cell plasmalemma in the root cortex is only increased by a factor of around 20 % by arbuscule formation in onion roots though, within any one cell, it may be increased by a factor of 2 or 3. The impression of enormous surface area can, therefore, be deceptive. The arbuscular inter-face has the advantage, in terms of material transfer, that there is no host cell wall between host and fungal plasmalemmas, but the fungal wall is still there (Cox & Sanders, 1974). Secondly, it is striking that ecto-mycorrhizas carry out the host–endophyte transfer without any such intracellular organs, the Hartig net being intercellular though specialized. No measurements of transfer rates equivalent to our *I* values have been made for intact plants with ecto-mycorrhizas, but the rate must clearly be comparable and yet take place without arbuscule formation. Thirdly, Dr Cooper (personal com-munication) has seen cases where a growth response in *Solanum laciniatum* was evident following infection, but before any appreciable number of arbuscules appeared in the roots. It therefore seems likely that much transfer, of both carbon compounds and minerals, takes place in the arbuscules, but that they may not be the sole site.

Growth decrease of host

Some authors have reported reductions in host plant growth following infection with v–a mycorrhizas. Thus Mosse (1973*b*) found that infection decreased growth of onions by up to 40 % when large amounts of soluble phosphorus were supplied in the soil, though growth improvements occurred when little or no phosphorus was supplied. This was attributed

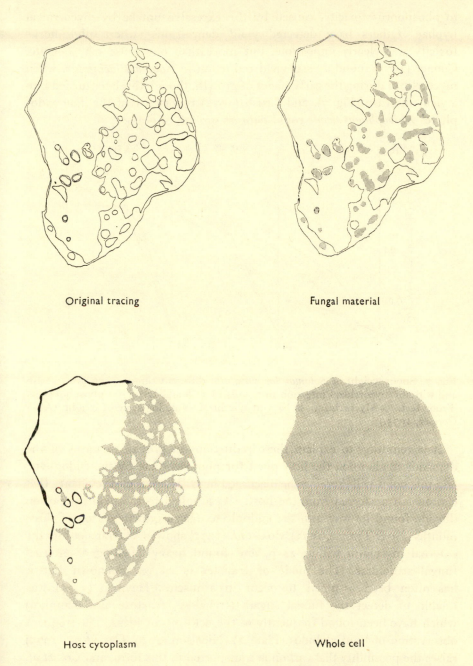

Original tracing

Fungal material

Host cytoplasm

Whole cell

Fig. 6. Tracing of electron micrograph mosaic of whole onion root cell containing arbuscule, showing relative volumes occupied by whole cell, fungal material, and host cytoplasm, and relative surface areas of host tonoplast and plasmalemma.

to phosphorus toxicity, caused by the excessive uptake by mycorrhizal hyphae. Others have detected yield depressions, where phosphorus toxicity was clearly impossible, but more than one cause may operate. Cooper (1975) found a clear yield reduction in *Solanum laciniatum* when mycorrhizal during the early stages of growth, though this later turned into a yield increase (Fig. 7), and Crush (1972) reported that the 'fine endophyte' *Rhizophagus tenuis* could depress growth of grasses.

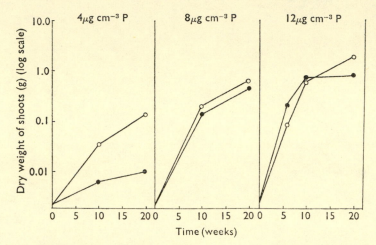

Fig. 7. Shoot weight of *Solanum laciniatum* at different times, when grown with and without mycorrhizal infection in 3 soils of 4, 8 and 12 μg g^{-1} phosphorus by Truog test. ○ Mycorrhizal, ● Non-mycorrhizal. Note log scale of weight. (After Cooper, 1975.)

It is tempting to explain these reductions in growth in terms of the demands made upon the host plant for photosynthate by the endophyte, and this point will therefore be discussed here. There is clear evidence that carbon is transferred from the host: ^{14}C supplied to the plant as carbon dioxide found its way into the endophyte over a growth period of three months (Ho & Trappe, 1973). Cox *et al.* (1975) showed ^{14}C transfer into the external mycelium within 24 h, and found heavy labelling of internal fungal structures. The mode of transfer is as yet unknown but it has often been assumed to occur in arbuscles. Hayman (1974) was unable to detect the fungal sugars (trehalose, fructose and mannitol) which have been found frequently in the ecto-mycorrhizas. The frequent observation of lipid globules (Plate 2) (Nicolson, 1959; Cox *et al.*, 1975) raises the possibility that carbon is transported in this form, and Cox *et al.* also reported some autoradiographic evidence suggesting that lipid in the endophyte was rapidly and heavily labelled with ^{14}C when the latter was supplied to the host.

The quantity of carbon required by the endophyte is also quite unknown at present. However, direct measurements of the volume of fungus by Cox (Cox & Tinker, unpublished) indicate only about 1–2 % of total root volume in onions with well-developed infection. The weight of external mycelium is small (some 0.04 mg per cm of infected root), and it appears strange that such small amounts of tissue should be able to divert significant amounts of photosynthate from the host. Possibly the formation of spores or of large amounts of host nuclear and cytoplasmic material acts as a sink. It is obvious that the possibility of yield reduction in crops due to mycorrhizal infection must be taken very seriously, and further and more detailed work is urgently needed.

Possibilities of practical application to crops

It is early to expect a practical use for studies on v–a mycorrhizas, but it is essential for serious thought to be given to the matter. The analogies with the uses of *Rhizobium*, the present shortages and high cost of phosphorus fertilizer and the possible limitations of phosphorus supplies from the present main deposits of phosphate rock, all suggest that the practical possibilities should be vigorously pursued, in the hope of improving phosphorus nutrition of crops cheaply.

It is occasionally felt that the ubiquity of the infection in natural vegetation and crops implies that all possible advantage is already being obtained. There is already considerable evidence that this is not the case. In arable crops there are two possibilities. Firstly, crops may not be infected, or may be infected too late because of insufficient inoculum in the soil. It is known (Hayman, 1975; Kruckelmann, 1973, 1975) that the spore population may depend strongly upon the preceding crops. Infected roots will also vary in quantity with crop and agronomic techniques, so that it appears possible that the level of inoculum may be insufficient. Khan (1972, 1975) has utilized this effect in establishing field experiments with wheat and maize on unfertilized land which had previously been occupied by weeds of the Chenopodiacae. He determined useful increases in growth (Fig. 8) when inoculated transplants were used. It is known that endo-mycorrhizas develop widely in fertilized crops, but there is still little evidence of the value of the infection. It seems quite inadmissable to assume that the endophyte must have a useful role merely because it is present, and more work with major field crops is urgently needed. In special cases where the soil is sterilized, as in citrus nurseries, there may be a major response to inoculation (Kleinschmidt & Gerdemann, 1972).

Secondly, there is the important possibility that improved strains of endophyte can be selected (Table 4). The work of Mosse & Hayman (1971),

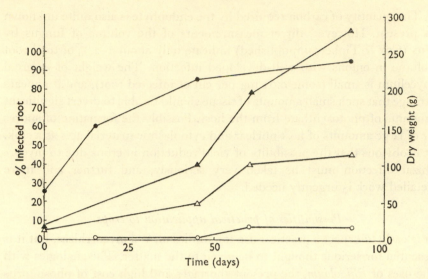

Fig. 8. Weight of shoots, and percentage infection of roots, for field grown maize
with and without artificial inoculation with mycorrhizal endophyte. (Redrawn
from Khan, 1975.)

○ , Per cent infection, non-inoculated; ● , per cent infection, inoculated; △ , yield,
non-inoculated; ▲ , yield, inoculated.

Mosse (1972), and Bevege, Bowen & Skinner (1975) shows clearly that
strains vary widely in their effectiveness.

In perennial vegetation, such as natural and improved grasslands,
orchards and forests it is unlikely that there will be any failure of infection
since there is ample time for inoculum to build up. No cases have yet been
reported of exotic species requiring the importation of the endophyte, as
has happened with the ecto-mycorrhizas, probably due to lack of host
specificity. The most interesting possibility here is again the selection of
new and improved strains of endophyte, but no comparative tests have been
reported so far.

Interactions with other organisms

There is no space here to discuss this matter, but it cannot be entirely
neglected, since important effects on higher plants are already being found.
Thus interactions between v–a endophytes and *Rhizobium, Thielaviopsis
basicola, Phytophora*, nematodes and viruses have already been reported
(Gerdemann, 1974; see Sanders, Mosse & Tinker, 1975). No clear pattern has yet emerged, but the subject seems likely to prove exceedingly
important, as is true for many other facets of study of the vesicular-
arbuscular mycorrhizas.

PLATE I

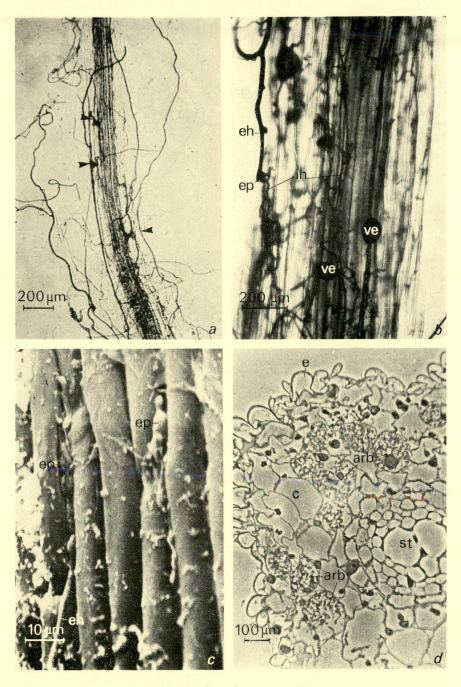

For explanation see p. 349

(*Facing p.* 344)

PLATE 2

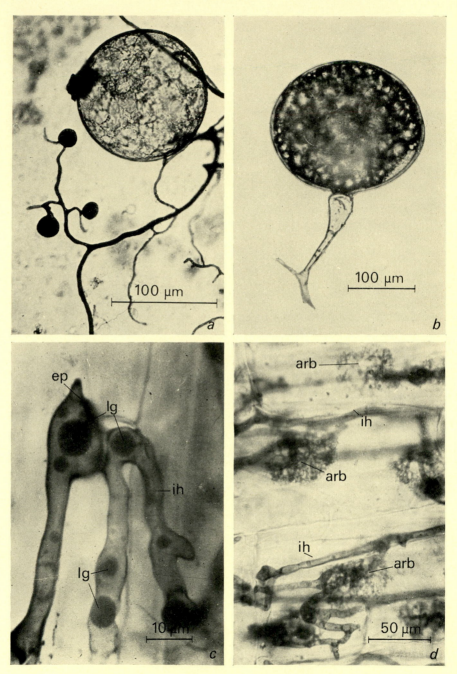

For explanation see p. 349

PLATE 3

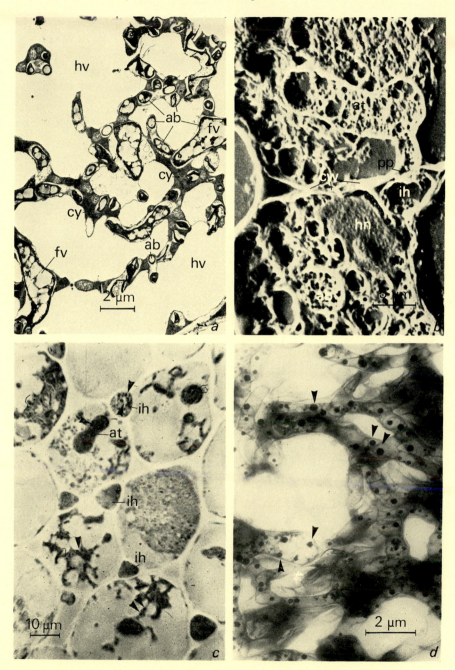

For explanation see p. 349

Table 4. *Dry weights of plants inoculated with different endophytes and grown in various soils (after Mosse, 1972)*

Plant	Soil	Endophyte	Dry wt (mg)	Degree of infection
Onion	No. 11, irradiated	Laminate	737	Moderate
Onion	No. 11, irradiated	E₃	708	Abundant
Onion	No. 11, irradiated	Honey-coloured	689	Moderate
Onion	No. 11, irradiated	Small brown clusters	649	Moderate
Onion	No. 11, irradiated	Yellow vacuolate	246	Slight
Onion	No. 11, irradiated	*E. microcarpa*	151	Slight
Onion	No. 11, irradiated	Bulbous reticulate	139	Very slight
Paspalum notatum	B, plus lime, irradiated	E₃	115	—
Paspalum notatum	B, plus lime, irradiated	Honey-coloured	41	—
Paspalum notatum	B, plus lime, irradiated	'Root inoculum'	79	—
Paspalum notatum	B, plus lime, irradiated	Laminate	33	—
Paspalum notatum	B, plus lime, irradiated	'Leachings'	10	—
Paspalum notatum	B, plus lime, non-sterilised		14	—

ACKNOWLEDGEMENTS

I wish to thank my collaborators at Leeds, and in particular Drs Cox and Sanders, for discussions, comments and ideas, and Professor Gerdemann and Dr Mosse for their comments.

REFERENCES

BALDWIN, J., TINKER, P. B. H. & NYE, P. H. (1971). Uptake of solutes by multiple root systems from soil. II. The theoretical effects of rooting density and pattern on uptake of nutrients from soil. *Pl. Soil*, **36**, 693–708.

BARTLETT, E. M. & LEWIS, D. H. (1973). Surface phosphatase activity of mycorrhizal roots of beech. *Soil Biol. Biochem.*, **5**, 249–257.

BAYLIS, G. T. S. (1969). Host treatment and spore production by *Endogone*. *N.Z. J. Bot.*, **7**, 173–174.

(1970). Root hairs and phycomycetous mycorrhizas in phosphate-deficient soils. *Pl. Soil*, **33**, 713–716.

(1975). The Magnolioid mycorrhiza and mycotrophy in root systems derived from it. In *Endomycorrhizas* (eds. F. E. T. Sanders, B. Mosse and P. B. H. Tinker). London: Academic Press (in press).

BENIANS, G. J. & BARBER, D. A. (1972). In *A. Rep. Letcombe Laboratory, 1971*, 7–9.

BEVEGE, D. I., BOWEN, G. D. & MOSSE B. (1975). Physiological studies in phosphate uptake by vesicular-arbuscular mycorrhizas. In *Endomycorrhizas* (eds. F. E. T. Sanders, B. Mosse and P. B. H. Tinker). London: Academic Press (in press).

BEVEGE, D. I., BOWEN, G. D. & SKINNER, M. F. (1975). Comparative carbohydrate physiology of ecto- and endo-mycorrhizas. In *Endomycorrhizas* (eds. F. E. T. Sanders, B. Mosse and P. B. H. Tinker). London: Academic Press (in press).

BIELESKI, R. L. (1973). Phosphate pools, phosphate transport and phosphate availability. *A. Rev. Pl. Physiol*, **24,** 225–252.

BOWEN, G. D. (1973). Mineral nutrition of ectomycorrhizae. In *Ectomycorrhizae*, 151–205 (eds. T. T. Kozlowski and G. C. Marks). New York: Academic Press.

BOWEN, G. D. & THEODOROU, C. (1967). Studies on phosphate uptake by mycorrhizas. *Proc. XIV Congr. Inst. U. For. Res. Org.* Munich, 116–138.

BURNETT, J. H. (1968). *Fundamentals of Mycology*. London: Edward Arnold.

COOPER, K. (1975). Growth responses to the formation of endotrophic mycorrhizas in *Solanum leptospermum* and New Zealand ferns. In *Endomycorrhizas* (eds. F. E. T. Sanders, B. Mosse and P. B. H. Tinker). London: Academic Press (in press).

COWAN, M. C., LEWIS, B. G. & THAIN, J. F. (1972). Mechanism of translocation of potassium in mycelium of *Phycomyces blakesleeanus* in an aqueous environment. *Trans. Brit. mycol. Soc.*, **58,** 103–112.

COX, G. C. & SANDERS, F. E. T. (1974). Ultrastructure of the host-fungus interface in a vesicular-arbuscular mycorrhiza. *New Phytol.*, **73,** 901–912.

COX, G. C., SANDERS, F. E. T., TINKER, P. B. H. & WILD, J. (1975). Ultrastructural evidence relating to nutrient transfer in vesicular-arbuscular mycorrhizas. In *Endomycorrhizas* (eds. F. E. T. Sanders, B. Mosse and P. B. H. Tinker). London: Academic Press (in press).

CRUSH, J. R. (1972). Mycorrhizas in some native and exotic grasses. Ph.D. thesis, University of Otago, New Zealand.

FURLAN, V. & FORTIN, J. A. (1973). Formation of endomycorrhizae by *Endogone calospora* on *Allium cepa* under three temperature regimes. *Naturaliste canadien*, **100,** 467–477.

GERDEMANN, J. W. (1968). Vesicular-arbuscular mycorrhizae and plant growth. *A. Rev. Phytopathol.*, **6,** 397–418.

(1974). Vesicular-arbuscular mycorrhizae. *Proc. Harvard Forest Symp.* (in press).

GERDEMANN, J. W. & TRAPPE, J. M. (1974). The Endogonaceae of the Pacific Northwest. *Mycologia Memoir*, no. **5** (in press).

GILMORE, A. E. (1971). The influence of endotrophic mycorrhizae on the growth of peach seedlings. *J. Am. Soc. hort. Sci.*, **96,** 35–38.

GRAY, L. E. & GERDEMANN, J. W. (1969). Uptake of phosphorus-32 by vesicular-arbuscular mycorrhizae. *Pl. Soil*, **30,** 415–422.

GRAY, L. E. & GERDEMANN, J. W. (1973). Uptake of sulphur-35 by vesicular-arbuscular mycorrhizae. *Pl. Soil*, **39,** 687–689.

HARLEY, J. L. (1969). *The Biology of Mycorrhiza*. London: Leonard Hill.

HAROLD, F. M. (1966). Inorganic polyphosphate in biology: structure, metabolism and function. *Bact. Rev.*, **30,** 772–794.

HATTINGH, M. J., GRAY, L. E. & GERDEMANN, J. W. (1973). Uptake and translocation of 32-P labelled phosphate to onion roots by endomycorrhizal fungi. *Soil Sci.*, **116,** 383–387.

HAYMAN, D. S. (1970). Endogone spore numbers in soil and vesicular-arbuscular mycorrhiza in wheat as influenced by season and soil treatment. *Trans. Br. mycol. Soc.*, **54**, 53–63.

(1974). Plant growth response to vesicular-arbuscular mycorrhiza. VI. Effect of light and temperature. *New Phytol.*, **73**, 71–80.

(1975). The occurrence of mycorrhizas in crops as affected by soil fertility. In *Endomycorrhizas* (eds. F. E. T. Sanders, B. Mosse and P. B. H. Tinker). London: Academic Press (in press).

HAYMAN, D. S. & MOSSE, B. (1971). Plant growth responses to vesicular-arbuscular mycorrhiza. I. Growth of *Endogone*-inoculated plants in phosphate-deficient soils. *New Phytol.*, **70**, 19–27.

(1972a). The role of vesicular-arbuscular mycorrhiza in the removal of phosphorus from soil by plant roots. *Rev. ecol. biol. Soc.*, **9**, 463–470.

(1972b). Plant growth responses to vesicular-arbuscular mycorrhiza. III. Increased uptake of labile P from soil. *New Phytol.*, **71**, 41–47.

HO, I. & TRAPPE, J. M. (1973). Translocation of ^{14}C from *Festuca* plants to their endomycorrhizal fungi. *Nature New Biol.*, **244**, 30–31.

ISAAC, P. K. (1964). Cytoplasmic streaming in filamentous fungi. *Can. J. Bot.*, **42**, 787–792.

JACKSON, H. E., MILLER, R. H. & FRANKLIN, R. E. (1973). The influence of vesicular-arbuscular mycorrhiza on uptake of ^{90}Sr from soil by soybeans. *Soil Biol. Biochem.*, **5**, 205–212.

JENNINGS, D. H. (1964). Changes in the size of the orthophosphate pools in mycorrhizal roots of beech with reference to absorption of the ion from the external medium. *New Phytol.*, **63**, 181–193.

JENNINGS, D. H., THORNTON, J. D., GALPIN, M. F. J. & COGGINS, C. R. (1974). Translocation in fungi. *Symp. Soc. exp. Biol.*, **28**, 137–154.

KHAN, A. G. (1972). The effect of v-a mycorrhizal associations on growth of cereals. I. Effects on maize growth. *New Phytol.*, **71**, 613–619.

(1975). Growth effects of vesicular-arbuscular mycorrhizas on crops in the field. In *Endomycorrhizas* (eds. F. E. T. Sanders, B. Mosse and P. B. H. Tinker). London: Academic Press (in press).

KLEINSCHMIDT, T. D. & GERDEMANN, J. W. (1972). Stunting of citrus seedlings in fumigated nursery soils related to the absence of endomycorrhiza. *Phytopathol.*, **62**, 1447–1453.

KRUCKELMANN, H. W. (1973). Die vesikular-arbusculären Mykorrizen und ihre Beeinflussung in landwirtschaftlichen Kulturen. Ph.D. thesis, Braunschweig.

(1975). Effect of plant species on the frequency of *Endogone* chlamydospore and mycorrhizal infection. In *Endomycorrhizas* (eds. F. E. T. Sanders, B. Mosse and P. B. H. Tinker). London: Academic Press (in press).

LEWIS, D. H. (1973). Concepts in fungal nutrition and the origin of biotrophy. *Biol. Rev.*, **48**, 261–278.

(1975). Comparative aspects of the carbon nutrition of mycorrhizas. In *Endomycorrhizas* (eds. F. E. T. Sanders, B. Mosse and P. B. H. Tinker). London: Academic Press (in press).

LYON, A. J. E. & LUCAS, R. L. (1969). The effect of temperature on the translocation of phosphorus by *Rhizopus stolonifer*. *New Phytol.*, **68**, 963–969.

MELIN, E. & NILLSON, H. (1958). Translocation of nutritive elements through mycorrhizal mycelia to pine seedlings. *Bot. Notiv.*, **111**, 251–256.

MOSSE, B. (1972). The influence of soil type and *Endogone* strain on the growth of mycorrhizal plants in phosphate deficient soils. *Rev. ecol. biol. Soc.*, **9**, 529–537.

(1973a). Advances in the study of vesicular-arbuscular mycorrhiza. *A. Rev. Phytopath.*, **11**, 171–195.

(1973b). Plant growth responses to vesicular-arbuscular mycorrhiza. IV. In soil given additional phosphate. *New Phytol.*, **72**, 127–136.

MOSSE, B. & HAYMAN, D. S. (1971). Plant growth responses to vesicular-arbuscular mycorrhiza. II. In unsterilized field soils. *New Phytol.*, **70**, 29–34.

MURDOCH, C. L., JAKOBS, J. A. & GERDEMANN, J. W. (1967). Utilization of phosphorus sources of different availability by mycorrhizal and non-mycorrhizal maize. *Pl. Soil.*, **27**, 329–334.

NICOLSON, T. H. (1959). Mycorrhiza in the Gramineae. I. Vesicular-arbuscular endophytes, with special reference to the external phase. *Trans. Br. mycol. Soc.*, **42**, 421–438.

(1967). Vesicular-arbuscular mycorrhizas – a universal plant symbiosis. *Sci. Prog. Oxford*, **55**, 561–581.

NYE, P. H. & TINKER, P. B. H. (1969). The concept of a root demand coefficient. *J. appl. Ecol.*, **6**, 293–300.

PEARSON, V. & TINKER, P. B. H. (1975). Measurement of phosphorus fluxes in the external hyphae of endomycorrhizas. In *Endomycorrhizas* (eds. F. E. T. Sanders, B. Mosse and P. B. H. Tinker). London: Academic Press (in press).

POWELL, C. (1975). Potassium uptake by endotrophic mycorrhizas. In *Endomycorrhizas* (eds. F. E. T. Sanders, B. Mosse and P. B. H. Tinker). London: Academic Press (in press).

REDHEAD, J. F. (1975). Endotrophic mycorrhizas in Nigeria: Some aspects of the ecology of endotrophic mycorrhizal association of *Khaya grandifolia* C. DC. In *Endomycorrhizas* (eds. F. E. T. Sanders, B. Mosse and P. B. H. Tinker). London: Academic Press (in press).

ROSS, J. P. & GILLIAM, J. W. (1973). Effect of *Endogone* mycorrhiza on phosphorus uptake by soybeans from inorganic sources. *Soil Sci. Soc. Am. Proc.*, **37**, 237–239.

RUSSELL, R. S. & SANDERSON, J. (1967). Nutrient uptake by different parts of the intact roots of plants. *J. exp. Bot.*, **18**, 491–508.

SAFIR, G. R., BOYER, J. S. & GERDEMANN, J. W. (1971). Mycorrhizal enhancement of water transport in soybean. *Science, Wash.*, **172**, 581–583.

SAFIR, G. R., BOYER, J. S. & GERDEMANN, J. W. (1972). Nutrient status and mycorrhizal enhancement of water transport in soybean. *Pl. Physiol., Lancaster*, **49**, 700–703.

SANDERS, F. E. T. (1975). The effect of phosphorus on the development and extent of endomycorrhizal infection of onion roots. In *Endomycorrhizas* (eds. F. E. T. Sanders, B. Mosse and P. B. H. Tinker). London: Academic Press (in press).

SANDERS, F. E. T., MOSSE, B. & TINKER, P. B. H. (eds.). (1975). *Endomycorrhizas*. London: Academic Press (in press).

SANDERS, F. E. T. & TINKER, P. B. H. (1971). Mechanism of absorption of phosphate from soil by *Endogone* mycorrhizas. *Nature, Lond.*, **233**, 278–279.

SANDERS, F. E. T. & TINKER, P. B. H. (1973). Phosphate flow into mycorrhizal roots. *Pestic. Sci.*, **4**, 385–395.

SPARLING, G. & TINKER, P. B. H. (1975). Mycorrhizas in Pennine grassland. In *Endomycorrhizas* (eds. F. E. T. Sanders, B. Mosse and P. B. H. Tinker). London: Academic Press (in press).

STRIBLEY, D. P. & READ, D. J. (1974). The biology of mycorrhizas in the Ericaceae, IV. The effect of mycorrhizal infection on uptake of ^{15}N from labelled soil by *Vaccinium macrocarpon* Ait. *New Phytol.*, **73** (in press).

STRZEMSKA, J. (1974). The occurrence and intensity of mycorrhizas in cultivated plants. *Proc. 10th Int. Cong. Soil Sci. Moscow*, **3**, 81–86.

TINKER, P. B. H. (1975). The soil chemistry of phosphorus and mycorrhizal effects on plant growth. In *Endomycorrhizas* (eds. F. E. T. Sanders, B. Mosse and P. B. H. Tinker). London: Academic Press (in press).

WILCOXON, R. D. & SUDIA, T. W. (1968). Translocation in fungi. *Bot. Rev.*, **34,** 32–50.

WOOLHOUSE, H. W. (1975). Turnover and transport of phosphorus in plants. In *Endomycorrhizas* (eds. F. E. T. Sanders, B. Mosse and P. B. H. Tinker). London: Academic Press (in press).

EXPLANATION OF PLATES

PLATE 1

(a) *Festuca rubra* infected with yellow vacuolate (*Glomus mosseae*) endophyte. Note external mycelium, entry points into root (arrowed), and 'runner' hyphae along root.

(b) Intercellular hyphae (ih) of yellow vacuolate endophyte in onion root, with entry point (ep), external hyphae (eh) and vesicles (ve).

(c) Scanning electron micrograph of surface of onion mycorrhiza after critical point drying. Note entry points (ep) and external hyphae (eh).

(d) Transverse section of onion mycorrhiza, showing stele (st), cortex (c), epidermis (e) and arbuscules in the cortical cells (arb), viewed under phase contrast.

(Pictures by G. C. Cox, F. E. T. Sanders and G. Sparling.)

PLATE 2

(a) Yellow vacuolate endophyte spore, with external mycelium and immature spores.

(b) Bulbous reticulate (*Gigaspora calospora*) spore.

(c) Entry point (ep) hypha with appressorium, stained with Trypan blue and Sudan IV, containing lipid globules (lg): (ih), intercellular hyphae.

(d) Onion root squash preparation stained with Trypan blue, showing internal hyphae (ih) and arbuscules (arb).

(Pictures by G. C. Cox and F. E. T. Sanders.)

PLATE 3

(a) Electron micrograph of mature arbuscule in onion root cell. Note host (hv) and fungal (fv) vacuoles, host cytoplasm (cy) and the fine branches of the arbuscules (ab). Thin section in TEM, stained with uranyl acetate and lead citrate.

(b) Scanning electron micrograph of onion root cells, with intercellular hyphae, (ih), penetration point into cell (pp), intracellular hyphae, probably arbuscular trunk (at), arbuscular branch (ab) and host cell wall (cw). Thick section in resin, extracted with sodium methoxide.

(c) Onion root section, stained with methylene blue. Note intercellular hyphae (ih), arbuscular trunk hyphae (at) and numerous (arrowed) heavily stained granules.

(d) Electron micrograph with ultra-high voltage microscope and thick section, of onion root cell with arbuscule branches. Material stained with lead nitrate. Note electron-dense granules in arbuscular branches (arrowed). (Courtesy of Professor B. Jouffrey, C.N.R.S. Toulouse.)

(Pictures by G. C. Cox and F. E. T. Sanders.)

SYMBIOTIC RELATIONSHIPS BETWEEN NEMATODES AND PLANTS

By A. F. BIRD

CSIRO, Division of Horticultural Research,
GPO Box 350, Adelaide, South Australia, Australia

INTRODUCTION

The term symbiosis is used here in its widest sense as originally defined by de Bary (1879) and used subsequently by Read (1970) and Smith, Muscatine & Lewis (1969) to mean the living together of two phylogenetically unrelated species of organism.

De Bary's original definition of the term appeared as a published address (de Bary, 1879) to a meeting of German physicians and naturalists at Cassel towards the end of 1878.

This paper was examined in some detail by a committee on terminology appointed by the American Society of Parasitologists (Hertig, Taliaferro & Schwartz, 1947) who stated that 'De Bary made it quite clear throughout that he was using symbiosis as a general term and that he included specifically all degrees of parasitism, commensalism and mutualism', and that authors who have used the term symbiosis in a more restricted sense to be synonymous with an obligatory form of mutualism 'have misinterpreted De Bary'.

The various categories of symbiosis include:

(1) Commensalism in which neither organism profits at the expense of the other, although one of them may benefit from the relationship.

(2) Parasitism in which one of the organisms benefits at the expense of the other.

(3) Mutualism in which both organisms benefit.

The symbiotic relationships between nematodes and plants can probably be classed under all three of these categories. Little or nothing is known about the relationships between the multitudes of free-living nematodes that exist in the soil, the bacteria on which they feed and the region of the rhizosphere around the roots of plants in which they abound. It would seem, however, that in these instances, commensalism or mutualism may occur, as bacteria are known to be able to utilize exudations from roots (Rovira, 1965; Bowen & Rovira, 1969), and many of these bacteria are devoured by nematodes (Overgaard-Neilsen, 1949) whose nitrogenous exudations and secretions (Weinstein, 1960) would serve as a source of nutrient for the plant. The bodies of dead and dying nematodes would in turn serve as a source of food for these bacteria so that in this instance we can readily imagine mutualism existing between the plant, bacteria and nematodes.

The type of symbiosis considered here is essentially a parasitic one involving a direct interaction between the nematode and its host plant in which the nematode is dependant on the plant's metabolism for its survival. Yeates (1971) has considered the concept of parasitism in plant nematodes and came to the conclusion that only in forms with sedentary females do we find a truly parasitic state. The various migratory, ecto- and endo-parasitic plant nematodes were regarded as being 'a highly adaptable group of plant browsers'.

These problems of nomenclature do not arise if we use a broader and more flexible definition of parasitism operating within the framework of symbiosis as outlined above.

Parasitism as Lincicome (1963) has pointed out 'has a chemical basis' and host and parasite exchange chemical substances which may at times be of mutual benefit. In fact it has been recorded in the literature on a number of occasions (Chitwood, Specht & Havis, 1952; Pitcher, 1965; Wallace, 1973) that small numbers of a sedentary plant parasitic nematode under a particular set of conditions can confer some benefit to its host. In these instances we get a well-recognized host–parasite relationship occurring more as a mutualistic relationship if only for a short time. Within the confines of a well-established host–parasite relationship, such as takes place between root-knot nematodes and their hosts we find, at the cell level, an obligatory form of mutualism existing as the normal pattern of events. The nematode stimulates metabolic processes in plant cells which lead to the induction of syncytia or giant cells on which the nematode is dependant for its survival and which in turn are dependant on the nematode for their development. I shall consider this type of situation in greater detail later in this paper.

Situations such as these in which two organisms can live together under more than one of the categories of symbiosis listed above draw attention to the usefulness of such an all-embracing term. It allows us to concern ourselves with degrees of integration between organisms rather than having to bridge rigid biological compartments.

ADAPTION TOWARDS A PARASITIC MODE OF LIFE

There are a number of characteristics which distinguish parasitic from non-parasitic nematodes. These include an increase in fecundity, pronounced sexual dimorphism and morphological change associated with feeding. Plant parasitic nematodes fit into this pattern in that those nematodes with a more complex and prolonged association with plant cells exhibit a greater egg-laying capacity and are more sexually dimorphic than those whose contact with the cells on which they feed is transitory by comparison. Plant parasitic nematodes all possess a hollow buccal stylet through which secretions may be injected into plant cells. The lumen of this stylet is usually less than 0.5 μm in diameter and appears to act as an effective bacterial filter, since micro-organisms have not been isolated and cultured from the digestive tract of plant parasitic nematodes. Micro-organisms have been detected intracellularly in *Heterodera rostochiensis* and *H. goettingiana* (Shepherd, Clark & Kempton, 1973) and *Xiphinema americanum* (Adams & Eichenmuller, 1963) where they are abundant in the ovaries, but infection by these particular micro-organisms is thought to take place through the egg.

Occasionally the buccal stylet of plant parasitic nematodes is absent in degenerate males of some species but it is always present in their larval forms.

BEHAVIOUR

The behaviour of plant parasitic nematodes is influenced in various ways by stimuli from plants. At the moment little is known about how these stimuli are received, transmitted and transformed into patterns of behaviour. However, studies on the fine structure of receptors and sensory neurones in a number of these nematodes, together with ciné photomicrographic studies on some aspects of their behaviour, are bridging these gaps in our knowledge.

Hatching

The hatching of the eggs of many plant parasitic nematodes is stimulated by exudates from the roots of host plants. The classic example of enhancement of hatching by root exudations is that of *Heterodera rostochiensis* in which exudations from the roots of solanaceous plants greatly enhance hatching of the eggs of this nematode. The precise nature and mode of action of this stimulus has not yet been completely resolved. The behaviour of second-stage larvae (L_2) of this nematode several days after being stimulated in this manner has been recorded by means of time lapse ciné photomicrography (Doncaster & Shepherd, 1967; Doncaster & Seymour, 1973) and provides an excellent illustration of the degree of co-ordination that can be achieved by a nematode. The L_2 cuts a slit in its egg shell by thrusting its stylet through it to form a linear series of perforations at the rate of approximately 20 per min. The egg shell eventually tears along these perforations and the L_2 emerges. This is one of the best-documented examples of co-ordinated behaviour in the Nematoda. The sense organs involved appear to be labial sensory structures and proprioceptors in the stylet retractor muscles and somatic muscles. These provide the L_2 with information about the inner surface texture of the egg shell and its deformability against thrust by the stylet as well as to information about change in body posture resulting from stylet thrusting.

Stylet activity is invariably increased prior to hatching but in some forms such as *Meloidogyne javanica* this appears to be associated with the secretion of lipases from the subventral oesophageal glands which hydrolyse the inner layer of the egg shell leading to changes in its fine structure and permeability (Bird, 1968a). It becomes plastic and can be distorted by movements of the L_2 which bursts through the weakened shell.

Orientation of nematodes towards plants

Plants emit both physical and chemical stimuli which attract nematodes (such as *Ditylenchus dipsaci* and *Aphelenchoides fragariae*) that feed on the stems and leaves of plants as well as the numerous species that feed on roots. In the natural state these stimuli probably consist of a combination of different factors (Bird, 1962). Two of these, namely heat and carbon dioxide, have received a more detailed experimental investigation than other substances (Klinger, 1972). This is a reflection of the difficulties involved in isolating, measuring and maintaining physiological gradients of other chemical attractants from plants.

The receptors which mediate this orientation towards non-specific food sources are located on the nematode's head (Ward, 1973) and are almost certainly the amphids.

Feeding

Doncaster & Seymour (1973) postulate that behaviour leading to feeding in eleven species of Tylenchida that they studied fell into three categories namely, widespread exploration, local exploration and stylet thrusting. The former involves locomotion coupled with searching movements of the head and lips and quick and intermittent stylet probing. Local exploration commences when the nematode stops moving and rubs its lips against the surface to be penetrated and this is briefly interspersed with stylet probing. This phase is marked in migratory ectoparasitic nematodes and less frequent in the L_2s of sedentary endoparasites. In the final stylet thrusting phase the nematode remains immobile with its lips pressed, but not rubbed, against the surface which the stylet rhythmically thrusts against in a highly co-ordinated manner similar to that described above for the hatching process. This rhythmical pattern of behaviour is often characteristic for a particular species, as has been shown using ciné film analysis of species of *Tylenchorhynchus*, *Ditylenchus* and *Aphelenchoides* feeding on epidermal root cells, stomatal guard cells and fungal hyphae respectively (Wyss, 1973; Doncaster & Seymour, 1973). Between feeding, movement is random and interspersed with quiescent periods.

In endoparasites such as *Heterodera* and *Nacobbus* the stylet is withdrawn after penetrating the host cell wall and another site is selected close by and the sequence is repeated, each perforation breaking into the previous one, and so making a slit through which the head eventually breaks. It takes the L_2 of *H. cruciferae* about 5 min to cut a slit in a cell wall large enough to pass through and an average of 15 min to enter a host root.

In forms which do not enter plants but feed on epidermal cells and root hairs, as do some species of *Trichodorus* and *Hemicycliophora*, an adhesive feeding tube or plug is formed (McElroy & Van Gundy, 1968; Wyss, 1971; Kisiel, Castillo & Zuckerman, 1971). This structure, which appears to be secreted by the nematode within its stoma, encircles its stylet and attaches the nematode firmly to the plant. It enables the nematode to maintain its feeding position for a relatively long time and this must be of considerable benefit to members of the genus *Hemicycliophora* which insert their buccal stylet as much as 70 μm into their host's tissues.

The feeding habits of plant parasitic nematodes vary. Some migratory forms feed with most of their bodies protruding from the root, others enter and emerge after feeding. Some feed for a few minutes on a single cell and others may feed for a day or two on the same cell (Klinkenberg, 1963; McElroy & Van Gundy, 1968).

Wyss (1973) has compared the feeding habits of two species of different genera namely, *Tylenchorhynchus dubius* and *Trichodorus similis* on similar cells of the same host *Brassica rapa*. Their feeding habits are completely different. *T. dubius*, for instance, fed for about 9 min during which it spent 3 % of its time penetrating the cell, 13 % salivating and 84 % ingesting its food whilst *Tr. similis* fed for about 3 min during which it spent 24 % of its time penetrating, 59 % salivating and 17 % ingesting its food before departure.

The best observations of nematode feeding behaviour under natural conditions are those which have been made on roots of trees growing in an orchard using time-lapse ciné photomicrography from an underground root-observation laboratory (Pitcher & Flegg, 1965; Pitcher, 1967). By means of this technique colonies of 10–100 *Trichodorus viruliferus* have been observed feeding on epidermal cells just behind the apical meristem of apple roots. Whilst this technique is limited by the degree of optical resolution that can be obtained, it is most valuable in determining the degree of involvement of parasites with roots, at their surface or in the rhizosphere. Conventional techniques following soil sampling can be very misleading in this respect because most of these nematodes are on the roots, as opposed to the surrounding soil, and are readily dislodged when the soil or roots are disturbed.

CHANGES INDUCED IN NEMATODES BY PLANTS

Salivation is associated with the secretion of granular material from the oesophageal glands. Typically these are three uninucleate structures, a dorsal and two subventrals, which connect with the lumen of the oesophagus through cuticularized ducts which may be swollen into ampullae (Bird, 1971).

Signs of oesophageal gland activity are first detectable just prior to hatching in some forms and it has been suggested (Doncaster & Seymour, 1973) that one of the first indications that a nematode has detected an attractive chemical gradient is that it starts to accumulate secretions in these structures.

In the L_2 of *M. javanica* these secretions can be observed just before hatching. They are granular and are readily seen in the two ducts of the subventral oesophageal glands if phase or Nomarski interference contrast optics are used (Plate 1*a*). It is tempting to suggest that various hydrolysing enzymes are produced by these glands (Bird, 1967) because the egg shell's inner layer appears to break down and the shell itself becomes quite plastic just before hatching (Bird, 1968*a*). Also the L_2 has to penetrate the cellulose cell walls of its host before entering it.

Little is known about enzymes exuded from plant parasitic nematodes and the methods used to determine their presence have been open to the criticism that bacterial contaminants could have been responsible. Furthermore, sensitive quantitative techniques have not been used. Recently we have been able to demonstrate that the L_2 of *M. javanica* do exude a cellulase (Bird, Downton & Hawker, 1975). By means of a calibration curve which related viscosity to formation of reducing-sugar hydrolysed from 0.5 % buffered carboxymethyl cellulose (CMC), it is possible to calculate the amount of reducing-sugar hydrolysed from the CMC per nematode per day.

In our experiments the activity of cellulase exuded from batches of from 6×10^3–45×10^3 freshly hatched L_2s per cm^3 of substrate at 27 °C brought about a decrease in viscosity which corresponded to from 0.5–2.5 picomoles of reducing-sugar per L_2 per day.

It was possible in every instance to show by means of subculturing and re-testing and by morphological and chemical identification of the organisms present that the hydrolysis of cellulose was not due to bacteria.

After penetration into the plant, a number of marked morphological and chemical changes take place in the L_2s of sedentary endoparasitic nematodes. In *M. javanica* the granules in the ducts of the subventral oesophageal glands, which are clearly visible in the preparasitic L_2, become indistinct within two to three days after entry into the root (Plate 1a) although their contents stain well when treated with the periodic acid–Schiff (PAS) technique in contrast to those structures in the preparasitic L_2s.

Furthermore, the different shapes of spectral absorption curves in similarly treated preparasitic and parasitic L_2s (Fig. 1) indicate that the protein component in the ducts of these glands changes with the onset of parasitism (Bird & Saurer, 1967). Clearly, marked chemical and morphological changes have occurred which are induced by the host. At higher resolution these morphological changes are even more pronounced (Plate 2a, b). The granules in the parasitic L_2 are smaller and the membrane that surrounds them appears to be breaking down. The contents of these

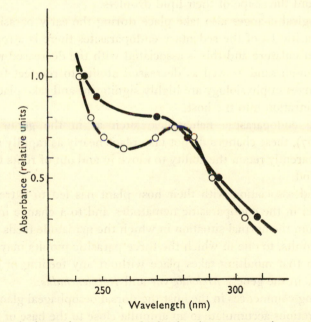

Fig. 1. Spectral absorption curves obtained using a u.v. microspectrograph of points taken through the region of the subventral oesophageal glands in preparasitic (○) and parasitic (●) L_2s of *M. javanica* (after Bird & Saurer, 1967).

granules stain more deeply and the lumen of the oesophagus becomes filled with material which resembles the internal contents of the granules. It is assumed that this substance has an important role in the establishment of the nematode-induced transfer cell or syncytium.

Another morphological change that takes place in the L_2 of sedentary nematodes shortly after their entry into the host plant is a change in the structure of the cuticle (Plate 2c, d, e). These changes can only be detected with the aid of the electron microscope because the entire cuticle of the L_2 of *M. javanica* is only 0.3 μm thick and so lies at the limit of resolution of the light microscope. Within a week of entry into the host plant the striated basal layer, a characteristic structure of the cuticles of pre-parasitic L_2s of many different species and genera, completely disappears.

Certain plant parasitic nematodes which cause galls in the leaves, stems and seeds of plants are able to survive in these galls in an anhydrobiotic state under extreme conditions. We (Bird & Buttrose, 1974) have detected a number of morphological differences between anhydrobiotic and active L_2s of *Anguina tritici*, using freeze-etch techniques. The anhydrobiotic forms can survive temperatures as low as -190 °C and as high as $+105$ °C for short periods of time and differ in the fine structure of the outer-most layers of their cuticles, the spatial arrangement of their muscle filaments and the shape of their lipid droplets.

Physiological changes also take place during the early parasitic period. In the parasitic L_2 of the sedentary endoparasites there is atrophy of the somatic musculature and this is associated with the decreased mobility of the L_2 through sand as well as decreased ability to reinfect fresh hosts. These changes in physiology are highly significant and take place within a day of penetration into the host.

In some endoparasitic nematodes, such as in the genus *Nacobbus* (Clark, 1967), these changes do not take place nearly as rapidly and the L_2 and L_3 apparently retain the ability to move in and out of roots for a much longer period.

Prolonged association with their host plant has led to extreme sexual dimorphism in the endoparasitic nematodes and to a change in moulting patterns from the normal situation in which the nematode feeds and grows between moults, to one in which the three parasitic moults may be super-imposed so that moulting takes place without any feeding or growth in between, as in the genera *Meloidogyne* and *Rotylenchulus*.

As feeding commences in the host the dorsal oesophageal gland enlarges and its secretions accumulate in an ampulla close to the base of the buccal stylet where it enters the alimentary tract (Plate 3). Gland secretions may be emitted by nematodes during feeding and a clear zone may form around the stylet tip. These secretions do not appear to be injected into the host cell in granular form. Observations with the electron microscope of material cut through the ampulla at the base of the stylet in *M. javanica* (Bird, 1969) indicate that a few of these granules may get squeezed out through the stylet but nearly all appear to break down close to the terminal duct which connects with the oesophageal duct (Plate 3*c*).

These granules in the oesophageal glands of plant parasitic nematodes have many features in common with micro-bodies (Frederick, Newcomb, Vigil & Wergin, 1968) that have been described in plant and animal cells. For instance, they are about the same diameter (0.4–0.8 μm), have a single bounding membrane, have a granular to fibrillar matrix of variable density, are associated with rough endoplasmic reticulum and probably contain peroxidase as this enzyme has been detected in the stylet exudations of adult female *M. incognita* (Hussey & Sasser, 1973).

Microbodies are thought to play an active role in cellular metabolism and it is thought that their enzyme complement and metabolic role may change during cellular differentiation. Certainly the number of enzymes synthesized by the oesophageal glands in nematodes is probably considerable. A basic protein which has some of the properties of a histone is also exuded from the buccal stylet of *M. javanica* (Bird, 1968*b*), and this could also have its origin in the microbodies.

CHANGES INDUCED IN PLANTS BY NEMATODES

Whole plants

Plants respond to nematodes in a variety of ways, depending on factors such as the balance between the numbers of nematodes and the food resources of the plant. This will vary depending on the species of nematode, the species of host, the nutritional status of the host, its age, and various environmental factors such as temperature, light and the composition of the soil, to mention a few of the variables.

The overall effect of a heavy infection of nematodes often is that the host plant becomes stunted and there is a considerable loss in yield. Seedlings which have germinated in heavily infected soil may die due to the physical disintegration of their root system caused by thousands of larvae. On the other hand, as mentioned above, very light infections may actually stimulate growth under certain conditions. In between these two levels, there is a whole range of symbiotic associations whose effect on the plant is difficult to measure with any degree of precision.

The photosynthetic rate of plants is significantly decreased by large numbers of *M. javanica* within one to two days of infection (Loveys &

Bird, 1973) and in experiments with the same nematode at later stages in the parasitic relationship, the photosynthetic rate of the plant has been shown to be reduced by as few as 2000 nematodes per plant (Wallace, 1974).

It is possible that root damage leading to water stress and partial closure of the stomata resulting in decreased carbon dioxide fixation could be responsible for this observed decline in the rate of photosynthesis brought about by the parasite. However, as there was no sign of wilting in the experimental plants, it is equally possible that photosynthesis may have been inhibited by interference with the synthesis and/or translocation of growth hormones in the host's roots. It has been shown that *Meloidogyne* can cause a decrease in cytokinins and gibberellins in root tissues and xylem exudates of its host (Brueske & Bergeson, 1972).

Growth-regulating substances are also associated with the degree of certain types of resistance in plants. For instance, cytokinins supplied to resistant tomatoes will make these plants more susceptible to *M. incognita* (Dropkin, Helgeson & Upper, 1969) and resistance in peaches to *M. javanica* is broken if these plants are wick-fed with kinetin or 1-naphthyl-acetic acid (Kochba & Samish, 1971). These workers also showed that cytokinin and auxin levels were significantly higher in susceptible root-stocks than in resistant ones (Kochba & Samish, 1972).

The situation is complicated by the wide range of plant physiological processes that are known to be influenced by cytokinins, such as control of cell division and cell enlargement, rate of enzyme activity and translocation of metabolites, lateral bud growth, development of inflorescences, the setting of fruit, responses to stress and the senescence of detached organs.

It is, perhaps, not surprising that little is known about the physiological processes associated with resistance of plants to nematodes for, as a pre-requisite, we require more information about the parasites' physiology to match that which the plant physiologists are supplying for the host.

Cells

Under corresponding conditions, similar numbers of nematodes will usually enter both susceptible and resistant plants; occasionally, in some species of plant, there is a barrier to penetration. Normally, resistance is manifest, once the parasite has entered its host, as a necrosis and darkening of cells. It is thought that both growth regulators and phenolic compounds may be important determinants of resistance and the outcome of these interactions often results in what is known as the hypersensitive reaction (HR) which is thought to be the same for both fungi and nematodes (Wallace, 1973).

Resistance to *H. rostochiensis* in various species of solanaceous plants has

PLATE I

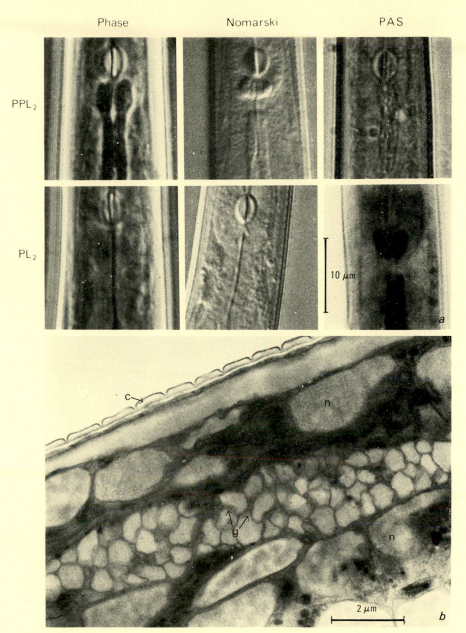

Phase   Nomarski   PAS

PPL₂

PL₂

10 μm

a

2 μm

b

For explanation see p. 370

PLATE 2

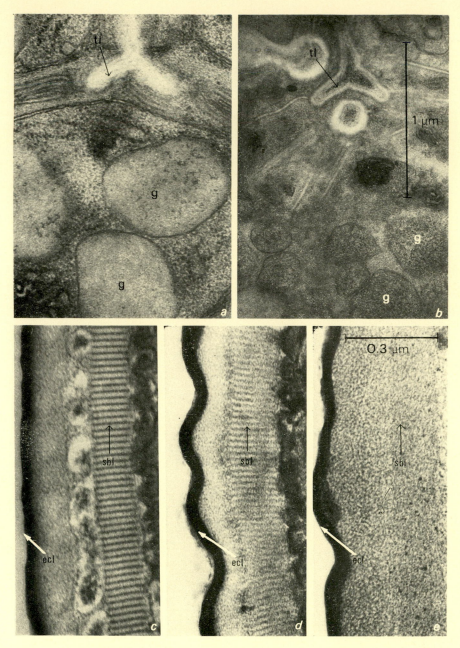

For explanation see p. 370

PLATE 3

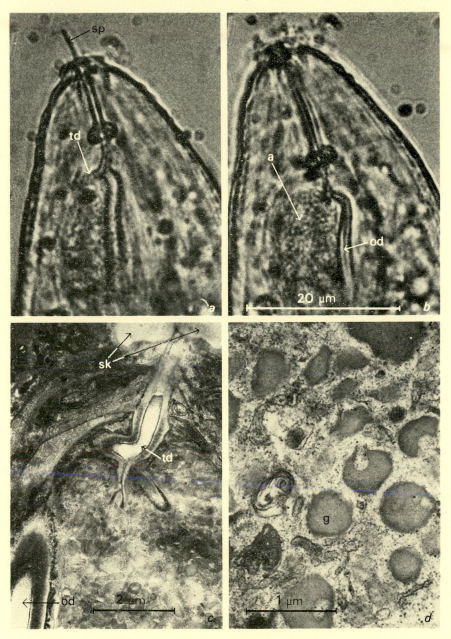

For explanation see p. 370

PLATE 4

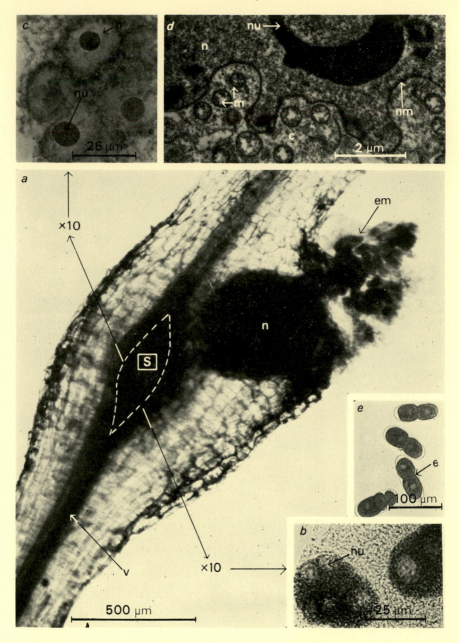

For explanation see p. 371

PLATE 5

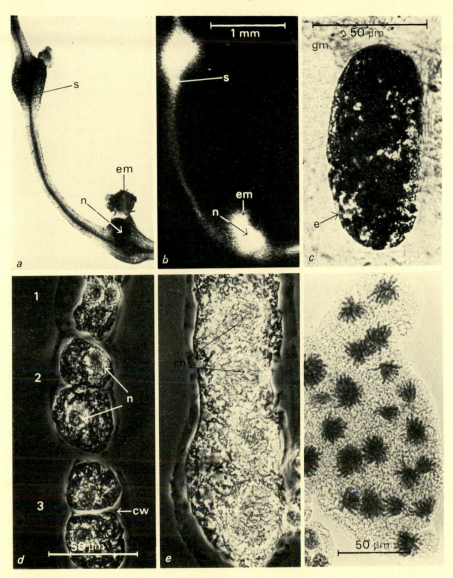

For explanation see p. 371

PLATE 6

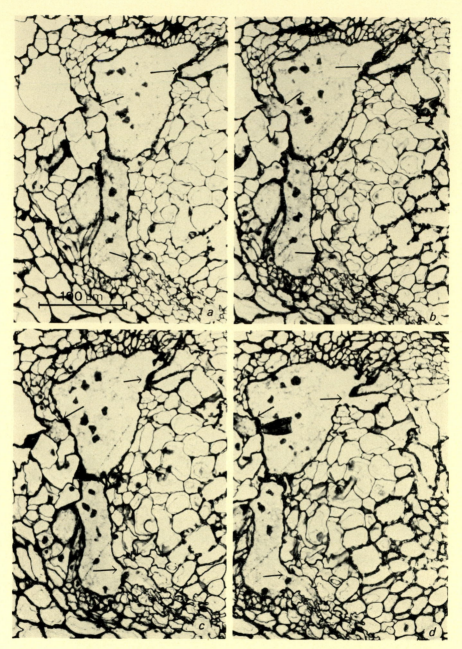

For explanation see p. 371

been correlated with a high ratio of mono- to polyphenolic compounds (Giebel, 1970), which are thought to act as precursors for lignin which is associated with necrotic spots and wall thickenings. The greatest concentrations of lignified tissues are found adjacent to the head and excretory pore of the parasite (Giebel, Krenz & Wilski, 1970). The fine structure of tissues of tomato roots involved in the HR to *M. incognita* has been examined (Paulson & Webster, 1972). These workers came to the conclusion that the death of the invading nematode was not due to the production of toxic substances but was caused by deteriorative changes in the cytoplasm of developing syncytia which prevented their formation. This deterioration was not enough to induce L_2s to move elsewhere, so they remained and died of starvation.

The cellular response to nematodes in resistant plants is clearly complex. In some instances there is no necrosis of the type usually associated with the HR (McClure, Ellis & Nigh, 1974) and in others the nematodes just migrate out of these roots several days after they have entered without inducing symptoms and are still capable of penetrating and maturing in susceptible plants (Reynolds, Carter & O'Bannon, 1970).

In comparison with the paucity of information available on cell responses in resistant plants, a considerable amount of work has been done on cell responses in susceptible plants, and there are numerous examples in the literature of various types of cell response to feeding. They range (Table 1) from a situation in which the parasite may obtain its food without any noticeable alteration to the cells' physiological mechanisms, to the formation of large, multinucleate syncytia which have a highly specialized function and which the nematode induces from undifferentiated tissue. In both these situations there is a balance in which the nematode obtains its nutrition with the minimum of disturbance to its host, provided that the numbers of parasites are not too great. When this does happen and there is competition for feeding sites, some nematodes can overcome this by means of adaptations which are self-regulatory. Thus, in *Meloidogyne* we can have sex reversal occurring during larval development resulting in the production of large numbers of males which are capable of moving to other feeding sites and whose nutritional requirements are much less than those of the female.

Between these two extremes of symbiosis, plant cells may undergo considerable physiological stress as a result of feeding by nematodes. This ranges from a decrease in cyclosis, with accompanying cytoplasmic coagulation which results in the gradual death of the cell, to an acceleration of these events leading in this case to a rapid death. In some instances, the cytoplasm may be sucked out of the cell by the parasite. These types of cell

Table 1. *Examples of host cell response to feeding by plant parasitic nematodes*

<table>
<tr><td></td><td>Cell response to nematode feeding</td><td>Example</td></tr>
<tr><td rowspan="6" style="writing-mode: vertical">INCREASING COMPLEXITY OF RESPONSE</td><td>1. Cyclosis normal – cell survives ⟶</td><td>*Paratylenchus projectus* feeding on root hair cell</td></tr>
<tr><td>2. Cyclosis decreases – cytoplasm coagulates – cells die gradually ⟶</td><td>*Tylenchorhynchus dubius* feeding on epidermal or root hair cells</td></tr>
<tr><td>3. Cytoplasmic coagulation and death immediately after feeding ⟶</td><td>*Trichodorus similis* feeding on epidermal or root hair cells</td></tr>
<tr><td>4. Hypertrophy of cells, loss of chloroplasts, lysis, cavities form ⟶</td><td>*Ditylenchus dipsaci* feeding on cortical parenchyma cells</td></tr>
<tr><td>5. Nuclear and nucleolar enlargement, increased cytoplasmic density, thickened cell walls – nurse cell, 6–10 of these make up a feeding site ⟶</td><td>*Tylenchulus semipenetrans* feeding on cortical parenchyma cells</td></tr>
<tr><td>6. Nuclear and nucleolar enlargement, increased cytoplasmic density, synchronous mitosis without cytokinesis – cell fusion ⟶</td><td>*Meloidogyne javanica* feeding on cells in the stele</td></tr>
</table>

response which lead to the death of the cell and to the formation of lesions and cracks in the host's epidermis or cortical parenchyma are associated with migratory parasites that can move from root to root. In sedentary forms such as *Rotylenchulus*, *Tylenchulus*, *Heterodera* and *Meloidogyne*, the host cells are stimulated rather than depleted, and respond to the nematode by an enlargement of their nuclei and nucleoli, an increase in cytoplasmic density and a thickening of their cell walls (Plate 4*a–d*). In forms such as *Rotylenchulus* and *Tylenchulus* the adult female will feed on a number of these cells located in the pericycle and the cortex respectively.

Members of the family Heteroderidae, to which *Heterodera* and *Meloidogyne* belong, induce the formation of multi-nucleate syncytia or giant cells, which are normally located in the vascular tissue. These structures have been examined with the aid of the electron microscope by workers from a number of different laboratories in recent years. Syncytia from both these genera have been examined by Jones & Northcote (1972*a*, *b*) and from *Meloidogyne* by Bird (1961), Huang & Maggenti (1969*b*) and by Paulson & Webster (1970). In their study on the syncytia induced by *H. rostochiensis* Jones & Northcote (1972*a*) have confirmed and reviewed the work of early investigators which showed that cell wall dissolution occurs during the formation of syncytia and that the cell walls of these structures have protuberances which are adjacent to conducting elements. Furthermore, they have drawn attention to the remarkable similarity in wall structure between syncytia and transfer cells (Pate &

Gunning, 1972) and have suggested that this differentiation of unspecialized cells to ones with a specific physiological function is induced by a demand from the feeding nematode for an increased flow of solutes. A similar pattern of events is thought to give rise to syncytia induced by *Meloidogyne* (Jones & Northcote, 1972*b*). The development of these structures is associated with a marked swelling or hyperplasia of the root cortex (Plate 4*a*), hence the parasite's common name of root-knot nematode. However, although this swelling is usually present, the nematode can grow normally and reproduce in its absence. The syncytia, on the other hand, are necessary for normal growth and reproduction.

The exact manner by which the syncytia or giant cells induced by *Meloidogyne* are formed, has not been clearly resolved and, indeed, may not be to everybody's satisfaction until suitable ciné time-lapse studies of the first few days of the induction of the structure are obtained. The question is whether or not there is cell wall breakdown and incorporation of adjacent cells, as with *Heterodera* in which case the structure should be termed a syncytium, or whether the cell just enlarges without ever incorporating adjacent cells and so should be termed a giant cell.

The evidence in favour of calling the structure a giant cell rests largely on the work of Huang & Maggenti (1969*a*), who consider that the nuclei of this structure are derived from repeated mitoses of the original diploid cell without subsequent cytokinesis so that the cell increases its size by swelling and the chromosome numbers fall into a 4, 8, 16, 32 and 64n ploidy sequence.

These workers observed some cell wall breakdown, but concluded that this was due to mechanical breakdown by migrating L_2s. Jones & Northcote (1972*b*) have also noticed occasional cell wall breakdown at the periphery of giant cells, but did not think that this contributed to their formation.

Alternatively, it has been stated (Christie, 1936) that 'at the beginning of giant-cell formation, it is usually several adjacent members of a row of undifferentiated cells in the central cylinder that first coalesce through the dissolution of the separating cell walls'. I have noticed this in roots during the early stages of syncytial formation (Plate 5*d, e*). It could easily be overlooked if only later stages of syncytial development were studied. The occurrence of synchronous mitoses in developing syncytia has been recorded by numerous investigators (Krusberg & Nielsen, 1958; Bird, 1961, 1973; Owens & Specht, 1964) and is an established fact (Plate 5*f*). It takes place in the earlier stages of syncytial development and there may be some asynchrony (Bird, 1973). Both direct observations of the chromosomes during mitosis (Bird, 1973) and photometric measurements of Feulgen-stained nuclei in syncytia (Fig. 2) have failed to support the notion that

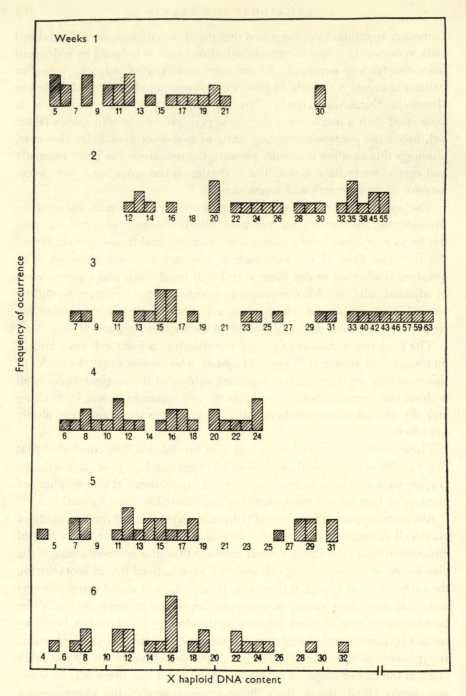

Fig. 2. Histograms depicting photometric measurements of Feulgen-stained nuclei in syncytia induced by *M. javanica* in *Vicia faba* during different stages of growth (Bird, 1972).

there is a regular ploidy sequence. These results are more consistent with the hypothesis that cells may be incorporated by the expanding syncytia. Furthermore, serial sections 2 μm thick through syncytia in tomato roots have revealed just such connections (Plate 6). Thus these structures can be regarded as true syncytia rather than giant cells because cell wall breakdown and cell fusion occur to a limited extent accompanied by synchronous mitosis and cell expansion. The concept of cell fusion in the early stages of syncytial development receives further support from the fact that a cell wall-degrading enzyme is secreted from the L_2 of *M. javanica* as mentioned above, and furthermore it is consistent with research that has been undertaken on artificial induction of cell fusion of plant protoplasts in which cellulase is used to degrade cell walls as part of the process (Power & Cocking, 1971). In fact, spontaneous fusion resulting from removal of cell wall constrictions on plasmodesmata has been known to give rise to large multinucleate protoplasts (Withers & Cocking, 1972).

Syncytia vary in size depending on the host and the rate of growth of the nematode within that host. The syncytia invariably contain numerous large nuclei and nucleoli (Plate 4) lying in dense cytoplasm surrounded by thick walls, which have numerous protuberances adjacent to xylem vessels and to sieve elements to a lesser extent. Here again the nematode is thought to function as a nutrient sink. We have obtained some physiological evidence for this notion by using radio tracer techniques (Bird & Loveys, unpublished results). Infected plants were allowed to photosynthesize in an atmosphere of $^{14}CO_2$ and were then left in the glasshouse for periods of time which varied from two hours to five days before being harvested. The uptake of ^{14}C into the nematodes and egg masses was measured using a liquid scintillation spectrometer (Fig. 3d) and uptake into the syncytia, nematodes and eggs was demonstrated using both macro- and microautoradiography (Plate 5a, b, c). The uptake of the label into the nematode is slow and not significant over periods of two and four hours. However, over periods of from one to five days, there was a much greater accumulation of translocated photosynthates in the galls than in adjacent tissues (Plate 5a, b). After five days exposure to $^{14}CO_2$ for instance, galls with egg masses contained about six times as much ^{14}C as did adjacent roots and about half of this had become incorporated into the eggs (Plate 5c).

The label apparently becomes incorporated into relatively insoluble compounds within 24 hours as we found that it was not removed from the nematode in significant amounts by aqueous fixatives, or by extraction with 80 % ethanol for 20 min at 75 °C. Clearly *Meloidogyne* acts as a nutrient sink as postulated by Jones & Northcote (1972b) and its organic nutrients

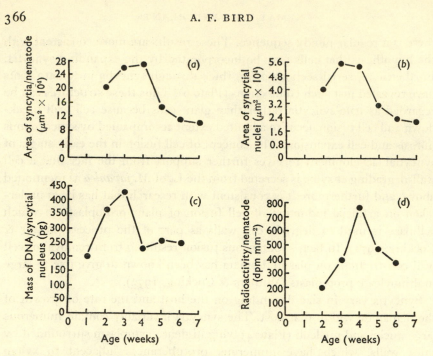

Fig. 3. Curves depicting morphological and physiological changes in syncytia induced by *M. javanica* in *Vicia faba* during their growth and decline (after Bird, 1972, and Bird and Loveys, unpublished work). (*a*) Changes in the size of syncytia; (*b*) changes in the size of syncytial nuclei; (*c*) changes in the amount of DNA per nucleus; (*d*) changes in the amount of ^{14}C incorporated into the female nematode from photosynthates after exposure to the label for 24 h.

appear to be derived, at least in part, from the products of current photosynthesis which are presumably translocated to the nematode via the phloem.

GROWTH AND DECLINE OF THE SYMBIOTIC RELATIONSHIP

Quantitative studies on the growth of symbiotic relationships between nematodes and plants at the cell level have, so far, only been done on syncytia induced by *Meloidogyne* (Bird, 1972). These experiments have shown that syncytia undergo a cycle of growth that is directly related to the physiological age of the nematode. Whether one measures size of syncytia, size of nucleus, weight of DNA per nucleus or uptake of ^{14}C per unit area of nematode, the answer is always much the same (Fig. 3*a–d*), namely that the syncytium has a relatively short cycle of activity and that it builds up to a peak around about the time the nematode has started to lay eggs and then it starts to decline. The rate of growth of the syncytium depends on the weather conditions and the species of host so, if comparative experiments are to be made, it is most important to compare nematodes of the same age under similar environmental conditions in the same species of host.

CONCLUSION

The exact manner by which the various nematode–plant symbiotic relationships are initiated and maintained remains unsolved. Perhaps one of the most interesting aspects is the response of plant cells to these nematodes. What factors emitted by the nematode are responsible for the complex changes in plant cell structure that are induced in the endoparasitic forms? How do these factors influence normal coding in cells of the developing root to give rise to the complex syncytia that have been described? What factors control synchronous mitoses in syncytia? Are virus-like microorganisms involved?

The answers to these and numerous other similar questions present a stimulating challenge to the biologist interested in understanding the nature of this symbiosis, and the forces that control cell differentiation.

ACKNOWLEDGEMENTS

I wish to thank my wife Jean and my colleagues in CSIRO for their advice and criticism and Mrs J. P. Milln for assistance in the preparation of the plates and figures.

REFERENCES

ADAMS, R. E. & EICHENMULLER, J. J. (1963). A bacterial infection of *Xiphinema americanum. Phytopathology*, **53,** 745.

DE BARY, A. (1879). *Die Erscheinung der Symbiose.* Strassburg: Karl J. Trübner.

BIRD, A. F. (1961). The ultrastructure and histochemistry of a nematode-induced giant cell. *J. biophys. biochem. Cytol.*, **11,** 701–715.

—— (1962). Orientation of the larvae of *Meloidogyne javanica* relative to roots. *Nematologica*, **8,** 275–287.

—— (1967). Changes associated with parasitism in nematodes. I. Morphology and physiology of preparasitic and parasitic larvae of *Meloidogyne javanica*. *J. Parasit.*, **53,** 768–776.

—— (1968a). Changes associated with parasitism in nematodes. III. Ultrastructure of the egg shell, larval cuticle, and contents of the subventral esophageal glands in *Meloidogyne javanica* with some observations on hatching. *J. Parasit.*, **54,** 475–489.

—— (1968b). Changes associated with parasitism in nematodes. IV. Cytochemical studies on the ampulla of the dorsal esophageal gland of *Meloidogyne javanica* and on exudations from the buccal stylet. *J. Parasit.*, **54,** 879–890.

—— (1969). Changes associated with parasitism in nematodes. V. Ultrastructure of the stylet exudation and dorsal esophageal gland contents of female *Meloidogyne javanica*. *J. Parasit.*, **55,** 337–345.

(1971). *The Structure of Nematodes*. New York: Academic Press.

(1972). Quantitative studies on the growth of syncytia induced in plants by root-knot nematodes. *Int. J. Parasit.*, **2**, 157–170.

(1973). Observations on chromosomes and nucleoli in syncytia induced by *Meloidogyne javanica*. *Physiol. Pl. Path.*, **3**, 387–391.

BIRD, A. F. & BUTTROSE, M. S. (1974). Ultrastructural changes in the nematode *Anguina tritici* associated with anhydrobiosis. *J. ultrastruct. Res.* **48**, 177–189.

BIRD, A. F., DOWNTON, W. J. S. & HAWKER, J. S. (1975). Cellulase secretion by second stage larvae of the root-knot nematode (*Meloidogyne javanica*). *Marcellia*, **38**, 165–169.

BIRD, A. F. & SAURER, W. (1967). Changes associated with parasitism in nematodes. II. Histochemical and microspectrophotometric analyses of preparasitic and parasitic larvae of *Meloidogyne javanica*. *J. Parasit.*, **53**, 1262–1269.

BOWEN, G. D. & ROVIRA, A. D. (1969). In *Root Growth*, 170–201 (ed. W. I. Whittington). London: Butterworths.

BRUESKE, C. H. & BERGESON, G. B. (1972). Investigation of growth hormones in xylem exudate and root tissue of tomato infected with root-knot nematode. *J. exp. Bot.*, **23**, 14–22.

CHITWOOD, B. G., SPECHT, A. W. & HAVIS, L. (1952). Root-knot nematodes. III. Effects of *Meloidogyne incognita* and *M. javanica* on some peach rootstocks. *Pl. Soil*, **4**, 77–95.

CHRISTIE, J. R. (1936). Development of root-knot nematode galls. *Phytopathology*, **26**, 1–22.

CLARK, S. A. (1967). The development and life history of the false root-knot nematode, *Naccobbus serendipiticus*. *Nematologica*, **13**, 91–101.

DONCASTER, C. C. & SEYMOUR, N. K. (1973). Exploration and selection of penetration site by Tylenchida. *Nematologica*, **19**, 137–145.

DONCASTER, C. C. & SHEPHERD, A. M. (1967). The behaviour of second stage *Heterodera rostochiensis* larvae leading to their emergence from the egg. *Nematologica*, **13**, 476–478.

DROPKIN, V. H., HELGESON, J. P. & UPPER, C. D. (1969). The hypersensitivity reaction of tomatoes resistant to *Meloidogyne incognita*: reversal by cytokinins. *J. Nematology.*, **1**, 55–61.

FREDERICK, S. E., NEWCOMB, E. H., VIGIL, E. L. & WERGIN, W. P. (1968). Fine-structural characterization of plant microbodies. *Planta*, **81**, 229–252.

GIEBEL, J. (1970). Phenolic content in roots of some Solanaceae and its influence on IAA-oxidase activity as an indicator of resistance to *Heterodera rostochiensis*. *Nematologica*, **16**, 22–32.

GIEBEL, J., KRENZ, J. & WILSKI, A. (1970). The formation of lignin-like substances in roots of resistant potatoes under the influence of *Heterodera rostochiensis* larvae. *Nematologica*, **16**, 601.

HERTIG, M., TALIAFERRO, W. H. & SCHWARTZ, B. (1947). Report of the Committee on Terminology. *J. Parasit.*, **23**, 325.

HUANG, C. S. & MAGGENTI, A. R. (1969a). Mitotic aberrations and nuclear changes of developing giant cells in *Vicia faba* caused by root knot nematode, *Meloidogyne javanica*. *Phytopathology*, **59**, 447–455.

(1969b). Wall modifications in developing giant cells of *Vicia faba* and *Cucumis sativus* induced by root-knot nematode, *Meloidogyne javanica*. *Phytopathology*, **59**, 931–937.

HUSSEY, R. S. & SASSER, J. N. (1973). Peroxidase from *Meloidogyne incognita*. *Physiol. Pl. Path.*, **3**, 223–229.

JONES, M. G. K. & NORTHCOTE, D. H. (1972a). Nematode-induced syncytium – a multinucleate transfer cell. *J. Cell Sci.*, **10**, 789–809.

(1972*b*). Multinucleate transfer cells induced in *Coleus* roots by the root-knot nematode, *Meloidogyne arenaria. Protoplasma,* **75,** 381–395.

KISIEL, M., CASTILLO, J. & ZUCKERMAN, B. M. (1971). An adhesive plug associated with the feeding of *Hemicycliophora similis* on cranberry. *J. Nematol.,* **3,** 296–298.

KLINGER, J. (1972). The effect of single and combined heat and CO_2 stimuli at different ambient temperatures on the behaviour of two plant parasitic nematodes. *J. Nematol.,* **4,** 95–100.

KLINKENBERG, C. H. (1963). Observations on the feeding habits of *Rotylenchus uniformis, Pratylenchus crenatus, P. penetrans, Tylenchorhynchus dubius* and *Hemicycliphora similis. Nematologica,* **9,** 502–506.

KOCHBA, J. & SAMISH, R. M. (1971). Effect of kinetin and 1-naphthylacetic acid on root-knot nematodes in resistant and susceptible peach rootstocks. *J. Am. Soc. hort. Sci.,* **96,** 458–461.

—— (1972). Level of endogenous cytokinins and auxin in roots of nematode resistant and susceptible peach rootstocks. *J. Am. Soc. hort. Sci.,* **97,** 115–119.

KRUSBERG, L. R. & NIELSEN, L. W. (1958). Pathogenesis of root-knot nematodes to the Porto Rico variety of sweet potato. *Phytopathology,* **48,** 30–39.

LINCICOME, D. R. (1963). Chemical basis of parasitism. *Ann. N.Y. Acad. Sci.,* **113,** 360–380.

LOVEYS, B. R. & BIRD, A. F. (1973). The influence of nematodes on photosynthesis in tomato plants. *Physiol. Pl. Path.,* **3,** 525–529.

McCLURE, M. A., ELLIS, K. C. & NIGH, E. L. (1974). Post-infection development and histopathology of *Meloidogyne incognita* in resistant cotton. *J. Nematol.,* **6,** 21–26.

McELROY, F. D. & VAN GUNDY, S. D. (1968). Observations on the feeding processes of *Hemicycliophora arenaria. Phytopathology,* **58,** 1558–1565.

OVERGAARD-NIELSEN, C. (1949). Studies on the soil microfauna. II. The soil inhabiting nematodes. *Nat. Jutland,* **2,** 1–131.

OWENS, R. G. & SPECHT, H. N. (1964). Root-knot histogenesis. *Contr. Boyce Thompson Inst. Pl. Res.,* **22,** 471–490.

PATE, J. S. & GUNNING, B. E. S. (1972). Transfer cells. *A. Rev. Pl. Physiol.,* **23,** 173–196.

PAULSON, R. E. & WEBSTER, J. M. (1970). Giant cell formation in tomato roots caused by *Meloidogyne incognita* and *Meloidogyne hapla* (Nematoda) infection. A light and electron microscope study. *Can. J. Bot.,* **48,** 271–276.

—— (1972). Ultrastructure of the hypersensitive reaction in roots of tomato, *Lycopersicon esculentum* L., to infection by the root-knot nematode, *Meloidogyne incognita. Physiol. Pl. Path.,* **2,** 227–234.

PITCHER, R. S. (1965). Interrelationships of nematodes and other pathogens of plants. *Helminth. Abstr.,* **34,** 1–17.

—— (1967). The host–parasite relations and ecology of *Trichodorus viruliferus* on apple roots, as observed from an underground laboratory. *Nematologica,* **13,** 547–557.

PITCHER, R. S. & FLEGG, J. J. M. (1965). Observation of root feeding by the nematode *Trichodorus viruliferus* Hooper. *Nature, Lond.,* **207,** 317.

POWER, J. B. & COCKING, E. C. (1971). Fusion of plant protoplasts. *Sci. Prog., Oxford,* **59,** 181–198.

READ, C. P. (1970). *Parasitism and Symbiology.* New York: Ronald Press.

REYNOLDS, H. W., CARTER, W. W. & O'BANNON, J. H. (1970). Symptomless resistance of alfalfa to *Meloidogyne incognita acrita. J. Nematol.,* **2,** 131–134.

ROVIRA, A. D. (1965). In *Ecology of Soil-borne Plant Pathogens,* 170–186 (eds. K. F. Baker and W. C. Snyder). Los Angeles: University of California Press.

SHEPHERD, A. M., CLARK, S. A. & KEMPTON, A. (1973). An intracellular micro-organism associated with tissues of *Heterodera spp. Nematologica*, **19**, 31–34.

SMITH, D., MUSCATINE, L. & LEWIS, D. (1969). Carbohydrate movement from autotrophs to heterotrophs in parasitic and mutualistic symbosis. *Biol. Rev.*, **44**, 17–90.

WALLACE, H. R. (1973). *Nematode Ecology and Plant Disease*. London: Edward Arnold.

(1974). The influence of root-knot nematode, *Meloidogyne javanica*, on photo-synthesis and on nutrient demand by roots of tomato plants. *Nematologica*, **20**, 27–33.

WARD, S. (1973). Chemotaxis by the nematode *Caenorhabditis elegans*: identifica-tion of attractants and analysis of the response by use of mutants. *Proc. nat. Acad. Sci. USA*, **70**, 817–821.

WEINSTEIN, P. P. (1960). In *Host Influence on Parasite Physiology*, 65–96 (ed. L. A. Stauber). New Brunswick: Rutgers University Press.

WITHERS, L. A. & COCKING, E. C. (1972). Fine-structural studies on spontaneous and induced fusion of higher plant protoplasts. *J. Cell Sci.*, **11**, 59–75.

WYSS, U. (1971). Der Mechanismus der Nahrungsaufnahme bei *Trichodorus similis*. *Nematologica*, **17**, 508–518.

(1973). Feeding of *Tylenchorhynchus dubius*. *Nematologica*, **19**, 125–136.

YEATES, G. W. (1971). Feeding types and feeding groups in plant and soil nema-todes. *Pedobiologica*, **11**, 173–179.

EXPLANATION OF PLATES

PLATE 1

(a) Ducts of subventral oesophageal glands in preparasitic (PPL_2) and parasitic (PL_2) second stage larvae of *M. javanica*. Examined alive under phase and Nomarski interference contrast microscopy and fixed and stained by the periodic acid–Schiff technique (PAS) (Bird, 1968a).

(b) Electron micrograph of a longitudinal section through the ducts of the subventral oesophageal glands of a PPL_2. Showing cuticle (c), nuclei (n) and granules (g) (Bird, 1968a).

PLATE 2

(a, b) Electron micrographs of transverse sections through the ducts of the sub-ventral oesophageal glands in (a) preparasitic L_2, and (b) a 2–3 day old parasitic L_2. Showing granules (g), triradiate lumen of oesophagus (tl) (Bird, 1968a).

(c, d, e) Electron micrographs of transverse sections through cuticles of L_2s of *M. javanica* showing changes that occur during the onset of the parasitic mode of life. (c) Preparasitic L_2; (d) 2–3 day old parasitic L_2; (e) a week old parasitic L_2. Note external cortical layer (ecl) and disappearance of striped basal layer (sbl) (Bird, 1968a).

PLATE 3

(a, b) Head region of living female *M. javanica* viewed under normal transmitted light. Both photographs of the same specimen were taken with electronic flash. Note granule-packed ampulla of dorsal oesophageal gland (a), terminal duct (td) connecting to oesophageal duct (od) and everted buccal stylet or spear (sp) in (a) and stylet withdrawn in (b) (Bird, 1968b).

(*c, d*) Electron micrographs of longitudinal sections through the ampulla of the dorsal oesophageal gland. (*c*) Stylet knobs (sk), terminal duct (td) and oesophageal duct (od). (*d*) Structure of granules (g) in the ampulla.

PLATE 4

(*a*) Whole gall showing female *M. javanica* (n), egg mass (em), vascular tissue (v) and syncytia (s) outlined by dotted white lines.

(*b, c*) Ten-fold enlargements of part of syncytia showing nucleus (n) and nucleolus (nu). (*b*) Feulgen-stained; (*c*) Sudan black B stained. (*b* is from Bird 1973; *c* is from A. F. Bird, 1961, *J. Cell Biol.*, **11**, 709, Fig. 8.)

(*d*) Electron micrograph of a section cut through part of a nucleus (n) showing part of nucleolus (nu) nuclear membrane (nm), cytoplasm (c) and mitochondria (m).

(*e*) Enlargement of eggs (e).

PLATE 5

(*a, b*) Tomato root infected with 7 week old *M. javanica* and harvested 5 days after exposure to $^{14}CO_2$. (*a*) Root removed from surface of X-ray film and treated at 60 °C for 18 h in lactophenol containing a trace of cotton blue. Nematode (n), egg mass (em) and syncytia (s). (*b*) Macroautoradiograph of freeze-dried root showing accumulation of labelled photosynthate in gall. Nematode (n), egg mass (em) and syncytia (s).

(*c*) Microautoradiograph of a single egg treated as for (*a*) and (*b*) but placed on a slide and then dipped in Ilford L4 nuclear emulsion and processed by standard techniques. Gelatinous matrix (gm), egg (e).

(*d, e*) Feulgen-stained bean roots showing young syncytia in the process of formation (Bird, 1973). (*d*) Cell fusion and cell wall breakdown. (1) Cells fused, (2) cell wall breaking down, and (3) cell wall still present. Cell wall (cw), nuclei (n). (*e*) Synchronous mitosis occuring in a newly formed syncytium. Chromosomes (ch). (*d* and *e* are from Bird, 1973.)

(*f*) Synchronous mitosis within a larger syncytium.

PLATE 6

(*a, b, c, d*) Serial sections 2 μm thick through syncytia in tomato root. Connections between syncytia and adjacent cells are shown by arrows. (*b* and *c* are from A. F. Bird, 1972, *Int. J. Parasit.*, **2**, 432.)

SYMBIOSIS AND THE BIOLOGY OF LICHENISED FUNGI

By D. C. SMITH

Department of Botany, The University, Bristol BS8 1UG

INTRODUCTION

Two outstanding characteristics of lichenised fungi are that they are symbiotic and that they are adapted to existence in habitats unfavourable to most other kinds of plants.

This article attempts a critical assessment of present understanding of these characteristics, and then considers the role of symbiosis in the ecological success of lichenised fungi. No comprehensive literature survey will be undertaken, especially in view of the recent or imminent publication of two substantial texts on lichens (Ahmadjian & Hale, 1973; Brown, Hawksworth and Bailey, 1975), and the excellent review of Farrar (1973a) on aspects of thallus physiology. Instead, the main emphasis will be on selected areas where current or future experimental research should be particularly useful in helping to understand problems of the biology of lichenised fungi.

The term 'lichenised fungi' is used synonymously with 'lichens', not only because this is in accord with increasing taxonomic practice but also to reflect the fact that by far the greater part of the mass of most lichen thalli is composed of fungal hyphae, with the algae restricted to a thin layer near the surface. Precise estimates of the proportion of alga to fungus have been made only for three species, all from the genus *Peltigera*, (Drew & Smith, 1966; Kershaw & Millbank, 1970; Millbank, 1972); the algae were reported to occupy 3–10 %, by weight, of the thallus.

It must also be stressed that the lichenised fungi are a large group: the total number of species in the various fungal groups listed in *Ainsworth & Bisby's Dictionary of Mycology* (Ainsworth, 1971) is 71 700, of which 18 000 or just over 25 % are lichenised. Lichenised fungi belong to a variety of taxonomic groups, and since the algae belong to 27 different genera, it is assumed that the symbiosis has evolved on a number of separate occasions.

[373]

THE LICHEN SYMBIOSIS:
INTERRELATIONSHIPS OF FUNGUS AND ALGA

Carbon transfer

It is well established that photosynthetically fixed carbon moves from alga to fungus, mainly as either glucose from blue-green symbionts, or polyols (ribitol, erythritol or sorbitol) from green symbionts (Smith, Muscatine & Lewis, 1969). Movement is substantial, and since methods of measuring transfer are likely to underestimate, Farrar (1975) considers that in lichens containing the two commonest symbionts, *Trebouxia* and *Nostoc*, at least 70–80 % of the carbon fixed by the alga probably passes to the fungus. When the algae are removed from the lichen, the massive release of carbon rapidly declines to a low value, and the carbohydrate released in symbiosis becomes much less prominent, being either absent or present in only small proportion amongst both the extracellular and intracellular carbohydrates (Green & Smith, 1974).

A central question is why existence in the lichen symbiosis causes a massive efflux of a single carbohydrate from the alga. Smith (1974) in a general review of transport in symbiotic systems concluded that there must be some effect of symbiosis upon certain membrane transport systems such that efflux was greatly stimulated relative to influx. In many alga/invertebrate associations it is known that the host animal tissue contains thermolabile 'factors' which cause this (Smith, 1974). Green (1970) made an exhaustive search for such factors in the lichen *Xanthoria aureola* but failed, and there is no other evidence to suggest that a continuous secretion of factors from the fungus stimulates efflux from the alga. Smith therefore proposed that the stimulus might be physical rather than chemical, and pointed to the conclusion of several authors that *physical* contact between the symbionts was of great importance. Since the direction of membrane transport of some molecules may be related to the transmembrane potential, he speculated that the fungus might be able to reduce the algal membrane potential – or even reverse it – so as to cause a net efflux of certain molecules. This might be achieved if the fungal membrane had a higher potential than the alga, and if the charge could be conducted through the matrix separating the symbiont membranes.

Because of the difficulty of making satisfactory measurements of membrane potentials within the thallus, he suggested indirect experimental approaches: partly studying the recommencement of the transport process when water is added to air-dry lichens and partly using various inhibitors. Such investigations might show, for example, whether the energy for the transport process was derived from the alga or the fungus. Investigations along these lines have now begun using the lichen *Peltigera polydactyla*, and results so far are discussed in the next two sections.

Carbon transfer in P. polydactyla: *recommencement of transport during re-wetting of dry thalli*

No metabolic activity is detectable in an air-dry lichen. It will be shown below that when water is added, various activities reappear, but at different rates. Fungal respiration and active uptake recover rapidly, but photosynthesis much more slowly. Outred (unpublished) and Smith (unpublished) have shown that fixation and transport of ^{14}C recover at closely similar rates in *Peltigera polydactyla* – but more slowly than the recovery of fungal glucose uptake. A distinctive feature in all lichens is that respiration rapidly rises to values well above undried controls for a period after addition of liquid water. This excess 'resaturation respiration' is distinct from basal respiration in being cyanide and azide sensitive (Smith & Molesworth, 1973; Farrar, 1973*b*). Smith & Molesworth suggested that it might provide additional energy for the restoration of cell functions such as membrane transport. However, treatment of *P. polydactyla* with azide to eliminate resaturation respiration has no effect on transport. In all experiments so far carried out on re-wetting, photosynthesis and transport out of the alga seem closely linked.

Carbon transfer in P. polydactyla: *effects of inhibitors*

In experiments with *Peltigera polydactyla*, a range of inhibitors has been tried to see if transport from alga to fungus could be prevented without affecting photosynthesis. In this lichen, the *Nostoc* symbiont releases glucose to the fungus which then converts it mainly to mannitol which accumulates. Since *Nostoc* does not contain mannitol, transport between the symbionts in intact thalli can be studied simply by measuring the degree of accumulation of photosynthetically fixed ^{14}C in mannitol (normally about 70 % of photosynthetically fixed ^{14}C accumulates in mannitol over an hour). The separate effects of an inhibitor on photosynthesis and transport can thus be rapidly investigated.

The main results of the investigations are summarised in Table 1. Inhibitors used included those affecting protein synthesis in prokaryotic and eukaryotic cells (chloramphenicol and cycloheximide), fungistatic agents (griseofulvin and nystatin), metabolic inhibitors (sodium azide and dinitrophenol), inhibitors of glucose uptake (phloridzin, sorbose and 2-deoxyglucose), Na/K ATPase (ouabain), hexokinase (*N*-acetylglucosamine), reagents affecting thiol groups [paramercuribenzoate (PCMB) and paramercuribenzosulphonate (PCMS)], and a detergent (digitonin).

Table 1. *Effects of inhibitors on transport between the symbionts of* Peltigera polydactyla

Compound	Treatment	Comments
Compounds with no detectable effect on transport		
[1]Cycloheximide	25 μg cm^{-3} 27 h pretreatment	Some stimulation of net ^{14}C fixation
[1]Chloramphenicol	25 μg cm^{-3} 1 h pretreatment	Some stimulation of net ^{14}C fixation; no effect on freshly isolated *Nostoc*
[1]Nystatin	Saturated solution 30 min pretreatment	
[1]Griseofulvin	Saturated solution 66 h pretreatment	
[2]Dinitrophenol	10^{-3} mol l^{-1} 30 min pretreatment	50 % reduction in net ^{14}C fixation
[2]Sodium Azide	10^{-3} mol l^{-1}	60 % reduction in net ^{14}C fixation
[3]Phloridzin	10 mmol l^{-1} 4 h pretreatment	No effect on [^{14}C]glucose uptake by discs in the dark
[3]Ouabain	10^{-2} mmol l^{-1} 4 h pretreatment	No effect on [^{14}C]glucose uptake by discs in the dark
[3]N-acetylglucosamine	45 mmol l^{-1}	No effect on [^{14}C]glucose uptake by discs in the dark
[3]Mannitol	18 h pretreatment in 110 mmol l^{-1}, followed by incubation with or without 110 mmol l^{-1} mannitol	Mannitol content of discs would normally increase by 200–300 % as a result of the pretreatment
Compounds with some detectable effect		
[1]p-Chloromercuri-benzsulphonate	10^{-3} mol l^{-1} 6 h pretreatment	Slight stimulation of net ^{14}C fixation accompanied by some reduction of ^{14}C incorporation into mannitol. No effect on ^{14}C incorporation into ethanol-insoluble material
[3]p-Chloromercuri-benzoate	10^{-1} mmol l^{-1} 14 h pretreatment	Net ^{14}C fixation reduced by 60 %, proportion incorporated into mannitol halved, but that into insoluble doubled. Some reduction of [^{14}C]glucose uptake in the dark
Compounds with distinct effects		
[3,4]Digitonin	0.01 % (mass/vol) 17 h pretreatment	See Table 2
[4]Sorbose	1 % (mass/vol) 4 h pretreatment	See Table 2

[1] Unpublished results of Sussman and Smith
[2] Unpublished results of Smith [3] Unpublished results of Morris
[4] Unpublished results of Chambers and Smith

[4]2-Deoxyglucose   2 % (mass/vol)   See Table 2
18 h pretreatment
then incubation
without compound

Samples of 7 *Peltigera* discs incubated on 2 ml media in 1″ diameter specimen tubes containing 10 μCi NaH^{14}CO$_3$ for 1 h at 19 °C with 5000 lx illumination. After killing and extraction in 80% ethanol, the proportion of fixed ^{14}C incorporated into mannitol estimated from scans of one-dimensional chromatograms developed in ethyl methyl ketone–acetic acid–borated water (9 : 1 : 1).

[4] Unpublished results of Chambers and Smith

The absence of any effect of cycloheximide and chloramphenicol implies that transport does not depend upon the continuous synthesis of enzymes [as might occur, for example, if the alga released a polysaccharide which was then broken down to glucose by extracellular fungal enzymes, a possible mechanism of transport once suggested by Smith *et al.* (1969) and Hill (1972)]. The stimulation of net ^{14}C fixation by these inhibitors is similar to the effects of antimycin-A on fixation by isolated spinach chloroplasts (Schacter & Bassham, 1972). No effect of chloramphenicol on photosynthesis by freshly isolated *Nostoc* cells was observed (Smith, unpublished).

The absence of any effect of the fungistatic agents griseofulvin and nystatin might be because these inhibitors act only on actively growing hyphae. Since phloridzin does not affect transport and fixation of ^{14}C, it might be concluded that algal glucose transport does not require energy, but the absence of an effect of this inhibitor on glucose uptake in the dark (shown to be an active process by Harley & Smith, 1956) suggests that lichen glucose transport systems are phloridzin insensitive. Likewise, it is surprising that N-acetylglucosamine, known as an in vitro competitive inhibitor of hexokinase, has no effect either on ^{14}C transport or glucose uptake. Pretreatment with mannitol over a period which would result in the internal mannitol pool being doubled or trebled in size also had no effect on transfer, showing there is no 'feed-back' effect or 'end-product inhibition' of transfer.

PCMB and PCMS caused some detectable reduction in the proportion of fixed ^{14}C transferred, presumably due to their supposed action on membranes, but the only compounds which have distinct effects on transport with only small effects on photosynthesis are 2-deoxyglucose, digitonin and sorbose. Each of these operates in a different way, so they will be considered separately. The unpublished work on these three compounds to be described below was carried out by Susan Chambers, M. Morris and D. C. Smith, and some of their results are summarised in Table 2 and Fig. 1.

Table 2. *Effects of 2-deoxyglucose pretreatment, digitonin and sorbose upon transport between the symbionts of* Peltigera polydactyla

	% Total ^{14}C fixed			^{14}C incorporation into mannitol	Effect on net ^{14}C fixation
	Medium	Soluble	Insoluble		
Water control	0.4	86.2	13.4	c. 70 % of soluble extract	—
18 h pretreatment on 2 % 2-deoxyglucose followed by incubation on water + NaH^{14}CO$_3$	51.8 (as glucose)	32.4	15.8	Trace	33 % reduction
17 h pretreatment then incubation in digitonin + NaH^{14}CO$_3$	53.8 (as glucose)	29.6	16.6	Trace	15–30 % reduction
Digitonin (0.01 % mass/vol) Discs incubated 1 h on water + NaH^{14}CO$_3$, chased 2 h on water, then 4 h on digitonin	42.3 (mainly mannitol)	45.0	12.7	—	—
1 % (mass/vol) sorbose (after 4 h pretreatment in sorbose)	3.2	68.7	28.1	Reduced by 33 %	10 % reduction

Consolidated results of experiments of Morris, Chambers and Smith (unpublished).

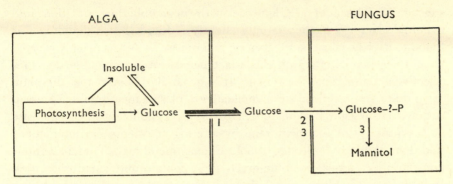

Fig. 1. Sites of action of inhibitors of transport between the symbionts of *Peltigera polydactyla*. 1 = Sorbose; 2 = digitonin; 3 = 2-deoxyglucose pretreatment.

2-Deoxyglucose

It was originally observed by Drew & Smith (1967b) that if *P. polydactyla* was incubated in the light in a solution of $NaH^{14}CO_3$ containing 1 % (mass/vol) 2-deoxyglucose, incorporation of fixed ^{14}C into mannitol was completely abolished, and there was a substantial release of [^{14}C]glucose to the medium in amounts almost equivalent to the ^{14}C accumulating in mannitol in control discs. The effect was originally explained on the basis that the 2-deoxyglucose competed at fungal uptake sites with the [^{14}C]-glucose released from the alga. However, it has been found more recently that if the lichen is pretreated for 17 h in 1 % (mass/vol) 2-deoxyglucose, thoroughly washed, and then incubated in $NaH^{14}CO_3$ alone, there was still a substantial release of [^{14}C]glucose to the medium with little incorporation into mannitol. Evidently, the effect of 2-deoxyglucose is not simply due to competition for uptake sites. Presumably it is blocking either the glucose transport system, or the metabolism of glucose immediately after entry into the cell. For example it might inhibit metabolism of glucose to mannitol, either by competitive inhibition of the glucose phosphorylation enzyme, or through competitive inhibition of phosphoglucoisomerase by 2-deoxyglucose-6-phosphate. Whatever the mechanism by which it operates, its effect shows clearly that glucose release from the alga occurs independently of glucose uptake by the fungus.

Digitonin

If the lichen is pretreated in a 0.01 % (mass/vol)* aqueous solution of digitonin for 4–17 h and then $NaH^{14}CO_3$ added to the medium, there is a

---

* Some digitonin preparations have subsequently been found to be less active, so that concentrations up to 0.05 % have had to be used.

substantial release of [^{14}C]glucose with negligible incorporation into mannitol. Net ^{14}C fixation is only slightly reduced. Since digitonin was found to have no effect on ^{14}C fixation and release by *Nostoc* freshly isolated from *Peltigera*, its effect on the intact lichen was presumably due to a selective action against the fungus. It is generally believed that digitonin can bind to non-esterified sterol molecules in plasma membranes (Lampen, Arnow, Borowska & Laskin, 1962; Kinsky, 1964) inducing a large increase in permeability of the membrane by altering sterol–protein interactions (Gordon, Wallach & Straus, 1969). Fungal membranes contain various non-esterified sterols, but blue-green algae contain only small quantities. Indeed Levin & Bloch (1964) showed that sterols are absent in *Nostoc muscorum* and that only a small proportion of the saponifiable lipid is precipitable by digitonin. Proof that fungal membrane permeability is affected in *Peltigera* was shown by experiments in which the lichen was first labelled with ^{14}C, chased for two hours while transport of ^{14}C to the fungus was completed, and then incubated in digitonin solutions. A massive loss of [^{14}C]mannitol to the medium was then observed. The effect of digitonin confirms that glucose release is independent from uptake by the fungus. Since nearly all the fixed ^{14}C released during fixation in the presence of digitonin is incorporated into glucose, this is additional evidence that fixed carbon is transferred almost entirely as glucose, and that no significant movement of carbon could be detected in other compounds such as amino acids. Finally Chambers has shown that [^{14}C]-glucose release in the presence of digitonin is almost unaffected by pH over the range 4.0–10.0.

Sorbose

Contrary to the preliminary report of Drew & Smith (1967*a*), independent experiments by Morris and by Chambers have shown clearly that *Nostoc* can accumulate glucose when inside the thallus or when freshly isolated from it. Morris suggested that the inward transport of glucose was mediated by the same carrier as outward transport. Since outward transport was insensitive to inhibitors such as phloridzin and ouabain, it was possible that the carrier system operated independently of an energy source – i.e. transport was by a carrier-mediated, facilitated-diffusion system. In yeast such carrier systems show an affinity for sorbose (Cirillo, 1961). Were this the case in *Nostoc*, then if the lichen was pretreated with a high concentration of sorbose and then NaH^{14}CO$_3$ added to the medium, no ^{14}C release should occur because the algal membrane carriers would be at least partly combined with sorbose. Table 2 shows this may be the case, and that in the presence of sorbose, scarcely any ^{14}C is released to the medium, but incorporation into mannitol is reduced by 33 %, ^{14}C fixation remains almost unaffected. The effect of sorbose still occurred if digitonin was included in the medium. If glucose efflux from *Nostoc* is by a carrier-mediated, facilitated-diffusion system, then efflux rates should be greatly affected by the internal glucose concentration. Experiments were therefore carried out in which the glucose concentration in the alga was artificially raised by incubating lichens in 112 mmol l^{-1} [^{14}C]glucose in the dark for 24 h, and then transferring to non-radioactive media in the light in the presence and absence of ^{12}[C]glucose. In both cases, efflux of [^{14}C]glucose to the medium was observed for a time, and it could be calculated that the initial rate of efflux was greater than the maximum velocity of the fungal uptake system.

Carbon transfer: conclusions

There is now substantial evidence that glucose transport out of the alga is by means of a carrier-mediated, facilitated-diffusion system in which efflux occurs at a much faster rate than influx. The rate of efflux seems little affected by fungal glucose uptake, which is an active process. Efflux is also little affected when the permeability of the fungal membrane is greatly increased by digitonin treatment. If this increased permeability was associated with a marked decrease in the fungal transmembrane potential, then the theory of Smith (1974) that a high transmembrane potential of the fungus is in some manner able to affect that of the alga becomes less easy to support. This is particularly so in view of the observation that glucose efflux from the alga in the presence of digitonin is insensitive to medium pH.

Nevertheless, the problem of explaining why efflux from the alga ceases as soon as physical contact with the fungus is broken still remains. Since the rate of efflux depends upon the concentration of glucose at the carrier site within the cell, and since incorporation of fixed ^{14}C into free glucose drops sharply when algae are isolated from the lichen, then perhaps one should investigate why existence in symbiosis induces free glucose to accumulate at the carrier sites. Indeed, it is observed for virtually all symbiotic algae that release of substantial fixed ^{14}C is associated with marked reduction of incorporation into insoluble material, and that the cessation of release after isolation is associated with marked increase of incorporation into insoluble material (Smith, 1974). This leads to the question of why existence in symbiosis represses incorporation of fixed carbon into insoluble material.

For lichens, physical attachment of the fungus to the alga ought to remain part of any theory advanced. One possibility might be that the fungus restricts the ability of the alga to grow, so that 'space' for storing

carbon in the cell is limited and a surplus available for transport is generated. However, in *Peltigera* this does not account for the apparent *absence*
of storage polyglucoside granules from *Nostoc* in symbiosis (Peat, 1968),
and as Table 2 shows, if release of fixed ^{14}C is reduced by adding $[^{12}C]$-
sorbose to the medium, there is a marked increase of incorporation into
insoluble material. It is as if there is some factor which either actively
represses incorporation into insoluble material or strongly promotes
formation of free glucose at the expense of reserve polysaccharide. Future
experimental work might therefore be directed to studying the enzymes
involved in the synthesis of glucose and the synthesis and breakdown of
polysaccharide. Studies of the polyol transport systems in the green
symbionts of lichens should also be carried out.

Nitrogen transfer

Lichens with blue-green symbionts show active nitrogen fixation, and,
where investigated, there is substantial transfer of the fixed nitrogen to the
fungus (Millbank & Kershaw, 1973). It was originally believed that existence in symbiosis enabled ordinary vegetative cells of *Nostoc* as well as
heterocysts to fix nitrogen. This was because early values for the rates of
nitrogen fixation in *Peltigera canina* were unusually high, but the heterocyst frequency was unusually low (Griffiths, Greenwood & Millbank,
1972). Further study of the problem using algae isolated from lichen homogenates was hampered by the fact that nitrogen fixation ceases immediately
the thallus is broken up.

More recent studies by Millbank (1975) using improved techniques for
measuring the amount of algal protein in the thallus show that some of the
earlier conclusions require revision. In particular, the fixation rates for
Peltigera polydactyla and *P. canina* (measured by acetylene reduction) were
of the order of 2–4 nmol C_2H_4 min^{-1} mg^{-1} protein not 10–12, though the
heterocyst frequency was confirmed by an additional method to be still of
the order of 3 %. Thus, there was no need to invoke vegetative cell fixation to explain high rates, provided that heterocysts function with reasonable efficiency. The loss of fixation on thallus disruption might simply
reflect disruption of the *Nostoc* filaments, rather than oxygen effects on
vegetative cells as was earlier thought. For cephalodia of *P. aphthosa* and
Lobaria pulmonaria, new techniques showed that the heterocyst frequency
was in fact higher than in cultures of free-living *Nostoc* so that again the
high rates observed in these structures do not require special explanations.

Thus, there is no longer convincing evidence that existence in symbiosis
increased the rate or efficiency of fixation by blue-green symbionts.

Although Millbank & Kershaw (1973) conclude that transfer of nitrogen from alga to fungus is substantial, the form in which fixed nitrogen is transferred is not yet clear. Millbank (1974) finds that excised cephalodia of *P. aphthosa* and dissected 'algal zones' of *P. polydactyla* leak a mixture of peptides to the medium over a 15 day incubation period which is similar in amino acid composition to that released by free-living *Nostoc* cultures. However, the amount of leakage in these experiments represented only about 15 % of the total nitrogen transferred as measured by other techniques.

On the basis of the general features of transport out of algae and chloroplasts in symbiosis described by Smith (1974), it might be expected that nitrogen would move in the form of neutral amino acids such as alanine and glutamine. But whether nitrogen moves mainly as peptides or as amino acids, movement of such molecules ought to be detectable in experiments using ^{14}C. Various 'inhibition' experiments carried out by Smith and his colleagues (unpublished) failed to demonstrate any movement of ^{14}C-labelled amino acids in *P. polydactyla*. When fungal uptake in this lichen was inhibited by digitonin treatment, virtually all of the photosynthetically fixed ^{14}C released to the medium by the algae was as glucose, and there was none in nitrogen-containing compounds. These experiments do not prove that nitrogen is not transferred in organic compounds, but they provide substantial circumstantial evidence against it.

The possibility must therefore be considered that the bulk of the nitrogen is transferred in inorganic form. Millbank (1974) reports the unpublished observation of Stewart and Rowlands that a substantial quantity of nitrogen moves in the form of ammonia in *P. aphthosa*. Ammonia uptake by *Peltigera* sp. has scarcely been studied since the investigations of Smith (1960). He showed that uptake was not particularly fast, except in the presence of glucose (Fig. 2). Since nitrogen fixation occurs mostly during photosynthesis (indeed, Millbank shows that fixation increases with light intensity), one might visualise that glucose released from the alga stimulates absorption of the ammonia that is simultaneously released.

Paradoxically, although about 10 % of lichenised fungi are associated with blue-green algae, they are not obviously more common than other types of lichen in nitrogen-poor habitats. Indeed some, such as *P. canina* and *P. polydactyla*, often occur in habitats where decaying vegetation should make it relatively nitrogen-rich.

Does the fungus 'benefit' the alga by supplying minerals and protection?

Because lichens have always been regarded as an outstanding example of mutualism, it has been automatically assumed by almost all biologists that the fungus must in some way 'benefit' the alga. Hence, there are two

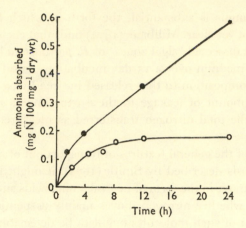

Fig. 2. Ammonia uptake by *Peltigera polydactyla* from 5 mmol l⁻¹ NH₄Cl with (●) and without (○) 14 mmol l⁻¹ glucose (Smith, 1960).

common assumptions that 'the fungus supplies the alga with minerals' and that 'the fungus protects the alga'.

With regard to mineral supply, there is as yet no published experimental evidence that the fungus actively supplies the alga with minerals. The only experiment which might have provided clear-cut evidence was that of Kershaw & Millbank (1970), who investigated how much ^{15}N fixed in the external cephalodia of *Peltigera aphthosa* and passed to the fungus subsequently appeared in the green *Coccomyxa* symbiont of the algal layer. After 25 days, only 3 % of the expected amount was found, suggesting that passage from fungus to green alga was negligible. In any case, Farrar (1973b) showed that phosphate absorbed from dilute solution by *Hypogymnia physodes* was virtually non-exchangeable, suggesting little tendency by fungi in moist thalli to release absorbed minerals.

Indeed, there is no necessity for substances entering the alga to have first passed into the fungal cytoplasm rather than diffusing from the exterior of the thallus along inert cell walls. In photosynthesis experiments with *P. polydactyla*, it has been shown that $H^{14}CO^-_3$ pass from the external solution to the photosynthetic centres of the alga within 15 seconds (Smith, 1973). In studies of the effects of pollution on lichens, it is widely assumed that lichen algae are particularly sensitive to sulphur dioxide – but nowhere is it assumed that the sulphur dioxide has first to traverse living fungal cells before being released to the alga (cf. articles in Ferry *et al.*, 1973).

Part of the reason for assuming that the fungus supplies minerals to the alga may be that in another mutualistic symbiosis, ectotrophic mycorrhizas, there is good experimental evidence for movement of phosphate and possibly other minerals from fungus to host. However, in this case the host autotroph is a large, actively growing tree with a high demand for nutrients. In the lichen, the autotrophic symbiont is small in volume compared to the fungus, grows very slowly and presumably has a low demand for nutrients.

On the other hand, it would be foolish to ignore the fact that lichen fungi have very efficient mechanisms of absorption, and that absorbed minerals do not in some way pass to the alga. A possible mechanism for this is suggested by the observations discussed below that when air-dry lichens are plunged into water, there is an initial brief period when a substantial loss of both inorganic and organic substances occurs from the thallus presumably because membrane integrity takes a short but significant time to become re-established after wetting. In nature, such large losses are unlikely to occur from intact thalli since they are usually not wetted by a sudden torrent of water, but more slowly by rain drops. However, if a rain drop hits a dry algal layer, one can visualise that for a short time in the localised area of contact between the symbionts, substances may be passively released from one symbiont to enter the other.

The concept that the fungus in some way 'protects' the alga is again not founded upon particularly solid experimental data, and is unconvincing since related free-living algae almost always occur in lichen habitats, frequently as epiphytes on thalli. Potentially misleading observations suggesting a 'protective' role for the fungus are that certain lichen algae such as *Trebouxia* and *Coccomyxa* grown in isolated culture will bleach irreversibly at light intensities above about 2000 lx. It is argued from this that the upper cortex of the lichen thallus 'protects' the alga from insolation. However, *Coccomyxa* only bleaches in culture if grown in an organic medium (Hill & Ahmadjian, 1972), showing yet again the danger of translating behaviour in culture to that in the thallus. *Trebouxia* is the commonest lichen alga, occurring in perhaps 70 % of all species; many are in exposed habitats where, even on rainy days in the summer, light intensities far in excess of 2000 lx occur. Even allowing for light interception by the upper cortex, it seems probable that they can tolerate, in nature, rather greater light intensities than in culture.

Since *Trebouxia* is reported not to be free-living, there is obviously some aspect of the lichen habitat for which it may be obligately dependent. As yet we have no concept of what this could be. When it is discovered, care will be required to assess whether the aspect is one which is 'beneficial' or 'protective' to *Trebouxia*, or whether it simply represents evolution from an independent to a dependent relationship.

Morphogenetic interactions

When grown in isolated culture, lichen fungi show none of the elaborate morphology and tissue differentiation of the intact thallus; they are little different from cultures of related free-living fungi except in having exceptionally slow growth and frequently forming rather compact, somewhat elevated, colonies on agar. The question therefore arises of the stimulus that evokes the remarkable morphogenetic response of thallus formation. Presumably it is partly the natural environmental conditions of the lichen, and partly association with the alga. There is no experimental evidence to enable a proper evaluation of the relative importance of these two factors and the degree of synergism between them. Although the algae occupy only a very small proportion of the thallus and are usually restricted to a limited region, there is little doubt that they can exert some profound influence. This is illustrated by certain lichen fungi which are known to be able to associate separately with different lichen algae and adopt strikingly different morphology with the different symbionts. An illustration of this is *Sticta filix*, containing the alga *Coccomyxa*, and *Dendriscocaulon*, the same fungus with *Nostoc* (Plate 1) as described by James & Henssen (1975). Less striking but equally important is the phenomenon whereby the same alga occurs in two different forms in different regions of the same thallus, as in *Stichococcus diplosphaera* (cf. Ahmadjian, 1973).

If the presence of the alga induces morphological responses, how is this achieved? As with the problem of the induction of carbon transfer described above, it is not known whether there is any secretion of morphogens or the involvement of phenomena such as catabolite repression, or whether it represents some aspect of physical contact between the symbionts.

Similar questions are again raised in considering the mechanism by which alga and fungus grow at compatible rates in the thallus, although they may show wide differences in culture. Presumably, the fungus exerts some influence on the rate of algal growth, but it is not known whether this is by physical restriction or secretion of substances.

Various studies of the area of contact between the symbionts have been reviewed by Peveling (1973). In all cases, there is close contact between the symbionts, but with some kind of matrix separating the membranes composed at least in part of cell wall material. There are various reports of limited penetration by projections of fungal hyphae into the algal cell to form haustoria. While such haustoria may often penetrate the algal cell wall, the membrane is rarely ruptured. However, despite the variety of reports of haustoria, it is still difficult in the case of the great majority of lichens to gain any clear picture of the frequency of haustoria in healthy thalli. There are some species, such as *Peltigera* spp. in which haustoria are definitely absent.

Synthesis of lichens and specificity of symbionts

Much experimental effort was devoted in the first part of this century to attempts to synthesise lichens from their isolated components. Until recently, these efforts met with little success. Indeed, there is only one case in which a taxonomically recognisable thallus has been artificially re-synthesised, *Endocarpon pusillum* (Ahmadjian & Heikkilä, 1970), and only very few in which even rudimentary thalli have been formed (Ahmadjian, 1973). In the great majority of cases, experimental synthesis has not proceeded much beyond initial stages of contact and occasional formation of pseudoparenchyma by the fungus. One reason for the absence of success in earlier experiments was lack of appreciation of the conditions favouring synthesis: nutrient-poor conditions, appropriate substrates and slow alternate wetting and drying.

However, now that these basic requirements are known, it should be possible to develop a laboratory system for studying at least the early stages of synthesis. This would also enable the study of the slow development of mechanisms of nutrient transfer, of which Hill & Ahmadjian (1972) made an interesting preliminary report. It should also help in determining the nature and extent of the need for physical contact between the symbionts to initiate morphogenetic responses such as pseudoparenchyma formation. Conceivably, it might be of value in studying the mechanisms of specificity of symbionts and how they recognise each other, although such studies might be severely hampered at present by inadequate understanding of the taxonomy of lichen algae.

PHYSIOLOGICAL ASPECTS OF THE ADAPTATION OF LICHENISED FUNGI TO THEIR HABITATS

This section considers those aspects of the physiology of lichens that seem of particular importance in their adaptation to habitat, and the principal aim will be to consider the intact lichen thallus as a functional unit. A central question to many of these aspects is the extent to which they explain why lichens grow so slowly.

Assimilation rates

Although, as shown above, lichen algae supply a high proportion of their fixed carbon to the fungus, none of the experiments on transfer gave measurements for the absolute rate of photosynthetic carbon fixation. It is important to know if the rate of algal photosynthesis is a major factor limiting the growth of lichens, especially since in isolated culture, lichen algae grow extremely slowly unless cultivated on an organic medium (but in which case they grow in the dark almost as efficiently as in the light) (Ahmadjian, 1967).

There have been a large number of measurements of the net photosynthetic rate of lichens made by a variety of conventional and standard techniques (cf. Richardson, 1973; Farrar, 1973a). Unfortunately, most results are expressed on a thallus dry weight basis. This is unsatisfactory, as pointed out by Farrar (1973a) and Richardson (1973) because over 90 % of the weight of the thallus is fungal, much of it presumably consisting of inert material in the thick cell walls of the medulla and cortices.

One of the most satisfactory methods of determining photosynthetic efficiency would be to determine the rate of carbon fixation per mg chlorophyll in the lichen, but only three such measurements have been published. Farrar (1973b) found rates of 160 μmol carbon fixed mg^{-1} chlorophyll h^{-1} in *Hypogymnia physodes* and calculated from the data of Schulze & Lange (1968) a value of 132 μmol for the same species, and Rundel (1972) obtained values of 41.8 in *Cladonia subtenuis*. For the experimental conditions used, these rates are within the range that might be expected from free-living algae such as *Chlorella*. If these two lichens are typical, then it could be assumed that algae in general function reasonably efficiently within the thallus (and it would be particularly interesting to see if the efficiency is markedly greater than that observed in culture). Indirect evidence for the efficiency of lichen algae is that rates of photosynthesis per unit area of thallus are only slightly below the range recorded for higher plant leaves (Richardson, 1973), but the chlorophyll content per unit area is also lower (Wilhelmsen, 1959).

The slow growth rates of lichens cannot therefore be explained simply in terms of the low photosynthetic efficiency of the algae or the low proportion of algae in the thallus.

Products of assimilation and their translocation in the thallus

Unlike many other photosynthetic tissues, much of the carbon fixed by the lichen initially accumulates in soluble compounds, especially fungal polyols such as mannitol and arabitol. The concentration of these polyols may be high. *Peltigera polydactyla* contains up to 10 % dry weight mannitol – this would be equivalent to a concentration of 0.13 mol l^{-1} in a saturated thallus assuming that all the water was held in the cytoplasm, which is unlikely; this is a remarkably high concentration bearing in mind that the maximum solubility of mannitol in distilled water at room temperature is 0.86 mol l^{-1}.

Since mannitol is known to be the principal respiratory substrate (Smith, 1961), the thallus is clearly rich in utilisable soluble carbohydrates, and not apparently starved of carbon. Farrar (1975) has made similar calculations for *Hypogymnia physodes*; the concentration of arabitol in the fugus was 0.014 mol l^{-1}, and of ribitol in the algal symbiont, 0.25 mol l^{-1}. Again, slow growth does not appear to be due to carbon shortage.

Although products of photosynthesis move rapidly from the algal layer to the underlying medulla, there is no evidence of any appreciable movement longitudinally through the thallus. Indeed, Armstrong (1975) suggests that since the radial growth rate of a lichen is far from being linearly dependent upon the surface area, there is little translocation along the thallus. Direct experimental investigation of this problem is almost completely lacking, although it should be technically simple. Even for ionic substances, studies of translocation seem limited almost entirely to vertical movements, especially in fruticose lichens such as *Cladina* spp. (cf. Tuominen & Jaakkola, 1973).

Accumulation of solutes: nutrition and protection from toxic cations

Lichens can accumulate remarkable amounts of certain solutes from the liquids passing over them in their habitat. There are two quite distinct mechanisms of solute accumulation: active uptake, and passive binding on to cell walls. Each of these mechanisms is very efficient, and each is of distinct ecological significance.

Active uptake

Active uptake of sugars, amino acids and phosphates has been studied in a range of lichens (e.g. Smith, 1962; Farrar, 1973b). Such uptake was described by Smith as 'very efficient' because very high rates of absorption from solutions of relatively high concentration (10^{-2} mol l^{-1} and above) were found. For example, the amount of glucose and of asparagine absorbed by *Peltigera polydactyla* in 24 hours from 1 % (mass/vol) solutions was higher than values obtained from yeast under comparable experimental conditions. It was argued that efficient uptake was advantageous to organisms which live in environments where nutrients are in short supply, but this argument had the drawback that it was based upon studies of uptake from high concentrations, while in the habitat nutrients would presumably be in very dilute solution. Recently, Farrar (1973b, 1975) showed that *Hypogymnia physodes* would indeed absorb phosphate efficiently from dilute solutions: when incubated in 10^{-6} mol l^{-1} KH$_2$PO$_4$, 4.37×10^{-9} mg phosphate was absorbed per mg dry wt per hour – a value higher than that recorded for excised barley roots under comparable conditions (Loughman, personal communication). Farrar also showed that at higher concentrations, phosphate uptake continued at a steady rate over 4 days from 10^{-4} mol 1^{-1} $KH_2{}^{32}PO_4$ – a total of 0.5 mg phosphate was absorbed per 100 mg dry weight, none of which was exchangeable when the lichen was then immersed in non-radioactive phosphate solutions.

While it may seem justifiable to describe active uptake by lichens as 'very efficient' and consequently of considerable ecological value, many questions remain unanswered. Do fungal hyphae have this high efficiency because they have a high membrane surface area for uptake (e.g. due to convolutions of the plasma membrane of the 'transfer cell' type), or because they have more transport sites per unit area of membrane, or because the transport systems work more rapidly than in other plants? While it is logical to assume that plants living in nutrient-poor habitats would evolve better uptake systems, it is far less clear how such evolution might occur.

Smith (1962) also suggested that while active uptake would be rapid, subsequent utilisation of absorbed substances was very slow. However, it is becoming clear that this conclusion applies only to situations where there is a massive uptake from solutions of high concentration. During uptake from more dilute solutions, metabolism of absorbed substances may be rapid. Farrar has shown that within five seconds of placing *H. physodes* in dilute solutions of $KH_2{}^{32}PO_4$, radioactivity could be detected in no fewer than 10 organic compounds. There was also rapid incorporation into polyphosphate in the acid-insoluble fraction. Rapid metabolism of both sugars (Drew, 1966) and amino acids (Farrar, 1973b) after uptake from dilute solutions has also been shown.

Passive binding

It is now clear that lichens can also accumulate substantial amounts of cations which can be potentially toxic by passive binding on to cell walls. Thus, *Acarospora sinopica* growing on iron slag has been reported to contain up to 55 000 ppm (by mass) iron and 1100 ppm (by mass) copper (Lange & Zeigler, 1963). Lack of appreciation that cations could be bound to cell walls led to some confusion in earlier literature and generated speculation that there may be 'plasmatic resistance' in lichens to such toxic elements, or that after entry to the cytoplasm they are then released into the extracellular matrix. Evidence for the mechanism of binding has been reviewed by Tuominen & Jaakkola (1973) who conclude that the ion exchange model of cation uptake explains most results of uptake studies and competition patterns for cation selectivity. Puckett, Neiboer, Gorzynski & Richardson (1973) conclude that the binding sites seem to be oxygen and oxygen–nitrogen donors, and ion exchange is of the strong field type.

Some cations are accumulated by both active and passive processes. Thus zinc is undoubtedly bound to cell walls, but in some experiments it seems to behave like potassium (Brown, 1975). In studies of uptake of radionuclides, differences in the uptake of strontium and caesium are also explained in this way (Tuominen & Jaakkola, 1973).

The ecological significance of strong cation binding onto lichen cell walls should be seen in the context of plants which are slow growing, long lived, and subject to frequent wetting and drying. As will be shown below, during the first stage of re-wetting a dry thallus, there is a short interval before membrane semi-permeability is re-established, so it is essential that there should be some 'purification' mechanism to remove toxic ions from the incoming liquid in this brief period. This is particularly important since the first raindrops coming through the atmosphere after a dry spell may contain over fifty times the concentration of dissolved substances of later rainfall (Allen, personal communication). Unfortunately, there has been little or no appreciation of this 'scouring' phenomenon in studies of the effects of atmospheric pollution on lichen physiology. The fact that the first raindrops, having an abnormally high solute content, arrive before membrane semi-permeability is re-established is certainly worth much more investigation than it has received.

Presumably, most binding sites are located on fungal hyphae since they have much thicker cell walls than the algae and since, in any case, the algae form only a small proportion of the thallus. Indeed, one aspect of the interaction between the symbionts which has received little attention from those who advocate that the fungus must 'benefit' the alga is the question of whether the abundance of cation binding sites in fungal hyphae 'protects' the alga from an influx of toxic substances during re-wetting.

Habitat sources of nutrients for lichens

Although the typical habitat of a lichen is traditionally described as being 'nutrient-poor', this does not necessarily mean either that nutrients are absent or that nutrient shortage limits growth. In so far as corticolous lichens are concerned, there are a number of studies showing that liquids flowing down tree trunks during rain have much higher concentrations of minerals such as potassium, calcium and magnesium, and of dissolved organic matter (including carbohydrates) than throughfall water (Brodo, 1973; Carlisle, Brown & White, 1967; Tamm, 1953). However, there is no evidence that epiphytic lichens derive significant quantities of nutrients from living tissues of their hosts. Trotet (1969) found that lichens on

Quercus suber did not accumulate any of the radioactive phosphate which had been introduced into its branches. With regard to lichens on exposed rocks, the 'scouring' effect of the early part of a rain shower may also concentrate nutrients to a high degree.

Farrar (1973*a*, 1975) concludes that in general, the level of nutrients in lichen thalli are not noticeably different from those in free-living algae and fungi. He believes that because lichens are slow growing they need only very small amounts of nutrients, rather than their growth being limited by low nutrient supply. It must be concluded that very little indeed is known about environmental sources of nutrients for lichens, but it is nevertheless of great importance for the understanding of their physiological ecology. Even simple questions, such as the extent to which lichen distribution is limited or affected by nutrient supply, cannot yet be properly answered.

Growth rates and longevity: is protein turnover slow?

It is axiomatic that lichens grow slowly, and this is borne out by numerous measurements (cf. Hale, 1973). Large thalli must therefore be very old, and it is obviously difficult to measure longevity precisely. In temperate climates, Linkola (1918) estimated that 'larger individuals' of several *Parmelia* spp. would be 30–80 years old; crustaceous lichens grow more slowly and so may reach greater ages – Hausman (1948) estimated a colony of *Rhizocarpon geographicum*, 8.6 cm in diameter, to be 120 years old. Much greater ages have been estimated for arctic lichens – some being well over 1000 years old (e.g. cf. Beschel, 1955).

Although a thallus may occupy a particular site for very long periods of time, it is not known how much turnover of living material occurs within that thallus. Some long-lived thalli – such as some types of *Rh. geographicum*, do not erode at the centre: to what extent has there been algal cell division in the centre region? There are some reports of variation in algal cell number per unit area of a lichen thallus at different times of the year (e.g. Harris, 1971) but information is extremely sketchy.

Smith (1962) originally suggested that rates of protein turnover in lichens were very slow, but the evidence upon which that statement was based is no longer acceptable by modern criteria. There have been no direct measurements of protein turnover by modern techniques, and there is but the solitary observation of Farrar (1973*b*) that only 2 % of the photosynthetically fixed carbon moved into protein in *Hypogymnia physodes* in 24 hours. In view of the recent suggestion that, in animals, there is a negative correlation between longevity and plasma protein turnover rate (Spector, 1974), corresponding comparative measurements between lichens and fast-growing free-living fungi might be of interest.

PLATE I

For explanation see p. 405

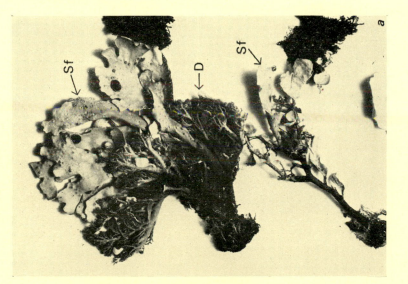

For explanation see p. 405

Why should lichens grow so slowly? A number of possibilities apart from low protein synthesis rates have been suggested. Smith (1962) speculated the low growth rates may be an adaptation to poor nutrient supply – but as seen earlier, there is no evidence that lichen thalli are any lower in nutrient content than many faster growing plants, or that their growth is severely limited by nutrient shortage. Further, field experiments on supplying lichens with additional nutrients do not cause particularly marked stimulations of growth. Hakulinen (1966) poured a 2 % (w/v) suspension of bird excrement over lichens at weekly intervals, but observed only a 10–30 % increase in growth over a two year period. Jones & Platt (1969) poured either distilled water or a mineral solution over lichens twice monthly and obtained a 40 % increase in rate – but this is still extremely slow compared to other types of plant. Also, as Tobler (1925) observed, the lichen symbiosis tends to break up if given excess nutrients.

Slow growth certainly permits the generation of a substantial excess fixed carbon in the alga for release to the fungus, but if this was the main factor limiting growth then it presumably could be easily remedied if the very small proportion of alga to fungus in a thallus increased.

Several (e.g. Ahmadjian, 1967) have commented that slow growth may be due to environmental conditions which only allow lichens brief periods for metabolic activities such as assimilation (i.e. when they are moist during the daytime). However, lichens which spend much or all of their time in a moist condition do not seem to have growth rates very much higher than those which frequently dry out. Anyway, as Farrar (1975) has observed, such measurements of assimilation rates as are available indicate that lichens have reasonable gross primary productivity – and that it is the net productivity which is low. Nearly all lichens are rich in polyols which form their principal respiratory substrate and presumably the main source of carbon for growth.

Farrar (1973b, 1975) suggested that the wide discrepancy between gross and net primary productivity in lichens was because a substantial part of the gross productivity was diverted to withstanding environmental stress – he termed this the concept of 'physiological buffering'. This concept will be discussed further, but first it is necessary to consider the effects of environmental stress on thallus physiology.

Resistance to drying and wetting, and other environmental factors

Almost all lichens in nature lose water and dry out, often to within 10 % or less of the dry weight of the thallus. Although drying may be infrequent in very moist habitats it is a frequent occurrence for the majority of species. If a crustaceous lichen on a boulder persists for fifty years it will probably undergo of the order of at least 10 000 cycles of drying and wetting during its life in a temperate climate.

This is significant because the re-wetting phase is a period of metabolic stress for a lichen. This is suggested by laboratory experiments in which distilled water is added to air-dry lichens. Two outstanding phenomena always observed are: (1) respiration rates rise rapidly to levels well above controls (Fig. 3), while photosynthesis starts slowly at levels below controls

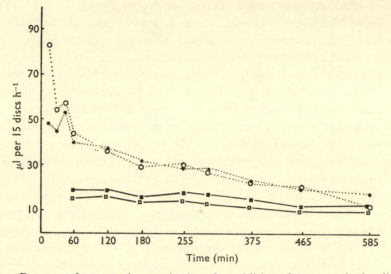

Fig. 3. Progress of resaturation respiration after addition of water to air-dry dics of *Peltigera polydactyla* (Smith & Molesworth, 1973). $\bullet$ = O_2 uptake; $\circ$ = CO_2 output. Control discs on water: $\blacksquare$ = O_2 uptake; $\square$ = CO_2 output.

(Fig. 4), so that there is a net carbon loss for a time; and (2) there is a brief period before membrane semi-permeability is re-established when both organic metabolites and mineral ions leak from the thallus. These are not phenomena caused by extreme desiccation, because in the lichen *Peltigera polydactyla* both are shown if the lichen dries out to only 40% of its saturated water content before re-wetting (Smith & Molesworth, 1973; Outred, unpublished) (Figs. 4 and 5).

The effects of re-wetting on gas exchange have been studied extensively (cf. review by Farrar, 1973a). The increased respiration is, unlike basal respiration, cyanide sensitive and has been termed 'resaturation respiration'. However, the carbon lost in resaturation respiration comes from the same polyol pool as basal respiration (Farrar, 1973b; Smith & Molesworth, 1973). Both the intensity and duration of resaturation respiration varies inversely with the moistness of the habitat (Table 3), but even in lichens from drier habitats, the loss of carbon dioxide is, during the period of resaturation respiration, appreciable and equivalent to at least one or two hours of basal respiration. During the very much slower process of absorption of water vapour, resaturation respiration does not seem to occur (Butin, 1954). Resaturation respiration is not a phenomenon confined to

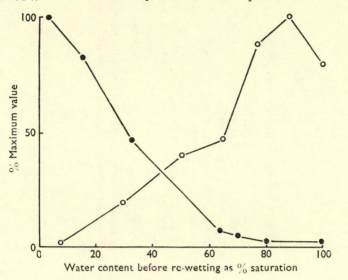

Fig. 4. Relationship between water content before re-wetting and effects of re-wetting *Peltigera polydactyla* discs. ● = leakage of ^{14}C fixed before drying out, measured during the first 5 min of re-wetting; ○ = ^{14}C fixation measured during 30 min after re-wetting (unpublished data of Heather Outred).

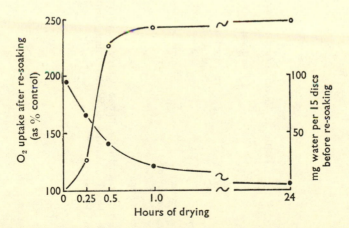

Fig. 5. Effect of water content of *Peltigera polydactyla* discs on their subsequent respiration rates after re-wetting (Smith & Molesworth, 1973). ● = water content before re-wetting, ○ = O_2 uptake during first hour after re-wetting.

resaturation respiration appreciable and equivalent to at least one or two hours of basal respiration. During the very rapid, slower phases of absorption of water vapour, net respiration/photosynthesis does not seem to occur during... Measurements are not so simple for a photosynthesis confined to

Table 3. *Resaturation respiration in lichens from different habitats (after Farrar, 1975)*

Species	Duration of resaturation respiration (h)	Total loss of CO_2 (cm^3 g^{-1})	Basal respiration: CO_2 loss per h (cm^3 g^{-1})	Assumed loss due to resaturation respiration (cm^3 g^{-1})	Habitat
Peltigera polydactyla	9·75	7.05	0.32	3.93	Moist woodland floor
Hypogymnia physodes	1	0.77	0.23	0.54	Branches of trees in open woodland
Xanthoria aureola	2	0.96	0.36	0.24	Tiled roof

All experiments carried out at 20 °C; respiration rates measured by Warburg manometry in which, for resaturation rates, air-dry lichen material placed in annulus of Warburg flask, distilled water tipped on from side-arm; further details given by Smith & Molesworth, 1973, and Farrar, 1975.

lichens, and is shown by a range of plant tissues. However, it is not known for any tissue whether resaturation respiration is an unavoidable consequence of metabolic disruption caused by desiccation, or whether it has a positive role in re-establishing cell organisation upon re-wetting.

Release of metabolites from dry tissues when immersed in water is again a phenomenon not restricted to lichens. Simon (1974) in a general review ascribed this leakage to changes in the nature of the cell membrane which renders it much more permeable until its normal configuration returns on hydration. The extent of leakage in lichens has been studied by first labelling moist thalli with isotopes (^{32}P or ^{14}C), drying them, and then measuring the rate and degree of loss of isotope upon subsequent immersion in distilled water. In *Hypogymnia physodes*, Farrar (1973b) observed losses of 11–23 % ^{32}P within the first five minutes of re-wetting. Although the lichen had been previously labelled by incubation in $KH_2{}^{32}PO_4$, the ^{32}P released to the medium was in hexose phosphates, phospholipids and other compounds besides inorganic phosphate. He also found losses of up to 5 % of the ^{14}C previously fixed by photosynthesis during the first hour of re-wetting, again distributed in a wide range of metabolites but especially polyols. Using *Peltigera polydactyla*, Outred (unpublished) found losses of up to 14 % of the ^{14}C previously fixed by photosynthesis within five minutes of immersion in water, again most of the ^{14}C released had been incorporated into mannitol. Although selective permeability may take several minutes to become fully re-established (Fig. 6) other processes may recover more

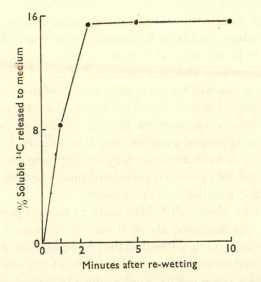

Fig. 6. Leakage of previously fixed ^{14}C from air-dry *Peltigera polydactyla* discs after re-wetting (unpublished data of Heather Outred).

quickly. Farrar found that when dry *H. physodes* was immersed in solutions of $KH_2{}^{32}PO_4$, organic phosphates were detectable after 20 seconds, and a measurable reduction in uptake in the presence of DNP after only 10 seconds.

Laboratory experiments in which lichens have been exposed to cycles of drying and wetting further illustrate the consequence to lichens of carbon loss both from resaturation respiration and leakage. Smith & Molesworth (1973) found that in four cycles of a regime in which *P. polydactyla* was soaked for two hours and dried for 22 in the dark, the mannitol content dropped sharply by 30 % to 40 % (Table 4). Farrar found that after only three cycles of this severe regime, over 50 % of the pre-applied ^{14}C had been lost from *H. physodes*. Seven cycles caused a substantial impairment of phosphate uptake and photosynthesis.

It is important to stress that the observations of Farrar, Smith & Molesworth, and Outred on the effects of re-wetting and of cycles of wetting and drying were carried out under laboratory conditions which may not be directly applicable to the field. For example, the substantial leakage of metabolites from thalli was obtained under conditions where dry thalli were rapidly plunged into water. Although such rapid re-wetting is unlikely to occur in nature, it is difficult to believe that there is not at least some loss during rainshowers after dry periods, and that this is of great importance to the lichen over long periods of time. Resaturation respiration presumably occurs each time even moderately dry thalli are exposed to rain.

A paradox of these observations is the apparent need of lichens to have fluctuating moisture conditions for healthy growth and development. As noted above, it is an essential requirement for successful synthesis in culture. Several authors who have attempted to grow lichens in the laboratory find that it is essential for the healthy growth of most species (Pearson, 1970; Dibben, 1971; Harris & Kershaw, 1971). Fletcher (1975) in studying the effects of salinity on maritime lichens, concluded that wetting and drying was more important a requirement than salinity. Finally, relatively few lichens grow in habitats where they are permanently submerged – or at least inundated for prolonged periods of time – perhaps not more than 2 % of the British flora fall into this category.

The reason why almost all lichens seem to require periodic drying remains obscure. As suggested above, it may be necessary for the flow of nutrients from fungus to alga. Possibly, the physical shrinkage and expansion may be an essential feature of substrate attachment: certainly, Mellor (1923) showed that this was the mechanism by which lichens etched stained glass windows of churches, and Fry (1926, 1927) came to similar conclusions

Table 4. *Effect of cycles of wetting and drying on mannitol content and resaturation respiration of Peltigera discs with and without prior incubation in glucose to increase initial mannitol content (after Smith & Molesworth, 1973)*

| | Incubation of discs before start of experiment | | | |
| | Water | | 1 % glucose | |
No. of previous cycles of wetting and drying	Mannitol content of dry discs before adding water (mg per 15 discs)	O_2 absorbed in first hour after re-wetting (cm^{-6} per 15 discs)	Mannitol content of dry discs before adding water (mg per 15 discs)	O_2 absorbed in first hour after re-wetting (cm^{-6} per 15 discs)
0	3.75	53	7.75	87
1	3.04	35	5.67	68
2	2.89	29	5.37	57
3	2.71	26	4.67	46
Basal respiration	—	20		46

Discs incubated on distilled water or 1 % (w/v) glucose for 17 h at room temperature before start of experiment. Each cycle consisted of 2 h soaking on distilled water followed by 22 h drying in a desiccator over $CaCl_2$, all in the dark at room temperature. Respiration measured at 25 °C by Warburg manometry.

about the effects of corticolous and saxicolous lichens on their substrates.

Of all ecological factors so far studied, habitat moisture and the frequency of wetting and drying may be one of the most important in determining lichen distribution. This should not, however, obscure the fact that lichens are extremely tolerant of low temperatures, and indeed there are several reports of lichens showing net carbon fixation at temperatures as low as -24 °C (Lange, 1962, 1965). As Farrar (1975) points out, it would be interesting to study assimilation at these low temperatures with ^{14}C. Most lichens are also resistant to high temperatures when dry; Kappen (1973) reports the observation of von Kerner (unpublished) that temperatures above 58 °C occurred for several hours in crustose thalli in the Karst of Istria; in South West Germany *Cladonia pyxidata* in its natural habitat withstood several hours at 66 °C.

The concept of 'physiological buffering'

Farrar (1975) has recently introduced the concept of 'physiological buffering' to explain the relationship between resistance to environmental stress and the problem of why lichens have such a low net primary productivity. This concept arose out of experiments where *Hypogymnia physodes* was first labelled with ^{14}C during photosynthesis and then exposed to a variety of harsh regimes including starvation in the dark at room temperature, cycles of wetting and drying, and prolonged soaking. The polyol levels showed marked and rapid changes during or in response to those regimes, but the level of insoluble material remained virtually constant. Similar phenomena were observed during starvation in *P. polydactyla* (Drew, 1966) when starvation in the dark produced rapid drops in polyols, but no detectable change in insoluble carbohydrate or protein even during seven days.

A central feature of Farrar's concept is that the inevitable disturbances due to environmental stress and fluctuation are mainly confined to the large pool of polyols, which appears to 'buffer' other metabolic processes, especially growth processes. The polyol pool fluctuates greatly in response to such disturbances, but other aspects of metabolism are much less disturbed. For example, when the protein fraction of *H. physodes* was labelled by externally supplied [^{14}C]leucine, the amount of ^{14}C in protein after 24 hours remained constant for the next seven days, although drastic changes occurred in ^{14}C in the ethanol soluble fraction (largely polyols) during this period.

The large pool of polyols in the thallus is envisaged as having a variety of roles. Firstly, polyols act as the substrate for both basal and resaturation respiration and they bear most of the carbon loss on re-wetting. Secondly, their high concentration in the cytoplasm presumably generates a high osmotic pressure – essential for plants normally living under conditions of water stress [hexoses never occur at similar high concentrations in plant cells, possibly because they might cause damage to cell contents (Scott, 1960)]. Thirdly, it is possible that polyols may protect macromolecules during drying out by direct substitution for water molecules in their hydration shells (Webb, 1963; Farrar, 1973b). Fourthly, as pointed out by Lewis & Smith (1967), they serve as convenient stores of reducing power.

A high internal concentration of polyols may therefore be essential for lichen existence – indeed, Smith & Molesworth (unpublished) observed that when the mannitol content of *Peltigera polydactyla* is depleted by cycles of wetting and drying, the lichen does not recover if the pool size drops below a certain level. The reason for the large discrepancy between gross and net primary productivity in lichens is that much of the fixed carbon is accumulated in the polyol pool to preserve its 'buffering' capacity, and there is a continual loss of carbon from this pool as it carries out this function.

CONCLUSIONS

At first sight, the role of the algal component of lichenised fungi might be seen simply as supplying the organic carbon necessary for growth and metabolism. However, since lichens grow so slowly and may spend appreciable periods in a dry and metabolically inactive condition, their demand for carbon for growth is extremely low, and indeed might be met in most habitats by the organic content of the rain and surface liquids. If Farrar's concept of 'physiological buffering' is valid, then the role of the alga may be seen as much in providing carbon for resistance to environmental stress and fluctuation as for growth.

Although the alga is very much the smaller component – and in some situations perhaps also the weaker component (e.g. it usually shows signs of sensitivity to pollution before the fungus) – there are some aspects of lichenisation where its influence may be profound. Particularly striking is the way in which association with the alga evokes profound morphogenetic responses from the fungus. It remains completely obscure why a specific alga is required to evoke the response, or how two different algae evoke different responses from the same fungus (e.g. cf. Plate 1).

Finally, there remains the question of the extent to which the lichenised fungi may be regarded as a typical example of mutualistic symbiosis, with the consequent implication that the algae in some way derive benefit from the association. It is still not possible to discern any clear and measurable benefit which the alga derives. If mutualism was defined in terms of mutual *exploitation* rather than *benefit*, then this small semantic change might make it easier to accept lichens as an example of mutualistic symbiosis. Certainly, if two-way flow of nutrients is to be regarded as an essential part of the definition of mutualism, lichens will have to be excluded until there is appropriate experimental evidence for a nutrient flow from fungus to alga.

But whether or not it is proper to term lichens 'mutualistic', they nevertheless represent a very good system for the experimental study of one of the central problems of the physiology of symbiosis, namely, how one symbiont is induced to release large quantities of nutrients to the other. So far, there is no reason to suppose that the mechanisms for transfer of fixed carbon in lichens will prove very different from those in most other autotroph/heterotroph associations.

REFERENCES

AHMADJIAN, V. (1967). *The Lichen Symbiosis*. Massachusetts: Blaisdell.
  (1973). Resynthesis of lichens. In *Lichens*, 565–580 (eds. V. Ahmadjian and M. E. Hale). London: Academic Press.
AHMADJIAN, V. & HALE, M. E. (1973). *Lichens*. London: Academic Press.
AHMADJIAN, V. & HEIKKILÄ, H. (1970). The culture and synthesis of *Endocarpon pusillum* and *Staurothele clopima*. *Lichenologist*, **4**, 259–267.
AINSWORTH, G. C. (1971). *Ainsworth & Bisby's Dictionary of Mycology*. London: Commonwealth Mycological Institute.
ARMSTRONG, R. A. (1975). Studies on the growth rate of lichens. In *Progress and Problems in Lichenology* (eds. D. H. Brown, D. Hawksworth and R. H. Bailey). London: Academic Press (in press).
BESCHEL, R. (1955). Individuum und Alter bei Flechten. *Phyton (Horn, N. O.)*, **6**, 60–68.
BRODO, I. M. (1973). Substrate ecology. In *Lichens*, 401–442 (eds. V. Ahmadjian and M. E. Hale). London: Academic Press.
BROWN, D. H. (1975). Mineral uptake by lichens. In *Progress and Problems in Lichenology*, (ed. D. H. Brown, D. Hawksworth and R. H. Bailey). London: Academic Press (in press).
BROWN, D. H., HAWKSWORTH, D. & BAILEY, R. H. (eds.) (1975). *Progress and Problems in Lichenology*. London: Academic Press (in press).
BUTIN, H. (1954). Physiologisch-ökologische Untersuchungen über den Wasserhaushalt und die Photosynthese bei Flechten. *Biol. Zbl.*, **73**, 459–502.
CARLISLE, A., BROWN, A. H. F. & WHITE, E. J. (1967). The nutrient content of tree stem flow and ground flora litter and leachates in a sessile oak (*Quercus petraea*) woodland. *J. Ecol.*, **55**, 615–627.
CIRILLO, V. P. (1961). Sugar transport in microorganisms. *A. Rev. Microbiol.*, **15**, 197–218.
DIBBEN, M. J. (1971). Whole-lichen culture in a phytotron. *Lichenologist*, **5**, 1–10.
DREW, E. A. (1966). Some aspects of the carbohydrate metabolism of lichens. D.Phil. thesis, University of Oxford.
DREW, E. A. & SMITH, D. C. (1966). The physiology of the symbiosis in *Peltigera polydactyla* (Neck.) Hoffm. *Lichenologist*, **3**, 197–201.
  (1967*a*). Studies in the physiology of lichens. VII. The physiology of the *Nostoc*

symbiont of *Peltigera polydactyla* compared with cultured and free-living forms. *New Phytol.*, **66**, 379–388.

(1967*b*). Studies in the physiology of lichens. VIII. Movement of glucose from alga to fungus during photosynthesis in the thallus of *Peltigera polydactyla*. *New Phytol.*, **66**, 389–400.

FARRAR, J. F. (1973*a*). Lichen physiology: progress and pitfalls. In *Air Pollution and Lichens*, 238–282 (ed. B. W. Ferry, M. S. Baddeley and D. L. Hawksworth). London: Athlone Press.

(1973*b*). Physiological lichen ecology. D.Phil. thesis, University of Oxford.

(1975). The lichen as an ecosystem: observation and experiment. In *Progress and Problems in Lichenology* (eds. D. H. Brown, D. Hawksworth and R. H. Bailey). London: Academic Press (in press).

FERRY, B. W., BADDELEY, M. S. & HAWKSWORTH, D. L. (1973). *Air Pollntion and Lichens*. London: Athlone Press.

FLETCHER, A. (1975). Nutritional aspects of marine and maritime lichen ecology. In *Progress and problems in Lichenology* (eds. D. H. Brown, D. Hawksworth and R. H. Bailey). London: Academic Press (in press).

FRY, E. J. (1926). The mechanical action of corticolous lichens. *Ann. Bot.*, **40**, 397–417.

(1927). The mechanical action of crustaceous lichens on substrate of shale, schist, gneiss, limestone and obsidian. *Ann. Bot.*, **41**, 437–460.

GORDON, A. S., WALLACH, D. F. H. & STRAUS, J. H. (1969). *Biochim. biophys. Acta*, **183**, 405.

GREEN, T. G. A. (1970). The biology of lichen symbionts. D.Phil. thesis, University of Oxford.

GREEN, T. G. A. & SMITH, D. C. (1974). Lichen physiology. XIV. Differences between lichen algae in symbiosis and in isolation. *New Phytol.*, **73**, 753–766.

GRIFFITHS, H. B., GREENWOOD, A. D. & MILLBANK, J. W. (1972). The frequency of heterocysts in the *Nostoc* phycobiont of the lichen *Peltigera canina* Willd. *New Phytol.*, **71**, 11–13.

HAKULINEN, R. (1966). Über die Wachstumsgeschwindigkeit einiger Laubflechten. *Ann. Bot. Fenn.*, **3**, 167–179.

HALE, M. E. (1973). Growth. In *Lichens*, 473–492 (eds. V. Ahmadjian and M. E. Hale). London: Academic Press.

HARLEY, J. L. & SMITH, D. C. (1956). Sugar absorption and surface carbohydrase activity of *Peltigera polydactyla* (Neck). Hoffm. *Ann. Bot.*, **20**, 513–543.

HARRIS, G. P. (1971). The ecology of corticolous lichens. II. The relation between physiology and the environment. *J. Ecol.*, **59**, 441–452.

HARRIS, G. P. & KERSHAW, K. A. (1971). Thallus growth and the distribution of stored metabolites in the phycobionts of the lichens *Parmelia sulcata* and *Parmelia physodes*. *Can. J. Bot.*, **49**, 1367–1372.

HAUSMAN, E. H. (1948). Measurements of the annual growth-rate of two species of rock lichens. *Bull. Torrey bot. Club*, **75**, 116–117.

HILL, D. J. (1972). The movement of carbohydrate from the alga to the fungus in the lichen *Peltigera polydactyla*. *New Phytol.*, **71**, 31–39.

HILL, D. J. & AHMADJIAN, V. (1972). Relationship between carbohydrate movement and the symbiosis in lichens with green algae. *Planta*, **103**, 267–277.

JAMES, P. W. & HENSSEN, A. (1975). Aspects of lichenisation. In *Progress and Problems in Lichenology* (eds. D. H. Brown, D. Hawksworth and R. H. Bailey). London: Academic Press (in press).

JONES, J. M. & PLATT, R. B. (1969). Effects of ionizing radiation, climate, and nutrition on growth and structure of a lichen *Parmelia conspersa* (Ach.) Ach. *Radioecol. Concent. Processes, Proc. 2nd Int. Symp.*, 111–119.

KAPPEN, L. (1973). Response to extreme environments. In *Lichens*, 310–380 (eds. V. Ahmadjian and M. E. Hale). London: Academic Press.

KERSHAW, K. A. & MILLBANK, J. W. (1970). Nitrogen metabolism in lichens. II. The partition of cephalodial-fixed nitrogen between the mycobionts and phycobionts of *Peltigera aphthosa*. *New Phytol.*, **69**, 75–79.

KINSKY, S. C. (1964). In *Antimicrobial Agents and Chemotherapy*, 387–394 (ed. J. C. Silvester). American Society for Microbiology.

LAMPEN, J. O., ARNOW, P. M., BOROWSKA, Z. & LASKIN, A. I. (1962). Location and role of sterol at nystatin-binding sites. *J. Bact.*, **84**, 1152–1160.

LANGE, O. L. (1962). Die Photosynthese der Flechten bei tiefen Temperaturen und nach Frostperioden. *Ber. Dt. Bot. Ges.*, **75**, 351–352.

(1965). Der CO_2-Gaswechsel von Flechten bei tiefen Temperaturen. *Planta*, **64**, 1–19.

LANGE, O. L. & ZEIGLER, H. (1963). Der Schwermetallgehalt von Flechten aus dem *Acarosporetum sinopicae* auf Erzschlackenhalden des Harzes. I. Eisen und Kupfer. *Mitt. Flor.-Soziol. Arbeitsgemeinsch.* (N.F.), **10**, 156–183.

LEVIN, E. Y. & BLOCH, K. (1964). Absence of sterols in blue-green algae. *Nature, Lond.*, **202**, 90–91.

LEWIS, D. H. & SMITH, D. C. (1967). Sugar alcohols (polyols) in fungi and green plants. I. Distribution, physiology and metabolism. *New Phytol.*, **66**, 143–184.

LINKOLA, K. (1918). Messungen über den jährlichen Langenzuwachs einiger *Parmelia*- und *Parmeliopsis*-Arten. *Medd. Soc. Fauna Flora Fenn.*, **44**, 153–158.

MELLOR, E. (1923). Lichens and their actions on the glass and leadings of church windows. *Nature, Lond.*, **112**, 299–301.

MILLBANK, J. W. (1972). Nitrogen metabolism in lichens. IV. The nitrogenase activity of the *Nostoc* phycobiont in *Peltigera canina*. *New Phytol.*, **71**, 1–10.

(1974). Nitrogen metabolism in lichens. V. The forms of nitrogen released by the blue-green phycobiont in *Peltigera* spp. *New Phytol.*, **73**, 1171–1181.

(1975). Aspects of nitrogen metabolism. In *Progress and Problems in Lichenology*, (eds. D. H. Brown, D. Hawksworth and R. H. Bailey). London: Academic Press.

MILLBANK, J. W. & KERSHAW, K. A. (1973). Nitrogen Metabolism. In *Lichens*, 289–307 (eds. V. Ahmadjian and M. E. Hale). London: Academic Press.

PEARSON, L. C. (1970). Varying environmental factors in order to grow lichens intact under laboratory conditions. *Am. J. Bot.*, **57**, 659–664.

PEAT, A. (1968). Fine structure of the vegetative thallus of the lichen *Peltigera polydactyla*. *Arch. Mikrobiol.*, **61**, 212–222.

PEVELING, E. (1973). Fine structure. In *Lichens*, 147–184 (eds. V. Ahmadjian and M. E. Hale). London: Academic Press.

PUCKETT, K. J., NEIBOER, E., GORZYNSKI, M. J. & RICHARDSON, D. H. S. (1973). The uptake of metal ions by lichens: a modified ion-exchange process. *New Phytol.*, **72**, 329–342.

RICHARDSON, D. H. S. (1973). Photosynthesis and carbohydrate movement. In *Lichens*, 249–288 (eds. V. Ahmadjian and M. E. Hale). London: Academic Press.

RUNDEL, P. W. (1972). CO_2 exchange in ecological races of *Cladonia subtenuis*. *Photosynthetica*, **6**, 13–17.

SCHACTER, B. & BASSHAM, J. A. (1972). Antimycin A stimulation of rate-limiting steps of photosynthesis in isolated spinach chloroplasts. *Pl. Physiol. Lancaster*, **49**, 411–416.

SCHULZE, E. D. & LANGE, O. L. (1968). CO_2-Gaswechsel der Flechten *Hypogymnia physodes* bei tiefen Temperaturen im Freiland. *Flora*, **158 B**, 180–184.

SCOTT, W. J. (1960). A mechanism causing death during storage of dried micro-organisms. In *Recent Researches in Freezing and Drying*, 188–202 (eds. A. S. Parkes and A. V. Smith). Oxford: Blackwell.

SIMON, E. W. (1974). Phospholipids and plant membrane permeability. *New Phytol.*, **73**, 377–420.

SMITH, D. C. (1960). Studies in the physiology of lichens. I. The effects of starvation and ammonia feeding upon the nitrogen content of *Peltigera polydactyla*. *Ann. Bot.*, **24**, 52–62.

(1961). The physiology of *Peltigera polydactyla* (Neck.) Hoffm. *Lichenologist*, **1**, 209–226.

(1962). The biology of lichen thalli. *Biol. Rev.*, **37**, 537–570.

(1973). *The Lichen Symbiosis*. Oxford Biology Readers No. 42. London: Oxford University Press.

(1974). Transport from symbiotic algae and symbiotic chloroplasts to host cells. *Symp. Soc. exp. Biol.*, **28**, 437–508.

SMITH, D. C. & MOLESWORTH, S. (1973). Lichen physiology. XIII. Effects of rewetting dry lichens. *New Phytol.*, **72**, 525–534.

SMITH, D. C., MUSCATINE, L. & LEWIS, D. H. (1969). Carbohydrate movement from autotrophs to heterotrophs in parasitic and mutualistic symbiosis. *Biol. Rev.*, **44**, 17–90.

SPECTOR, I. M. (1974). Animal longevity and protein turnover rate. *Nature, Lond.*, **249**, 66.

TAMM, C. O. (1953). Growth, yield and nutrition in carpets of a forest moss (*Hylocomium splendens*). *Medd. Statens Skogsforskn. Inst. Stockholm*, **43**, 1–140.

TOBLER, F. (1925). *Biologie der Flechten*. Berlin: Borntraeger.

TROTET, G. (1969). Recherches sur la nutrition des lichens. *Rev. bryol. lichen.*, **36**, 733–736.

TUOMINEN, Y. & JAAKKOLA, T. (1973). Absorption and accumulation of mineral elements and radioactive nuclides. In *Lichens*, 185–224 (eds. V. Ahmadjian and M. E. Hale). London: Academic Press.

WEBB, S. J. (1963). Possible role for water and inositol in the structure of nucleoproteins. *Nature, Lond.*, **198**, 785–787.

WILHELMSEN, J. B. (1959). Chlorophylls in the lichens *Peltigera*, *Parmelia* and *Xanthoria*. *Bot. Tidsskr.*, **55**, 30–39.

EXPLANATION OF PLATE

The lichen *Sticta filix* (Sf) contains the alga *Coccomyxa*, and the lichen *Dendriscocaulon* (D) contains the alga *Nostoc*, but both lichens have the same fungus [as shown by specimens in which *Dendriscocaulon* can also be found growing out of *S. filix* (*a*); *Dendriscocaulon* can also be found growing independently (*b*). This illustrates how different algae evoke different morphogenetic responses from the same fungus. (Pictures by P. W. James of samples collected from South Island, New Zealand, on rocks by a stream.)

HAEMATOPHAGOUS INSECT AND HAEMOFLAGELLATE AS HOSTS FOR PROKARYOTIC ENDOSYMBIONTS

By K.-P. CHANG

The Rockefeller University, New York, N.Y. 10021, USA

INTRODUCTION

Intracellular entities are frequently found in invertebrates including protozoa (Ball, 1969; Kirby, 1941; Sonneborn, 1959) and insects (Buchner, 1965; Steinhaus, 1967). Some of the entities in the latter have long been considered micro-organisms, though their precise status has been a matter of controversy. The work of Preer, Jurand, Preer & Rudman (1972) definitely established the bacterial nature of certain inclusions in protozoa; that of Pant & Fraenkel (1954), the occurrence of yeasts in certain Coleoptera; that of Brecher & Wigglesworth (1944), the occurrence of actinomyces in *Rhodnius*; that of Grinyer & Musgrave (1966), Musgrave & Grinyer (1968), Bush & Chapman (1961) and Brooks (1963), the occurrence of bacteroids in certain weevils and cockroaches. Some of the entities in insects are, however, still controversial (Chang & Musgrave, 1969) in that no DNA can be detected. Some investigators are prepared to accept the organismal nature of the entities in the absence of evidence of DNA (e.g. Buchner, 1965; Koch, 1960; Körner, 1969, 1972).

With few exceptions (e.g. Brecher & Wigglesworth, 1944; Pant & Fraenkel, 1954; van Wagtendonk, Clark & Godoy, 1963), these micro-organisms have been refractory to in-vitro cultures (e.g. Musgrave & McDermott, 1961; Brooks & Richards, 1966). Presumably, they have adapted to an environment, which we have not been able to simulate *in vitro* (Lanham, 1968; Trager, 1970).

The micro-organisms and hosts appear to co-operate for mutual benefit (Buchner, 1965; Brooks, 1963; Koch, 1956; Musgrave, 1964; Trager, 1970). Again, clear evidence of this is limited to a few cases, in which symbionts have been found to benefit hosts by supplying them with essential nutrients (Brecher & Wigglesworth, 1944; Pant & Fraenkel, 1954; van Wagtendonk *et al.*, 1963).

The regulatory mechanisms whereby the hosts maintain their micro-organisms are still largely speculative (Brooks, 1963; Chang & Musgrave, 1973; Trager, 1970). This paper explores the relationships between

[407]

prokaryotic micro-organisms and three species of invertebrates, *Blasto-crithidia culicis*, *Crithidia oncopelti* and *Cimex lectularius*. The prokaryotic micro-organisms will, in this paper, be referred to as symbionts for convenience.

MYCETOMAL SYMBIONTS OF HAEMATOPHAGOUS INSECTS

Prokaryotic symbionts are often sheltered by insects in specialized cells, the mycetocytes, or in the mycetome, which is an aggregation of cells or a syncytium (Buchner, 1965). Mycetomal symbionts are often inherited maternally through ovarial infection (Buchner, 1965).

Well-developed mycetomes can be found frequently in insects living on restricted diets including avian and mammalian blood, which are low in vitamins (Koch, 1956). That mycetomal symbionts may supplement this dietary deficiency has been well illustrated in the studies of blood-sucking lice (*Pediculus corporis*) (Aschner & Ries, 1933; Aschner, 1932). Removal of mycetomes from these insects by physical means results in the failure of reproduction and growth, which can be partially restored by a rectal injection of a yeast extract rich in B vitamins. The blood-sucking insect, *Rhodnius prolixus*, also derives similar growth factors from its endo-symbionts in the alimentary canal and these are essential for adequate growth and reproduction of hosts (Brecher & Wigglesworth, 1944).

These types of insects all live on a blood diet throughout their lives. Blood-sucking species without symbionts invariably feed on micro-organisms during their young stadia, thereby acquiring sufficient growth factors. Exceptional cases are the blood-sucking ticks, which harbour certain intracellular, rickettsia-like micro-organisms in ovaries or mal-pighian tubules (Roshdy, 1968) or other organs (Reinhardt, Aeschlimann & Hecker, 1972). Experimental evidence seems to suggest that these rickettsial symbionts are commensal micro-organisms (Suitor, 1964) and that they may become pathogenic when inoculated into the haemocoele of the host (Burgdorfer, Brinton & Hughes, 1973).

The bedbug, *Cimex lectularius*, which lives exclusively on blood through-out its life cycle, has paired mycetomes consisting of syncytial chambers and containing pleomorphic symbionts (see Buchner, 1965). The nature of the mycetomal symbionts has been a matter of controversy: Buchner considered them as bacteria, whereas others regarded them as a rickettsia (Hertig & Wolbach, 1924; Philip, 1957) identical to *Symbiotes lectularius* found in bedbugs by Arkwright, Atkin & Bacot (1921). Clarification of this and of symbiont–host relationships has been offered in several published observations (Chang, 1972; Chang & Musgrave, 1973; Chang, 1974a) and in some recent findings, which are mentioned briefly below.

SYMBIONTS OF *CIMEX LECTULARIUS*

Mycetomal symbionts have been studied in two different strains, here referred to as Hamilton strain, collected at Hamilton, Ontario, Canada, and Orlando strain obtained through the courtesy of Dr G. S. Burden, Orlando, Florida. Examination of the smear preparations of mycetomes by means of light microscopy reveals a variety of entities showing little consistency with regard to their form and stainability, which may explain how different interpretations of their identity could arise in earlier studies. With the aid of the electron microscope, it has become clear that three different kinds of micro-organisms may be associated with the mycetome. They differ markedly from one another in ultrastructural characteristics and no intermediate forms occur. Based on the intimacy of their associations with the mycetome, they are here categorized into primary, secondary and guest symbionts.

Primary symbionts

These are small pleomorphic entities, previously termed descriptively pleomorphic symbionts (Chang & Musgrave, 1973). They occur in the syncytial cytoplasm of the mycetomes (Plate 1a). Each symbiont has dense cytoplasm with dispersed DNA-like fibres and is located in a host vacuole (Plate 2c). The symbionts divide by binary fission different from that of conventional bacteria. Division septa are first formed by invaginations of the plasma membrane (Plate 2d), which segments the symbiont into protoplasts enclosed by a common cell wall. The latter then follows the invaginations of the division furrows, so that the symbiont becomes divided into daughter cells sharing a common host vacuole (Plate 2e). In their general ultrastructure and the unusual way of cell division, the primary symbionts show resemblance to the intracellular 'reticulate bodies' of the rickettsia *Chlamydia* (Mitsui, Fugimoto & Kajuma, 1964; Tamura, Matsumoto, Manire & Higashi, 1971). These similarities led us (Chang & Musgrave, 1973) to suggest that the primary symbionts may be identified as rickettsiae.

Ovarial ampullae also harbour primary symbionts in saccular enclaves distributed among nurse cells in Hamilton and Orlando strains of *C. lectularius*. In Europe, Louis, Laporte, Carayon & Vago (1973) found in ovaries the mycetomal 'typical symbionts', which are similar in ultrastructure to the primary symbionts described here. Possibly, they are common to the mycetomes of all strains collected in different geographic areas.

Secondary symbionts

These have been named rod-shaped symbionts based on their general appearance (Chang & Musgrave, 1973). When present in the mycetome, they are randomly mingled with the primary symbionts (Plate 1*a*). Each secondary symbiont has a dense granular cytoplasm, electron-lucid nuclear areas, and multilayered envelopes composed of an inner trilaminar plasma membrane, an intermediate zone and a tripartite outer cell wall that is coated with a thin layer of amorphous dense material (Plate 2*a*). Symbionts divide by binary fission in the usual manner and lie in the cytoplasm free from host-derived membranes (Plate 2*b*). Occasionally, they infect the nuclei (Plate 1*b*), seemingly without adverse effect. A similar phenomenon has also been reported in *Paramecium*, which often contains endosymbionts in macro- or micronuclei (Preer, 1969; Beale, Jurand & Preer, 1969). The intranuclear occurrence of the secondary symbionts indicates that they are rickettsiae and may be *Symbiotes lectularius*, described by Philip (1957) as occurring in mycetomes and other tissues.

These symbionts are strain-specific. They are absent in the Orlando strain (Plate 5*b*), but are associated with the mycetomes and ovaries of the Hamilton strain and probably of others as well, as judged by the descriptions of Buchner (1965) and Arkwright *et al.* (1921).

Guest symbionts

These are filamentous motile rodlets, infrequently encountered in the mycetome. They occur in the cytoplasm and are not enclosed by host-derived membranes. Each symbiont has a typical bacterial cytoplasm and a Gram-negative profile: an inner plasma membrane and an outer cell wall (Plate 5*a*). Motility is probably achieved by means of peritrichous flagella.

Buchner (1965) considered these micro-organisms as the motile variants of true mycetomal symbionts. In Hamilton and Orlando strains these motile rods are more numerous in ovarial tissues than in mycetomes. Hertig & Wolbach (1924) found similar organisms chiefly in the malpighian tubules. In the ovaries and haemolymph of *C. lectularius*, Louis *et al.* (1973) observed flagellated rodlets, which they considered to be 'symbiotes RC' or *Wolbachia lectularius* (see Krieg, 1963). Further study is needed to determine whether or not all these symbionts are the same species of micro-organism. However, it seems likely that only the primary symbionts are the true mycetomal symbionts and the secondary symbionts may be *S. lectularius*.

Regulation of symbionts by host lysosomal activity

To maintain a symbiotic association, host organisms must be able to curtail the excessive proliferation of symbionts by certain means. In *C. lectularius*, ultrastructural evidence indicates that the regulatory mechanism is a controlled lysosomal degradation of symbionts in the mycetomes. In the Hamilton strain, reconstructed events from different degradation stages of mycetomal symbionts suggest that the process is similar to cellular autophagy (Chang & Musgrave, 1973). Initially, groups of symbionts, along with mitochondria, become segregated in cytoplasmic portions, surrounded by double cytoplasmic membranes of the mycetome (Plate 3a). The resulting structures are analogous to the isolated bodies formed in the early stage of autophagic vacuoles in insect fat cells (Locke & Collins, 1965). The sequestered symbionts appear normal initially, suggesting that it is not a process for clearing dead symbionts. Fusion of these cytoplasmic bodies with lysosomes seems to occur, as mycetomes often contain large, conspicuous cytolysomes in which degenerate symbionts are present among membrane fragments (Plate 3b). Degradation of symbionts appears to continue until they are beyond recognition and the cytolysomes only contain swirls of membranes and dense granular material (Plate 4a). Eventually, cytolysomes become residue bodies (Plate 4b, c). Degradation of symbionts in cytolysomes is a constant feature of mycetomes, irrespective of insect stadia or metamorphosis when lysosomal activity of cells may be high. In the mycetomes of the Orlando strain, symbionts are similarly lysed in cytolysomes, in which acid phosphatase activity can be detected by electron microscopy, indicating that symbiont degradation is indeed mediated by the host lysosomal system (Plate 5b). It should be noted that host cells only suppress some of the symbionts, others survive and multiply, without causing damage to the host. Cytolysomes and residue bodies associated with lysing symbionts have been observed in the mycetomes and mycetocytes of other insects (Griffiths & Beck, 1973; Hinde, 1971; see Müller, 1972; Musgrave & Grinyer, 1968). It seems likely that the lysosomal digestion may be one of the general mechanisms employed by insects for the regulation of their symbiont population.

Aposymbiotic insects

Insects can be rendered symbiont-free, or aposymbiotic, by various treatments (see Brooks, 1963; Koch, 1956). DeMeillon & Goldberg (1947) were unsuccessful when using penicillin treatment in their efforts to obtain symbiont-free *C. lectularius*. Raising the ambient temperature of the insects leads to greater success (Chang, 1974a). When the Hamilton strain of *C. lectularius* was subjected to 36 °C for two weeks, their mycetomes became smaller and almost symbiont-free. Examination of these mycetomes by electron microscopy showed that they contain the full complement of normal cell organelles (Plate 6*a*, *b*), but very few symbionts and these invariably showed signs of degeneration (Plate 6*c*).

The fecundity of treated insects has been studied at the normal temperature of 27 °C for more than ten weeks. The results showed that treated colonies produced far fewer eggs and hatchlings than did the untreated insects. A somewhat similar finding has recently been reported for the aposymbiotic tse-tse flies, *Glossina morsitans*, obtained by antibiotic treatment (Hill, Saunders & Campbell, 1973). Perhaps the low fecundity of these insects may be related to the loss of symbionts. Such a correlation has been well demonstrated in several species of insects rendered symbiont-free by other means (Brooks, 1963; Koch, 1956). In these insects, fecundity is partially restored by supplementing their diets with B vitamins or other essential nutrients. Possibly, the low fecundity of treated *C. lectularius* is due to the lack of similar factors normally supplied by their mycetomal symbionts. The success of obtaining aposymbiotic bedbugs by elevated temperature should facilitate further investigation of this possibility.

PROKARYOTIC ENDOSYMBIONTS OF HAEMOFLAGELLATES

Intracellular endosymbiosis is not as widespread a phenomenon in these flagellates as in ciliates, such as *Paramecium*. It may prove useful to introduce briefly some general characteristics of the host organisms. Haemoflagellates, or trypanosomatid flagellates, are best known as parasitic protozoa which cause diseases in man and domestic animals. Many other species are, however, non-pathogenic and are frequently associated with insects. Different species of these flagellates are very similar in their general ultrastructure, for which Plate 8*a* gives a representative picture. They are mononucleate cells, bound by a rigid pellicle or micro-tubule-associated plasma membrane. They contain a single ramified mitochondrion, in which a mass of DNA fibrillae is confined in a special structure called the kinetoplast. The nutritional requirements of these organisms have been extensively studied in several parasitic species in insects and a cold-blooded vertebrate (Trager, 1968). A unique feature in this aspect is that most of these species require haemin for growth.

Intracellular entities resembling bacteria have been observed in several species of haemoflagellates, among which *Blastocrithidia culicis* and *Crithidia oncopelti* are originally parasites of insect guts (see Wallace, 1966).

The original authors who discovered these entities were reluctant to speculate on their nature and, on morphological grounds, termed them bipolar bodies in *C. oncopelti* (Newton & Horne, 1957) and diplosomes in *B. culicis* (Novey, McNeal & Torrey, 1907). Suggestions have been made later that these entities are symbiotic bacteria which confer nutritional benefit on their host flagellates. However, supporting evidence was based mainly on fragmentary and often controversial experiments (Gill & Vogel, 1963; Guttman & Eisenman, 1965), which aroused opposing views (Newton, 1968). Recently, we have reinvestigated these entities and confirmed their prokaryotic nature and nutritional significance to host flagellates (Chang & Trager, 1974). Supporting evidence for this view is described below.

Morphology and ultrastructure of symbionts

In the original paper of Novey *et al.* (1907), their excellent micrographs vividly showed the bacterial appearance of diplosomes in *B. culicis.*

Bipolar bodies and diplosomes are readily detectable in Giemsa-stained smears, especially after Tween-80 treatment (Plate 7*a*, *c*). Both entities often occur in pairs, like bacteria in division. Flagellates usually harbour a single pair of symbionts; these duplicate before flagellate division and are passed into each daughter cell. It seems likely that the population and multiplication of symbionts are under the strict control of host flagellates.

Ultrastructural studies of bipolar bodies and diplosomes by previous authors did not permit a definite conclusion as to their nature (Newton & Horne, 1957; Gill & Vogel, 1963; Gutteridge & Macadam, 1971; Indurkar, 1965). Recent work provides more convincing evidence that the symbionts are indeed prokaryotic organisms (Chang, 1974*b*). Each symbiont is enclosed by two unit membranes separated by an irregular, light zone and consists of dense matrix with electron-lucid areas (Plate 8*b*). The matrix contains ribosome-like particles, which are smaller in size than flagellate ribosomes; the lucid zones have a network of fibrillae, resembling bacterial DNA fibres. In general, bipolar bodies and diplosomes have the same ultrastructure as bacteria, with the exception of their peripheral membranes. While the inner envelopes are clearly the plasma membranes of the symbionts, their outer envelopes differ from bacterial cell walls. Moreover, continuation of the outer envelope with the vesicular membrane of host flagellates occurs, suggesting that it is a host-derived membrane (Chang, 1974*b*). Furthermore, the electron lucidity and irregular spacing between the two envelopes indicate that the cell wall material of bacterial type may be absent or greatly reduced. In this aspect, bipolar bodies and diplosomes differ from *Paramecium* endosymbionts which closely resemble conventional bacteria (Beale *et al.*, 1969) but are more like certain insect symbionts which have been considered as L-phase bacteria or bacteria with defective cell walls (Chang & Musgrave, 1972).

Effects of antibiotics on symbionts

The prokaryotic nature of bipolar bodies and diplosomes is further indicated by their sensitivity to antibiotic treatment (Chang, 1974*b*). They are similar to *Escherichia coli* (Morgan, Rosenkranz & Rose, 1967) in their response to chloramphenicol (CAP). The response is manifested first by an expansion of nuclear zones with a concomitant reduction of cytoplasmic areas (Plate 9*a*). With further CAP treatment, the inner membrane of the symbionts becomes deformed and draws away from the outer membrane, which remains unchanged (Plate 9*b*). The symbionts eventually degenerate to become dense bodies surrounded by membrane fragments (Plate 9*c*).

Similar effects could not be observed by using penicillin G at a concentration of 2.4 %. After repeated treatment with penicillin, bipolar bodies remain unchanged, while diplosomes become pleomorphic bodies. The latter, however, revert to normal forms when transferred back to the penicillin-free medium. The change of diplosomes after the penicillin treatment suggests that they may have remnants of cell wall material, i.e. mucopeptide.

Thus, these symbionts may be classed as bacterial sphaeroplasts or bacteria with a partial cell wall.

Isolation of symbionts

The bacterial symbionts have been isolated from host flagellates of both species and partially purified for further studies (Tuan & Chang, unpublished). Previously, disruption of protozoa for isolation of endosymbionts has relied solely on mechanical means (e.g. Newton & Horne, 1957; Stevenson, 1969), but these were found to be unsatisfactory when applied to *B. culicis* and *C. oncopelti*. Thus, a new method was developed involving a sequential treatment of flagellates by hypotonic shock, complement-dependent 'immune lysis' and needle passage. Since the isolation method produced very satisfactory results, the procedure will be described in some detail.

Flagellates (10^{10} cells) were pretreated to bring about hypotonic swelling in 1/10 strength Trager's buffer (Trager, 1959). (Further experiments proved that this procedure is not essential.) Flagellates were subsequently incubated for one hour in 10 cm³ of the buffer containing 2 % guinea pig serum as a source of complement and 25 % rabbit antiserum against flagellates. Symbiont-free flagellates (see the following section) were used to prepare the immune sera, which presumably contained antibody against only host components. Although most flagellates remained intact after the treatment, their plasma membranes were evidently weakened so that a gentle force provided by the needle passage ($27G\frac{1}{2}''$ for four times) disrupted the flagellates. Microscopic examination of these preparations revealed numerous intact symbionts together with unbroken flagellates and their cellular organelles. Very few symbionts were liberated after the flagellates underwent a similar procedure without immune sera and/or complement, indicating that such treatments are essential for the ultimate disruption of the flagellates. Subsequent treatment of the samples by DNAase digestion, followed by differential centrifugation, yielded a final symbiont fraction essentially free of nuclei, kinetoplasts and unbroken flagellates. Recovery of symbionts in the final fractions was determined to be 20 %. Structural integrity and numerical abundance of symbionts were verified by light and electron microscopy (Plate 10*a*).

By means of electron microscopy, the isolated symbionts were shown to retain their double envelopes in their entirety and to have granular cytoplasm with electron-lucid nuclear zones, essentially identical to those *in situ* (Plate 10*b*). Isolated symbionts retain their outer envelopes, which are considered to be host-derived membranes (see above). Similar findings for intracellular parasitic protozoa suggest that they may have integrated host-derived membranes into their own economy (Trager, 1974). This seems even more likely for symbionts, which are more intimately and permanently associated with host cells. Although the symbiont fractions obtained still contained minor contaminants, the absence of major cellular organelles and unbroken flagellates makes them useful for the study of symbiont DNA and soluble enzymes. Enrichment of symbionts in the final fractions is further indicated by these biochemical analyses. These will be discussed further.

Production of symbiont-free flagellates

For producing symbiont-free flagellates, Brueske (1967) has tried a variety of antibiotics, among which only CAP was found to be effective at a concentration of 800 μg cm^{-3}. The influence of CAP on the ultrastructure of bipolar bodies and diplosomes indicates that they are indeed susceptible to the treatment of this drug (see above). We have also confirmed Brueske's finding by producing symbiont-free strains of *B. culicis* and *C. oncopelti* (Chang & Trager, 1974). Further experiments have been undertaken to find a minimal effective dose of CAP by culturing flagellates in enriched blood media containing different drug concentrations. Flagellates are able to grow in the presence of CAP at concentrations ranging from 50–800 μg cm^{-3}. However, the growth rate was reduced in proportion to the drug concentration used. The reduction of growth is manifested by a lengthened lag-phase and generation time, and smaller peak populations of the flagellates. Electron microscopic observations of CAP-treated flagellates showed that they often contain dense granules and become binucleated, which suggests metabolic disturbance and impairment of cell division. However, these are temporary effects, as flagellates become normal again when transferred back to CAP-free media.

Symbionts could no longer be seen by light and electron microscopy in flagellates of *B. culicis* treated with 200–800 μg cm^{-3} of CAP for two weeks or more and in those of *C. oncopelti* with 800 μg cm^{-3} for one month, in most cases. These flagellates correspond to symbiont-free strains, which remain stable after repeated subculture in CAP-free media (Plate 7*b*, *d*). Evidently, at these concentrations, CAP shows a selective killing property by which flagellates are not significantly affected and symbionts are irreversibly destroyed. This effect agrees with the general properties of CAP (Malik, 1972).

Evidence for the identity of symbiont-containing and symbiont-free flagellates

A comparative study between symbiont-containing and symbiont-free flagellates should shed some light on the functional roles of the symbionts. This study has now been made possible by the procuration of symbiont-free flagellates by CAP treatment. However, it may be questioned whether the results are due to the treatment itself or the loss of symbionts. Answers to this question are already provided by the fact that CAP is generally considered to be non-mutagenic (Malik, 1972) and produces only temporary reversible effects on mitochondria, even at very high concentrations (Kellerman, Bigg & Linnane, 1968). Furthermore, by serological and DNA studies we have been able to show that symbiont-free strains of both species have not changed their species specificity, i.e. genetic characteristics.

By the slide agglutination method, cross-reactions were detected between the two strains of each species at 10^{-4} to 10^{-6} dilution of antisera. Specific agglutination indicates that the serological characteristics of these flagellates remain unchanged when made symbiont-free by CAP treatment.

Total DNAs of flagellates and of symbiont fractions were isolated and purified according to the procedure reported elsewhere (Tuan & Chang, 1974). Isopycnic caesium chloride gradient centrifugation of flagellate

DNAs yielded two and three bands respectively for symbiont-free and symbiont-containing strains of *C. oncopelti*, and two bands for both strains of *B. culicis*.

In *C. oncopelti*, the DNA band, present in symbiont-containing flagellates and absent in those freed of symbionts, clearly represents symbiont DNA. This is further confirmed by its buoyant density of 1.693 g cm^{-3}, which is in total agreement with the value determined from single homogeneous DNA isolated from symbiont fractions (see above). The buoyant density of symbiont DNA also agrees with that previously determined in these organisms by other workers (Marmur, Cahoon, Shimura & Vogel, 1963). The remaining two DNA bands for symbiont-containing flagellates are identical to those for symbiont-free flagellates in their buoyant densities of 1.712 and 1.702 g cm^{-3} respectively, which are very similar to the values determined for nuclear and kinetoplast DNAs of the flagellates (Newton & Burnett, 1972).

In *B. culicis*, the two DNA bands obtained from both symbiont-containing and symbiont-free flagellates are similar in pattern. Their buoyant densities were determined to be 1.696 and 1.717 g cm^{-3} respectively, which accord well with previously determined values for DNAs, presumably derived from kinetoplasts and nuclei of the symbiont-containing flagellates (Newton & Burnett, 1972). The absence of a third band in the symbiont-containing flagellates raises the question of whether their symbionts are bacteria containing DNA. However, the presence of DNA in symbionts has been suggested as they contain DNA-like fibrillae (see above) and are able to incorporate tritiated thymidine (Indurkar, 1965). This seeming controversy becomes clarified by the further study of the DNA of symbiont-fractions isolated from the flagellates (see above). Isopycnic caesium chloride gradient centrifugation of the symbiont fractions yielded a single homogeneous band of DNA, whereas an identical treatment of comparable fractions obtained from symbiont-free flagellates produced no banding at all. Thus, the DNA is indeed derived from symbionts, and not from contaminating host DNA in the symbiont fractions. The buoyant density of symbiont DNA was determined to be 1.695 – a value very close to that of kinetoplast DNA. Evidently, this similarity of symbiont and kinetoplast DNAs is enough to prevent the appearance of two distinct bands after caesium chloride density gradient centrifugation. A parallel finding has been reported in a *Paramecium*–endosymbiont association where the DNAs of macronuclei and mu particles have a similar buoyant density and are thus not separable by centrifugation (Stevenson, 1969).

The most significant aspect of this DNA study is perhaps the finding that, when flagellates have lost their symbiont DNA after CAP treatment.

the characteristics of their own DNA do not change. Thus, the symbiont-free strains produced in this way differ from the Guttman's 'cured' strain (Guttman & Eisenman, 1965), presumably derived from *C. oncopelti*, but appear to resemble *C. fasciculata* in their DNA and serological characteristics (Newton, 1968). The present finding also constitutes additional proof for the absence of symbionts in symbiont-free flagellates.

Nutritional comparisons between symbiont-containing and symbiont-free strains

During continuous cultivation in CAP-free blood media for one year, symbiont-free flagellates never attained the same level of growth as that of the normal symbiont-containing flagellates. This suggests that symbionts may have nutritional value to their hosts. Further evidence is provided by the facts that symbiont-free flagellates fail to grow in Grace's medium supplemented with 10 % foetal bovine serum (Grace, 1962) or Trager's defined medium (Trager, 1957), both of which support the growth of normal symbiont-containing strains well (Chang & Trager, 1974). To explore the implications of this finding further, the defined medium was modified by quantitative increases of its specific ingredients or by the addition of growth-promoting materials. The additives included liver fractions, yeast powder, extract of symbiont-containing flagellates, and whole blood or its fractions. The results showed that symbiont-free flagellates grew only when the defined medium was supplemented with 0.25 % liver extract concentrate (Nutritional Biochemical Co.). Although this semi-defined medium supported the growth of symbiont-free strains, they did not grow as well as the normal symbiont-containing flagellates in the defined medium. This suggests that the nutritional contribution of symbionts to their host flagellates is probably not fully compensated for by the addition of the liver fraction. Further characterization of the liver concentrate showed that the growth factor is water-soluble, heat-stable, dialysable and is probably not a lipid fraction in the liver. While the chemical nature of the growth factor in the liver concentrate awaits further investigation, its use with the defined medium made possible a comparison of haemin requirement between symbiont-containing and symbiont-free strains. The results demonstrated that the deletion of haemin from the defined medium does not impair the growth of symbiont-containing flagellates, but fails to support that of symbiont-free strains even in the presence of liver concentrate (Chang & Trager, 1974). Thus, symbiont-free strains of *B. culicis* and *C. oncopelti* are similar in their haemin requirement to other species of haemoflagellates which normally lack endosymbionts (Trager, 1968). The nutritional study shows that when the symbiont is present in the host flagellate there is no longer the need for exogenous haemin or other nutrient(s) which are present in the liver concentrate.

Haem biosynthetic capability of symbiont-containing and symbiont-free flagellates

The haemin requirement of symbiont-free strains indicates that they are probably incapable of adequate haem biosynthesis. Conversely, symbiont-containing flagellates seem to possess this capability, as they do not require haemin for growth. Here, symbionts apparently play an essential role as a determining factor mediating haem biosynthetic capability of these flagellates. To explore this further, flagellates of symbiont-containing and symbiont-free strains and isolated symbionts have been studied with respect to their haem biosynthetic activity (Chang, Chang, Tuan & Sassa, 1974).

Porphyrins and haem

By fluorophotometric method, porphyrins and haem could be detected in flagellates of both symbiont-containing and symbiont-free strains. The finding of haem compounds in symbiont-free flagellates is of little significance, as they can be cultured only in the presence of haem or proto-porphyrin IX. On the other hand, since symbiont-containing flagellates have been cultured in haemin-free defined medium, the presence of haem and porphyrins in them is an indication of haem biosynthesis by these flagellates. A substantial accumulation of porphyrins was found in symbiont-containing flagellates of *C. oncopelti*. The cause for this phenomenon is unknown. However, it is probably not a result of iron deficiency, as the addition of excessive chelated iron into the culture media did not reduce the porphyrin level.

[^{14}C]Glycine incorporation

Comparison of haem biosynthetic ability between symbiont-containing and symbiont-free flagellates has been made by isotope incorporation study using [$2-^{14}C$]glycine – an initial substrate for haem biosynthesis (Granick & Sassa, 1971). For this experiment, flagellates of both strains were cultured in defined or semi-defined media, to which [^{14}C]glycine was added. After appropriate incubation periods, the total haem of the flagellates was extracted, crystallized and assayed for radioactivity (Takaku, Wada, Sassa & Nakao, 1968). Incorporation of the isotope occurred in both strains, but differed in the rate of incorporation; it was three to 13 times higher in symbiont-containing flagellates than in those freed of symbionts. Apparently, symbiont-containing flagellates have a higher haem biosynthetic activity than do the symbiont-free strains.

Uroporphyrinogen I synthetase (URO-S)

Different haem biosynthetic capability between symbiont-containing and symbiont-free flagellates is further indicated by the study of URO-S – an enzyme in the haem biosynthetic pathway (Granick & Sassa, 1971). By using a highly sensitive micro-assay method (Sassa *et al.*, 1974), the specific activity of URO-S was found to be 115–190 nmole URO g^{-1} protein h^{-1} in symbiont-containing flagellates, while little or no activity could be detected in symbiont-free strains of both species (Chang *et al.*, 1974). Neither could URO-S activity be detected in two reference species of haemoflagellates, *Leishmania tarentolae* and *Trypanosoma conorhini* which, like symbiont-free strains of *B. culicis* and *C. oncopelti*, require haemin for growth and possess no symbionts. Evidently, URO-S activity of symbiont-free flagellates is too low to be detected, though they are capable of a limited haem biosynthesis, as indicated by the isotope incorporation study. The specific activity of URO-S in symbiont-containing flagellates is comparable to that in mammalian cells (Sassa *et al.*, 1974), suggesting that it should support adequate haem biosynthesis.

The presence of URO-S activity only in symbiont-containing flagellates suggests that it is associated with symbionts. The URO-S study of isolated symbionts tends to confirm that the enzyme is localized in symbionts but not in flagellates. Specific activity of URO-S was found to be two- to three-fold higher in isolated symbiont fractions than in homogenates of comparable symbiont-containing flagellates (Chang *et al.*, 1974). Since URO-S is a soluble enzyme present in the cytosol of eukaryotic cells (Granick & Sassa, 1971) and since the symbiont fractions are free of unbroken flagellates (see above), the URO-S activity found in the fractions is likely to be entirely of symbiont origin. Symbiont numbers and URO-S activity were found to correspond well both before and after isolation of the symbionts, which suggests that no URO-S activity is lost during the isolation. It seems likely that symbionts supply URO-S to support adequate haem biosynthesis of these flagellates.

Utilization of haem biosynthetic intermediates

The lack of URO-S activity represents one aspect of defective machinery for haem biosynthesis in symbiont-free flagellates. Other possible defects have been investigated by testing the growth of these flagellates in media containing haem biosynthesis intermediates instead of haemin. It was found that haemin could be replaced by protoporphyrin IX (Proto.), but not by porphyrin precursors like δ-amino-levulinic acid or porphobilinogen for the growth of symbiont-free flagellates. The results suggest that the symbiont-free flagellates can synthesize haem by utilizing Proto. and iron present in the culture medium. In other haemoflagellates which are naturally symbiont-free, a similar culture condition suffices for the growth of *C. fasciculata* (see Trager, 1968), but not for that of *L. tarentolae* (Gaughan & Krassner, 1971). The similarity of symbiont-free strains and *C. fasciculata* in their ability to utilize Proto. may be explained by their closer phylogenetic relationships.

Utilization of Proto. and iron by symbiont-free strains for haem formation was further studied by using ^{59}Fe. For this study, symbiont-free flagellates were cultured in the presence of either haemin or Proto. The procedure was like that for [^{14}C]glycine incorporation study. The uptake of ^{59}Fe into haem by symbiont-free flagellates occurred under both culture conditions, but was found to be higher in the presence of Proto. This confirms the finding of the culture experiment in that symbiont-free flagellates are able to assemble haem from Proto. and iron.

From the above findings, it is evident that symbionts are essential for an adequate haem biosynthesis of *B. culicis* and *C. oncopelti*. As indicated by the incorporation of [^{14}C]glycine into haem, the flagellates themselves are not completely incapable of haem biosynthesis. It is, however, limited in capacity as manifested by the haemin requirement of the symbiont-free flagellates. Symbionts greatly augment this limited haem biosynthetic capability, thereby rendering them independent of exogenous supplies of haem. Though it is unclear whether symbionts supply host flagellates with haem or just remedy defects in their haem biosynthetic pathway, there is no doubt that symbionts make a significant contribution by supplying URO-S, which is essential for haem biosynthesis and is absent or very low in host flagellates. A similar situation has been suggested to account for the increase of haem biosynthetic activity in rhizobium–legume associations (Jordan, 1962). However, a more recent investigation indicates that the haem moiety of leghaemoglobins in the effective nodules is synthesized entirely by rhizobial bacteroids (Cutting & Schulman, 1972). The ability of symbiont-free flagellates to utilize Proto. and to incorporate ^{59}Fe into haem indicates that they can synthesize the enzyme, ferrochelatase; thus the host flagellates may not depend on the symbionts for this enzyme. Whether symbionts are capable of producing the enzyme is worthy of further investigation. A negative finding would be of great interest, as it implies that the haem biosynthesis of symbiont-containing flagellates depends on a symbiont–host co-operation and, here, flagellates could have, by controlling the availability of ferrochelatase, a convenient mechanism for the regulation of their endosymbionts.

DISCUSSION

In concluding this paper, it seems appropriate to make a comparison between the two symbioses described above. Common to both are symbionts which resemble prokaryotic micro-organisms and which are essential for the well-being of the hosts, at least under certain conditions. While there is no doubt that symbionts of *B. culicis* and *C. oncopelti* are important in host nutrition, further work is needed on the symbionts of *C. lectularius*.

A population balance between symbionts and hosts seems to be maintained by the lysosome-mediated autophagy of symbionts in the mycetomes of *C. lectularius*. Such a regulatory mechanism has not been found in haemoflagellates (Chang, unpublished), which probably resort to metabolic or nutritional regulation. Of the two invertebrate hosts as organisms for the study of symbiosis, the flagellate protozoa has the advantage of structural simplicity, short life-span and ready axenic cultivation. Insects clearly present more difficulties for the detailed analysis of the symbiont–host relationships. One apparent solution – the cultivation of symbionts in artificial media – has been largely unsuccessful (e.g. Brooks & Richards, 1966; Musgrave & McDermott, 1961), though Landureau (1966), for example, has used cultured cells of cockroaches for growing endosymbionts and their identification by fluorescent antibody technique. If symbionts could be further maintained continuously in insect cells, as in the unicellular flagellates, then study of insect endosymbiosis could be extended beyond the more morphological and traditional approaches.

ACKNOWLEDGEMENTS

I wish to thank Professor A. J. Musgrave, The University of Guelph, and Professor W. Trager, The Rockefeller University, for invaluable suggestions at different phases of this study and for critical review of the manuscript. Thanks are further extended to my collaborators at The Rockefeller University, Dr S. Sassa with whom the haem biosynthesis study was made in his laboratory; and Mr R. S. Tuan who did the DNA study and participated in the experiments of symbiont isolation. I am also very grateful to Dr Sassa for preparing excellent graphs for the oral presentation and to Miss R. Klatt and my wife, Chin, for their skillful help. This work received financial aids mainly from a Post-doctoral Training Grant (AI-00192-13) of US Public Health Service given to Professor Trager, but also from The National Research Council of Canada given to Professor Musgrave for part of the *Cimex* study.

REFERENCES

ARKWRIGHT, J. A., ATKIN, E. E. & BACOT, A. (1921). An hereditary rickettsia-like parasite on the bed-bug (*Cimex lectularius*). *Parasitology*, **13**, 27–36.

ASCHNER, M. (1932). Experimentelle Untersuchungen über die Symbiose der Kleiderlaus. *Naturwissenschaften*, **20**, 501–505.

ASCHNER, M. & RIES, E. (1933). Das Verhalten der Kleiderlaus bei Ausschaltung ihrer Symbionten. Eine experimentelle Symbiosestudie. *Z. Morph. Ökol. Tiere*, **26**, 529–590.

BALL, G. H. (1969). Organisms living on and in protozoa. In *Research in Protozoology*, vol. **3**, 565–718 (ed. T.-T. Cheng). London: Pergamon Press.

BEALE, G. H., JURAND, A. & PREER, J. R. (1969). The classes of endosymbiont of *Paramecium aurelia*. *J. Cell Sci.*, **5**, 65–91.

BRECHER, G. & WIGGLESWORTH, V. B. (1944). The transmission of *Actinomyces rhodnies erikson* in *Rhodnius prolixus* and its influence on the growth of the host. *Parasitology*, **35**, 220–224.

BROOKS, M. A. (1963). Symbiosis and aposymbiosis in arthropods. *Symp. Soc. gen. Microbiol.*, **13**, 200–231.

BROOKS, M. A. & RICHARDS, K. (1966). On the *in vitro* culture of intracellular symbiotes of cockroaches. *J. Invertebr. Path.*, **8**, 150–157.

BRUESKE, W. A. (1967). The diplosome of *Blastocrithidia culicis*. (Novy, McNeal & Torrey, 1907) (Mastigophora: Trypanosomatidae). Ph.D. Dissertation, University of Minnesota.

BUCHNER, P. (1965). *Endosymbiosis of Animals with Plant Micro-organisms*. New York: Wiley.

BURGDORFER, W., BRINTON, L. P. & HUGHES, L. E. (1973). Isolation and characterization of symbiotes from the Rocky Mountain wood tick, *Dermacentor andersoni*. *J. Invertebr. Path.*, **22**, 424–434.

BUSH, G. L. & CHAPMAN, G. B. (1961). Electron microscopy of symbiotic bacteria in developing oocytes of the American cockroach, *Periplaneta americana*. *J. Bact.*, **81**, 267–276.

CHANG, K.-P. (1972). Endosymbiosis of prokaryotic micro-organisms with phytophagous Homoptera and haematophagous Heteroptera. Ph.D. Thesis, The University of Guelph, Ontario, Canada.

CHANG, K.-P. (1974a). Effects of elevated temperature on mycetome and its symbiotes of *Cimex lectularius* (L.). *J. Invertebr. Path.*, **23**, 333–340.

(1974b). Ultrastructure of symbiotic bacteria in normal and antibiotic treated *Blastocrithidia culicis* and *Crithidia oncopelti*. *J. Protozool.*, **21**, 699–707.

CHANG, K.-P., CHANG, J. C., TUAN, R. S. & SASSA, S. (1974). Ultrastructure and heme biosynthesis of symbiotic bacteria in two species of haemoflagellates. *Proc. 3rd Int. Cong. Parasit.*, **1**, 26–27.

CHANG, K.-P., & MUSGRAVE A. J. (1969). Histochemistry and ultrastructure of the mycetome and its 'symbiotes' in the pear Psylla, *Psylla pyricola* Foerster (Homoptera). *Tissue & Cell*, **1**, 597–606.

(1972). Multiple symbiosis in a leafhopper, *Helochara communis* Fitch (Cicadellidae: Homoptera): envelopes, nucleoids and inclusions of the symbiotes. *J. Cell Sci.*, **11**, 275–293.

(1973). Morphology, histochemistry and ultrastructure of mycetome and its rickettsial symbiotes in *Cimex lectularius* L. *Can. J. Microbiol.*, **19**, 1075–1081.

CHANG, K.-P. & TRAGER, W. (1974). Nutritional significance of symbiotic bacteria in two species of hemoflagellates. *Science, Wash.*, **183**, 531–532.

CUTTING, J. A. & SCHULMAN, H. M. (1972). The site of haem synthesis in soyabean root nodules. *Biochim. biophys. Acta*, **192**, 486–493.

DeMeillon, B. & Goldberg, L. (1947). Preliminary studies on the nutritional requirements of the bed bugs (*Cimex lectularius* L.) and the tick (*Ornithodorus moubata* Murray). *J. exp. Biol.*, **24**, 41–63.

Gaughan, P. L. Z. & Krassner, S. M. (1971). Haemin deprivation in culture stages of the haemoflagellate, *Leishmania tarentolae*. *Comp. Biochem. Physiol.*, **39B**, 5–18.

Gill, J. W. & Vogel, H. J. (1963). A bacterial endosymbiont in *Crithidia* (*Strigomonas*) *oncopelti*: biochemical and morphological aspects. *J. Protozool.*, **10**, 148–152.

Grace, T. D. C. (1962). Establishment of a line of mosquito (*Aedes aegypti* L.) cells grown *in vitro*. *Nature, Lond.*, **211**, 366–367.

Granick, S. & Sassa, S. (1971). δ-Aminolevulinic acid synthetase and the control of haem and chlorophyll synthesis. In *Metabolic Regulation – Metabolic Pathways*, vol. **5**, 79–95 (ed. H. J. Vogel). New York: Academic Press.

Griffiths, G. W. & Beck, D. S. (1973). Intracellular symbiotes of the pea aphid *Acrythosiphon pisum*. *J. Insect Physiol.*, **19**, 75–85.

Grinyer, I. & Musgrave, A. J. (1966). Ultrastructure and peripheral membranes of the mycetomal micro-organisms of *Sitophilus granarius* (L.) Coleoptera. *J. Cell Sci.*, **1**, 181–186.

Gutteridge, W. E. & Macadam, R. F. (1971). An electron microscopic study of the bipolar bodies in *Crithidia oncopelti*. *J. Protozool.*, **18**, 637–640.

Guttman, H. N. & Eisenman, R. N. (1965). 'Cure' of *Crithidia* (*Strigomonas*) *oncopelti* of its bacterial endosymbionts. *Nature, Lond.*, **206**, 113–114.

Hertig, M. & Wolbach, S. B. (1924). Studies on rickettsia-like micro-organisms in insects. *J. Med. Res.*, **44**, 328–374.

Hill, P., Saunders, D. S. & Campbell, J. A. (1973). The production of 'symbiote-free' *Glossina moristans* and an associated loss of female fertility. *Trans. R. Soc. trop. Med. Hyg.*, **67**, 727–728.

Hinde, R. (1971). The control of the mycetome symbiotes of the aphids *Brevicoryne brassicae*, *Myzus persicae* and *Macrosiphum rosae*. *J. Insect Physiol.*, **17**, 1791–1800.

Indurkar, A. K. (1965). Studies on the kinetoplast of three species of flagellates. Ph.D. Dissertation, The University of Wisconsin.

Jordan, D. C. (1962). The bacteroids of the genus *Rhizobium*. *Bact. Rev.*, **26**, 119–141.

Kellerman, G. M., Bigg, D. R. & Linnane, A. W. (1968). Biogenesis of mitochondria. XI. A comparison of the effects of growth-limiting oxygen tension, intercalating agents and antibiotics on the obligate aerobe, *Candida parapsilosis*. *J. Cell Biol.*, **42**, 378–391.

Kirby, H. Jr (1941). Organisms living on and in protozoa. In *Protozoa in Biological Research*, 1009–1113 (eds. G. N. Calkins and F. M. Summers). New York: Columbia University Press.

Koch, A. (1956). The experimental elimination of symbionts and its consequences. *Exptl Parasit.*, **5**, 481–518.

— (1960). Intracellular symbiosis in insects. *A. Rev. Microbiol.*, **14**, 121–140.

Körner, H. K. (1969). Zur Ultrastruktur der intrazellularen Symbionten im Embryo der Kleinzikade *Euscelis plebejus* Fall. (Homoptera, Cicadina). *Z. Zellforsch. mikrosk. Anat.*, **100**, 466–473.

— (1972). Elektronenmikroscopische Untersuchungen an embryonalen Mycetom der Kleinzikade *Euscelis plebejus* Fall. (Homoptera, Cicadina). *Zentbl. Bakt. ParasitKde Abt. II*, **40**, 203–226.

Krieg, A. (1963). Rickettsiae and rickettsioses. In *Insect Pathology*, vol. **1**, 577–617 (ed. E. A. Steinhaus). New York: Academic Press.

PLATE I

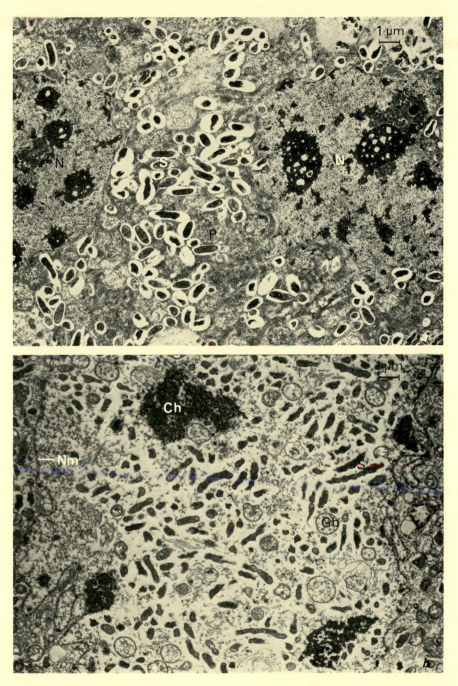

For explanation see p. 427

PLATE 2

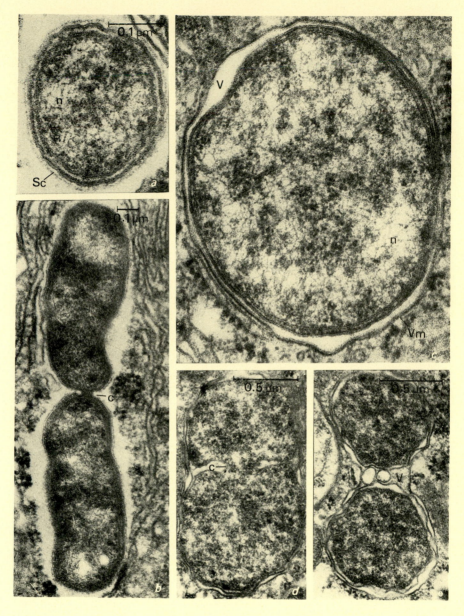

For explanation see p. 427

PLATE 3

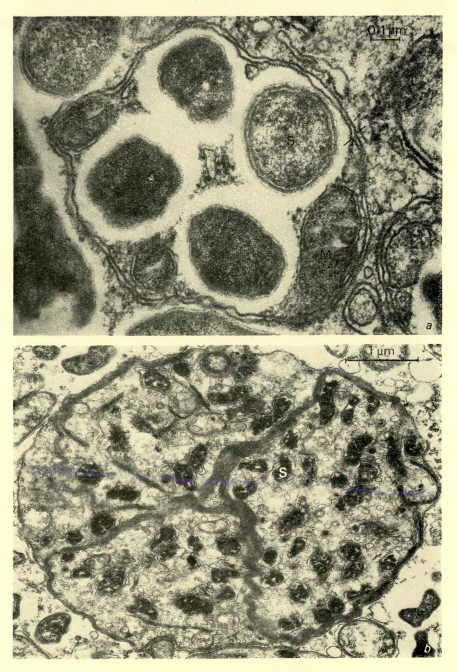

For explanation see p. 427

PLATE 4

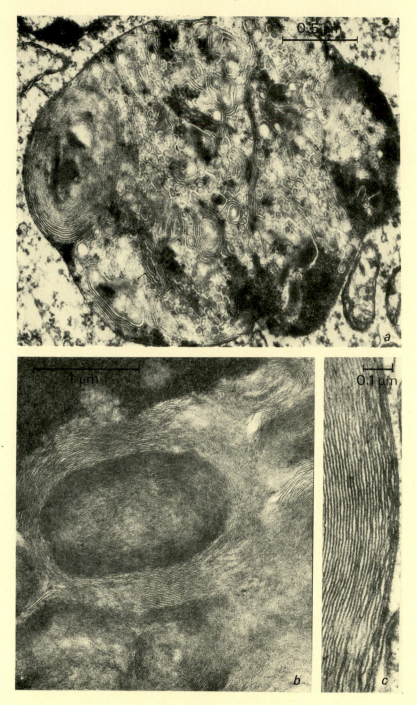

For explanation see p. 427

PLATE 5

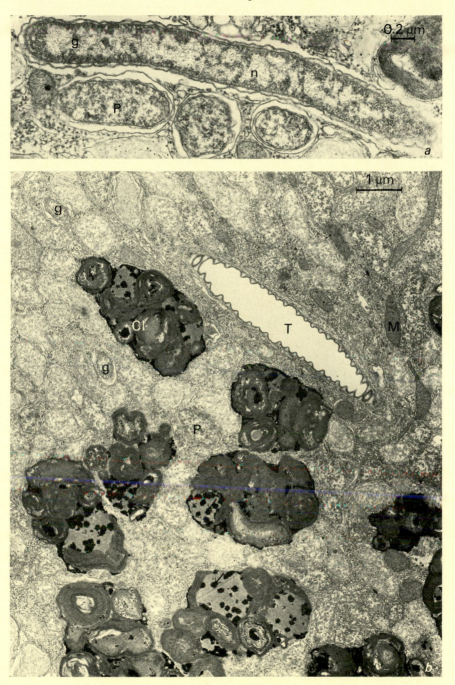

For explanation see p. 427

PLATE 6

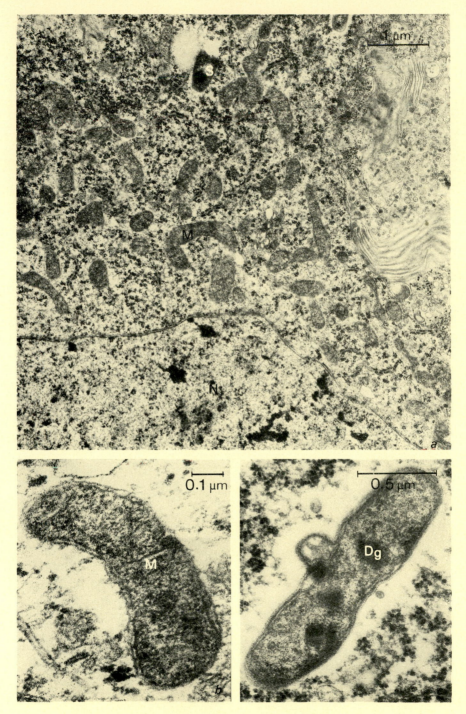

For explanation see p. 428

PLATE 7

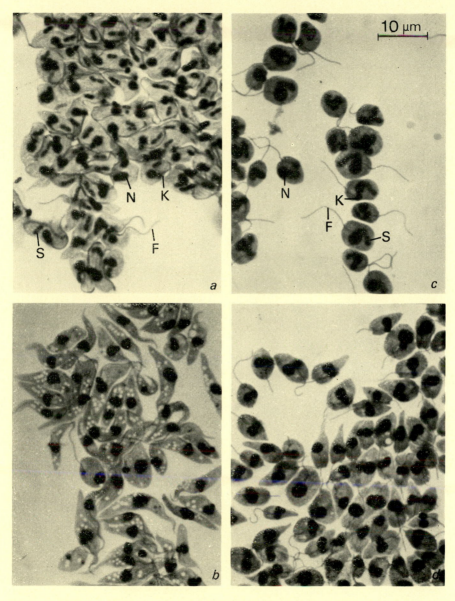

For explanation see p. 428

PLATE 8

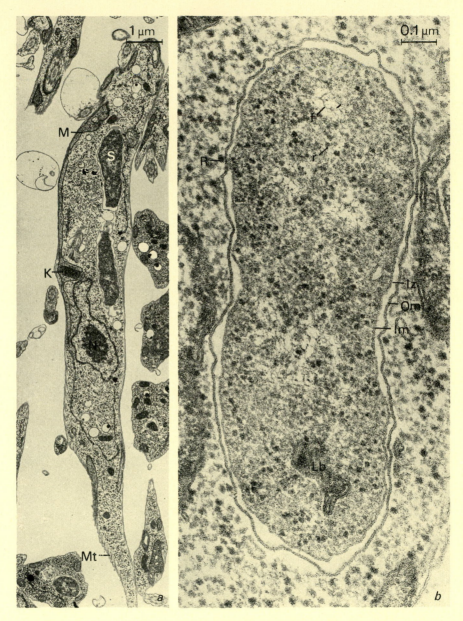

For explanation see p. 428

PLATE 9

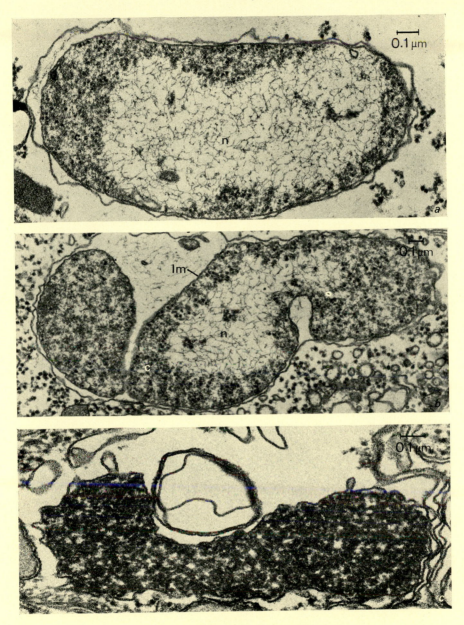

For explanation see p. 428

PLATE 10

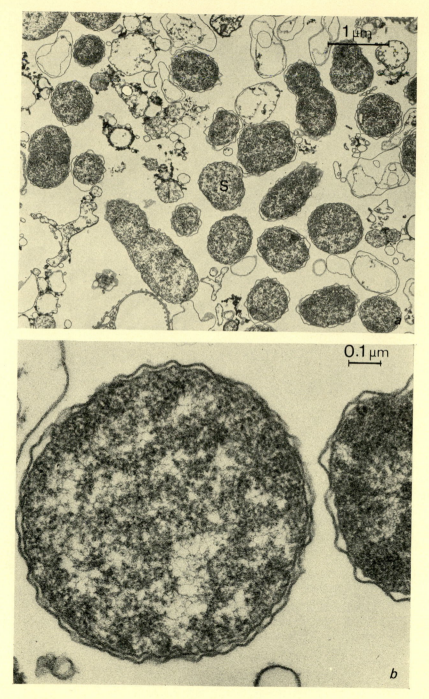

For explanation see p. 428

LANDUREAU, J.-C. (1966). Des cultures de cellules embryonnaires de Blattes permettent d'obtenir la multiplication in vitro des bactéries symbiotiques. *C. r. hebd. Séanc. Acad. Sci., Paris*, **262**, 1484–1487.

LANHAM, U. N. (1968). The Blochmann bodies: hereditary intracellular symbionts in insects. *Biol. Rev.*, **43**, 269–286.

LOCKE, M. & COLLINS, J. V. (1965). The structure and formation of protein granules in the fat body of an insect. *J. Cell Biol.*, **26**, 857–884.

LOUIS, C., LAPORTE, M., CARAYON, J. & VAGO, C. (1973). Mobilité, ciliatères ultrastructuraux des micro-organismes symbiotiques endo- et exocellulaires de *Cimex lectularius* L. (Hemiptera Cimicidae). *C. r. hebd. Séanc. Acad. Sci., Paris*, **277**, 607–611.

MALIK, U. S. (1972). Chloramphenicol. *Adv. appl. Microbiol.*, **15**, 297–331.

MARMUR, J., CAHOON, M. E., SHIMURA, Y. & VOGEL, H. J. (1963). DNA type attributable to a bacterial endosymbiote in the protozoon, *Crithidia* (*Strigomonas*) *oncopelti*. *Nature, Lond.*, **197**, 1228–1229.

MILLER, F. & PALADE, G. E. (1964). Lytic activities in renal protein absorption droplets. An electron microscopical cytochemical study. *J. Cell Biol.*, **23**, 519–525.

MITSUI, Y., FUGIMOTO, M. & KAJUMA, M. (1964). Development and morphology of trachoma agent in the yolk sac cell as revealed by electron microscopy. *Virology*, **23**, 30–45.

MORGAN, C., ROSENKRANZ, H. & ROSE, H. M. (1967). Electron microscopy of chloramphenicol treated *Escherichia coli*. *J. Bact.*, **93**, 1987–2003.

MÜLLER, V. J. (1972). Die intrazellulare Symbiose der Zikaden mit Mikroorganismen. *Biol. Rdsch.*, **10**, 46–57.

MUSGRAVE, A. J. (1964). Insect mycetomes. *Can. Ent.*, **96**, 377–389.

MUSGRAVE, A. J. & GRINYER, I. (1968). Membranes associated with the distingration of mycetomal micro-organisms in *Sitophilus zea-mais* (Mots.) (Coleoptera). *J. Cell Sci.*, **3**, 65–70.

MUSGRAVE, A. J. & McDERMOTT, L. A. (1961). Some media used in attempts to isolate and culture the mycetomal micro-organisms of *Sitophilus* weevils. *Can. J. Microbiol.*, **7**, 842–843.

NEWTON, B. A. (1968). Biochemical peculiarities of trypanosomatid flagellates. *A. Rev. Microbiol.*, **22**, 109–130.

NEWTON, B. A. & BURNETT, J. K. (1972). DNA of kinetoplastidae: a comparative study. In *Comparative Biochemistry of Parasites*, 185–198 (ed. H. van den Bossche). New York: Academic Press.

NEWTON, B. A. & HORNE, R. W. (1957). Intracellular structures in *Strigomonas oncopleti*. 1. Cytoplasmic structures containing ribonucleoprotein. *Exptl Cell Res.*, **13**, 563–574.

NOVEY, F. G., McNEAL, W. J. & TORREY, H. N. (1907). The trypanosomes of mosquitoes and other insects. *J. infect. Dis.*, **4**, 223–276.

PANT, N. C. & FRAENKEL, G. (1954). Studies on the symbiotic yeasts of two insect species, *Lasioderma serricorne* F. and *Stegobium* (*Sitodrepa*) *paniceum* L. *Science, Wash.*, **112**, 498–500.

PHILIP, C. B. (1957). Rickettsiales. In *Bergey's Manual of Determinative Bacteriology*, 7th edn, 937–979 (eds. R. S. Breed, R. E. G. Murray, and N. R. Smith). Baltimore: The Williams & Wilkins Co.

PREER, L. B. (1969). Alpha, an infectious macronuclear symbiont of *Paramecium aurelia*. *J. Protozool.*, **16**, 570.

PREER, L. B., JURAND, A., PREER, J. R. JR & RUDMAN, B. M. (1972). The classes of Kappa in *Paramecium aurelia*. *J. Cell Sci.*, **11**, 581–600.

REINHARDT, C., AESCHLIMANN, X. & HECKER, H. (1972). Distribution of rickettsia-like micro-organisms in various organs of an *Ornithodorus moubata* laboratory strain (Ixodoidea, Argasidae) as revealed by electron microscopy. *Z. Parasit-Kde*, **39**, 201–209.

ROSHDY, M. A. (1968). A rickettsia-like micro-organism in the tick, *Ornithodorus savignyi*; observations on its structure and distribution in the tissues of the tick. *J. Invertebr. Path.*, **11**, 155–169.

SASSA, S., GRANICK, S., BICKERS, D. R., BRADLOW, H. L. & KAPPAS, A. (1974). A microassay for uroporphyrinogen I synthase, one of three abnormal enzyme activities in acute intermittent porphyria, and its application to the study of the genetics of this disease. *Proc. nat. Acad. Sci. USA*, **71**, 732–736.

SONNEBORN, T. M. (1959). Kappa and related particles in *Paramecium*. In *Advances in Virus Research*, vol. IV, 231–353 (eds. K. M. Smith and M. A. Lauffer). New York: Academic Press.

STEINHAUS, E. A. (1967). *Insect Microbiology*. (3rd printing). New York: Hafner Publishing Co.

STEVENSON, I. (1969). The biochemical status of μ particles in *Paramecium aurelia*. *J. gen. Microbiol.*, **57**, 61–75.

SUITOR, E. J. Jr (1964). The relationship of *Wolbachia persica* Suitor and Weiss to its host. *J. Insect Path.*, **6**, 111–124.

TAKAKU, F., WADA, O., SASSA, S. & NAKAO, K. (1968). Haem synthesis in normal and leukemic leukocytes. *Cancer Res.*, **28**, 1250–1255.

TAMURA, A., MATSUMOTO, A., MANIRE, G. P. & HIGASHI, N. (1971). Electron microscopic observations on the structure of the envelopes of mature elementary bodies and developmental reticulate forms of *Chlamydia psittaci*. *J. Bact.*, **105**, 355–360.

TRAGER, W. (1957). Nutrition of a haemoflagellate (*Leishmania tarentolae*) having an interchangeable requirement for choline or pyridoxal. *J. Protozool.*, **4**, 269–276.

(1959). The enhanced folic and folinic acid contents of erythrocytes infected with malaria parasites. *Exptl Parasit.*, **8**, 265–273.

(1968). Cultivation and nutritional requirements. In *Infectious Blood Diseases of Man and Animals*, vol. 1, 149–171 (eds. D. Weinman and M. Ristic). New York: Academic Press.

(1970). *Symbiosis – Selected Topics in Modern Biology*. New York: van Nostrand Reinhold.

(1974). Some aspects of intracellular parasitism. *Science, Wash.*, **183**, 269–273.

TUAN, R. S. & CHANG, K. P. (1974). Isolation and characterization of DNA from symbiotic and aposymbiotic strains of *Blastocrithidia culicis*. *Proc. 3rd Int. Cong. Parasit.*, **3**, 1453–1454.

VAN WAGTENDONK, W. J., CLARK, J. A. D. & GODOY, G. A. (1963). The biological status of lambda and related particles in *Paramecium aurelia*. *Proc. natn. Acad. Sci. USA*, **50**, 835–883.

WALLACE, F. G. (1966). The trypanosomatid parasites of insects and arachnids. *Exptl Parasit.*, **18**, 124–193.

Note added in proof: The genome complexity of symbionts in *B. culicis* was found to be 6.7×10^9 daltons (Tuan & Chang, *J. Cell Biol.*, **65**, 309–323).

EXPLANATION OF PLATES

Plates 1–6. Thin sections of mycetomes from the Hamilton and Orlando Strains of *Cimex lectularius* Plates 1*a*, 2*a*, *c*, and 3*a*, *b* reproduced by permission of the National Research Council of Canada from the *Canadian Journal of Microbiology*, **19**, 1973, 1075–1081.

PLATE 1

(a) The occurrence of the primary (P) and secondary (S) symbionts in the syncytial cytoplasm (Chang & Musgrave, 1973).

(b) The intranuclear occurrence of the secondary symbionts (S); the globular bodies (Gb) in the nucleus are variants of secondary symbionts. Ch, chromatin granules; Nm, nuclear membrane; N, nucleus.

PLATE 2

(a) Cross-section of a secondary symbiont. The symbiont contains dense cytoplasm, an electron-lucid nuclear zone (n) and multilayered envelopes resembling the profile of a Gram-negative bacterium with a surface coat (Sc) (Chang & Musgrave, 1973).

(b) Longitudinal section of a secondary symbiont undergoing binary fission. Note the absence of host-derived membrane: c, division furrow.

(c) Section of a primary symbiont. The symbiont contains dense granular cytoplasm and electron-lucid nuclear zones (n) with DNA-like filaments. The peripheral envelope is of Gram-negative type. The symbiont occurs in a host vacuole (V) limited by a vacuolar membrane (Vm) (Chang & Musgrave, 1973).

(d) A dividing primary symbiont with a central division furrow (c) formed by the invagination of the inner plasma membrane of the symbiont.

(e) Two primary symbionts in a common host vacuole (V). Presumably, they represent two daughter cells which have just completed cell division. The host vacuole seems to be dividing as well, supplying each symbiont with a separate vacuole.

PLATE 3

(a) Segregation of several symbionts (S) and mitochondria (M) with double cytoplasmic membranes (between arrows) in a mycetome. The structure resembles an isolated body or early cytosegresome formed before cellular autophagy.

(b) A large cytolysome in the mycetomal syncytium containing many small vesicles, membrane bundles and very dense symbionts (S) apparently degenerating (Chang & Musgrave, 1973).

PLATE 4

(a) A mycetomal cytolysome at a later stage to that shown in Plate 3b containing dense material, vesicles and concentric membranes but no recognizable symbionts.

(b) Portion of a large residue body in a mycetome consisting of membrane whirls and dense granular materials.

(c) Portion of a residue body made up of parallel trilaminar membranes.

PLATE 5

(a) Longitudinal section of a guest symbiont (g) in the mycetome. The symbiont contains dense cytoplasm, electron-lucid nuclear zones (n), a cell envelope typical of Gram-negative bacteria, and flagella (not shown). Note the absence of host-derived membrane. P, primary symbiont.

(b) Portion of a mycetome from the Orlando strain of *Cimex lectularius* prepared for showing acid phosphatase activity, a marker enzyme for lysosomes, by a modified procedure after Miller & Palade (1964). The enzyme activity, manifested as heavy lead deposits (arrows), is evident in the cytolysomes (Cl) of mycetomal syncytia. Note the absence of the secondary symbionts in this strain of bedbugs. g, Guest symbiont; M, mitochondrion; T, tracheole; P, primary symbiont.

PLATE 6

(a) Portion of a mycetome from the Hamilton Strain of *C. lectularius* reared at 36 °C for two weeks. Symbionts are absent in the mycetome, except for very few degenerating ones (S). The mycetome contains normal cellular organelles, e.g. nuclei (N), mitochondria (M), Golgi apparatus, endoplasmic reticulum and ribosomes.

(b) A mitochondrion in the aposymbiotic mycetome.

(c) A secondary symbiont at elevated temperature apparently degenerating, as suggested by the presence of dense granules (Dg) in its cytoplasm.

PLATE 7

Light micrographs of smear preparations of haemoflagellates treated with 4 % (v/v) Tween 80 in phosphate buffer, pH 7.0 for four minutes and stained with Giemsa. (All at the same magnification.)

(a, c) Normal symbiont-containing flagellates of *Blastocrithidia culicis* (a) and *Crithidia oncopelti* (c). Each flagellate possesses a pair of symbionts (S), besides a nucleus (N), a kinetoplast (K) and a flagellum (F).

(b, d) Symbiont-free flagellates of *B. culicis* (b) and *C. oncopelti* (d) obtained by chloramphenicol treatment. The flagellate contains normal cellular organelles but no symbionts.

PLATE 8

(a) Electron micrograph of *B. culicis* showing the general ultrastructure of the flagellate and its symbionts. K, kinetoplast; M, mitochondrion; Mt, subpellicular microtubule; N, nucleus.

(b) Ultrathin section of a symbiont in *B. culicis*. The symbiont possesses nuclear areas containing DNA-like fibres (f) and dense cytoplasm with ribosome-like particles (r). The peripheral envelope of the symbiont consists of an inner (Im) and an outer (Om) membrane separated by an irregular intermediate zone (Iz) of very low density. Lb, lamellar body (mesosome?) of the symbiont; R, ribosome of flagellate.

PLATE 9

Effects of chloramphenicol treatment at different intervals on the ultrastructure of symbionts in the flagellate of *B. culicis*.

(a) The symbiont appears to have an expanded electron-lucent nuclear zone (n) with a concomitant reduction of cytoplasmic areas (c) four days after the treatment.

(b) On day 8, the symbiont becomes an irregular entity due to the distortion of its inner membrane (Im) (a and b from Chang, 1974b).

(c) The symbiont degenerates to become a dense body surrounded by membrane fragments on day 12.

PLATE 10

(a) A survey electron micrograph of the symbiont fraction obtained from flagellates of *B. culicis* (see text for detail) showing the abundance of symbionts (S) and the absence of intact cellular organelles of host flagellates.

(b) Electron micrograph of isolated symbionts from *B. culicis* at higher magnification showing the structural integrity of their peripheral envelopes and cytoplasmic contents.

FACTORS AFFECTING INFECTIONS OF MAMMALS WITH INTRAERYTHROCYTIC PROTOZOA

By F. E. G. COX

Department of Zoology, King's College, Strand, London, WC2R 2LS

INTRODUCTION

The red blood cell of a mammal provides a suitable habitat for parasites and has been utilized by a number of protozoa. The best known of these are the malaria parasites and the related piroplasms, species of which cause red water fever and East Coast fever in cattle. These economically important parasites have received considerable attention and, fortunately for experimental biologists, representative species can be induced to develop in laboratory rodents. This paper will be restricted to those parasites which can be maintained in the erythrocytes of laboratory mice but the principles that emerge from the experiments described are applicable to other intraerythrocytic parasites in other hosts.

The outcome of any infection is the result of the parasite's ability to live in its host and the ability of the host to support the infection. It is well known that factors such as the age, sex or nutritional state of the host, or less clearly defined factors such as stress, may modify the course of an infection. It is not so well known that prior or simultaneous exposure to other infectious agents may also influence the outcome of an infection and the subject of this paper will be the effects of other infections on those caused by intraerythrocytic protozoa.

In general, it has been found that the normal infection with a malaria parasite or piroplasm may be altered under the following circumstances:

1. Previous exposure to the same species of parasite.
2. Previous exposure to a different species of parasite with common antigens.
3. Previous exposure to a different species of parasite without antigens in common.
4. Previous exposure of the mother to the same species of parasite.
5. Previous exposure of the mother to a different species of parasite with common antigens.
6. Previous exposure of the mother to a different species of parasite without antigens in common.

[429]

The outcomes of some of these interactions are due to immunological effects while others are not and for some years we have been investigating the relative importance of immune responses in determining the outcome of any particular infection. The experiments which follow have mostly been carried out at King's College as part of our research programme into the interactions between parasites but the results have been integrated with those obtained in other laboratories.

THE PARASITES USED

The intraerythrocytic protozoa used in these investigations are listed in Table 1. They belong to two classes and three genera. Normally the

Table 1. *Characteristics of the parasites used in the experiments described in this paper*

Family	Genus and species	Characteristics of infection in mice	
		Virulence	Type of cell preferred
Plasmodiidae	*Plasmodium vinckei vinckei*	Fatal	Mature erythrocytes
(malaria	*Plasmodium vinckei chabaudi*	Non-fatal	Mature erythrocytes
parasites)	*Plasmodium berghei berghei*	Fatal	Reticulocytes
	Plasmodium berghei yoelii	Non-fatal	Reticulocytes
	Plasmodium atheruri	Fatal or non-fatal	Mature erythrocytes or reticulocytes
Babesiidae	*Babesia rodhaini*	Fatal	Mature erythrocytes
(piroplasms)	*Babesia microti*	Non-fatal	Mature erythrocytes
Dactylosomidae	*Anthemosoma garnhami*[a]	Non-fatal	Mature erythrocytes or reticulocytes

[a] It is not always possible to infect mice with *A. garnhami* and infections range from transient ones, with a maximum parasitaemia at which less than 1 % of red cells are infected, to peaks of 60 % or more.

members of the Plasmodiidae undergo one or more cycles of division in the liver before entering red cells but the members of the Babesiidae spend all their lives in red cells. It is not known whether or not *Anthemosoma* develop outside erythrocytes. In the experiments described below, mice were infected with blood from another infected host and under these circumstances only the stages in the blood occur in the recipient. The cell cycles of three representative parasites are shown in Fig. 1 and they are basically very similar. The young parasite penetrates an erythrocyte and grows within it. Eventually nuclear division occurs and daughter parasites are produced and liberated when the host cell bursts. Eight daughters are

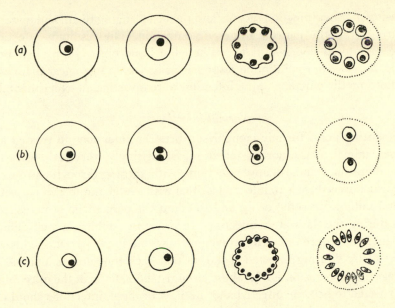

Fig. 1. The cell cycles of (*a*) *Plasmodium vinckei vinckei*, (*b*) *Babesia rodhaini* and (*c*) *Anthemosoma garnhami*. The cycles take place within the erythrocytes of a mouse and the end-products (merozoites) are liberated into the plasma to invade fresh cells.

produced in each cell cycle of the plasmodia used, two in the babesias and eight or 16 in *Anthemosoma*. The young daughter parasites are called merozoites in the case of the Plasmodiidae and, by analogy, although this is by no means widely accepted, it is convenient to refer to the daughters of *Babesia* and *Anthemosoma* as merozoites also. The merozoites penetrate fresh cells and the cycle is repeated.

The characteristics of the eight parasites used are summarized in Table 1. Briefly, there are three pairs of morphologically similar parasites, one of each producing fatal infections in mice and the other milder infections. In addition there is a malaria parasite, *Plasmodium atheruri* which is quite different from the other four, and a piroplasm *Anthemosoma garnhami* which has only remote affinities with the other parasites. Together these parasites provide a range of affinities, infections and ability to develop in red cells of different types. All these parasites can be maintained in a constant state frozen in liquid nitrogen and when put into mice of a standard strain produce reproducible infections.

In general, the experiments consisted of infecting mice with these parasites or other infectious agents and then challenging them, and matched controls, with the parasite under investigation. The resultant infections were monitored and compared. When the basic observations had been made, the roles of the immune response, or non-immunological changes in the host, on the outcome of the infection were investigated experimentally.

Normal infections

The starting point for any comparison between infections in treated and control animals is the normal infection. Surprisingly, very little work has been done on standardizing infections with intraerythrocytic protozoa. The usual procedure is to infect an animal with a large number of parasites, several millions in some cases, and to count the percentage of parasitized cells during the days after infection. In order to establish a well-defined base line, mice were infected intravenously with a single parasite. This was achieved by diluting infected blood so that each inoculum contained an average of one parasite. Under these circumstances, 60 % of the mice might be expected to become infected and, in 60 % of these, infections should be initiated from single parasites. In practice, the actual numbers of infections achieved were very close to the estimated ones. The results obtained for infections with *P. b. berghei*, *P. b. yoelii*, *P. v. vinckei*, *P. v. chabaudi*, *B. rodhaini* and *B. microti* are shown in Fig. 2. In all cases the increase in the number of parasites was exponential until about 30 % of the cells were infected. At this stage, most of the infected cells contained single parasites but, as the infections progressed, random invasion of cells occurred resulting in cells containing 1–5 parasites during the final stages of the infection. The four malaria parasites multiplied at a rate of eight every 24 hours and this rate did not slow down during the course of the infection. *B. rodhaini* multiplied at a rate of two every eight hours and therefore achieved the same growth rate as the malaria parasites. *B. microti* multiplied much more slowly than any of the other parasites, 2.7 times every 24 hours. Single parasite infections could not be established with *A. garnhami*.

With increased numbers of parasites in the inoculum the rate of multiplication was unaltered but the time taken for the parasites to reach the peak of infection was decreased in direct proportion to the number of parasites introduced into the host. After the level of infection passes 30 % various complications, such as multiple invasion and non-availability of red cells, occur. These make the analysis of the further progress of the infection difficult but, under normal circumstances, there are only two possible outcomes of infections of the kind described here; the parasites occupy every possible cell, or the infection declines. In the former, the host in fact dies before this stage is reached.

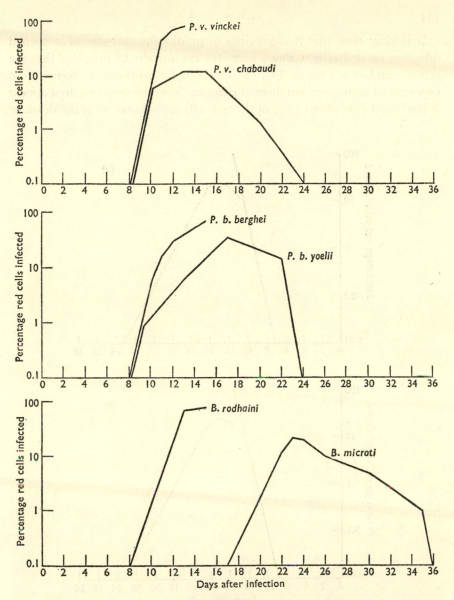

Fig. 2. The patterns of infections of *Plasmodium vinckei vinckei, P. v. chabaudi,*
P. berghei berghei, P. b. yoelii, Babesia rodhaini and *B. microti* in mice. For convenience these graphs have been drawn in pairs on the same axes but each graph
represents a single infection. These infections were initiated from a single parasite
injected intravenously on day 0. The parasitaemia is expressed as the percentage of
erythrocytes infected and is shown on a logarithmic scale. A parasitaemia of
1 % represents a parasite load of approximately 2×10^8. The pattern of infection
with *P. b. yeolii* is often quite different from its theoretical pattern because the
infection is markedly influenced by the anaemia and reticulocytosis it induces in its
host.

It is clear then that it is possible to predict the outcome of a normal infection no matter how many parasites are introduced provided that the actual number is known. The outcome can be expressed in terms of an exponential increase in numbers of parasites from the time they first appear in the blood until about 30 % of the red cells are infected or in the death or

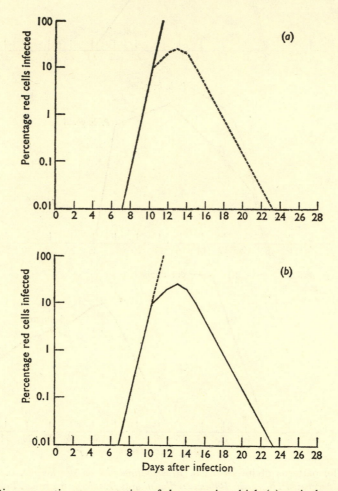

Fig. 3. Diagrammatic representation of the ways in which (a) a virulent infection (e.g. *P. v. vinckei*) and (b) an avirulent infection (e.g. *P. v. chabaudi*) may be altered by the presence of other infections. In (a) the infection is restrained and cannot proceed to its normal fatal conclusion. In (b) a constraint is lifted and the infection proceeds in an unrestrained way. The slope of the graph is determined by the intrinsic rate of increase of the parasite (× 8 every 24 hours in this case) and therefore cannot move to the left. The more the graph slopes to the right the more chance the animal has of overcoming the infection. Solid line, normal infection; dashed line, modified infection.

recovery of the infected animals (see Fig. 3). Any deviation from the expected outcome in animals previously or simultaneously infected with another infectious agent can be attributed directly or indirectly to that agent.

FACTORS AFFECTING INFECTIONS

Previous exposure to the same species of parasite

It has been shown repeatedly that after recovery from infections with the parasites listed above (and many others) rodents do not become infected when challenged with the same parasite (Table 2). In the case of potentially fatal infections it is necessary to control the primary infection with drugs or by maintaining the animals on a diet deficient in *para*-aminobenzoic acid which malaria parasites require. Solid resistance to a challenge infection following recovery from an infection with the same parasite is known as homologous immunity and has been demonstrated in mice infected with *P. b. berghei* (Kretschmar, 1962), *P. b. yoelii* (Cox & Voller, 1966), *P. v. vinckei* (Cox, 1966), *P. v. chabaudi* (Cox, 1970c), *P. atheruri* (Cox, 1972a), *B. rodhaini* and *B. microti* (Cox & Young, 1969), and *A. garnhami* (Cox, 1972b). Immunity of this kind is strong and often life-long although some animals may suffer from mild infections after challenge. The sera of animals that have recovered from such infections contain antibodies against the homologous parasite and these can easily be detected by the fluorescent antibody technique (Cox, Crandall & Turner, 1969; Cox, 1970b; Cox & Turner, 1970; Cox, 1972a). In general, IgM antibody is produced early in the infection while IgG replaces it and persists during convalescence. In a long-term experiment it has been found that mice which have recovered from infections with *B. microti* have antibody titres of between 1/10 240 and 1/20 480 for up to nine months after recovery and there is no indication that this antibody level falls or that the mice lose their resistance to infection during their life-time.

Previous exposure to different species of parasites with common antigens

It might be expected that related parasites such as *P. v. vinckei* and *P. v. chabaudi* or *P. b. berghei* and *P. b. yoelii* can protect against each other and indeed this is the case (Cox & Voller, 1966). However, what is unexpected is that prior exposure to distantly related parasites sometimes also results in immunity. Mice were infected with the eight species or subspecies of blood parasites listed on p. 430 and allowed to recover either naturally or after treatment with appropriate drugs. They were then challenged with the

Table 2. *Percentage of mice protected by prior infections with homologous or heterologous blood parasites*

Immunizing infection	Challenged with							
	P. v. vinckei	*P. v. chabaudi*	*P. b. berghei*	*P. b. yoelii*	*P. atheruri*	*B. rodhaini*	*B. microti*	*A. garnhami*
P. v. vinckei	100	70	7	100	50	100	100	92
P. v. chabaudi	80	100	0	34	100	60	100	100
P. b. berghei	90	40	80	56	50	66	60	58
P. b. yoelii	56	34	34	100	17	75	56	58
P. atheruri	17	83	0	67	100	75	75	50
B. rodhaini	95	40	0	0	83	100	100	100
B. microti	90	75	7	0	67	100	100	100
A. garnhami	53	67	0	20	67	100	100	100

Data from Cox, 1970a, 1972a,b, and additional experiments.

Table 3. *Fluorescent antibody titres obtained using blood parasites and homologous and heterologous antisera*

Antiserum to	Antigen						
	P. v. vinckei	*P. v. chabaudi*	*P. b. berghei*	*P. b. yoelii*	*P. atheruri*	*B. rodhaini*	*B. microti*
P. v. vinckei	2560	640	640	320	—	160	160
P. v. chabaudi	640	2560	320	160	80	10	10
P. b. berghei	320	160	2560	320	—	160	80
P. b. yoelii	40	80	160	2560	—	0	0
P. atheruri	—	160	—	—	2560	—	—
B. rodhaini	20	20	40	20	—	2560	80
B. microti	40	40	40	20	—	40	2560

Data from Cox & Turner, 1970; Cox, 1970b, 1972a, and additional experiments.

homologous or heterologous parasites and the resultant infections, if any, were monitored. The results obtained are shown in Table 2. Briefly, these show that out of the 56 heterologous combinations, immunity occurred in the majority of cases.

One of the many ways by which antigens can be identified is by the use of the fluorescent antibody technique. Parasites are first exposed to dilutions of homologous and heterologous antisera. The antibodies in the serum bind to antigens common to those of the parasite against which they were formed. The parasites coated with antibody are then exposed to fluorescein-labelled anti-immunoglobulin and the intensity of the fluorescence produced is an index of the amount of antibody bound. In the case of a parasite in the presence of the homologous antiserum, fluorescence occurs at dilutions of 1/2560 and this is less in the presence of heterologous antisera. The reciprocal of the dilution of antiserum required to produce fluorescence is called the antibody titres and a comparison of the antibody titres produced by exposing different parasites to a variety of heterologous antisera gives some idea of the antigens common to different parasites (Cox & Turner, 1970). Table 3 shows the antibody titres obtained as a result of exposing some of the parasites described above to antisera against themselves and against other parasites. These results indicate that, broadly, heterologous immunity can be correlated with the presence of common antigens. The actual relationship between antigens and protection is not a simple one, because in some cases, for example *B. microti* and *P. v. chabaudi*, there is considerable cross-protection but few common antigens, whereas *P. v. vinckei* and *P. b. berghei* appear to have a number of antigens in common but *P. v. vinckei* does not protect against *P. b. berghei*.

The actual role of the antigens identified by the fluorescent antibody technique in protection is not at all clear but, in general, it seems that heterologous immunity between malaria parasites and piroplasms can be explained in terms of common antigens. The detection and measurement of other antigens using other techniques might provide further proof of this hypothesis.

Previous exposure to different species of parasite without antigens in common

Rickettsiae

Eperythrozoon and *Haemobartonella* spp. are common rickettsial parasites of mammals (see Kreier & Ristic, 1968). They live in or on red blood cells and interact with a variety of other organisms including blood parasites. *Eperythrozoon coccoides* is the best known of the rickettsiae of mice. In general, latent infections with eperythrozoa are potentiated by malaria infections and such infections result in a reduction of the severity of the malaria infections. We have found that mice infected with *E. coccoides* recover from infections with the virulent and usually fatal *P. v. vinckei* (Cox, 1966). Peters (1965) found that *E. coccoides* reduced the severity of *P. b. berghei* infections and he obtained similar results with *B. rodhaini*. Ott, Astin & Stauber (1967) and Voller & Bidwell (1968) showed that their strains of *P. v. chabaudi* were fatal in eperythrozoon-free mice but relatively benign in animals infected with this organism. Young (1970) has shown that infections with *B. microti* are lower in *Eperythrozoon*-infected mice than in controls. Experiments with *Haemobartonella muris* have been less extensive. In a series of experiments with mice infected with *Haemobartonella muris*, one, two, four or six weeks before infection with various blood parasites we found that infections with *P. v. vinckei*, *B. rodhaini* and *B. microti* were less intense in mice that had been infected with the rickettsia, one, two or four weeks before infection than they were in control mice or mice infected six weeks earlier. These results suggest that *Haemobartonella* has a moderating effect on the infections caused by these parasites. Young (1970) has made similar observations. On the other hand Hsu & Geiman (1952), working with *P. b. berghei* in rats, found that *Haemobartonella* enhanced the malaria infection. Kretschmar (1963) made similar observations using mice.

It is clear from a considerable amount of experimental work that *Eperythrozoon* and *Haemobartonella* exert a variety of different kinds of effects that might influence infections with blood parasites (see the review by Baker, Cassell & Lindsey, 1971). The most likely possibilities are the following:

1. The possession of antigens in common with the blood parasites resulting in cross-immunity.
2. Competition for metabolites (suggested by Peters, 1965).
3. Increased reticuloendothelial activity as a result of the rickettsial infection acting non-specifically to remove parasites from the blood.
4. An alteration in the ratio between immature and mature red blood cells making it impossible for parasites to find sufficient suitable cells for their development.

The possession of common antigens can be dismissed for lack of evidence. Finerty, Evans & Hyde (1973), in a detailed study of the immune responses in mice infected with *P. b. berghei* and *E. coccoides*, found no cross-reacting antigens using the fluorescent antibody technique. This has also been our experience with all the parasites used in our studies.

Peters (1965) suggested that competition for essential metabolites might result in lowered infections of blood parasites in mice infected with rickettsiae. This suggestion has never really been tested and our present state of knowledge of the metabolism of blood parasites is insufficient to provide proof or otherwise of this suggestion.

Eperythrozoon coccoides causes a marked increase of over 150% in reticuloendothelial activity in infected mice (Gledhill, Bilbey & Niven, 1965). Theoretically, this increased activity could be responsible for the removal of merozoites, or infected cells, from the blood. However, various attempts to simulate the effects of *Eperythrozoon* using substances that enhance reticuloendothelial activity have been unsuccessful.

P. b. berghei and *P. b. yoelii* preferentially invade immature red blood cells whereas *P. v. chabaudi*, *P. v. vinckei* and *Babesia* spp. invade mature ones. It has been shown on a number of occasions that the outcome of these infections depends more on the availability of suitable cells than on any other factor (Cox, 1974). Substances that increase the number of reticulocytes favour *P. b. berghei*-type parasites and inhibit *P. v. vinckei*-type parasites and *Babesia*. Phenylhydrazine, for example, causes enhancement of infections with *P. b. berghei* (Singer, 1954; Ott, 1968) but restricts infections with *P. v. chabaudi* (Ott, 1968) and *B. rodhaini* (McHardy, 1973). Anaemia induced in various ways also causes an increase in the number of reticulocytes. Thus anti-erythrocyte serum enhances *P. b. berghei* infections (Schwink, 1960) but reduces the severity of *B. rodhaini* (McHardy, 1972).

Both *Eperythrozoon* and *Haemobartonella* cause severe anaemias accompanied by reticulocytosis. The anaemia only persists for a short time but reticulocytosis is marked for 6–7 days after infection. The factors affecting the balance between mature and immature cells are complex and have been discussed in some detail by Ott *et al.* (1967). In general, it is possible to correlate infections with various blood parasites following rickettsial infections with the availability of suitable blood cells. Thus *Haemobartonella* has its most inhibitory effect on *B. rodhaini* infections if animals are infected with both agents at about the same time. The *Babesia* finds itself with insufficient mature red cells when they are required and is unable to fulfil its potential growth before the immune response becomes operative (Fig. 4). When mice are infected with *B. rodhaini* after the number of reticulocytes in the blood has fallen to normal the rickettsiae have little or no effect on the *Babesia* infection. Observations on other parasites could be explained in a similar way and it should be possible to interpret discrepancies in the results obtained by different observers in terms of the timing of the infections with the interacting organisms.

It would be naive to claim that the varied interactions between rickettsiae

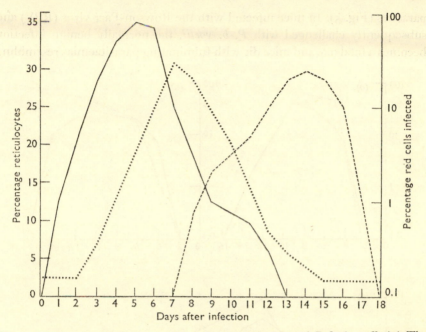

Fig. 4. The interaction between *Haemobartonella muris* and *Babesia rodhaini*. The *Haemobartonella* infection causes anaemia and reticulocytosis. If the mouse is infected with a large inoculum of *B. rodhaini* five days after the *Haemobartonella*, there are insufficient mature erythrocytes available to allow the *Babesia* infection to proceed to its normal fatal conclusion at which 80–90 % of the erythrocytes are usually infected. Solid line, *Haemobartonella* infection; dashed line, *B. rodhaini* infection; dotted line, reticulocytosis.

and blood parasites are entirely due to the availability of suitable cells. The final outcome of the infection must be governed by the host's immune response. This often cannot eliminate the infection unless the numbers of parasites are kept down in some other way; in this case the availability of suitable cells keeps the infection within the limits that can be coped with by the immune response.

Viruses

Virus infections of mammals often go undetected unless they cause overt disease yet several viruses of minimal pathogenicity markedly influence infections with blood parasites. The intensity of *P. b. berghei* infections is reduced in mice exposed to West Nile Virus (Yoeli, Becker & Bernkopf, 1955) and Newcastle Disease Virus (Jahiel, Nussenzweig, Vanderberg & Vilček, 1968; Schultz, Huang & Gordon, 1968) but the effect on the malaria infections is very slight. On the other hand, exposure to certain mouse leukaemia viruses causes enhancement of infections caused by blood parasites (Fig. 5). In mice infected with the Rowson-Parr virus (RPV) and subsequently challenged with *P. b. yoelii*, the normally benign infection becomes a fatal one and mice die with fulminating parasitaemias resembling

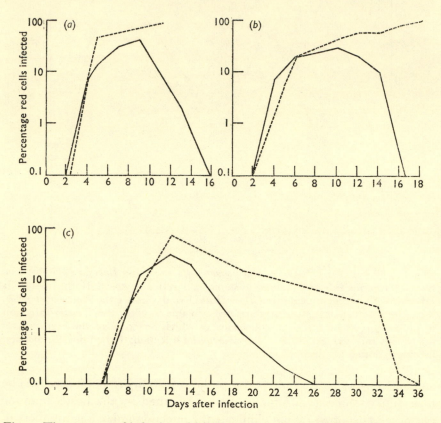

Fig. 5. The patterns of infection of (*a*) *Plasmodium vinckei chabaudi*, (*b*) *P. berghei yoelii* and (*c*) *Babesia microti* in mice simultaneously infected with the Rowson-Parr Virus. Each mouse received approximately 1×10^5 parasites. (Compare these graphs with those shown in Figs. 2 and 3. The slopes of the graphs are similar but the parasitaemias reach their peaks earlier in these experiments because of the large number of parasites given initially.) Solid line, blood infection alone; dashed line, blood infection with virus.

those due to *P. b. berghei* (Salaman & Wedderburn, 1969; Cox, Wedderburn & Salaman, 1974). RPV enhances infections with *P. v. chabaudi*, turning them into fatal ones like those caused by *P. v. vinckei* (Cox *et al.*, 1974). RPV also enhances infections with *B. microti* (Cox & Wedderburn, 1972). In this case the infections are never fatal but are much more intense than those in control animals. Parasitaemias persist for six months or more whereas, in controls, parasites are seldom seen in the blood after a month.

RPV exerts its effect if given to mice between seven days before and seven days after infection with the blood parasite. The enhanced infections in the virus-infected mice can be correlated with lowered antibody levels as detected by the fluorescent antibody technique (Cox & Wedderburn, 1972; Cox et al., 1974) and reduced numbers of antibody-forming cells are detected by the plaque-forming technique (Cox et al., 1974). It is likely, then, that the enhanced infections of blood parasites seen in mice infected with RPV is due to immunodepression caused by the virus. The fact that infections with both P. b. yoelii and P. v. chabaudi are similarly affected by RPV eliminates the possibility that changes in the red cell balance are important in determining the outcome of the infection as is the case with rickettsiae.

RPV is not the only virus to exert an immunodepressive effect resulting in enhanced infections with blood parasites. The urethane leukaemia virus (ULV) also exerts such an effect on P. b. yoelii (Salaman, Wedderburn & Bruce-Chwatt, 1969) and B. microti (Cox & Wedderburn, 1972). As a number of viruses are immunodepressive (Salaman, 1969) it is likely that several, as yet unidentified, are capable of enhancing infections with a variety of blood parasites.

Other infectious agents

Depression of the immune response to other antigens is a characteristic of a number of parasitic infections including malaria (Salaman et al., 1969), trypanosomiasis (Goodwin, Green, Guy & Voller, 1972) and toxoplasmosis (Strickland, Voller, Pettit & Fleck, 1972). The effects of this kind of immunodepression on blood infections have been investigated by Strickland et al. (1972). They found that mice infected with the ubiquitous sporozoan Toxoplasma gondii suffered from more intense infections with P. b. yoelii than did controls. As more investigations are carried out it is likely that more interactions of this kind will be discovered.

Previous exposure of the mother to the same species of parasite

It is a well-established fact that a mother immune to an infection often passes protective antibodies to her offspring (Brambell, 1970). This passive transfer of immunity may occur before birth, after birth with the milk or in both of these ways. The blood parasites are no exception to this general rule. Most work has been done with P. b. berghei in rats. Adult rats recover from infections with this parasite and the usual experiment has been to infect a female rat and to challenge her offspring with the same parasite some time after birth. This kind of experiment has shown that the offspring are immune to challenge, that the protection is relatively short-lived (about four weeks) and that antibodies are involved. The early work has been reviewed by Bruce-Chwatt (1963) and more detailed studies have subsequently been carried out by Zuckerman, Spira & Shor (1969), Desowitz (1971) and Smalley (1973). The passive transfer of immunity to *P. b. berghei* in mice has been demonstrated by Bruce-Chwatt & Gibson (1955), to *P. v. vinckei* in mice by Adler & Foner (1965) and to *P. v. chabaudi*, *P. b. yoelii* and *B. microti* in mice in our laboratory. It can be accepted, then, that the transfer of passive immunity to the offspring of animals infected with blood parasites is a commonplace phenomenon. Antibodies received from the mother affect subsequent infections in various ways. If too little antibody is present the protection is minimal. If too much is present, the young animal either does not have an opportunity to acquire an immunity, because all the parasites with which it subsequently becomes infected are immediately destroyed, or its own immune response is inhibited by passively acquired antibodies (see for example, Uhr & Möller, 1968). Passively acquired antibody, therefore, influences blood infections but the actual effect depends on a whole range of different factors and the actual outcome of the infection under such conditions may be unpredictable.

Antibody is not the only thing that passes from a mother to her offspring and the possibility exists that antigens may be transferred to young animals *in utero*. Gill, Kunz & Bernard (1971) were able to detect labelled antigen (a synthetic polypeptide aggregated with methylated bovine serum albumin) in offspring born to immunized rats and there is considerable evidence to suggest that this is not an uncommon phenomenon. The immunity shown by young animals born from immune mothers and infected with blood parasites is sometimes too good and lasts for too long to be wholly explained in terms of passively transferred antibody. This has led several workers to ask whether antigens do pass across to the offspring in blood infections. Bruce-Chwatt (1956) suggested that some component that had a stimulating effect on the immune system passed to children from mothers infected with malaria. This suggestion was investigated by McGregor (1971) and the same theme has been taken up by Desowitz (1971) and Smalley (1973) using *P. b. berghei* in rats. It is not at all clear from the evidence available whether or not antigens pass to the offspring during infections with blood parasites or, if they do, that they have any role in priming the immunological system in preparation for subsequent challenge. It is unlikely that this question will be answered until the antigens responsible for inducing the protective antibody response to malaria have been isolated and identified.

The passage of antigen from mothers to their offspring has been shown to extend to a second generation of offspring by Gill *et al.* (1971). They demonstrated that female rats born to immunized mothers passed the antigen, methylated bovine serum albumin, on to their offspring and they on to theirs but that there was no further transfer after the third generation. The passage of parasite antigens in this way has been examined by Smalley (1973) using *P. b. berghei* in rats. He found, however, that in rats of the second generation the infections were longer and more intense than in control animals. Smalley suggests that the immune response in these second generation animals is in some way different from that in normal rats but from the evidence available the passage of protective immunity to second and third generation animals is unlikely to be of any importance under natural conditions.

Previous exposure of the mother to a different species of parasites with common antigens

The investigation of the passage of antibodies against blood parasites from a mother to her offspring is difficult enough using homologous parasites but the use of heterologous parasites produces even more difficulties. In theory, the fact that parasites of different species may have cross-reacting antigens which play a role in protection means that the protective antibodies could pass from a mother to her offspring. This possibility has been investigated using mice born of mothers immune to *B. microti* and challenging them with *P. v. vinckei*. Fourteen out of 36 young mice born to mothers immune to *B. microti* died from the malaria infection while all 24 matched controls died. This kind of experiment needs to be repeated but it does indicate that heterologous immunity passes from a mother to young animals in the same way that homologous immunity does. This also implies that antibodies are involved in this kind of response.

Previous exposure of the mother to a different species of parasites without antigens in common

There is no evidence to suggest that the exposure of a mother to parasites without cross-reacting antigens has any ameliorating effect on subsequent blood infections in her offspring. However, a number of infectious diseases can be passed on to the unborn animal and some of these may exert direct effects on infections with blood parasites. The urethane leukaemia virus, for example, can be passed vertically, that is from mother to offspring, and enhances infections with blood parasites by depressing the immune response (see p. 443). Similarly, *Toxoplasma gondii* can be passed to unborn animals and this also enhances blood infections (see p. 443). There are probably a number of other pathogens that produce similar effects which are likely to be the same as if the animal in question was infected after birth.

DISCUSSION

Intensive studies on blood parasites have been carried out using laboratory animals and considerable attention has been paid to factors such as stress and nutrition that might affect the course of any particular infection (see for example, Kretschmar, 1965). Little attention, however, has been paid to the fact that, in the wild, the situation is basically different from that in the laboratory because animals are usually infected with a number of agents which may interact with each other in various ways. In the largest survey of its kind, Young (1970) examined a number of British small mammals for blood parasites and found that of the 1215 voles and field mice that he studied 403 had patent infections and 118 (29 %) were infected with more than one parasite. In fact, patent infections detected in blood films probably represent only part of the true incidence of such parasites and, of the apparently uninfected animals, about one third harbour sub-patent infections. This means that over half the animals in the sample had blood infections and that about 16 % had multiple infections. This ignores viruses and other infections which do not appear in the blood and also ignores past infections represented only by residual immune responses. Young's survey is typical of many others so it can probably be accepted that in most natural situations a number of infections co-exist in particular animals.

The interactions of these infections cannot be studied except by isolating them and examining them in laboratory animals free from extraneous infections. Thus the controlled experiment in a laboratory animal does have an essential role in any understanding of field situations and this approach has formed the basis of the experiments described in this paper. From the experiments performed it is clear that the outcome of any blood infection is influenced by prior or concurrent infections with other infectious agents and also by the mother's experience of other agents. The actual observable effects of these interactions are becoming clear and, in many cases, so is the biological basis of the effect.

Prior exposure to the same species of parasite, or to a parasite with cross-reacting antigens, results in an amelioration of the infection. This is due to the presence of antibodies and if these come from the mother the young animal is protected without experiencing any infection itself. Homologous immunity, in which an animal is protected against a second or subsequent infection with the same species of parasite, can be regarded as a normal situation and the passage of antibodies from the mother to offspring is also normal. Heterologous immunity, in which recovery from one species of parasite is followed by protection against another, is relatively uncommon and it is surprising that so many examples exist among the blood parasites of small mammals. Heterologous immunity appears to differ from homologous immunity in that it is not always possible to correlate cross-reacting antigens with protection. It may be that this is because the antigens identified using techniques such as the fluorescent antibody technique cannot differentiate between antigens that are important in protection and those that are not. It is likely that, when the antigens associated with protection have been identified, correlation between antigens and protection will become the rule and that there will be no real difference between homologous and heterologous immunity. The passage of heterologous immunity from mother to offspring has been demonstrated but not as clearly as for homologous immunity. This situation is probably due to the lack of suitable experiments and does not indicate any real difference between homologous and heterologous immunity.

Infections with blood parasites are also often ameliorated by prior or simultaneous infections with rickettsiae. The most likely explanation of this effect is that the anaemia, and subsequent reticulocytosis, associated with the rickettsial infection interferes with the balance between mature and immature red cells and deprives the parasites of their preferred cells. This explanation is a better one than competition for metabolites or non-specific stimulation of phagocytic activity.

Mouse leukaemia viruses enhance infections with blood infections and this is apparently due to the immunodepressive effect that these viruses exert. The enhancement recorded is parallel with that seen when animals infected with the same parasites are treated with immunodepressive drugs. This situation has been well studied and it is probably a general rule that infectious agents that are immunodepressive enhance infections with blood parasites.

The outcome of any infection in an animal in the field results from the intrinsic factors, such as the potential of the parasite to develop in a particular host and the ability of the host to support the parasite's growth, and extrinsic factors. The intrinsic factors are usually regarded as aspects of host specificity and therefore determine a base line for susceptibility to infection. The extrinsic factors include nutrition and stress and also the interactions with other organisms described in this paper. Some of these interactions lessen the effects of a particular infection while others enhance them. The actual effect of these interactions on the epidemiology of any particular infection is difficult to predict. An agent such as a virus that enhances an infection may increase the chances of its spread or it may decrease them by killing an animal before it has a chance to pass the infection on to another one. Similarly, an agent that depresses an infection may increase the chances of the transmission of a virulent and potentially fatal infection and decrease the chances of the transmission of a mild one.

The experiments described in this paper point to a series of interactions between infectious agents which may be as complex as those that occur in any ecological situation. These interactions are not confined to blood parasites or to small mammals and an appreciation of the interactions between parasites and other infectious agents in man, domesticated and wild animals may be a pre-requisite for any future advance in our understanding of the branch of symbiosis known as parasitism.

It is notoriously difficult to define symbioses even when only two organisms are involved. Starr (p. 1, this volume) has chosen to ignore three organism partnerships but these raise major problems when attempting any kind of general definition of symbiosis or parasitism. A parasite that depends on its host but evokes an immune response that ultimately protects that host from another unrelated parasite can be said to exert a beneficial effect on the host. Its relationship with the other parasite is, however, an antagonistic one. Interactions involving only two parasites in a host are, presumably, rare in nature as many parasites themselves harbour other infectious agents. These may in turn affect both the parasite in which they live and later other parasites and eventually the host. It may well be that situations such as these will make it impossible to produce any overall definition of parasitism.

ACKNOWLEDGEMENTS

I would like to express my gratitude to my postgraduate students, Dr A. S. Young, Dr N. McHardy and Dr M. E. Smalley, for their help in the work that has been described in this paper. I would like also to thank the World Health Organization for financial support.

REFERENCES

ADLER, S. & FONER, A. (1965). Transfer of antibodies to *P. vinckei* through milk of immune mice. *Israel J. Med. Sci.*, **1**, 988–993.

BAKER, H. J., CASSELL, G. H. & LINDSEY, J. R. (1971). Research complications due to *Haemobartonella* and *Eperythrozoon* infections in experimental animals. *Am. J. Path.*, **64**, 625–652.

BRAMBELL, F. W. R. (1970). *The Transmission of Passive Immunity from Mother to Young*. Amsterdam and London: North-Holland.

BRUCE-CHWATT, L. J. (1956). Biometric study of spleen and liver weights in Africans and Europeans, with special reference to endemic malaria. *Bull. Wld Hlth Org.*, **15**, 513–548.

(1963). Congenital transmission of immunity in malaria. In *Immunity to Protozoa*, 89–108 (eds. P. C. C. Garnham, A. E. Pierce and I. Roitt). Oxford: Blackwell Scientific Publications.

BRUCE-CHWATT, L. J. & GIBSON, F. D. (1955). Transplacental passage of *Plasmodium berghei* and passive transfer of immunity in rats and mice. *Trans. R. Soc. trop. Med. Hyg.*, **50**, 47–53.

COX, F. E. G. (1966). Acquired immunity to *Plasmodium vinckei* in mice. *Parasitology*, **56**, 719–732.

(1970a). Protective immunity between malaria parasites and piroplasms in mice. *Bull. Wld Hlth Org.*, **43**, 325–336.

(1970b). The specificity of Immunoglobulin G and Immunoglobulin M in the fluorescent-antibody test for malaria parasites in mice. *Bull. Wld Hlth Org.*, **43**, 341–344.

(1970c). Acquired immunity to *Plasmodium chabaudi* in Swiss TO mice. *Ann. trop. Med. Parasit.*, **64**, 309–314.

(1972a). Protective heterologous immunity between *Plasmodium atheruri* and other *Plasmodium* spp. and *Babesia* spp. in mice. *Parasitology*, **65**, 379–387.

(1972b). Immunity to malaria and piroplasmosis in mice following low level infections with *Athemosoma garnhami* (Piroplasmea: Dactylosomidae). *Parasitology*, **65**, 389–398.

(1974). A comparative account of the effects of betamethasone on mice infected with *Plasmodium vinckei chabaudi* and *Plasmodium berghei yoelii*. *Parasitology*, **68**, 19–26.

COX, F. E. G., CRANDALL, C. A. & TURNER, S. A. (1969). Antibody levels detected by the fluorescent antibody technique in mice infected with *Plasmodium vinckei* and *P. chabaudi*. *Bull. Wld Hlth Org.*, **41**, 251–260.

COX, F. E. G. & TURNER, S. A. (1970). Antigenic relationships between the malaria parasites and piroplasms of mice as determined by the fluorescent-antibody technique. *Bull. Wld Hlth Org.*, **43**, 337–340.

COX, F. E. G. & VOLLER, A. (1966). Cross-immunity between the malaria parasites of rodents. *Ann. trop. Med. Parasit.*, **60**, 297–303.

COX, F. E. G. & WEDDERBURN, N. (1972). Enhancement and prolongation of *Babesia microti* infections in mice infected with oncogenic viruses. *J. gen. Microbiol.*, **72**, 79–85.

COX, F. E. G., WEDDERBURN, N. & SALAMAN, M. H. (1974). The effect of Rowson-Parr Virus on the severity of malaria in mice. *J. gen. Microbiol.*, **85**, 358–364

COX, F. E. G. & YOUNG, A. S. (1969). Acquired immunity to *Babesia microti* and *Babesia rodhaini* in mice. *Parasitology*, **59**, 257–268.

DESOWITZ, R. S. (1971). *Plasmodium berghei*: enhanced protective immunity after vaccination of white rats born of immune mothers. *Science, Wash.*, **172**, 1151–1152.

FINERTY, J. F., EVANS, C. B. & HYDE, C. L. (1973). *Plasmodium berghei* and *Eperythrozoon coccoides*: antibody and immunoglobulin synthesis in germ-free and conventional mice simultaneously infected. *Expl Parasit.*, **34**, 76–84.

GILL, T. J., KUNZ, H. W. & BERNARD, C. F. (1971). Maternal-foetal interaction and immunological memory. *Science, Wash.*, **172**, 1346–1348.

GLEDHILL, A. W., BILBEY, D. L. J. & NIVEN, J. S. F. (1965). Effect of certain murine pathogens on phagocytic activity. *Br. J. exp. Path.*, **46**, 433–442.

GOODWIN, L. G., GREEN, D. G., GUY, M. W. & VOLLER, A. (1972). Immunosuppression during trypanosomiasis. *Br. J. exp. Path.*, **53**, 40–43.

Hsu, D. V. M. & Geiman, Q. M. (1952). Synergistic effect of *Haemobartonella muris* on *Plasmodium berghei* in white rats. *Am. J. trop. Med. Hyg.*, **1**, 747–760.

Jahiel, R. I., Nussenzweig, R. S., Vanderberg, J. & Vilček, J. (1968). Anti-malarial effect of interferon inducers at different stages of development of *Plasmodium berghei* in the mouse. *Nature, Lond.*, **220**, 710–711.

Kreier, J. P. & Ristic, M. (1968). Haemobartonellosis, eperythrozoonosis, grahamellosis and ehrlichiosis. In *Infectious Blood Diseases of Man and Animals*, vol. **2**, 387–472 (eds. D. Weinman and M. Ristic). New York and London: Academic Press.

Kretschmar, W. (1962). Resistenz und Immunität bei mit *Plasmodium berghei* infizierten Mäusen. *Z. Tropenmed. Parasit.*, **13**, 159–175.

— (1963). Die Abhängigkeit des Verlaufs der Nagetiermalaria (*Plasmodium berghei*) in der Maus von exogenen Faktoren und der Wahl des Mäusestammes. I. Interferierende Bartonellosen. *Z. Versuchstierk.*, **3**, 151–166.

— (1965). The effects of stress and diet on resistance to *Plasmodium berghei* and malarial immunity in the mouse. *Annls Soc. belge Méd. trop.*, **45**, 325–344.

McGregor, I. A. (1971). Immunity to plasmodial infections; consideration of factors relevant to malaria in man. *Int. Rev. trop. Med.*, **4**, 1–52.

McHardy, N. (1972). Protective effect of haemolytic serum on mice infected with *Babesia rodhaini*. *Ann. trop. Med. Parasit.*, **66**, 1–5.

— (1973). Effects of stimulating erythropoiesis in mice infected with *Babesia rodhaini*. *Ann. trop. Med. Parasit.*, **67**, 301–306.

Ott, K. J. (1968). Influence of reticulocytosis on the course of infection of *Plasmodium chabaudi* and *P. berghei*. *J. Protozool.*, **15**, 365–369.

Ott, K. J., Astin, J. K. & Stauber, L. A. (1967). *Eperythrozoon coccoides* and rodent malaria: *Plasmodium chabaudi* and *Plasmodium berghei*. *Expl Parasit.*, **21**, 68–77.

Peters, W. (1965). Competitive relationship between *Eperythrozoon coccoides* and *Plasmodium berghei* in the mouse. *Expl Parasit.*, **16**, 158–166.

Salaman, M. H. (1969). Immunodepression by viruses. *Antibiotic Chemother.*, **15**, 393–406.

Salaman, M. H. & Wedderburn, N. (1969). The immunodepressive effect of a virus of minimal pathogenicity derived from Friend virus infected mice and its interaction with other agents. *Proceedings of the International Conference on Immunity and Tolerance in Oncogenesis*, 613–621. University of Perugia: Perugia.

Salaman, M. H., Wedderburn, N. & Bruce-Chwatt, L. J. (1969). The immuno-depressive effect of a murine plasmodium and its interaction with murine oncogenic viruses. *J. gen. Microbiol.*, **59**, 383–391.

Schultz, W. W., Huang, K. Y. & Gordon, F. B. (1968). Role of interferon in experimental mouse malaria. *Nature, Lond.*, **220**, 709–710.

Schwink, T. M. (1960). The effect of anti-erythrocytic antibodies upon *Plasmodium berghei* infections in white mice. *Am. J. trop. Med. Hyg.*, **9**, 293–296.

Singer, I. (1954). The effect of splenectomy or phenylhydrazine on infections with *Plasmodium berghei* in white mice. *J. infect. Dis.*, **94**, 159–163.

Smalley, M. (1973). Factors affecting the course of intraerythrocytic protozoan infections in laboratory rats. Ph.D. thesis, University of London.

Strickland, G. T., Voller, A., Pettit, L. E. & Fleck, D. G. (1972). Immuno-depression associated with concomitant toxoplasma and malarial infections in mice. *J. infect. Dis.*, **126**, 54–60.

Uhr, J. W. & Möller, G. (1968). Regulatory effect of antibody on the immune response. *Adv. Immunol.*, **8**, 81–127.

VOLLER, A. & BIDWELL, D. E. (1968). The effect of *Eperythrozoon coccoides* infection in mice on superimposed *Plasmodium chabaudi* and Semliki Forest Virus. *Ann. trop. Med. Parasit.*, **62**, 342–348.

YOELI, M., BECKER, Y. & BERNKOPF, H. (1955). The effect of West Nile Virus on experimental malaria infections (*Plasmodium berghei*) in mice. *Harefuah, Jerusalem*, **49**, 116–119.

YOUNG, A. S. (1970). Investigations on the epidemiology of blood parasites of small mammals with special reference to piroplasms. Ph.D. thesis, University of London.

ZUCKERMAN, A., SPIRA, D. T. & SHOR, A. (1969). Partial protection and precipitin passively transferred to their litters by mother rats infected or superinfected with *Plasmodium berghei. Milit. Med.*, **134**, 1249–1257.

EVASION OF THE IMMUNE RESPONSE BY PARASITES

By R. J. TERRY* and S. R. SMITHERS†

* Brunel University, Uxbridge, Middlesex

† National Institute of Medical Research, London NW7

THE VERTEBRATE IMMUNE RESPONSE

The vertebrate immune response has, as its cellular basis, the lymphocyte and, as its molecular basis, the immunoglobulin molecule. Lymphocytes and immunoglobulins have now been demonstrated in all living vertebrates but not conclusively in any invertebrate species.

Over the past few years it has been shown that lymphocytes are not a homogeneous population. Many of the stem cells arising in the foetal liver, or later in the bone marrow, pass to the thymus gland. Here they proliferate and acquire immunocompetence, perhaps under the influence of a thymus hormone. These 'thymus influenced' lymphocytes then pass to the peripheral lymphoid organs, spleen, lymph nodes, and gut-associated lymphoid tissues, and are now referred to as T cells. Other stem cells do not migrate to the thymus. In birds, they acquire competence by migrating to the Bursa of Fabricius. No bursal analogue has been found in mammals and it may be that competence is acquired in the bone marrow. In any event these lymphocytes also eventually arrive in the peripheral lymphoid organs, forming part of the circulating pool of effector cells, and are known as B cells. Resting T and B cells are morphologically indistinguishable, at least at the light microscope level, but may be identified by their possession of surface markers and other physiological properties (see Raff, 1973, for an excellent short review).

The clonal selection hypothesis of Burnett (1959) holds that, during ontogeny, clones of lymphocytes arise which are pre-committed to respond to one (or a small number) of antigenic determinants. This hypothesis has the support of most immunologists who further believe that this commitment is expressed by the lymphocytes having antigen-specific receptors at their surface. Any introduced antigen then selects out those lymphocytes having specific complementary surface receptors and induces the proliferation of these cells, also committing them to effector pathways. It is generally agreed that antigen receptors on B cells are specific immunoglobulin (antibody) molecules. The nature of T cell receptors is a matter of continuing controversy.

T and B lymphocytes have different functions in the immune response. When B cells are stimulated by antigen they divide and many of them differentiate into cells with abundant rough endoplasmic reticulum. These plasma cells secrete immunoglobulins which have the same antigen specificity as the original receptor immunoglobulin, although the molecules may not be identical in all respects. The immunoglobulins (antibodies) circulate in the blood and tissue fluids and, acting in concert with the non-specific components of complement, and cells such as macrophages, play a major role in protective immunity.

When T cells are stimulated by antigen they also transform and pro-liferate but do not secrete large amounts of antibody. Instead they secrete non-specific factors (lymphokines) which activate other cells, particularly macrophages, to take part in the series of reactions known as cell-mediated immunity. In this way, target cells bearing offending antigens may be destroyed. Cell-mediated, as opposed to circulating-antibody-mediated reactions, are of paramount importance in allograft rejection. They are also known to play a major role in tumour rejection and in the immune response to many infectious agents.

It is now known that although T cells do not secrete large amounts of antibody, their co-operation is often required in order to enable B cells to do so. The exact mechanisms of co-operation are still to be determined but interference with T–B cell interactions results in a markedly reduced antibody response to many antigens.

In summary, the vertebrate immune response is a complex multi-component system of cells and soluble factors, capable of the rapid recognition and destruction of foreign invading organisms. Its importance in preserving individual integrity is perhaps best appreciated by considering those unfortunate individuals with genetic or acquired impairment of the response. Whether the defect affects mainly T cells or mainly B cells, these patients suffer from a series of life-threatening infections with agents which are easily overcome by those with intact responses (Good, 1972). Never-theless, there are a number of parasites – viruses, bacteria, protozoa and helminths – which are able to produce chronic and long-lasting infections in individuals which seem to have no basic defect in immunity. Clearly, these agents must have evolved ways of evading host immunity. In many instances we know nothing of these evasion mechanisms, but we would like to discuss two examples where we have at least some superficial under-standing. These are the salivarian trypanosomes of Africa, best exemplified by *Trypanosoma brucei brucei* and the human blood fluke *Schistosoma mansoni*.

THE AFRICAN TRYPANOSOMES

At first sight the salivarian trypanosomes would seem to be particularly vulnerable to acquired immunity. They are present in the bloodstream; they stimulate, at least initially, a high level of antibody production; and both *in vitro* and *in vivo* they are susceptible to antibody-mediated complement-dependent lysis. Yet, in man and his domestic animals, they produce either fulminating or insidiously chronic disease which in most untreated infections leads to the death of the host. Trypanosomes must therefore possess mechanisms which enable them to evade host immunity. One such mechanism, 'antigenic variation', has been known for many years. Two further possible mechanisms, the occupation of immunologically privileged sites and the immunodepressive properties of trypanosome infections, have been studied more recently. These mechanisms will be considered in turn.

Antigenic variation

It was recognized early on that infection of domestic and experimental animals with *T. b. brucei* and its relatives follows a characteristic course of successive waves of parasitaemia (Fig. 1). Initial infection is followed by increasing parasitaemia as the trypanosomes multiply by binary fission. This parasitaemia terminates in a crisis with a marked reduction in trypanosome numbers, coinciding with the appearance of trypanolytic

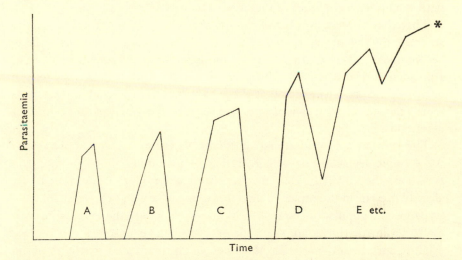

Fig. 1. Diagrammatic representation of the course of blood parasitaemia in *T. b. brucei* infections. Each succeeding wave of parasitaemia, A, B, C, is due to the development of an antigenically distinct population and its subsequent destruction by specific antibody. The eventual failure to control the infection may be brought about by the immunodepressive properties of the infection.

antibodies in the blood, but within a day or so parasitaemia again increases. This new population is unaffected by the existing antibody and is of a new antigenic constitution. This pattern is repeated with new antigenic variants appearing every few days and with succeeding crises and relapses until the animal dies. The number of possible variants is unknown but at least 22 variants arising from a single trypanosome have been demonstrated (Ritz, 1916).

Classical studies on antigenic variation are those of Lourie & O'Connor (1937) and Gray (1965a, b). A given trypanosome strain produces similar variants in different hosts but not necessarily in the same sequence, and there is no convincing evidence that a variant ever appears twice during an infection. Perhaps important in our understanding of the biology of antigenic variation is the tendency for trypanosomes to revert to a 'parental' or 'basic' strain. This may occur if the variants are rapidly passaged through mice, avoiding their exposure to specific antibody, or on cyclical passage through tsetse flies.

The variant antigens are situated in an outer surface coat, 12–15 nm thick, which is characteristic of bloodstream forms but not stages within the tsetse (Vickerman, 1969; Vickerman & Luckins, 1969; Steiger, 1973). Chemically, there is evidence that the antigens may comprise a polydisperse group of glycoproteins (Allsop, Njogu & Humphryes, 1971; Njogu & Humphryes, 1972); their nature is being explored more fully with the newer techniques of membrane biology (Cross, 1973).

There have been a number of reports that these variant antigens may be secreted by the trypanosomes into the blood stream, either as soluble 'exoantigens' (Weitz, 1970) or as filopodia or plasmonemes (Wright, Lumsden & Hales, 1970). Allsop et al. (1971) dispute that this release occurs in vivo and consider that these exoantigens represent surface proteins released when trypanosomes are kept in vitro in less than optimal conditions.

The mechanisms underlying antigenic variation are not understood. As a single trypanosome can give rise to infections showing antigenic variation, the diversity cannot arise from a number of variants being already present in the infective inoculum. Either, during the growth of the trypanosome population, there is random mutation and viable mutants with novel surface antigens are selected; or certain trypanosomes can replace existing surface antigens with new ones when threatened by the immune response. Cantrell (1958) and Watkins (1964) have calculated that multiplication rates and imputed mutation rates are great enough to allow of variation through mutation and selection; but other workers, particularly Gray (1967, 1969), support an adaptive process. His main argument is concerned with the relatively fixed pattern of variation and the tendency of strains to persist in or revert to a basic variant.

There is less argument about the force driving antigenic variation. New variants follow the appearance of antibodies to existing variants and there is evidence that variation does not occur in hosts incapable of immune responses (Goedbloed, 1971; Luckins, 1971). It is easy to envisage a role for antibody in the selection of new antigenic mutants but more difficult to see it acting as a signal in an adaptive process. It must be presumed that some trypanosomes are able to detect the presence of antibody without being destroyed as the majority of bloodstream trypanosomes are destroyed. Consideration of the other possible evasion mechanisms might suggest how this might be achieved.

Privileged sites

The 'adaptive' hypothesis of antigenic variation requires the presence of a subpopulation of trypanosomes which switch, on some signal associated with antibody, to the synthesis of new variant antigens before they are destroyed by antibody reacting with their original variant antigens. The survival of such a subpopulation might occur in specially privileged sites where they are exempt from the full effect of the immune response. There is no conclusive evidence that salivarian trypanosomes do occupy such sites in their mammalian hosts but certain findings may be significant.

Viens, Targett, Wilson & Edwards (1972) have demonstrated the survival of the stercorarian *T. (Herpetosoma) musculi* in the immune mouse host. These trypanosomes may be observed in the vasa recta of the kidneys of mice with no detectable parasitaemia and which are completely refractory to reinfection. The trypanosomes from the vasa recta are capable of producing infections on inoculation into non-immune mice and so are fully viable. It is postulated that the vasa recta affords a privileged site, which enables a subpopulation of trypanosomes to evade the antibodies, and a presumed cellular component which characterize immunity to this infection. *T. musculi* has not been observed to undergo antigenic variation and as yet no biological advantage for this survival mechanism has been shown.

On the other hand, occupation of such sites by salivarian trypanosomes showing antigenic variation might explain a great deal. Ormerod & Venkatesan (1971) have suggested that the choroid plexus may act as a privileged site to *T. b. brucei* in the rat and more recently Ssenyonga (1975) has shown that the numbers of *T. b. brucei* in the tissues of mice do not show the marked fluctuations characteristic of the blood populations. One may speculate that even partially protected sites in the tissues may allow some trypanosomes to undergo variation and re-seed the blood with new forms. Such a process would be aided if the immune responses were not operating at full efficiency and, in this context, the generalized immuno-depression associated with trypanosome infections may be significant.

Immunodepression

There are now a number of reports that trypanosome infections may bring about depression of immune responses to unrelated antigens. Thus mice infected with *T. brucei* make poor antibody responses to sheep red blood cells (Goodwin, Green, Guy & Voller, 1972; Longstaffe, Freeman & Hudson, 1973; Murray, Urquhart, Murray & Jennings, 1973). A reduced immune response to the nematode *Nippostrongylus brasiliensis* has been reported by Urquhart, Murray & Jennings (1972) for trypanosome-infected rats; and there seems to be a reduction in the symptoms of experimental allergic neuritis in rabbits infected with *T. brucei* (Allt, Evans, Evans & Targett, 1971).

The mechanism of this immunodepression is obscure, particularly when account must be taken of the generally high levels of immunoglobulins found in trypanosomiasis. Hudson, Freeman, Byner & Terry (1975) find that the 'background' antibody response (in the absence of specific antigen) to a number of unrelated antigens rises in trypanosome-infected animals. They favour the idea that trypanosome infections may somehow commit the clones of B lymphocytes to limited antibody production and to exhaustion before the antigenic stimulus is presented. This would at the same time explain the generally elevated immunoglobulin levels, the immuno-depression to specific antigens and the disruption of lymphoid architecture seen in trypanosome infection (Hudson & Byner, 1973; Murray, Murray, Jennings, Fisher & Urquhart, 1973).

Although not proven, it is likely that this immunodepression will also relate to trypanosome antigens, particularly to the later variant antigens and may result in less and less effective responses. There is also some evidence that immunoglobulin G (IgG) responses may be depressed sooner than the immunoglobulin M (IgM) responses. The larger IgM antibodies are usually the first to be produced and are mainly confined to the bloodstream, whereas the smaller IgG molecules appear later and easily enter the tissues. If the synthesis of IgG antibodies against variant antigens is defective this may help to explain the survival of trypanosomes in the tissues at a time when they are being destroyed in the blood.

Thus, through antigenic variation, and perhaps helped by their occupation of privileged sites and their immunodepressive properties, trypano-somes of the *brucei* group are able to evade any effective immune response mounted by the host. These evasion mechanisms constitute a formidable barrier to vaccination against the disease. All experience suggests that whilst it is relatively easy to vaccinate against one particular variant, this affords no protection against others (Herbert & Lumsden, 1968; Lanham & Taylor, 1972). We must have more knowledge of the mechanisms underlying antigenic variation and immunodepression before there is any prospect of being able to tilt the host–parasite balance in favour of the host. At present, the scale weighs heavily in favour of the parasite.

SCHISTOSOMA MANSONI

It is estimated that over 200 million people are infected with the human blood flukes or schistosomes. The adults live in the mesenteric or pelvic veins, and eggs are passed out with the faeces or urine. On contact with fresh water, the eggs hatch, giving rise to miracidia. The miracidia infect various mollusc intermediate hosts and, following a phase of asexual multiplication, large numbers of infective cercariae are released into the water. These penetrate the skin of man, and the juvenile forms (now known as schistosomula) migrate via the lungs to the liver and from thence into the portal vessels. Here the flukes mature and the female commences egg-laying about five or six weeks after penetration.

Schistosomiasis is a chronic disease and the adult worms may live for many years in the blood (anything but a privileged site!). This would seem to argue that the host is unable to mount an immune response against the parasite but there is now much evidence that this is not the case. For example, Smithers & Terry (1965) showed that following a small initial infection, rhesus monkeys developed a powerful resistance to reinfection with large numbers of cercariae. It was further demonstrated by inter-host transfer that the presence of adult worms provided the major stimulus to this immunity but, paradoxically, the adult worms were not affected by this immunity (Smithers & Terry, 1967). This survival of adult worms in the presence of immunity was termed 'concomitant immunity' (Smithers & Terry, 1969) and there is now some evidence that it may also occur in human infections (McCullough & Bradley, 1973; Bradley & McCullough, 1973). It has the obvious biological advantage that the worms of a primary infection, in conjunction with host immunity, create a barrier against continual reinfection which might otherwise lead to the death of both host and parasite.

Attention was now turned to the mechanism of protection of the adult worm and the possibility of some kind of antigenic disguise was considered. The existence of shared antigens between schistosomes and their hosts had been demonstrated by Capron, Biguet, Rose & Vernes (1965) and Damian (1967). The hypothesis was made that if these 'host antigens' were present at the surface of the adult schistosomes, they might serve to protect the worms against antibodies, and perhaps immune cells, capable of destroying unprotected juveniles. The hypothesis has now been tested in an extensive series of experiments, all of which provide support for its validity. Smithers, Terry & Hockley (1969) transferred worms grown to maturity in mice directly into the portal circulation of monkeys immunized against mouse tissues. Such 'mouse worms' were rapidly destroyed by an antibody-mediated response directed against the surface of the worm; mouse worms survived well on transfer to normal monkeys. This could only mean that the mouse worms possessed antigens shared by mouse tissues at their surface and that these host antigens were the target of the immune response. The host antigens appeared to be species specific as worms grown in monkeys or in Libyan jirds survived normally on transfer into anti-mouse-red-blood-cell monkeys.

Interestingly, the mouse antigens were not permanently expressed in the worms; they could not be detected by the transfer system when mouse worms had been in the circulation of normal monkeys for seven days. Nevertheless, good evidence was obtained that the antigens were not merely gross contaminants of host origin but were more intimately incorporated into the surface (Clegg, Smithers & Terry, 1970). They did not seem to be mouse immunoglobulin, attaching specifically but in-effectively to the worm surface, neither were they the widely-shared heterophile Forsmann antigen. Rather, they seem to be antigenically identical with components of the mouse erythrocyte membrane.

In order to facilitate identification of these host antigens, an in-vitro cultivation system has been devised whereby schistosomes may be grown in human blood. Human blood group antigens have been well studied and specific reagents for their identification are available. Clegg et al. (1971a) used this system to demonstrate that worms grown in human blood shared antigens with human erythrocytes and that among these appeared to be antigens showing ABO specificity. This has now been confirmed (Gold-ring, Clegg, Smithers & Terry, 1975) by immunizing monkeys with purified A blood group substance; these monkeys rejected worms grown in A group blood but not worms grown in B group blood. Further, fluorescent antibody and mixed agglutination reactions have shown that worms grown in human blood acquire antigens showing AB and H specificity but not antigens with other blood group specificities such as Lewis, rhesus (D), Kell & Duffy.

The origin of these host antigens has attracted much interest. Sprent (1959) originally suggested that parasites might, during the course of evolution, have developed antigens resembling those of their host; this might prevent the host from mounting an effective response. Both Capron *et al.* (1965) and Damian (1967) generally agreed with these ideas of antigenic mimicry. The system would be expected to work well where the parasite showed fairly rigid host specificity but would be less effective where the parasites, such as schistosomes, can grow normally in a range of hosts. Antigens adapted to one host species would be less appropriate if expressed in another. In order to overcome this difficulty Capron, Biguet, Vernes & Afchain (1968) suggested that parasites such as schistosomes might have a series of genes coding for host antigens; the appropriate gene would then be turned on by some kind of induction process when a new host was infected.

Smithers *et al.* (1969) took the opposite view believing that the antigens were truly of host origin and were incorporated by the parasite into its surface membrane. Grounds for this were the high degree of specificity of the antigens and their seemingly rapid acquisition and turn-over. More recently, Goldring, Clegg & Kusel (personal communication) have investigated the acquisition of host antigens *in vitro* using radio-labelled precursors. They found no evidence for synthesis by the parasite, but a small, yet significant, incorporation of pre-existing erythrocyte substances into the worms. The host antigens have yet to be conclusively identified chemically, but all the evidence so far points to their being glycolipids. These investigations do not of course deny the existence of other antigens synthesized by the parasite but cross-reacting with antigens of the host. A good example is the schistosome antigen which cross-reacts with mouse α_2 macroglobulin (Damian, Greene & Hubbard, 1973). This antigen is present not only in schistosomes grown in mice but also in worms grown in monkeys, and therefore cannot be acquired from the host.

Undoubtedly then schistosomes have host antigens at their surface and these are probably synthesized by the host and acquired by the parasite. The question remains – do these in fact protect the worm from the immune reponse that destroys juveniles? A major difficulty in conclusively demonstrating that this is so, is that we are still unsure of the exact mechanism of immunity. Clegg *et al.* (1971*b*) obtained preliminary evidence that juveniles were not fully expressing host antigens and Clegg & Smithers (1972) found that these juveniles were susceptible to the action of antibody from the serum of hyperimmune monkeys in an in-vitro assay system. More recently, McLaren, Clegg & Smithers (1975) using peroxidase-labelled antibodies and electron microscopy have explored the distribution of schistosome antigens and mouse host antigens on the tegument of schistosomula of various ages. Immunoglobulins from rhesus monkeys, hyperimmune to schistosome infection, readily attached to three hour old schistosomula but failed to bind to the surface of schistosomula which had been grown for four days in a mouse. As a corollary, mouse erythrocyte antigens were not detected on the surface of the younger worms but were easily demonstrable on the older worms. It seemed that the presence of host antigens at the surface prevented the attachment of specific anti-schistosome antibodies.

Although we have yet to obtain conclusive proof that antigens, acquired by schistosomes from host erythrocytes, serve to protect them from host immunity, there is much circumstantial evidence that this is the case. Our views of the immunological relationships between a schistosome and its hosts is summarized in Fig. 2. It may be that when we understand the

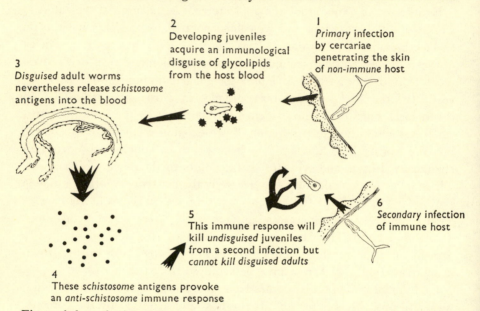

2
Developing juveniles acquire an immunological disguise of glycolipids from the host blood

1
Primary infection by cercariae penetrating the skin of *non-immune* host

3
Disguised adult worms nevertheless release *schistosome* antigens into the blood

5
This immune response will kill *undisguised* juveniles from a second infection but *cannot kill disguised adults*

6
Secondary infection of immune host

4
These *schistosome* antigens provoke an *anti-schistosome* immune response

Fig. 2. A hypothesis concerning the immunological relationships of *Schistosoma mansoni* and its host. These relationships would favour the long-term survival of both host and parasite.

acquisition of host antigens more fully we might interfere with the process, making the prospect of immunological control of schistosomes more hopeful than it presently appears.

GENERAL CONCLUSIONS

The vertebrate immune response functions to detect and destroy invading organisms that are genetically foreign to the host. Against African trypanosomes and schistosomes the immune response is not fully effective as the parasites possess means of evading the efferent arm of the response. Because these parasites are important pathogens of man, they have been extensively studied and we have at least some inkling of the mechanisms of evasion. Many other organisms, parasites, commensals and symbionts, must have evolved similar or different mechanisms of evasion, and of these we know little. We believe that studies aimed to discover the evasion mechanisms of other organisms would not only provide intellectual stimulation but also practical benefits to mankind.

REFERENCES

ALLSOPP, B. A., NJOGU, A. R. & HUMPHRYES, K. C. (1971). Nature and location of *Trypanosoma brucei* subgroup exoantigen and its relationship to 4 S antigen. *Exptl Parasit.*, **29**, 271–284.

ALLT, G., EVANS, E. M. E., EVANS, D. M. L. & TARGETT, G. A. T. (1971). Effect of infection with trypanosomes on the development of experimental allergic neuritis in rabbits. *Nature, Lond.*, **233**, 197–199.

BRADLEY, D. J. & McCULLOUGH, F. S. (1973). Egg output stability and the epidemiology of endemic *Schistosoma haematobium*. II. An analysis of the epidemiology of endemic *S. haematobium. Trans. R. Soc. trop. Med. Hyg.*, **67**, 491–500.

BURNETT, M. F. (1959). *The Clonal Selection Theory of Acquired Immunity*. London: Cambridge University Press.

CANTRELL, W. (1958). Mutation rate and antigenic variation in *Trypanosoma equiperdum. J. infect. Dis.*, **103**, 263–271.

CAPRON, A., BIGUET, J., ROSE, F. & VERNES, A. (1965). Les antigènes de *Schistosoma mansoni*. II. Étude immunoélectrophorique comparée de divers stades larvaires et des adultes des deux sexes. Aspects immunologiques des relations hôte–parasite de la cercaire et de l'adulte de *S. mansoni. Annls Inst. Pasteur Paris*, **109**, 798–810.

CAPRON, A., BIGUET, J., VERNES, A. & AFCHAIN, D. (1968). Structure antigénique des helminthes. Aspects immunologiques des relations hôte–parasite. *Path. Biol., Paris*, **16**, 121–138.

CLEGG, J. A. & SMITHERS, S. R. (1972). The effects of immune rhesus monkey serum on schistosomula of *Schistosoma mansoni* during cultivation *in vitro. Int. J. Parasit.*, **2**, 79–98.

CLEGG, J. A., SMITHERS, S. R. & TERRY, R. J. (1970). 'Host' antigens associated with schistosomes: observations on their attachment and nature. *Parasitology*, **61**, 87–94.

(1971a). Acquisition of human antigens by *Schistosoma mansoni* during cultivation *in vitro. Nature, Lond.*, **232**, 653–654.

(1971b). Concomitant immunity and host antigens associated with schistosomiasis. *Int. J. Parasit.*, **1**, 43–49.

CROSS, G. A. M. (1973). Identification and purification of a class of soluble surface proteins from *Trypanosoma brucei. Trans. R. Soc. trop. Med. Hyg.*, **67**, 261.

DAMIAN, R. T. (1967). Common antigens between adult *Schistosoma mansoni* and the laboratory mouse. *J. Parasit.*, **53**, 60–64.

DAMIAN, R. T., GREENE, N. D. & HUBBARD, W. J. (1973). Occurrence of mouse

α_2-macroglobulin antigenic determinants on *Schistosoma mansoni* adults with evidence on their nature. *J. Parasit.*, **59**, 64–73.

GOEDBLOED, E. (1971). *Trypanosoma rhodesiense*: antigenic stability in embryonated chicken eggs. *Exptl Parasit.*, **30**, 257–259.

GOLDRING, O. L., CLEGG, J. A., SMITHERS, S. R. & TERRY, R. J. (1975). Acquisition of human blood group antigens by *Schistosoma mansoni*. *Clin. exp. Immun.* (in press).

GOOD, R. A. (1972). Recent studies on the immunodeficiencies of man. *Am. J. Path.*, **69**, 484–490.

GOODWIN, L. G., GREEN, D. G., GUY, M. W. & VOLLER, A. (1972). Immuno-suppression during trypanosomiasis. *Br. J. exp. Path.*, **53**, 40–43.

GRAY, A. R. (1965a). Antigenic variation in clones of *Trypanosoma brucei*. 1. Immunological relationships of the clones. *Ann. trop. Med. Parasit.*, **59**, 27–36.

(1965b). Antigenic variation in a strain of *Trypanosoma brucei* transmitted by *Glossina morsitans* and *Glossina palpalis*. *J. gen. Microbiol.*, **41**, 195–214.

(1967). Some principles of the immunology of trypanosomes. *Bull. Wld Hlth Org.*, **37**, 177–193.

(1969). The epidemiological significance of some recent finds from research on antigenic variation in trypanosomes. *Bull. Wld Hlth Org.*, **41**, 805–813.

HERBERT, W. J. & LUMSDEN, W. H. R. (1968). Single-dose vaccination of mice against experimental infection with *Trypanosoma (Trypanozoon) brucei*. *J. med. Microbiol.*, **1**, 23–32.

HUDSON, K. M. & BYNER, C. (1973). Changes in the lymphoid architecture of trypanosome infected mice. *Trans. R. Soc. trop. Med. Hyg.*, **67**, 265.

HUDSON, K. M., FREEMAN, J., BYNER, C. & TERRY, R. J. (1975). Immunodepression in mice infected with *Trypanosoma brucei brucei*. *Immunology* (in press).

LANHAM, S. M. & TAYLOR, A. E. R. (1972). Some properties of the immunogens (protective antigens) of a single variant of *Trypanosoma brucei brucei*. *J. gen. Microbiol.*, **72**, 101–116.

LONGSTAFFE, J. A., FREEMAN, J. & HUDSON, K. M. (1937). Immunosuppression in trypanosomiasis: some thymus dependent and thymus independent responses. *Trans. R. Soc. trop. Med. Hyg.*, **67**, 264–265.

LOURIE, E. M. & O'CONNOR, R. J. (1973). A study of *Trypanosoma rhodesiense* relapse strains *in vitro*. *Ann. trop. Med. Parasit.*, **31**, 319–340.

LUCKINS, A. G. (1971). Infectivity and antigenicity of *Trypanosoma rhodesiense* in albino rats exposed to whole body X-irradiation. *Trans. R. Soc. trop. Med. Hyg.*, **65**, 232–234.

McCULLOUGH, F. S. & BRADLEY, D. J. (1973). Egg output stability and the epidemiology of *Schistosoma haematobium*. Part I. Variation and stability in *Schistosoma haematobium* egg counts. *Trans. R. Soc. trop. Med. Hyg.*, **67**, 475–490.

McLAREN, D. J., CLEGG, J. A. & SMITHERS, S. R. (1975). Acquisition of host antigens by young *Schistosoma mansoni* in mice: correlation with failure to bind antibody *in vitro*. *Parasitology*, **70**, 67–75.

MURRAY, M., MURRAY, P. K., JENNINGS, F. W., FISHER, E. W. & URQUHART, G. M. (1973). The pathology of *Trypanosoma brucei* in the rat. *Trans. R. Soc. trop. Med. Hyg.*, **67**, 276–277.

MURRAY, P. K., URQUHART, G. M., MURRAY, M. & JENNINGS, F. W. (1973). The response of mice infected with *T. brucei* to the administration of sheep erythrocytes. *Trans. R. Soc. trop. Med. Hyg.*, **67**, 267.

NJOGU, A. R. & HUMPHRYES, K. C. (1972). The nature of the 4 S antigens of the *brucei* subgroup trypanosomes. *Exptl Parasit.*, **31**, 178–187.

ORMEROD, W. E. & VENKATESAN, S. (1971). The occult visceral phase of mammalian trypanosomes with special reference to the life cycle of *Trypanosoma* (*Trypanozoon*) *brucei*. *Trans. R. Soc. trop. Med. Hyg.*, **65**, 722–741.

RAFF, M. C. (1973). T and B lymphocytes and immune responses. *Nature, Lond.*, **242**, 19–23.

RITZ, H. (1916). Über Rezidive bei experimenteller Trypanosomiasis, II. Mitteilung. *Arch. Schiffs- u. Tropenhyg.*, **20**, 397.

SMITHERS, S. R. & TERRY, R. J. (1965). Naturally acquired resistance to experimental infections of *Schistosoma mansoni* in the rhesus monkey (*Macaca mulatta*). *Parasitology*, **55**, 701–710.

— (1967). Resistance to experimental infection with *Schistosoma mansoni* in rhesus monkeys induced by the transfer of adult worms. *Trans. R. Soc. trop. Med. Hyg.*, **61**, 517–533.

— (1969). Immunity in schistosomiasis. *Ann. N. Y. Acad. Sci.*, **160**, 826–840.

SMITHERS, S. R., TERRY, R. J. & HOCKLEY, D. J. (1969). Host antigens in schistosomiasis. *Proc. R. Soc. B*, **171**, 483–494.

SPRENT, J. F. A. (1959). Parasitism, immunity and evolution. In *The Evolution of Living Organisms*, 149–165 (ed. G. W. Leaper). Melbourne: Melbourne University Press.

SSENYONGA, G. S. (1975). The distribution of *Trypanosoma brucei* and *T. congolense* in tissues of mice. *Parasitology*, **69**, XXV.

STEIGER, R. F. (1973). On the ultrastructure of *Trypanosoma* (*Trypanozoon*) *brucei* in the course of its life cycle and some related aspects. *Acta Trop.*, **30**, 64–168.

URQUHART, G. M., MURRAY, M. & JENNINGS, F. W. (1972). The immune response to helminth infection in trypanosome-infected animals. *Trans. R. Soc. trop. Med. Hyg.*, **66**, 342–343.

VICKERMAN, K. (1969). On the surface coat and flagellar adhesion in trypanosomes. *J. Cell Sci.*, **5**, 163–193.

VICKERMAN, K. & LUCKINS, A. G. (1969). Localisation of variable antigens in the surface coat of *Trypanosoma brucei* using ferritin conjugated antibody. *Nature, Lond.*, **224**, 1125–1126.

VIENS, P., TARGETT, G. A. T., WILSON, V. C. L. C. & EDWARDS, C. I. (1972). The persistence of *Trypanosoma* (*Herpetosoma*) *musculi* in the kidneys of immune mice. *Trans. R. Soc. trop. Med. Hyg.*, **66**, 669–670.

WATKINS, J. F. (1964). Observation on antigenic variation in a strain of *Trypanosoma brucei* growing in mice. *J. Hyg., Camb.*, **62**, 69–80.

WEITZ, B. G. F. (1970). Infection and resistance. In *The African Trypanosomiases*, 97–124 (ed. H. W. Mulligan). London: Allen & Unwin.

WRIGHT, K. A., LUMSDEN, W. H. R. & HALES, H. (1970). The formation of filopodium-like processes by *Trypanosoma* (*Trypanozoon*) *brucei*. *J. Cell. Sci.*, **6**, 285–297.

RELATIONSHIPS BETWEEN ACANTHOCEPHALA AND THEIR HOSTS

By D. W. T. CROMPTON

The Molteno Institute, University of Cambridge,
Downing Street, Cambridge CB2 3EE

Since their recognition by seventeenth-century microscopists, about a thousand species of acanthocephalan worm have been described (Schmidt, 1969) and all are endoparasites which attain sexual maturity in the alimentary tracts of vertebrates. Their life cycles, where known, have been found to involve an arthropod intermediate host. Adult acanthocephalans may be recognized by the possession of a retractile, hook-bearing proboscis (Plate 1a), by which they become attached to the host's intestinal wall (Hammond, 1967a), a muscular proboscis sheath or receptacle, a pair of lemnisci and a characteristic body wall (compare Fig. 1 with Plate 1d). There is a body cavity, but no alimentary tract has been observed at any stage of the life-cycle. The adults are dioecious and males are usually smaller than females of the same species, although male *Corynosoma hamanni* are larger than the females (Holloway & Nickol, 1970). The reproductive organs and reproductive processes have special features which have been described recently by Whitfield (1968) and Crompton & Whitfield (1974). Most acanthocephalans measure a few centimetres in

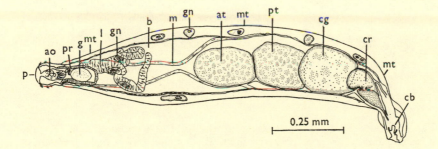

Fig. 1. The morphology of an adult male *Octospiniferoides chandleri* (after Bullock, 1966, 1969). ao, Apical organ; at, anterior testis; b, body cavity; cb, copulatory bursa; cg, cement gland; cr, cement reservoir; g, ganglion; gn, giant nucleus; l, lemniscus; m, muscle; mt, metasoma; p, proboscis; pr, proboscis receptaculum; pt, posterior testis.

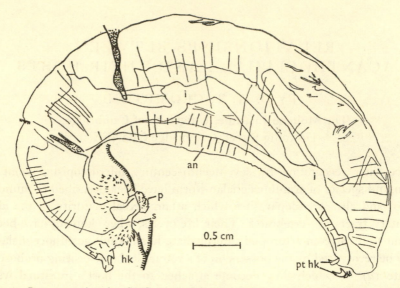

Fig. 2. Interpretation by S. Conway-Morris of the specimen of *Ottoia prolifica* shown in Plate 3a. an, annulation; hk, hook; i, intestine; p, proboscis; pt hk, posterior hook; s, spine.

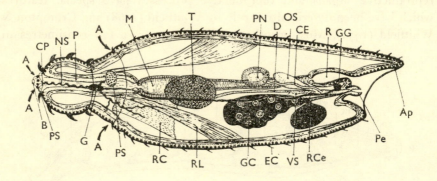

Fig. 3. Coup longitudinal dorso-ventral du Protacanthocéphale. A = points d'invagination; Ap = appendice terminal; B = bouche; CE = canal excréteur de la protonéphridie; CP = crochet du proboscis; D = canal déférent; EC = épine cuticulaire du tronc; G = ganglion cérébroide; GC = gland cémentaire syncytiale; GG = ganglion génital; M = intestin moyen (mésentéron); NS = nerf sensoriel; OS = organe de Saefftigen; P = pharynx (stomodeum); Pe = pénis; PN = protonéphridie; PS = papille sensorielle; R = rectum (proctodeum); RC = rétracteurs du cou; RCe = réservoir cémentaire; RL = rétracteurs longs; T = testicule; VS = vésicule séminale (after Fig. 10 of Golvan, 1958).

length; the longest specimens on record appear to be individuals of *Nephridiacanthus longissimus* from the aardvark, *Orycteropus afer* (Golvan, 1962; Schmidt & Canaris, 1967).

After copulation and development within the body cavity, the shelled acanthors, hereafter referred to as eggs, are released by the female worm and are discharged in the faeces of the definitive host. On ingestion by the appropriate arthropod host, the egg hatches and the acanthor bores through the intestinal wall into the haemocoele, where development proceeds, through a series of recognizable stages known as acanthellae (Plate 1c; King & Robinson, 1967; Butterworth, 1969a; Lackie, 1972a), until the cystacanth stage is reached (Plate 1b). In the simplest type of life-cycle, for example those of *Moniliformis dubius* (Moore, 1946) and *Polymorphus minutus* (Hynes & Nicholas, 1957), the definitive host acquires acanthocephalan parasites by eating arthropods containing infective cystacanths. The life-cycle may be prolonged and made more complicated by the inclusion of transport or paratenic hosts (Van Cleave, 1953), which may be indispensible for the completion of the life-cycles with which they are associated (Baer, 1961). For example, the definitive host of *Neoechinorhynchus cylindratus* is the bass, *Huro* (= Micropterus) *salmoides*, which, in nature, is unlikely to eat the ostracod intermediate host. The bluegill, *Lepomis pallidus*, does eat infected ostracods in the laboratory, with the result that encapsulated parasites are found in its abdominal tissues (Ward, 1940). When such bluegills are devoured by bass, mature *N. cylindratus* are eventually recovered from the intestine. In the few cases which have been studied, acanthocephalans in vertebrate transport hosts are usually encapsulated by a process reminiscent of the inflammatory response (Bogitsh, 1961). Interesting, but still incomplete, evidence has been obtained by Hopp (1954) to suggest that the gastropod *Campeloma rufum* serves as a second intermediate host for *N. emydis*, an acanthocephalan from freshwater turtles. Larger specimens of *N. emydis*, which undoubtedly develops in the ostracod *Cypria maculata*, were found in the gastropod rather than in the ostracod. An attempt to establish *N. emydis* in turtles by feeding them infected ostracods failed, whereas an attempt with infected gastropod tissue was successful.

The works of Meyer (1933), Van Cleave (1948, 1953), Petroschenko (1956, 1958), Baer (1961), Yamaguti (1963) and Bullock (1969) and the publications of Golvan cited by Bullock (1969), indicate that the Acanthocephala have a world-wide distribution and that their taxonomy is fascinating and difficult. In this article, I have adopted the view of Bullock that the group is a phylum composed of three orders, the Palaeacanthocephala, the Archiacanthocephala and the Eoacanthocephala. In addition to those references, the recent reviews of Nicholas (1967, 1973), Crompton (1970) and von Brand (1973) provide an introduction to many aspects of acanthocephalan biology, including pathology, reproduction and biochemistry.

I have chosen to discuss, in relative isolation, observations on some of the physiological interactions which I hope will illustrate something of the delicate equilibrium of acanthocephalan parasitism. This approach, however, may give an oversimplified impression of the complexity of the host–parasite relationship. The many factors affecting the course of controlled laboratory infections of acanthocephalans in their intermediate and definitive hosts are varied and incompletely understood (Burlingame & Chandler, 1941; Kates, 1944; Awachie, 1966; Crompton & Whitfield, 1968; Crofton, 1971; Crompton & Walters, 1972; Lackie, 1972a; Kennedy, 1972, 1974; Harris, 1972). The course of infection in nature is even more complex and difficult to analyse not only because of the effects on the parasite in question of intra- and interspecific reactions and of the responses of the host's digestive physiology to seasonal, dietary and other environmental changes, but also because of the natural ecological interactions between the intermediate and definitive hosts.

RELATIONSHIPS WITH INTERMEDIATE HOSTS

The process of infection

The mature eggs of acanthocephalan worms are very resistant to adverse environmental conditions. Representatives of each of the orders of the phylum have been found to retain their infectivity to the intermediate host for periods varying from 1–10 months if stored in water at temperatures of 4–17 °C (Crompton, 1970). Some eggs of *Macracanthorhynchus hirudinaceus* contained originally in pig faeces spread on soil in Maryland, USA, retained their infectivity for $3\frac{1}{2}$ years (Spindler & Kates, 1940), and in the laboratory other eggs of *Mac. hirudinaceus* retained their infectivity after various drastic treatments, sometimes extending over 550 days, but not after direct exposure to uv light for 10 min (Kates, 1942). Eggs of *Mac. hirudinaceus* (Glasgow & Deporte, 1939) and *Moniliformis clarki* (Crook & Grundmann, 1964) retained their infectivity after exposure to the entire digestive processes of pigeons and deer mice, *Peromyscus maniculatus sonoriensis*, respectively. The structure and composition of the envelopes or shells surrounding the acanthors probably account for the resistant properties of acanthocephalan eggs (Fig. 4). In addition to the references cited in the legend of this figure and by Crompton (1970), recent informa-

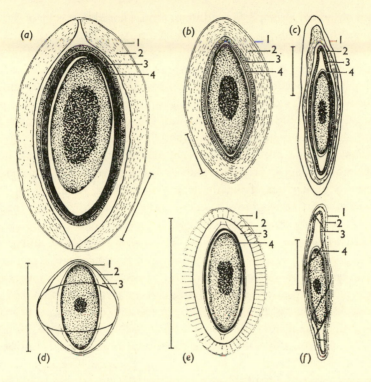

Fig. 4. Eggs of six species of Acanthocephala. (*a*) *Macracanthorhynchus hirudinaceus* (after Meyer, 1933, Fig. 369); (*b*) *Moniliformis dubius* (after Nicholas, 1967, Fig. 3E); (*c*) *Echinorhynchus gadi* (after West, 1964, Fig. 1); (*d*) *Neoechinorhynchus emydis* (after Hopp, 1954, Fig. 3); (*e*) *Neoechinorhynchus rutili* (after Meyer, 1931, Fig. 27); (*f*) *Acanthocephalus jacksoni* (after West, 1964, Fig. 2). The egg shells are numbered 1, 2, 3 and 4 for convenience only and not to indicate homologies between eggs. Scale bars represent 25 μm (reproduced from Crompton, 1970).

tion about the egg envelopes is given by Wright (1971) and Whitfield (1973) who studied *M. dubius* and *P. minutus* respectively.

Observations and comments made on the fate of eggs ingested by the appropriate intermediate host may be interpreted to suggest that both the digestive physiology of the host and the activity of the acanthor are necessary if the liberation of the acanthor from the egg is to occur before the egg is carried out of the host in the faeces. For example, DeGiusti (1949*a*) recorded that, when an egg of *Leptorhynchoides thecatus* (Palaeacanthocephala) is eaten by the amphipod *Hyalella azteca*, the acanthor within the egg becomes active and orientated against the inner envelope, the host's digestive enzymes weaken the outer envelopes and the acanthor's movements then enable it to penetrate the envelopes, usually within about $\frac{3}{4}$ h of the ingestion of the egg. Evidence for the functioning of the host's enzymes was not obtained. Crook & Grundmann (1964) noted that eggs of *M. clarki* (Archiacanthocephala) hatch in three stages with the acanthor actively participating in the final stage. Harms (1965) observed that, within 4 h of ingestion by the ostracod host, acanthors of *Octospinifer macilentis* (Eoacanthocephala) become motile while enclosed within the middle and inner envelopes after the outer envelope has split transversely at its equator. Most papers on the development and life cycles of acanthocephalans include observations like those described above.

Various artificial procedures have been used to study the hatching of acanthocephalan eggs. After alternate drying and wetting of eggs collected from the body cavities of female *Mac. hirudinaceus*, *Mac. ingens*, *Mediorhynchus grandis*, *Hamanniella tortuosa* and *M. dubius*, Manter (1928) and Moore (1942) observed that acanthors were liberated *in vitro*. Eggs of *Centrorhynchus* spp., *N. cylindratus*, *N. emydis*, *P. minutus* and *Prosthorhynchus formosus* tested in this way did not respond, nor did eggs of *Mac. hirudinaceus* and *M. dubius* if these were collected from the faeces of the definitive host (Manter, 1928; Moore, 1942; Hynes & Nicholas, 1957; Schmidt & Olsen, 1964). Eggs of *Pros. formosus* were induced to hatch in crushed digestive glands from isopod hosts (Schmidt & Olsen, 1964), but eggs of *P. minutus* did not hatch in macerated portions of intestine from amphipod hosts (Hynes & Nicholas, 1957). This diversity of results illustrates the specificity of this aspect of the relationship between acanthocephalans and their intermediate hosts.

A detailed investigation has been made into the factors involved in the hatching of eggs removed from the body cavities of female *M. dubius* (Edmonds, 1966). His results show that some of the eggs hatch *in vitro* when the concentration of the incubation medium is greater than about 0.2 mol l^{-1} (Fig. 5a), that hatching is increased by the presence of bicarbonate ions (Fig. 5b), and that hatching was minimal below pH 7.0 (Fig. 5c) except when the gas phase contained 1 % carbon dioxide. The acanthors were observed to become active during hatching and to release chitinase, presumably in response to the external conditions which favoured hatching *in vitro*. Similar conditions to these exist in the foregut and crop of the intermediate host, *Periplaneta americana* (Wigglesworth, 1927; Treherne, 1957) and evidence indicating that chitin is a component of one of the inner egg envelopes of *M. dubius* (Fig. 4b) was obtained by Edmonds (1966). Later, eggs of *Mac. hirudinaceus* were stimulated to hatch *in vitro* by application of Edmonds's technique (Robinson & Strickland, 1969). Edmonds's work provides the best support for the hypothesis that the process of infection depends on contributions from both host and parasite.

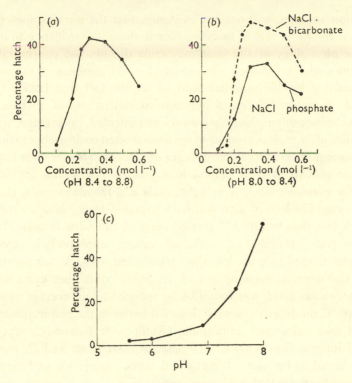

Fig. 5. The hatching of eggs of *Moniliformis dubius in vitro*. (a) The effect of molarity on percentage hatch; (b) the effect of bicarbonate ions on percentage hatch; (c) the effect of pH on percentage hatch. Ordinate represents the percentage hatch (after Edmonds, 1966, Figs. 1–4, modified from Crompton, 1970).

Effects on intermediate hosts

It is difficult to draw general conclusions about the metabolic relationships that must exist between acanthocephalans and their arthropod hosts, although the environmental temperature to which the poikilothermic host is exposed undoubtedly affects the rate of development of the parasites in the haemocoele (see Crompton, 1970; Lackie, 1972a). From measurements of the utilization of glucose by developing stages of *P. minutus in vitro* and estimations of the dry weights of the parasites (Butterworth, 1969b) and from measurements of the volume (Butterworth, 1968a; Lockwood & Inman, 1973) and glucose concentration of the haemolymph (Butterworth, 1968b) of the intermediate host, *Gammarus pulex*, it is possible to calculate that the rate of glucose utilization by an early acanthella, a middle acanthella, a late acanthella and a cystacanth, is about 0.885 µg, 0.525 µg, 1.12 µg and 0.315 µg per hour respectively from an average amount of about 3.85 µg available in the haemocoele of a host of average size. This calculation assumes, without any evidence, that the parasite uses glucose *in vivo* at the same rate as *in vitro*. Nor is there any evidence to indicate how the physiology of the host maintains the glucose concentration of the haemocoele in order to compensate for the presence of the parasite. Compensation by the host may account for the fact that Lackie (1972a) detected no major differences between concentrations of 16 free amino acids in the haemolymph of *Peri. americana* infected, on average, with 120 acanthellae of *M. dubius*, compared with uninfected cockroaches maintained in the same manner. Developing stages of *Acanthocephalus dirus* appear to affect the growth of their isopod host, *Asellus intermedius* (Seidenberg, 1973); for example, in October 1967, male and female isopods, including their average burdens of 4.70 and 6.08 parasites respectively, were found to weigh less than uninfected counterparts of the same length. Hynes & Nicholas (1963) stated that, in the laboratory, amphipods infected with *P. minutus* tended to grow less than uninfected ones. At temperatures of 20 °C, the oxygen consumption of *G. pulex* containing cystacanths of *Pomphorhynchus laevis* was found to be reduced by an average of 19.3 %, but at 10 °C no differences were detected between the respiration rates of infected and uninfected amphipods (Rumpus & Kennedy, 1974). The infected intermediate hosts of *Corynosoma constrictum* and *P. paradoxus* were believed to be more sluggish and more susceptible to fatigue than uninfected hosts (Bethel & Homes, 1973).

Some effects of developing acanthocephalans appear to be very detrimental to the intermediate hosts. Occasionally, hosts have been reported to be killed by mass penetration of the intestinal wall by acanthors under laboratory conditions in which no control was exercised over the number of eggs eaten (DeGiusti, 1949a; Crompton, 1964a; Crook & Grundmann, 1964). *Polymorphus minutus* is considered to interfere with the reproductive activity of mature female *G. pulex* and to prevent immature females from reaching maturity (Le Roux, 1931a; Hynes, 1955). The effect is characterized by the disappearance of the setae from the oostegites, the cessation of ovarian development and the termination of mating activity; the same signs and symptoms can be produced by the irradiation of female *G. pulex* (Le Roux, 1931b). No major effects of *P. minutus* on the sexual development of male *G. pulex* have been observed. Thus, in nature, the parasite may act as a population regulator. Hynes (1955) obtained evidence to suggest that isolated populations of *G. lacustris* in lakes could be depleted or eliminated by *P. minutus*. Hynes & Nicholas (1963) examined many thousands of infected *Gammarus* and found only six females carrying eggs, whereas normally about 30–50 % of the adult females would have been breeding. Another, as yet unidentified, acanthocephalan has been reported to alter the sexual development of the isopod, *Asellus aquaticus*. In a sample of 80 infected isopods, Munro (1953) found 28 intersexes, which appeared to be modified females, since they possessed oostegites with immature setae. The ovaries of terrestrial isopods infected with *Pros. formosus* were observed to become vestigial or occasionally to hypertrophy and be filled with an oily substance (Schmidt & Olsen, 1964). Recently, Rumpus concluded that *Pom. laevis* also reduces the reproductive capacity of some female, though not male, *G. pulex* (see Rumpus & Kennedy, 1974). Munro also noticed that most infected isopods were darker in colour than uninfected specimens. Hindsbo (1972) observed that *G. lacustris* containing *P. minutus* were blue in colour, while uninfected amphipods were brown, and he demonstrated that the haemolymph of the infected amphipods contained significantly less carotenoid pigment (Barrett & Butterworth, 1968) than uninfected amphipods.

In some cases, the presence of an acanthocephalan alters the normal behaviour of the intermediate host. Hindsbo (1972) found that the blue specimens of *G. lacustris*, infected with *P. minutus*, were more easily detected in surface water than the brown uninfected amphipods. The infected hosts showed greater positive phototropism, and one test, involving a domestic duckling, indicated that the altered behaviour of the infected amphipods resulted in their being captured and eaten more readily than the uninfected amphipods. By means of experimental lighting and aquaria offering a choice between light and dark backgrounds, Bethel & Holmes (1973) demonstrated significant differences between the responses of *G. lacustris* infected with cystacanths of *P. paradoxus* or *P. marilis* and *H. azteca* infected with cystacanths of *C. constrictum* compared with those of uninfected amphipods or those containing acanthellae of the parasites (Fig. 6). In response to disturbance, the evasive behaviour of *G. lacustris* and *H. azteca* infected with cystacanths of *P. paradoxus* and *C. constrictum* also differed significantly from that of uninfected amphipods. After enticing uninfected amphipods to the water surface with food, they were touched, whereupon all dived to the bottom of the aquaria. Hosts containing cystacanths needed no enticement to the surface; on prodding, *G. lacustris* was observed to skim on the water and to cling to flotsam, while *H. azteca* continued to occupy the surface water despite repeated stimulation. Bethel & Holmes (1974) discovered that the onset of the altered evasive behaviour of *G. lacustris* infected with *P. paradoxus* coincided with the time when the parasite became infective to hamsters, an experimental definitive host. The significance of these results cannot be appreciated without reference to the papers of Bethel & Holmes (1973, 1974) and their analyses of their observations and experiments. They made

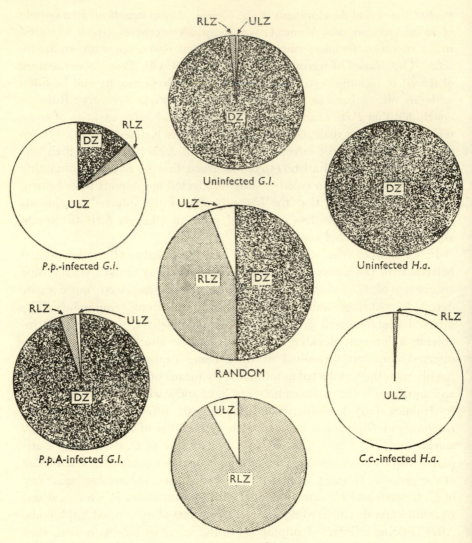

Fig. 6. Proportion of time spent by infected and uninfected amphipods in different areas of a light-dark choice aquarium. ULZ, upper 3 cm of light zone; RLZ, remainder of light zone; DZ, dark zone; *G.l.*, *Gammarus lacustris*; *H.a.*, *Hyalella azteca*; *P.p.*, *Polymorphus paradoxus*; *P.p.*A., *Paradoxus* acanthellae; *P.m.*, *P. marilis*; *C.c.*, *Corynosoma constrictum*. Random values calculated from relative volume of each zone (after Fig. 1 of Bethel & Holmes, 1973).

similar qualitative observations under field conditions and it now appears that the presence of some acanthocephalans increases the chances of the ingestion of their intermediate hosts by definitive hosts.

Development and host resistance

Much evidence has established that haemocytic encapsulation is the principle defence reaction of arthropods to foreign objects which enter the haemocoele and are too large to be neutralized by phagocytosis (Salt, 1963, 1970). During development in the haemocoele, several species of acantho-cephalan are observed to be enclosed within a thin, transparent envelope (Plate 1b,c), the origin of which was studied with the light microscope by Crompton (1964b) for *P. minutus* developing in *G. pulex*. Once the acanthor of *P. minutus* has penetrated the intestinal tissues, it often remains associated with the haemocoelic surface of the intestine and becomes surrounded by haemocytes, but after some degree of development relatively few haemocytes are associated with the healthy parasite. Crompton interpreted his observations on tissue sections of these events as showing that the parasite did not enter the haemocoele directly, but that it grew for a time beneath the connective tissue sheath or serosa which encloses the host's midgut. He proposed that the envelope of *P. minutus* originated as a result of a wound-healing response of the host to its stretched or damaged connective tissue. This suggestion was attractive because it provided an explanation for the development of a parasite despite the presence of haemocytes, the inference being that the connective tissue, which would not be expected to elicit haemocytic encapsulation (Salt, 1961), was keeping the parasite's surface and the haemocytes apart. On the basis of histochemical evidence, however, Crompton concluded that material produced by the parasite was added to the inner surface of the envelope during development in the haemocoele. A survey of the relevant literature revealed that, in cases where an envelope had been described, the authors had often recorded or depicted incidental observations which could be interpreted as suggesting that all envelopes originated in the manner proposed for that around *P. minutus*. More recent observations on acantho-cephalan development can still be interpreted in favour of this suggestion. For example, Bowen (1967) found that, during development, all stages of *Mac. ingens* within the haemocoeles of the millipedes *Narceus gordanus* and *Floridobolus penneri* were encapsulated by a thin layer of cells and fibres which were acquired as the acanthors penetrated the wall of the posterior part of the midgut. Denny (1969) recorded that he found acanthors of *P. marilis* bound to the intestinal wall of *G. lacustris* by a membrane of connective tissue, probably the serosa, and associated with a small number of smaller host cells, probably haemocytes. Robinson & Strickland (1969) published photomicrographs of sections through developing *M. dubius* and the midgut of *Peri. americana*; there is general similarity between these illustrations and the drawings of Crompton (1964*b*). Podesta & Holmes (1970), in a study of the development of *C. constrictum* in *H. azteca*, stated that hatched acanthors were observed within the intestinal wall of amphipods and by the tenth day all the observable acanthors were contained in small pouches formed between the muscle and serosal layer of the intestinal wall. Awachie (1966), however, concluded that the envelope of *Echinorhynchus truttae* from *G. pulex* was probably of parasitic origin.

An important discovery, which undermines Crompton's interpretation, was made by Robinson & Strickland (1969) who obtained acanthors of *M. dubius* by the method of Edmonds (1966) and injected them directly into the haemocoeles of *Peri. americana*. Such acanthors were encapsulated by haemocytes, but the numbers of haemocytes gradually decreased and the parasites developed normally within an envelope to form cystacanths. In this case, the formation of the envelope could not have involved haemocytic repair of the connective tissue serosa of the intestine. Evidence obtained by means of transmission electron microscopy strongly supported the conclusion that acanthors of *M. dubius* are encapsulated on entering the haemocoele of *Peri. americana* by either natural or artificial routes (Rotheram & Crompton, 1972, Plate 2*a*). The dispersal of the haemocytes, however, was accompanied by the presence of a coat of membranous material which appeared to be connected to the acanthor's surface (Plate 2*b,c*; Fig. 7). The membranous material was found to become more complex in structure during the parasite's development and eventually to become detached from the parasite's surface (Lackie & Rotheram, 1972). When acanthors, obtained from eggs hatched *in vitro*, were maintained in culture at 24 °C in the absence of host haemocytes, some form of membranous material was observed to be associated with the surface of about a third of the acanthors (Plate 2*d*). These observations suggest that the envelope surrounding *M. dubius* is largely parasitic in origin and composition. The envelope may be found, however, to contain components from the host. The electron micrographs of Mercer & Nicholas (1967) show haemocytes enmeshed in the membranous, vesicular material surrounding an acanthella of *M. dubius*, and the adsorption of materials from the haemolymph may occur.

These findings mean that the conclusion of Crompton (1964*b*) about the mode of origin of acanthocephalan envelopes should be set aside because the most intensively studied example, *M. dubius*, appears to be an exception. Furthermore, the ultrastructure of the envelope around

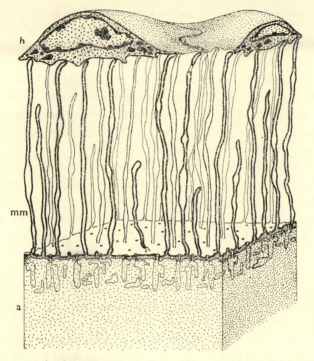

Fig. 7. Diagrammatic interpretation of a phase in the relationship between *Moniliformis dubius* and the haemocytes of *Periplaneta americana* (see Plate 2*a–c*). (Illustration supplied by Susan Rotheram.) a, Acanthor; h, haemocyte; mm, membranous material.

acanthellae of *P. minutus* (Butterworth, 1969*a*) is similar to that around equivalent acanthellae of *M. dubius* (Mercer & Nicholas, 1967). The envelopes around *Pros. formosus* and *Prosthenorchis elegans* have been studied by light microscopy by Wanson & Nickol (1973), who proposed that these, as well as the envelope around *M. dubius*, contained much material from the parasite. They interpreted their observations as showing that, after entry into the haemocoele, the acanthor begins to swell and transform into a covering for the acanthella developing from the nuclear mass at its centre (Plate 1*f*). Then, as development proceeds, the tissues of the acanthor begin to atrophy and eventually its body covering is left as a fluid-filled membranous envelope. Thus, the results of recent studies indicate that a major component of some acanthocephalan envelopes is material derived in some manner from the acanthor.

Some species, for example *Paulisentis fractus* (Cable & Dill, 1967), appear not to possess an envelope even though they develop in the haemocoele of an arthropod. Although the experiments were marred by technical difficulties, when envelope-free *P. minutus* were introduced directly into the haemocoeles of *G. pulex*, the parasites were found to be dead and encapsulated on retrieval some days later (Crompton, 1967). Similarly, acanthellae and cystacanths of *M. dubius*, from which the envelopes had been removed *in vitro*, were encapsulated and killed on transplantation into *Peri. americana*, whereas specimens with their envelopes intact were not encapsulated (Robinson & Strickland, 1969). The results of these experiments and the account of observations on the formation of the envelope suggest that it is associated in some manner with host resistance.

Not all acanthocephalan parasites complete their development after being ingested by their natural intermediate hosts. Lackie (1972*a*) exposed *Peri. americana* to known numbers of mature eggs of *M. dubius* and found an exponential decline in the numbers of cystacanths recovered as the dose was increased. Apart from eggs which fail to hatch in the intestine, some acanthors seem to perish during, or shortly after, their entry into the haemocoele and they are found in a melanized and encapsulated state on the haemocoelic surface of the intestinal wall (Lackie, 1972*a*). Individuals of *Leptorhynchoides thecatus* (DeGiusti, 1949*a*), *Mac. hirudinaceus* (Miller, 1943), *P. minutus* (Hynes & Nicholas, 1958) and *Pros. formosus* (Schmidt & Olsen, 1964), have also been found in this condition. The cause of death of the acanthors is unknown. In addition to its potential to destroy foreign objects, encapsulation prevents necrotic and moribund tissues from contaminating the haemocoele (Salt, 1961).

In some cases, the equilibrium between the early development of the parasite and the resistance of the intermediate host appears to be influenced by the fact that the range of ambient temperatures tolerated by the parasite and its host are different. DeGiusti (1949*a*) observed more individuals of *L. thecatus* associated with haemocytes and the intestinal wall of *H. azteca* maintained at 13–15 °C than in amphipods maintained at 20–25 °C and he concluded that such parasites soon died and degenerated. Lackie (1972*b*) found that the development of *M. dubius* in *Peri. americana* kept at 37 °C could be completed, but only if the hosts were kept at a lower temperature during the process of infection and penetration of the intestinal wall. Attempts to infect hosts kept at 37 °C were mainly unsuccessful and many melanized and encapsulated parasites were retrieved from the wall of the lower midgut.

The literature also reveals that acanthocephalans which have successfully entered the haemocoele of the intermediate host may be encapsulated, presumably because their development or metabolism has become abnormal (Crompton, 1967; Schaefer, 1970; Bethel & Holmes, 1973), and that acanthocephalans in unnatural hosts often elicit haemocytic encapsulation whether they enter the haemocoele by natural or artificial means (DeGiusti, 1949*b*; Robinson & Strickland, 1969; Lackie, 1973; Plate 2*e*).

RELATIONSHIPS WITH DEFINITIVE HOSTS

The distribution of Acanthocephala in the vertebrate alimentary tract

Observations on the distribution of a selection of adult acanthocephalans in the tracts of their definitive hosts are given in Table 1. This information has been compiled in the manner described by Crompton (1973), whose terminology for site and emigration have also been adopted, and the references given in the table should be consulted for detailed descriptions of the sites. Although there is little doubt that different physico-chemical conditions prevail in different regions of the alimentary tract during digestion (see Crompton, 1973), there is little direct evidence at present to indicate how acanthocephalans respond to these conditions and whether they influence the observed distribution. Some species must be exposed simultaneously to the varying conditions of the alimentary tract. *Neoechinorhynchus topseyi* from mango fish (Podder, 1937) and *M. clarki* from pine mice (Benton, 1954) were observed to extend for almost the entire length of the small and large intestines.

Some acanthocephalans appear to emigrate from one part of the tract to another during the establishment of the adults in their site. For example, immature *M. dubius* emigrate from the posterior half of the small intestine of the rat to the anterior half during the first few weeks of the infection (Burlingame & Chandler, 1941; Holmes, 1961; Crompton & Nesheim, 1973; Fig. 8). *Acanthocephalus jacksoni* (Bullock, 1963) and *Echinorhynchus truttae* (Awachie, 1966) appear to achieve maturity after an extended period of emigration in the direction of gastro-intestinal flow. Crompton (1973) discussed some of the difficulties which are encountered in trying to assess the significance of observations on emigrations. A further difficulty, which handicaps the interpretation of many studies of the distribution of parasites in their hosts, is the fact that the results nearly always consist of a series of observations, which have to be made on separate populations of parasites of different ages, but are assumed to apply to the parasite in question as if it had not been disturbed during its association with the host.

No satisfactory explanation, supported by experimental evidence, exists for the emigration of *M. dubius* against the direction of gastro-intestinal flow. The observation that the emigration seems to occur in rats fed at least 10 diets (Burlingame & Chandler, 1941; Holmes, 1961; Crompton & Nesheim, 1973, and unpublished observations) suggests that the emigration may be associated either with some factor present in the tract irrespective of

Table 1. *The distribution of sites of a selection of adult Acanthocephala in the alimentary tract of vertebrates*
(after Crompton, 1973)

Parasite	Hosts	Site	References
FISH HOSTS			
Palaeacanthocephala			
Acanthocephalus jacksoni	*Salvelinus fontinalis*	Ileo-rectal regions of intestine	Bullock (1963)
	Salmo gairdneri		
A. lucii	*Perca fluviatilis*	Stomach	Rawson (1952)
	Lota lota lota	Pyloric appendices	Ejsymont (1970a)
Echinorhynchus borealis	*Lota lota lota*	Pyloric appendices, stomach and intestines	Ejsymont (1970a)
E. gadi	*Silurus glanis*	Stomach and middle portion of intestine; some in posterior intestine	Ejsymont (1970b)
	Oncorhynchus kisutch	Lower intestine	Ekbaum (1938)
	O. tschawytscha		
E. lagenuiformis	*Platichthys stellatus*	Proximal loop of intestine	Prakash & Adams (1960)
E. salmonis	*Perca fluviatilis*	Anterior half of intestine	Barnes (1968)
E. truttae	*Salmo trutta*	Posterior part of intestine	Tedla & Fernando (1970)
Leptorhynchoides thecatus	*Ambloplites rupestris Huro (=Micropterus) salmoides*	Posterior part of intestine	Awachie (1966)
		Pyloric caeca	DeGiusti (1949a)
Micracanthocephalus hemirhamphus	*Hemirhamphus intermedius*	Stomach	Venard & Warfel (1953); Baylis (1944)
Pomphorhynchus bulbocolli	*Catostomus commersoni*	Between middle and posterior intestine	Chaicharn & Bullock (1967)
P. laevis	*Squalius cephalus*	Anterior half of intestine	Hine (1970); personal communication
Pomphorhynchus sp.	*Platichthys flesus*	Rectum	MacKenzie & Gibson (1970)
Telosentis tenuicornis	*Leiostomus xanthurus*	Hind gut	Huizinga & Haley (1962)

Eoacanthocephala		
Neoechinorhynchus carpiodi	Anterior small intestine	Dechtiar (1968)
N. cristatus	Posterior part of intestine	Chaicharn & Bullock (1967)
N. prolixoides	Posterior part of intestine	Chaicharn & Bullock (1967)
N. rutili	Pyloric appendices, stomach and rarely in anterior intestine	Eisymont (1970a)
Carpoides cyprinus		
Catostomus commersoni		
Erimyzon oblongus		
Lota lota lota		
Semotilus atromaculatus	First flexus of intestine below stomach	Uglem & Larson (1969)
N. saginatus	Middle of intestine	Cross (1934)
'Neoiacanthorinchus' sp.	Posterior part of intestine	Chaicharn & Bullock (1967)
Octospinifer macilentis	Middle third of intestine	Van Cleave & Haderlie (1950
O. torosus	Posterior part of intestine	Bullock (1967)
Octospiniferoides chandleri	Duodenum and intestine	Bhalerao (1931)
Pallisentis nagpurensis	First flexus of intestine below stomach	Cable & Dill (1967)
Paulisentis fractus		

AMPHIBIAN HOSTS

Palaeacanthocephala		
Acanthocephalus ranae	Anterior part of small intestine	Crompton (1970)
Bufo bufo	Upper part of small intestine	Pflugfelder (1949)
Rana temporaria		

REPTILIAN HOSTS

Palaeacanthocephala		
Sphaerechinorhynchus rotudocapitatus	Lower part of intestine and rectum	Johnston & Deland (1929)
Pseudechis prophyriacus		

Eoacanthocephala		
Neoechinorhynchus emydis	Attached to duodenal wall between entry of bile duct and tail of adjacent pancreas	Dunagan (1962)
Graptemys geographica		

Table 1 continued

Parasite	Hosts	Site	References
AVIAN HOSTS			
Palaeacanthocephala			
Apororhynchus amphistomi	*Wilsonia canadensis*	Cloacal region	Byrd & Denton (1949)
Arhythmorhynchus capellae	*Capella galinago delicata*	Small intestine and caeca	Schmidt (1963)
A. longicollis	*Larus canus*	Posterior half of small intestine	Bakke (1973)
Corynosoma bipapillum	*Larus philadelphia*	Lower intestine	Schmidt (1965)
Filicollis anatis	Ducks	Posterior two-thirds of small intestine	Soliman (1955)
Polymorphus contortus	Ducks	Within 1 or 2 inches above and below the ileo-caecal junction	Podesta & Holmes (1970)
P. minutus	Ducks	Posterior part of intestine	Crompton & Whitfield (1968)
Profilicollis botulus	*Somateria mollissima*	Posterior small intestine	Threlfall (1968)
Prosthorhynchus formosus	*Turdus migratorius*	Posterior part of intestine	Schmidt & Olsen (1964)
Archiacanthocephala			
Mediorhynchus grandis	*Quiscalus quiscala*	Lower portion of small intestine	Moore (1962)
MAMMALIAN HOSTS			
Palaeacanthocephala			
Bolbosoma balaenae	*Balaenoptera acutorostrata*	Duodenum	Barker & Macalister (1865)
Corynosoma strumosum	*Phoca richardii*	Anterior small intestine	Ball (1928)
	Erignathus barbatus	Small intestine	Lyster (1940)
Polymorphus paradoxus	*Castor canadensis*	Jejunum	Connell & Corner (1957)
Archiacanthocephala			
Macracanthorhynchus hirudinaceus	Domestic pigs	Duodenum	Schwartz (1929)
Moniliformis dubius	Rats	Jejunum	Kates (1944)
		Anterior small intestine	Burlingame & Chandler (1941); Holmes (1961)
Oncicola canis	Dogs	Jejunum and ileum	Van Cleave (1920)
Prosthenorchis sp.	*Saimiri sciurea*	Posterior small intestine and ileocaecal junction	Crompton (1973)

the composition of the diet or with some endogenous behaviour pattern of the parasite itself. Crompton & Nesheim (1973) recorded the sex, attachment position and length of individual *M. dubius* from rats fed on diets of known composition for different periods. The distributions of the parasites in two groups of four male rats each fed diet D (Crompton & Nesheim, 1973) for four and eight weeks respectively are represented in Fig. 8*a*, *c*. By considering the distribution in individual rats, it is possible to see the overlap of the parasites at the time of death of the host and to estimate the numbers of worms which were in potential contact with any individual. The results of this type of analysis for the first nine weeks of the course of infection of *M. dubius* in male rats fed diet D are shown in Fig. 8*b*. It appears that during their occupation of the anterior part of the host's small intestine, each female *M. dubius* is, on average, in contact with the maximum number of males and the highest percentage of the females present in the host are experiencing contact with at least one male. The preliminary observation that each female *M. dubius* is in contact with most males from about three to six weeks after the start of the infection (Fig. 8*b*) agrees with the conclusion that copulatory activity is necessary for this length of time if a female worm is to undergo a maximum patent period (Crompton, Arnold & Barnard, 1972; Crompton, 1974). Copulation between acanthocephalans, which are relatively large compared with the dimensions of the host's tract, may be achieved more easily in the anterior part of the tract where roughage and ingesta neither collect nor occlude the intestinal lumen of the healthy host. One consequence of the anterior emigration of *M. dubius* is that roughage is avoided for the major period of copulation. After female worms have been inseminated sufficiently to expend their full egg-producing potential (Crompton, 1974; Crompton & Whitfield, 1974), the result of the spreading of *M. dubius* along the small intestine (Fig. 8*c*) will tend to reduce mutual contamination of environments. It is interesting to note that the worms from rats fed diets A and B (Crompton & Nesheim, 1973) also emigrate anteriorly, but there is no evidence of reproduction. Other observations indicate that male and female *M. dubius* can find each other and mate after surgical transplantation and manipulations bearing no regard for the anterior emigration (Crompton, 1974).

The process of infection

The degree of invagination of the proboscis and other tissues of a cystacanth can probably be correlated with the amount of mechanical pressure and grinding activity to which cystacanths are exposed during their passage through the anterior part of the definitive host's alimentary tract (Crompton, 1970). Cystacanths of *P. minutus* (Plate 1*b*) must withstand the

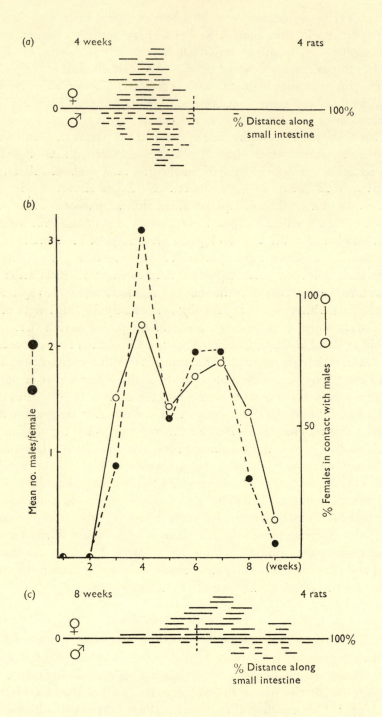

(a) 4 weeks 4 rats

♀
♂
0 100%
% Distance along
small intestine

(b)

Mean no. males/female

% Females in contact with males

(weeks)

(c) 8 weeks 4 rats

♀
♂
0 100%
% Distance along
small intestine

squeezing action of a duck's bill and pressures of about 178 mm Hg generated as the ventriculus contracts four times per min (Farner, 1960). In contrast, cystacanths of *E. truttae*, which are ingested by trout, have the appearance of small adult worms in which the proboscis has been retracted (Awachie, 1966). In some hosts, for example, the common toad, *Bufo bufo*, adverse mechanical factors in the anterior part of the tract must be minimal, since adult *Acanthocephalus ranae* can be recovered from the small intestine of one toad and introduced orally into another toad without any apparent damage (Hammond, 1968*a*). In simple acanthocephalan life-cycles, the process of infection involves the release of the cystacanth from the tissues of the intermediate host and the activation of the cystacanth from its semi-dormant state followed by the eversion of the proboscis.

The release of the cystacanths from their intermediate hosts appears to occur in the portion of the tract anterior to the pylorus, the duration of this process depending on whether the host is poikilothermic or homoio-thermic (Awachie, 1966; Lingard & Crompton, 1972). The ventriculus of ducks represents a hazard to *P. minutus*, and some crushed cystacanths were recovered from the lumen of this organ shortly after infected amphi-pods had been swallowed (Lingard & Crompton, 1972). The envelope surrounding cystacanths is readily removed by the action of acid pepsin, but this stage does not appear to be an essential step in the process of infection (Nicholas & Hynes, 1958; Crompton, 1964*a*; Graff & Kitzman, 1965).

The infection of rats with *M. dubius* (Graff & Kitzman, 1965) and ducks with *P. minutus* (Lingard & Crompton, 1972; Lackie, 1974*a*, *b*) have been studied in detail. Graff & Kitzman orally infected each of five rats with 10 cystacanths of *M. dubius* which had been removed from the intermediate host. The rats were killed at intervals of 10, 45, 90, 160 and 310 min and the condition and location of the parasites were observed. There was no evidence of activation having occurred until the parasites had passed the pylorus. Activation and establishment of *M. dubius* also occurred when

---

Fig. 8. Observations on the distribution of *Moniliformis dubius* in the small intestine of male rats fed diet D of Crompton & Nesheim (1973). (*a*) The distribution of worms in 4 male rats fed diet D for 4 weeks. The lengths of the worms in this figure and in (*c*) have been converted to the same scale as that used for the small intestine.

(*b*) Graphical representation of the results of an analysis of potential contacts between worms during the first 9 weeks of the course of the infection. The observations were made on the exact positions and lengths of worms in the small intestine within about 10 min of the death of the hosts between 0900 and 1230 h.

(*c*) The distribution of worms in 4 male rats fed diet D for 8 weeks. (*a*, *b* and *c* from Crompton & Nesheim, unpublished observations.)

cystacanths were introduced directly into the jejunum of rats. Similar observations were made by Lingard & Crompton (1972), who allowed domestic ducks, that had been starved for 1 h, to ingest known numbers of cystacanths of *P. minutus* in *G. pulex*. No sign of activation was detected until the parasites had entered the small intestine. When cystacanths of *P. minutus* were introduced directly into the small intestine of ducks by means of cannulae, placed in the duodenum or region of the yolk stalk, activation and successful evagination of the proboscides occurred.

The results of investigations of activation *in vitro* generally agree with the observations made *in vivo*. Bile salts were found to be essential for the activation of cystacanths of *M. dubius in vitro* and an increased rate was observed at concentrations greater than 0.005 % (Graff & Kitzman, 1965). Surgical alteration of the point of entry of bile from the duodenum to the apex of the caecum of the rat rendered the host insusceptible to oral infection with *M. dubius*; this surgical procedure appeared to have little effect on established worms during the experimental period. The optimum temperature and pH for activation *in vitro* were found to be similar to conditions believed to exist in the small intestine, and the presence of carbon dioxide in the gas phase *in vitro* had a stimulatory effect on activation.

Lackie (1974*a*, *b*) found that *in vitro* the optimum temperature for the activation of cystacanths of *P. minutus* was 42–44 °C (Fig. 9*a*) and the optimum pH was 7.0 (Fig. 9*b*) with no evidence of activation occurring at pH 5.0 and below. Duck bile, bile salts and chromatographically pure samples of sodium taurocholate (NaTC), at appropriate concentrations and under the optimum conditions of temperature and pH, had a marked enhancing effect on the rate of activation (Fig. 9*c*). The effect was achieved when cystacanths were exposed to the bile salt for 5 min before being transferred to physiological saline from which the bile salt was omitted. As expected, the osmotic pressure of the incubation medium affected the rate

Fig. 9. Observations on the activation of cystacanths of *Polymorphus minutus in vitro*. (*a*) the effect of temperature on cystacanth activation. □ — □, 49 °C; ● - - - ●, 48 °C; ○ — ○, 47 °C; ▲ — ▲, 44 °C; ● — ●, 42 °C; △ — △, 40 °C; ■ — ■, 39 °C.

   (*b*) Effect of pH on cystacanth activation. ○, pH 7.4; ▲, pH 7.0; ●, pH 6.7; ■, pH 6.0.

   (*c*) Effect of different concentrations of NaTC on cystacanth activation. ▲, 0.05 % NaTC; ○, 0.01 % NaTC; ■, 0.002 % NaTC; ●, control.

   (*d*) Effect of osmotic pressure on activation. ■ — ■, 100 mmol l⁻¹; ○ — ○, 150 mmol l⁻¹; ● - - - ●, BSS; □ — □, 180 mmol l⁻¹; ■ - - - ■, 190 mmol l⁻¹; △ — △, 220 mmol l⁻¹; ▲ — ▲, 250 mmol l⁻¹. Osmolarity of solutions expressed as equivalent to mmol l⁻¹ NaCl. Bars show standard deviation in all figures. BSS represents balanced salt solution (from Figs. 1, 2, 3 and 6 of Lackie, 1974*a*).

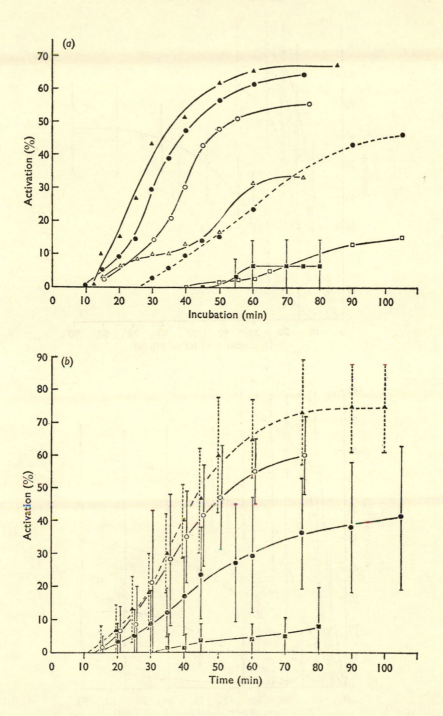

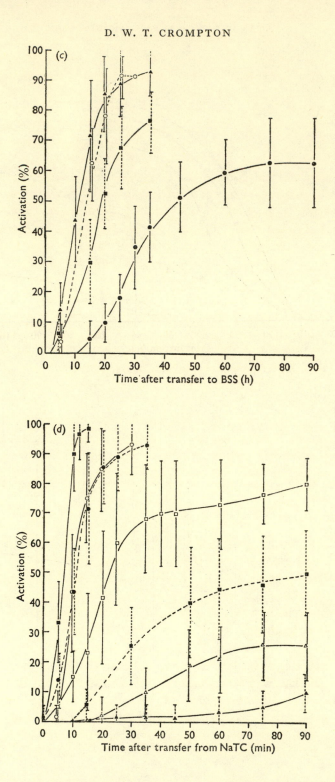

of activation (Fig. 9d). Various observations indicate that these conditions probably occur in the small intestine of domestic ducks. The body temperature is about 42 °C, the pH near the yolk stalk is about 7.4, the osmotic pressure of luminal fluid from the environment of *P. minutus* is about 170 mmol l^{-1} NaCl and bile salts are present in concentrations similar to those used *in vitro* by Lackie (Crompton, 1969; Crompton & Edmonds, 1969; Crompton & Nesheim, 1970). In the case of *P. minutus*, the enhancing effect of the bile salts may be considered as an adaptation which results in the evagination of the proboscis before the majority of the parasites have been propelled by intestinal motility past their normal attachment zone in the duck's small intestine (Table 1). This conclusion is based on the assumption that intestinal motility is one of the factors which affects the distribution of acanthocephalans in the small intestine. Further details of the physiology of infection and the possible action of the bile salts are discussed by Graff & Kitzman (1965) and Lackie (1974a, b).

Observations on the course of infection of acanthocephalans may be explained in terms of the relationship between the process of infection and the digestive physiology of the host. Kennedy (1972) infected gold-fish, *Carassius auratus*, by introducing *G. pulex* containing cystacanths of *Pom. laevis* into their tanks. Although the immediate process of infection was not studied and *Pom. laevis* is not known to reach maturity in *C. auratus* (Kennedy, personal communication), the observed variations in the distribution and populations of parasites could have been related to the effects, on intestinal motility and digestion, of having given different numbers of *Gammarus* to the fish, on having fed some and fasted others at the time of infection, and on having kept the hosts at different temperatures.

Feeding and nutrition

An adult acanthocephalan feeds by absorbing nutrients from the contents of that part of the host's intestine in which the worm lives. Convincing evidence for this statement was obtained by Edmonds (1965), who gave oral doses of either ^{32}P disodium hydrogen phosphate or $[^{14}C]$L-leucine to some rats infected with *M. dubius* and intraperitoneal doses to others. The recovery of radioactivity from worms infecting rats given the oral doses was far greater than that from rats given intraperitoneal injections. Most of an acanthocephalan's absorptive surface is associated with the metasoma (Fig. 1) which is the part exposed to the intestinal lumen. There is no evidence for the implication of Van Cleave (1952) that absorption from the mucosa may occur through the surface of the proboscis and praesoma. By means of autoradiographic techniques applied to worms incubated with nutrients *in vitro*, Hammond (1968a) concluded that the praesoma of *A.*

ranae was not involved in the absorption of radioactive glyceryl trioleate, and Hibbard & Cable (1968) reached a similar conclusion about the absorption of radioactive glucose, tyrosine and thymidine by *Paulisentis fractus*. Nor is there any direct evidence to support the suggestion that some acanthocephalans may have mouths and stomachs (Weinland, 1856) or that *P. boschadis* (= *P. minutus*) sucks blood (Lapage, 1968).

Investigations of the ultrastructure of the body walls of *P. minutus* (Crompton & Lee, 1965; Crompton, 1970), *M. dubius* (Nicholas & Mercer, 1965; Edmonds & Dixon, 1966; Rothman, 1967; Wright & Lumsden, 1968, 1969), *Pom. laevis* (Stranack, Woodhouse & Griffin, 1966; Hammond, 1968b), *A. ranae* (Hammond, 1967b, 1968b), *E. truttae* (Hammond, 1968b), *E. gadi* (Lange, 1970) and *Filicollis anatis* and *Mac. hirudinaceus* (Baraba-shova, 1965) have made observations from which absorption could be adduced as an obvious funtcion of the surface layers of palae- and archi-acanthocephalans (Plate 1d). Where comparisons have been made, the praesoma appears to possess fewer pores per unit of surface area and a simpler layer of ducts and canals than the metasoma (Crompton & Lee, 1965; Hammond, 1967b).

Acanthocephalans have been reported to absorb glucose, galactose, mannose, fructose, maltose, trehalose and a variety of amino acids in addition to glyceryl trioleate and various dyes (Crompton, 1970). The nutrient molecules of this list are all likely to be present in the host's small intestine during digestion and the observed distribution of adult worms (Table 1) and current opinion about their feeding activity illustrates their dependence on the digestive physiology of their hosts (Crompton, 1973). By incubating worms *in vitro* with neutral amino acids labelled with ^{14}C, the active transport of leucine by *M. dubius* has been demonstrated (Rothman & Fisher, 1964; Branch, 1970; Uglem & Read, 1973) and evidence has been obtained by the same workers, from *M. dubius* and *Mac. hirudinaceus*, indicating that the uptake of methionine, alanine, serine, leucine and isoleucine may involve either mediated or diffusion processes and, in some cases, both mechanisms for the same amino acid.

The results from another line of work suggest that the relationships between the parasite's feeding processes and the host's digestive physiology are even more intricate. When adult *M. dubius* were incubated *in vitro* with the peptides, leucylleucine, leucylglycylglycine, glycylglycine, alanyl-alanine, tri-alanine and tetra-alanine, amino acids or mixtures of amino acids and peptide subunits were liberated into the incubation medium (Uglem, Pappas & Read, 1973). This hydrolytic activity of *M. dubius* is related to an aminopeptidase associated with the parasite's surface; washing the worms repeatedly with Krebs–Ringer saline did not reduce the activity.

The probable function of the enzyme is that it facilitates the uptake of amino acids by liberating these molecules from the peptides in the vicinity of the amino acid transport system. The enzyme appears to be a component of the parasite and not an adsorbed secretion of the host. Cystacanths of *M. dubius*, which have not been ingested by rats, show surface aminopeptidase activity after being stimulated to excyst *in vitro*. Adult *M. dubius* also possess considerable surface amylase activity which can be removed by successive washings with saline, partially restored by incubation of washed worms with luminal contents from the small intestine of rats and inhibited by the presence of rabbit antiserum raised against extracts of rat pancreas (Ruff, Uglem & Read, 1973). Thus, this surface enzyme appears to be an adsorbed product of the host; the possession of a filamentous epicuticle of mucopolysaccharide (Wright & Lumsden, 1968; Rothman & Elder, 1969; Plate 1e) would probably provide a suitable adsorptive surface.

Relatively little information is available about the nutritional requirements of acanthocephalans, but the withholding of food for 48 h from rats and ducks harbouring established infections of *M. dubius* (Burlingame & Chandler, 1941) and *P. minutus* (Nicholas & Hynes, 1958; Hynes & Nicholas, 1963) respectively, greatly reduces the worm burden. Read & Rothman (1958) detected considerable reductions in the glycogen content and wet weights of *M. dubius* which survived the starvation of their hosts from 24–80 h, compared with those from feeding hosts. They also observed that the growth of *M. dubius* stopped and that many worms were lost after infected rats were transferred from a standard to a carbohydrate-free diet. Crompton & Nesheim (1973) fed male rats on four isocaloric diets which varied in carbohydrate content from 0 to about 59 %. Infections of *M. dubius* became established irrespective of diet, but there was little increase in dry weight and no sign of reproductive activity in worms from rats fed the carbohydrate-free diets, whereas worms from rats fed diets containing 3.6 % or about 59 % carbohydrate gained in dry weight and contained fully developed eggs after five weeks. Thus, it has been demonstrated that the growth of *M. dubius* in rats appears to be dependent on the ingestion of carbohydrate by the host. There is also evidence which indicates that the cholesterol requirement of *Mac. hirudinaceus* and *M. dubius* must be satisfied by the diet or secretions of the host (Barrett, Caid & Fairbairn, 1970).

Not all species of acanthocephalan from poikilotherms respond in the same manner as those from homoiotherms when their hosts are fasting. Although starvation of gold-fish has a deleterious effect on *Pom. laevis* (Kennedy, 1972), withholding food from brown trout, *Salmo trutta*, for up to 24 days did not appear to result in the loss of *E. truttae* (Awachie, 1972), unless the hungry fish ate any worms they passed. Specimens of *A. ranae* were recovered from frogs which had not fed for four months (Pflugfelder, 1949) and adult *N. emydis* have been observed to live for many months in the intestines of turtles which are not feeding (Van Cleave & Ross, 1944).

EVOLUTION OF THE HOST–PARASITE RELATIONSHIP

Most authorities tend to agree that the Acanthocephala have phylogenetic affinities with the Aschelminthes (Meyer, 1933; Haffner, 1950; Hyman, 1951; Golvan, 1958; Lang, 1963; Remane, 1963; Nicholas, 1970; Whitfield, 1971). From anatomical evidence obtained by light microscopy, the Acanthocephala have been considered to have a phylogenetic relationship with the Priapulida (Meyer, 1933; Golvan, 1958), which are a small group of free-living, benthic, marine aschelminths having an alimentary tract and an introversible praesoma (Hyman, 1951). Meyer and Golvan appear to have been influenced by the similarities between the external morphology of priapulids, acanthocephalans and the fossil worms found by Walcott (1911) in the mid-Cambrian Burgess Shale of British Columbia (Whittington, 1971). I am most grateful to Mr S. Conway-Morris who took new photographs of the specimens of *Ottoia prolifica*, which were first published by Walcott and then reproduced by Meyer (Plate 3*a*, *b*), and to Professor H. B. Whittington for Plate 3*c* which depicts the praesoma of a hitherto unstudied specimen of *O. prolifica*, which appears to have died with a partially inverted praesoma. Comparison of an interpretation of the morphology of these fossils (Fig. 2) with the morphology of a modern acanthocephalan (Fig. 1) and with the diagrammatic hypotheses of either Haffner (1950) or Golvan (1958) about the morphology of an ancestral relative of the Acanthocephala (Fig. 3) does not contradict the view that acanthocephalans or their ancestors may have evolved from worms not unlike *Ottoia*, although it now appears that modern priapulids may not be pseudocoelomates (Shapeero, 1961). Perhaps judgement should be further reserved, however, until the observations of Banta & Rice (Whittington, personal communication) and Conway-Morris, Department of Geology, University of Cambridge, on other fossil worms from the Burgess Shale have been published, particularly since Whitfield (1971) concluded that the ultrastructure of acanthocephalan spermatozoa, body walls and musculature was more like that of the nematodes and rotifers than the priapulids.

If one accepts the opinion that endoparasitic acanthocephalans may have had their origins in the seas of the mid-Cambrian era, in association with a flourishing fauna of arthropods (Van Cleave, 1947; Whittington, 1971), it follows that the evolution of acanthocephalan host–parasite relationships has taken about 600 million years to reach the present state of complexity. It is not surprising, therefore, that the present state of our knowledge of these relationships is superficial and incomplete.

ACKNOWLEDGEMENTS

I am very grateful to Miss Susan Arnold, Mr David Barnard, Mr S. Conway-Morris, Mrs Y. Gardner, Dr C. R. Kennedy, Dr Ann M. Lackie, Dr J. M. Lackie, Professor M. C. Nesheim, Dr B. B. Nickol, Dr Susan Rotheram, Mrs Jan Saunders, Dr G. D. Schmidt, Dr W. W. Wanson, Dr P. J. Whitfield and Professor H. B. Whittington, FRS, for much assistance and helpful criticism. I also thank the Agricultural Research Council for generous financial support.

REFERENCES

AWACHIE, J. B. E. (1966). The development and life history of *Echinorhynchus truttae* Schrank, 1788 (Acanthocephala). *J. Helminth.*, **40**, 11–32.

—— (1972). Experimental studies on some host–parasite relationships of the *Acanthocephala*. Effect of host size and starvation on *Echinorhynchus truttae* Schrank, 1788, in its definitive host. *Acta parasit. pol.*, **20**, 383–388.

BAER, J. C. (1961). Embranchement des Acanthocéphales. In *Traité de Zoologie*, vol. IV (ed. P.-P. Grassé). Paris: Masson et Cie, 733–782.

BAKKE, T. A. (1973). Studies of the helminth fauna of Norway XXV: The common gull, *Larus canus* L. as final host for Acanthocephala in a Norwegian locality. *Norw. J. Zool.*, **20**, 1–5.

BALL, G. (1928). An acanthocephalid, *Corynosoma strumosum* (Rudolphi), in the California harbor seal. *Anat. Rec.*, **41**, 82.

BARABASHOVA, V. N. (1965). Fine structure of the epidermis of two species of Acanthocephala – *Filicollis anatis* and *Macracanthorhynchus hirudinaceus* (in Russian). *Mater. nauch. Konf. vses. Obshch. Gelmint.*, part III, 25–29.

BARKER, J. & MACALISTER, A. (1865). On *Echinorhynchus porrigens*. *Proc. nat. Hist. Soc. Dublin*, **4**, 293–294.

BARNES, J. R. (1968). The morphology and ecology of *Echinorhynchus lageniformis* Ekbaum, 1938 (Acanthocephala). Master's thesis, Oregon State University.

BARRETT, J. & BUTTERWORTH, P. E. (1968). The carotenoids of *Polymorphus minutus* (Acanthocephala) and its intermediate host, *Gammarus pulex*. *Comp. Biochem. Physiol.*, **27**, 575–581.

BARRETT, J., CAID, G. D. & FAIRBAIRN, D. (1970). Sterols in *Ascaris lumbricoides* (Nematoda), *Macracanthorhynchus hirudinaceus* and *Moniliformis dubius* (Acanthocephala), and *Echinostoma revolutum* (Trematoda). *J. Parasit.*, **56**, 1004–1008.

BAYLIS, H. A. (1944). Three new Acanthocephala from marine fishes of Australasia. *Ann. Mag. nat. Hist.*, 11th series, **11**, 462–472.

BENTON, A. H. (1954). Notes on *Moniliformis clarki* (Ward) in Eastern New York (Moniliformidae: Acanthocephala). *J. Parasit.*, **40**, 102–103.

BETHEL, W. M. & HOLMES, J. C. (1973). Altered evasive behavior and responses to light in amphipods harboring acanthocephalan cystacanths. *J. Parasit.*, **59**, 945–956.

(1974). Correlation of altered evasive behavior in *Gammarus lacustris* (Amphipoda) harboring cystacanths of *Polymorphus paradoxus* (Acanthocephala) with the infectivity to the definitive host. *J. Parasit.*, **60**, 272–274.

BHALERAO, G. D. (1931). On a new species of Acanthocephala from *Ophiocephalus striatus. Ann. Mag. nat. Hist.*, 10th series, **7**, 569–573.

BOGITSH, B. (1961). Histological and histochemical observations on the nature of the cyst of *Neoechinorhynchus cylindratus* in *Lepomis* sp. *Proc. helminth. Soc. Wash.*, **28**, 75–81.

BOWEN, R. C. (1967). Defense reactions of certain spirobolid millipedes to larval *Macracanthorhynchus ingens. J. Parasit.*, **53**, 1092–1095.

BRANCH, S. I. (1970). Accumulation of amino acids by *Moniliformis dubius* (Acanthocephala). *Expl Parasit.*, **27**, 95–99.

BRAND, T. VON (1973). *Biochemistry of Parasites*, 2nd edn. New York and London: Academic Press.

BULLOCK, W. L. (1963). Intestinal histology of some salmonid fishes with particular reference to the histopathology of acanthocephalan infections. *J. Morph.*, **112**, 23–44.

(1966). A redescription of *Octospiniferoides chandleri* Bullock, 1957. *J. Parasit.*, **52**, 735–738.

(1967). The intestinal histology of the mosquito fish, *Gambusia affinis* (Baird and Girard). *Acta zool. Stockh.*, **48**, 1–17.

(1969). Morphological features as tools and as pitfalls in acanthocephalan systematics. In *Problems in Systematics of Parasites*, 9–45 (ed. G. D. Schmidt). Baltimore, Maryland and Manchester, England: University Park Press.

BURLINGAME, P. L. & CHANDLER, A. C. (1941). Host–parasite relations of *Moniliformis dubius* (Acanthocephala) in albino rats, and the environmental nature of resistance to single and superimposed infections with this parasite. *Am. J. Hyg.*, **33**, 1–21.

BUTTERWORTH, P. E. (1968a). An estimation of the haemolymph volume in *Gammarus pulex. Comp. Biochem. Physiol.*, **26**, 1123–1125.

(1968b). Aspects of the nutrition of *Polymorphus minutus* during its development. *Parasitology*, **58**, 3P.

(1969a). The development of the body wall of *Polymorphus minutus* (Acanthocephala) in its intermediate host, *Gammarus pulex. Parasitology*, **59**, 373–388.

(1969b). Studies on the physiology and development of *Polymorphus minutus*. Ph.D. Dissertation, University of Cambridge.

BYRD, E. E. & DENTON, F. J. (1949). The helminth parasites of birds. II. A new species of Acanthocephala from North American birds. *J. Parasit.*, **35**, 391–410.

CABLE, R. M. & DILL, W. T. (1967). The morphology and life history of *Paulisentis fractus* Van Cleave and Bangham, 1949 (Acanthocephala: Neoechinorhynchidae). *J. Parasit.*, **53**, 810–817.

CHAICHARN, A. & BULLOCK, W. L. (1967). The histopathology of acanthocephalan infections in suckers with observations on the intestinal histology of two species of catostomid fishes. *Acta zool. Stockh.*, **48**, 19–42.

CONNELL, R. & CORNER, A. H. (1957). *Polymorphus paradoxus* sp. nov. (Acanthocephala) parasitizing beavers and muskrats in Alberta, Canada. *Can. J. Zool.*, **35**, 525–533.

CROFTON, H. D. (1971). A quantitative approach to parasitism. *Parasitology*, **62**, 179–193.

CROMPTON, D. W. T. (1964a). Studies on Acanthocephala, with special reference to *Polymorphus minutus*. Ph.D. Dissertation, University of Cambridge.

(1964b). The envelope surrounding *Polymorphus minutus* (Goeze, 1782) (Acanthocephala) during its development in the intermediate host, *Gammarus pulex*. *Parasitology*, **54**, 721–735.

(1967). Studies on the haemocytic reaction of *Gammarus* spp., and its relationship to *Polymorphus minutus* (Acanthocephala). *Parasitology*, **57**, 389–401.

(1969). On the environment of *Polymorphus minutus* (Acanthocephala) in ducks. *Parasitology*, **59**, 19–28.

(1970). *An Ecological Approach to Acanthocephalan Physiology*. London: Cambridge University Press.

(1973). The sites occupied by some parasitic helminths in the alimentary tract of vertebrates. *Biol. Rev.*, **47**, 27–83.

(1974). Experiments on insemination in *Moniliformis dubius* (Acanthocephala). *Parasitology*, **68**, 229–238.

CROMPTON, D. W. T., ARNOLD, S. & BARNARD, D. (1972). The patent period and production of eggs of *Moniliformis dubius* (Acanthocephala) in the small intestine of male rats. *Int. J. Parasit.*, **2**, 319–326.

CROMPTON, D. W. T. & EDMONDS, S. J. (1969). Measurements of the osmotic pressure in the habitat of *Polymorphus minutus* (Acanthocephala) in the intestine of domestic ducks. *J. exp. Biol.*, **50**, 69–77.

CROMPTON, D. W. T. & LEE, D. L. (1965). The fine structure of the body wall of *Polymorphus minutus* (Goeze, 1782) (Acanthocephala). *Parasitology*, **55**, 357–364.

CROMPTON, D. W. T. & NESHEIM, M. C. (1970). Lipid, bile acid, water and dry matter content of the intestinal tract of domestic ducks with reference to the habitat of *Polymorphus minutus* (Acanthocephala). *J. exp. Biol.*, **52**, 437–445.

(1973). Relationships between *Moniliformis dubius* (Acanthocephala) and the carbohydrate intake of rats. *Parasitology*, **67**, ii.

CROMPTON, D. W. T. & WALTERS, D. E. (1972). An analysis of the course of infection of *Moniliformis dubius* (Acanthocephala) in rats. *Parasitology*, **64**, 517–523.

CROMPTON, D. W. T. & WHITFIELD, P. J. (1968). The course of infection and egg production of *Polymorphus minutus* (Acanthocephala) in domestic ducks. *Parasitology*, **58**, 231–246.

(1974). Observations on the functional organization of the ovarian balls of *Moniliformis* and *Polymorphus* (Acanthocephala). *Parasitology*, **69**, 429–443.

CROOK, J. R. & GRUNDMANN, A. W. (1964). The life history and larval development of *Moniliformis clarki* (Ward, 1917). *J. Parasit.*, **50**, 689–693.

CROSS, S. X. (1934). A probable case of non-specific immunity between two parasites of ciscoes of the trout lake region of Northern Wisconsin. *J. Parasit.*, **20**, 244–245.

DECHTIAR, A. O. (1968). *Neoechinorhynchus carpiodi* n. sp. (Acanthocephala: Neoechinorhynchidae) from quillback of Lake Erie. *Can. J. Zool.*, **46**, 201–204.

DEGIUSTI, D. L. (1949a). The life cycle of *Leptorhynchoides thecatus* (Linton), an acanthocephalan of fish. *J. Parasit.*, **35**, 437–460.

(1949b). Partial development of *Echinorhynchus coregoni* in *Hyalella azteca* and the cellular reaction of the amphipod to the parasite. *J. Parasit.*, **35**, suppl., 31.

DENNY, M. (1969). Life-cycles of helminth parasites using *Gammarus lacustris* as an intermediate host in a Canadian lake. *Parasitology*, **59**, 795–827.

DUNAGAN, T. T. (1962). Studies on *in vitro* survival of Acanthocephala. *Proc. helminth. Soc. Wash.*, **29**, 131–135.

EDMONDS, S. J. (1965). Some experiments on the nutrition of *Moniliformis dubius* Meyer (Acanthocephala). *Parasitology*, **55**, 337–347.

— (1966). Hatching of the eggs of *Moniliformis dubius*. *Expl Parasit.*, **19**, 216–226.

EDMONDS, S. J. & DIXON, B. R. (1966). Uptake of small particles by *Moniliformis dubius* (Acanthocephala). *Nature, Lond.*, **209**, 99.

EJSYMONT, L. (1970a). Parasites of common burbot, *Lota lota lota* (L.), from the river Biebrza. *Acta parasit. pol.*, **17**, 195–201.

— (1970b). Parasites of the sheatfish, *Silurus glanis* L., from the river Biebrza and its tributaries. *Acta parasit. pol.*, **17**, 203–216.

EKBAUM, E. (1938). Notes on the occurrence of Acanthocephala in Pacific fishes. I. *Echinorhynchus gadi* (Zoega) Müller in salmon and *E. lageniformis* sp. nov. and *Corynosoma strumosum* (Rudolphi) in two species of flounder. *Parasitology*, **30**, 267–273.

FARNER, D. S. (1960). Digestion and the digestive system. In *Biology and Comparative Physiology of Birds*, vol. I (ed. A. J. Marshall). New York and London: Academic Press.

GLASGOW, R. D. & DEPORTE, P. (1939). Recovery from excreta of the pigeon of viable eggs of the giant thorny-headed worm of swine. *J. econ. Ent.*, **32**, 882.

GOLVAN, Y. J. (1958). Le phylum des Acanthocephala. Première note. Sa place dans l'échelle zoologique. *Ann. Parasit. hum. et comp.*, **33**, 538–602.

— (1962). Acanthocéphales parasites des mammifères. Not seen, abstracted in *Helminth. Abstr.*, **33**, 210.

GRAFF, D. J. & KITZMAN, W. B. (1965). Factors influencing the activation of acanthocephalan cystacanths. *J. Parasit.*, **51**, 424–429.

HAFFNER, K. VON (1950). Organisation und systematische Stellung der Acanthocephalen. In *Festschrift für Dr Berthold Klatt. Zool. Anz.*, **145**, 243–274.

HAMMOND, R. A. (1967a). The mode of attachment within the host of *Acanthocephalus ranae* (Schrank, 1788), Lühe, 1911. *J. Helminth.*, **41**, 321–328.

— (1967b). The fine structure of the trunk and praesoma wall of *Acanthocephalus ranae* (Schrank, 1788), Lühe, 1911. *Parasitology*, **57**, 475–486.

— (1968a). Some observations on the role of the body wall of *Acanthocephalus ranae* in lipid uptake. *J. exp. Biol.*, **48**, 217–225.

— (1968b). Observations on the body surface of some acanthocephalans. *Nature, Lond.*, **218**, 872–873.

HARMS, C. E. (1965). The life cycle and larval development of *Octospinifer macilentis* (Acanthocephala: Neoechinorhynchidae). *J. Parasit.*, **51**, 286–293.

HARRIS, J. E. (1972). The immune response of a cyprinid fish to infections of the acanthocephalan *Pomphorhynchus laevis*. *Int. J. Parasit.*, **2**, 459–469.

HIBBARD, K. M. & CABLE, R. M. (1968). The uptake and metabolism of tritiated glucose, tyrosine, and thymidine by adult *Paulisentis fractus* Van Cleave and Bangham, 1949 (Acanthocephala: Neoechinorhynchidae). *J. Parasit.*, **54**, 517–523.

HINDSBO, O. (1972). Effects of *Polymorphus* (Acanthocephala) on colour and behavior of *Gammarus lacustris*. *Nature, Lond.*, **238**, 333.

HINE, P. M. (1970). Studies on the parasites of some freshwater fish. Ph.D. Dissertation, University of Exeter.

HOLLOWAY, H. L. & NICKOL, B. B. (1970). Morphology of the trunk of *Corynosoma hamanni* (*Acanthocephala*: Polymorphidae). *J. Morph.*, **130**, 151–162.

HOLMES, J. C. (1961). Effects of concurrent infections on *Hymenolepis diminuta* (Cestoda) and *Moniliformis dubius* (Acanthocephala). I. General effects and comparison with crowding. *J. Parasit.*, **47**, 209–216.

HOPP, W. B. (1954). Studies on the morphology and life cycle of *Neoechinorhynchus emydis* (Leidy), an acanthocephalan parasite of the map turtle, *Graptemys geographica* (La Sueur). *J. Parasit.*, **40**, 284–299.

HUIZINGA, H. W. & HALEY, A. J. (1962). The occurrence of the acanthocephalan parasite, *Telosentis tenuicornis*, in the spot, *Leiostomus xanthurus*, in Chesapeake Bay. *Chesapeake Sci.*, **3**, 35–42.

HYMAN, L. H. (1951). *The Invertebrates*, vol. III. New York, Toronto and London: McGraw-Hill Book Company, Inc.

HYNES, H. B. N. (1955). The reproductive cycle of some British freshwater Gammaridae. *J. Anim. Ecol.*, **24**, 352–387.

HYNES, H. B. N. & NICHOLAS, W. L. (1957). The development of *Polymorphus minutus* (Goeze, 1782) (Acanthocephala) in the intermediate host. *Ann. trop. Med. Parasit.*, **51**, 380–391.

(1958). The resistance of *Gammarus* spp. to infection by *Polymorphus minutus* (Goeze, 1782) (Acanthocephala). *Ann. trop. Med. Parasit.*, **52**, 376–383.

(1963). The importance of the acanthocephalan *Polymorphus minutus* as a parasite of domestic ducks in the United Kingdom. *J. Helminth.*, **37**, 185–198.

JOHNSTON, T. H. & DELAND, E. W. (1929). Australian Acanthocephala, No. 2. *Trans. R. Soc. S. Aust.*, **53**, 155–166.

KATES, K. C. (1942). Viability of the eggs of the swine thorn-headed worm (*Macracanthorhynchus hirudinaceus*). *J. agric. Res.*, **64**, 93–100.

(1944). Some observations on experimental infections of pigs with the thorn-headed worm, *Macracanthorhynchus hirudinaceus*. *Am. J. vet. Res.*, **5**, 166–172.

KENNEDY, C. R. (1972). The effects of temperature and other factors upon the establishment and survival of *Pomphorhynchus laevis* (Acanthocephala) in goldfish, *Carassius auratus*. *Parasitology*, **65**, 283–294.

(1974). The importance of parasite mortality in regulating the population size of the Acanthocephalan *Pomphorhynchus laevis* in goldfish. *Parasitology*, **68**, 93–101.

KING, D. & ROBINSON, E. S. (1967). Aspects of the development of *Moniliformis dubius*. *J. Parasit.*, **53**, 142–149.

LACKIE, A. M. (1974a). The activation of cystacanths of *Polymorphus minutus* (Acanthocephala) *in vitro*. *Parasitology*, **68**, 135–146.

(1974b). The activation of cystacanths of *Polymorphus minutus* (Acanthocephala). Ph.D. Dissertation, University of Cambridge.

LACKIE, J. M. (1972a). The course of infection and growth of *Moniliformis dubius* (Acanthocephala) in the intermediate host *Periplaneta americana*. *Parasitology*, **64**, 95–106.

(1972b). The effect of temperature on the development of *Moniliformis dubius* (Acanthocephala) in the intermediate host, *Periplaneta americana*. *Parasitology*, **65**, 371–377.

(1973). Studies on interactions between *Moniliformis* (Acanthocephala) and its intermediate hosts. Ph.D. Dissertation, University of Cambridge.

LACKIE, J. M. & ROTHERAM, S. (1972). Observations on the envelope surrounding *Moniliformis dubius* (Acanthocephala) in the intermediate host, *Periplaneta americana*. *Parasitology*, **65**, 303–308.

LANG, K. (1963). The relationship between the Kinorhyncha and Priapulida and their connection with the Aschelminthes. In *The Lower Metazoa*, 256–263 (eds. E. C. Dougherty, Z. N. Brown, E. D. Hanson and W. D. Hartmann). Berkeley: University of California Press.

LANGE, H. (1970). Über Struktur und Histochemie des Integuments von *Echinorhynchus gadi* Müller (Acanthocephala). *Z. Zellforsch. mikrosk. Anat.*, **104**, 149–164.

LAPAGE, G. (1968). *Veterinary Parasitology*, 2nd edn. Edinburgh and London: Oliver & Boyd.

LE ROUX, M. L. (1931*a*). Castration parasitaire et caractères sexuels secondaires chez les Gammariens. *C. r. hebd. Séanc. Acad. Sci. Paris*, **192**, 889–891.

(1931*b*). La castration expérimentale de femelles de Gammariens et sa répercusion sur l'évolution des oostégites. *C. r. hebd. Séanc. Acad. Sci. Paris*, **193**, 885–887.

LINGARD, A. M. & CROMPTON, D. W. T. (1972). Observations on the establishment of *Polymorphus minutus* (Acanthocephala) in the intestines of domestic ducks. *Parasitology*, **65**, 159–165.

LOCKWOOD, A. P. M. & INMAN, C. B. E. (1973). The blood volume of some amphipod crustaceans in relation to the salinity of the environment they inhabit. *Comp. Biochem. Physiol.*, **44A**, 935–941.

LYSTER, L. L. (1940). Parasites of some Canadian sea mammals. *Can. J. Res.*, **18**, 395–409.

MACKENZIE, K. & GIBSON, D. I. (1970). Ecological studies of some parasites of plaice, *Pleuronectes platessa* L., and flounder, *Platichthys flesus* (L.). *Symp. Br. Soc. Parasit.*, **8**, 1–42.

MANTER, H. W. (1928). Notes on the eggs and larvae of the thorny-headed worm of hogs. *Trans. Am. microsc. Soc.*, **47**, 342–347.

MERCER, E. H. & NICHOLAS, W. L. (1967). The ultrastructure of the capsule of the larval stages of *Moniliformis dubius* (Acanthocephala) in the cockroach *Periplaneta americana*. *Parasitology*, **57**, 169–174.

MEYER, A. (1931). Urhautzelle, Hautbahn und plasmodiale Entwicklung der Larve von *Neoechinorhynchus rutili* (Acanthocephala). *Zool. Jb.*, **53**, 103–126.

(1933). *Acanthocephala. Bronn's Klassen und Ordnungen des Tierreichs*, **4**. Leipzig: Akademische Verlagsgesellschaft m.b.H.

MILLER, M. A. (1943). Studies on the developmental stages and glycogen metabolism of *Macracanthorhynchus hirudinaceus* in the Japanese beetle larva. *J. Morph.*, **73**, 19–42.

MOORE, D. V. (1942). An improved technique for the study of the acanthor stage in certain acanthocephalan life histories. *J. Parasit.*, **28**, 495.

(1946). Studies on the life history and development of *Moniliformis dubius* Meyer, 1933. *J. Parasit.*, **32**, 257–271.

(1962). Morphology, life history and development of the acanthocephalan *Mediorhynchus grandis* Van Cleave, 1916. *J. Parasit.*, **48**, 76–86.

MUNRO, W. R. (1953). Intersexuality in *Asellus aquaticus* L. parasitized by a larval acanthocephalan. *Nature, Lond.*, **172**, 313.

NICHOLAS, W. L. (1967). The biology of the Acanthocephala. *Adv. Parasit.*, **5**, 205–246.

(1970). Taxonomy: genetics and evolution of parasites (Acanthocephala). *J. Parasit.*, **56**, 506–507.

(1973). The biology of the Acanthocephala. *Adv. Parasit.*, **11**, 671–706.

NICHOLAS, W. L. & HYNES, H. B. N. (1958). Studies on *Polymorphus minutus* (Goeze, 1782) (Acanthocephala) as a parasite of the domestic duck. *Ann. trop. Med. Parasit.*, **52**, 36–47.

NICHOLAS, W. L. & MERCER, E. H. (1965). The ultrastructure of the tegument of *Moniliformis dubius* (Acanthocephala). *Q. Jl microsc. Sci.*, **106**, 137–146.

PETROSCHENKO, V. I. (1956). *Acanthocephala of Domestic and Wild Animals*. I. (In Russian). Moskva: Akad. Nauk. SSSR.

(1958). *Acanthocephala of Domestic and Wild Animals*. II. (In Russian). Moskva: Akad. Nauk. SSSR.

PFLUGFELDER, O. (1949). Histophysiologische Untersuchungen über die Fettresorption darmloser Parasiten: die Funktion der Lemnisken der Acanthocephala. *Z. ParasitKde.*, **14,** 274–280.

PODDER, T. N. (1937). A new species of Acanthocephala, *Neoechinorhynchus topseyi* n. sp., from a Calcutta fish, *Polynemus heptadactylus* (Cuv. & Val.). *Parasitology*, **29,** 365–369.

PODESTA, R. B. & HOLMES, J. C. (1970). The life cycles of three polymorphids (Acanthocephala) occurring as juveniles in *Hyalella azteca* (Amphipoda) at Cooking Lake, Alberta. *J. Parasit.*, **56,** 1118–1123.

PRAKASH, A. & ADAMS, J. R. (1960). A histopathological study of the intestinal lesions induced by *Echinorhynchus lageniformis* (Acanthocephala–Echinorhynchidae) in the starry flounder. *Can. J. Zool.*, **38,** 895–897.

RAWSON, D. (1952). The occurrence of parasitic worms in British freshwater fishes. *Ann. Mag. nat. Hist.*, 12th series, **5,** 877–887.

READ, C. P. & ROTHMAN, A. H. (1958). The carbohydrate requirement of *Moniliformis* (Acanthocephala). *Expl Parasit.*, **7,** 191–197.

REMANE, A. (1963). The systematic position and phylogeny of the pseudocoelomates. In *The Lower Metazoa*, 247–255 (eds. E. C. Dougherty, Z. N. Brown, E. D. Hanson and W. D. Hartmann). Berkeley: University of California Press.

ROBINSON, E. S. & STRICKLAND, B. C. (1969). Cellular responses of *Periplaneta americana* to acanthocephalan larvae. *Expl Parasit.*, **26,** 384–392.

ROTHERAM, S. & CROMPTON, D. W. T. (1972). Observations on the early relationship between *Moniliformis dubius* (Acanthocephala) and the haemocytes of the intermediate host, *Periplaneta americana. Parasitology*, **64,** 15–21.

ROTHMAN, A. H. (1967). Ultrastructural enzyme localization in the surface of *Moniliformis dubius* (Acanthocephala). *Expl Parasit.*, **21,** 42–46.

ROTHMAN, A. H. & ELDER, J. E. (1969). Histochemical nature of an acanthocephalan, a cestode and a trematode absorbing surface. *Comp. Biochem. Physiol.*, **33,** 745–762.

ROTHMAN, A. H. & FISHER, F. M. (1964). Permeation of amino acids in *Moniliformis* and *Macracanthorhynchus* (Acanthocephala). *J. Parasit.*, **50,** 410–414.

RUFF, M. D., UGLEM, G. L. & READ, C. P. (1973). Interactions of *Moniliformis dubius* with pancreatic enzymes. *J. Parasit.*, **59,** 839–843.

RUMPUS, A. E. & KENNEDY, C. R. (1974). The effect of the acanthocephalan *Pomphorhynchus laevis* upon the respiration of its intermediate host, *Gammarus pulex. Parasitology*, **68,** 271–284.

SALT, G. (1961). The haemocytic reaction of insects to foreign bodies. In *The Cell and the Organism*, (eds. J. A. Ramsay and V. B. Wigglesworth). London: Cambridge University Press.

(1963). The defence reactions of insects to metazoan parasites. *Parasitology*, **53,** 527–642.

(1970). *The Cellular Defence Reactions of Insects.* London: Cambridge University Press.

SCHAEFER, P. W. (1970). *Periplaneta americana* (L.) as an intermediate host of *Moniliformis moniliformis* (Bremser) in Honolulu, Hawaii. *Proc. helminth. Soc. Wash.*, **37,** 204–207.

SCHMIDT, G. D. (1963). *Arhythmorhynchus capellae* sp. n. (Polymorphidae: Acanthocephala), a parasite of the common snipe *Capella gallinago delicata. J. Parasit.*, **49,** 483–484.

(1965). *Corynosoma bipapillum* sp. n. from Bonaparte's gull *Larus philadelphia* in Alaska, with a note on *C. constrictum* Van Cleave, 1918. *J. Parasit.*, **51,** 814–816.

(1969). Acanthocephala as agents of disease in wild mammals. *Wildl. Dis.*, **53,** 10 pp. (microfiche, Wildlife Disease Association, Ames, Iowa, USA).

SCHMIDT, G. D. & CANARIS, A. G. (1967). Acanthocephala from Kenya with descriptions of two new species. *J. Parasit.*, **53,** 634–637.

SCHMIDT, G. D. & OLSEN, O. W. (1964). Life cycle and development of *Prosthorhynchus formosus* (Van Cleave, 1918) Travassos, 1926, an acanthocephalan parasite of birds. *J. Parasit.*, **50,** 721–730.

SCHWARTZ, B. (1929). Important internal parasites of livestock. *Vet. Med.*, **24,** 336–346.

SEIDENBERG, A. J. (1973). Ecology of the acanthocephalan, *Acanthocephalus dirus* (Van Cleave, 1931), in its intermediate host, *Asellus intermedius* Forbes (Crustacea: Isopoda). *J. Parasit.*, **59,** 957–962.

SHAPEERO, W. L. (1961). Phylogeny of priapulida. *Science, Wash.*, **133,** 879–880.

SOLIMAN, K. N. (1955). Observations on some helminth parasites from ducks in Southern England. *J. Helminth.*, **29,** 17–26.

SPINDLER, L. A. & KATES, K. C. (1940). Survival on soil of eggs of the swine thornheaded worm *Macracanthorhynchus hirudinaceus*. *J. Parasit.*, **26,** suppl., 19.

STRANACK, F. R., WOODHOUSE, M. A. & GRIFFIN, R. L. (1966). Preliminary observations on the ultrastructure of the body wall of *Pomphorhynchus laevis* (Acanthocephala). *J. Helminth.*, **40,** 395–402.

TEDLA, S. & FERNANDO, C. H. (1970). Some remarks on the ecology of *Echinorhynchus salmonis* Muller, 1784. *Can. J. Zool.*, **48,** 317–321.

THRELFALL, W. (1968). Helminth parasites of some birds in Newfoundland. *Can. J. Zool.*, **46,** 909–913.

TREHERNE, J. E. (1957). Glucose absorption in the cockroach. *J. exp. Biol.*, **34,** 478–485.

UGLEM, G. L. & LARSON, O. R. (1969). The life history and larval development of *Neoechinorhynchus saginatus* Van Cleave and Bangham, 1949 (Acanthocephala: Neoechinorhynchidae). *J. Parasit.*, **55,** 1212–1217.

UGLEM, G. L., PAPPAS, P. W. & READ, C. P. (1973). Surface aminopeptidase in *Moniliformis dubius* and its relation to amino acid uptake. *Parasitology*, **67,** 185–195.

UGLEM, G. L. & READ, C. P. (1973). *Moniliformis dubius*: uptake of leucine and alanine by adults. *Expl Parasit.*, **34,** 148–153.

VAN CLEAVE, H. J. (1920). Acanthocephala parasitic in the dog. *J. Parasit.*, **7,** 91–94.

(1947). The Eoacanthocephala of North America, including the description of *Eocollis arcanus*, new genus and new species, superficially resembling the genus *Pomphorhynchus*. *J. Parasit.*, **33,** 285–296.

(1948). Expanding horizons in the recognition of a phylum. *J. Parasit.*, **34,** 1–20.

(1952). Some host–parasite relationships of the Acanthocephala, with special reference to the organs of attachment. *Expl Parasit.*, **1,** 305–330.

(1953). Acanthocephala of North American Mammals. *Illinois biol. Monogr.*, **23,** 1–179.

VAN CLEAVE, H. J. & HADERLIE, E. C. (1950). A new species of the acanthocephalan genus *Octospinifer* from California. *J. Parasit.*, **36,** 169–173.

VAN CLEAVE, H. J. & ROSS, E. L. (1944). Physiological responses of *Neoechinorhynchus emydis* (Acanthocephala) to various solutions. *J. Parasit.*, **30,** 369–372.

VENARD, C. E. & WARFEL, J. H. (1953). Some effects of two species of Acanthocephala on the alimentary canal of the large mouth bass. *J. Parasit.*, **39,** 187–190.

WALCOTT, C. D. (1911). Middle Cambrian Annelids. *Cambrian Geol. and Paleont. II. Smithsonian Misc. Coll.*, **57,** 109–144.

WANSON, W. W. & NICKOL, B. B. (1973). Origin of the envelope surrounding larval acanthocephalans. *J. Parasit.*, **59**, 1147.

WARD, H. (1940). Studies on the life-history of *Neoechinorhynchus cylindratus* (Van Cleave, 1913) (Acanthocephala). *Trans. Am. microsc. Soc.*, **59**, 327–347.

WEINLAND, D. F. (1856). On the digestive apparatus of the Acanthocephala. *Proc. Am. Ass. Advmt Sci.*, **10**, 197–201.

WEST, A. J. (1964). The acanthor membranes of two species of Acanthocephala. *J. Parasit.*, **50**, 731–734.

WHITFIELD, P. J. (1968). A histological description of the uterine bell of *Polymorphus minutus* (Acanthocephala). *Parasitology*, **58**, 671–682.

(1971). Phylogenetic affinities of Acanthocephala: an assessment of ultrastructural evidence. *Parasitology*, **63**, 49–58.

(1973). The egg envelopes of *Polymorphus minutus* (Acanthocephala). *Parasitology*, **66**, 387–403.

WHITTINGTON, H. B. (1971). The Burgess Shale: History of research and preservation of fossils. *North Am. Paleont. Convention, Chicago, 1969, Proc. I*, 1170–1201.

WIGGLESWORTH, V. B. (1927). Digestion in the cockroach. I. The hydrogen ion concentration in the alimentary canal. *Biochem. J.*, **21**, 791–796.

WRIGHT, R. D. (1971). The egg envelopes of *Moniliformis dubius*. *J. Parasit.*, **57**, 122–131.

WRIGHT, R. D. & LUMSDEN, R. D. (1968). Ultrastructural and histochemical properties of the acanthocephalan epicuticle. *J. Parasit.*, **54**, 1111–1123.

(1969). Ultrastructure of the tegumentary pore-canal system of the acanthocephalan *Moniliformis dubius*. *J. Parasit.*, **55**, 993–1003.

YAMAGUTI, S. (1963). *Systema Helminthum*, vol. V. *Acanthocephala*. New York and London: John Wiley and Sons.

EXPLANATION OF PLATES

PLATE 1

(*a*) The proboscis of an adult acanthocephalan worm.

(*b*) Longitudinal section through a cystacanth of *Polymorphus minutus* in the haemocoele of *Gammarus pulex*. The material was fixed in Carnoy's fluid, dehydrated in a series of solutions of *n*-butanol, embedded in paraffin wax (56 °C), sectioned at 10 μm and stained with Mallory's triple stain. The arrow indicates the envelope.

(*c*) Longitudinal section through a late acanthella of *P. minutus* in *G. pulex*. See (*b*) for details. The arrows indicate the envelope.

(*d*) Electron micrograph of the surface layers of the body wall of a 6 day old *Moniliformis dubius* from the small intestine of a rat. The material was fixed and prepared as described by Rotheram & Crompton (1972).

(*e*) Electron micrograph of the surface of a 106 day old *M. dubius*. See (*d*) for details. Note the filamentous epicuticle (Wright & Lumsden, 1968).

(*f*) Photomicrograph of *M. dubius* after 25 days of development in the haemocoele of *Periplaneta americana*. Note the haemocytes around the parasite. (Illustration supplied by W. W. Wanson.)

PLATE 2

(*a*) Electron micrograph of a section through a 1 day old acanthor of *Moniliformis dubius* (a) surrounded by haemocytes (h) of *Peri. americana*. Illustration prepared from Plate 1 of Rotheram & Crompton (1972).

(*b*) Electron micrograph of a section through an 8 day old acanthor of *M. dubius* (a) in *Peri. americana*. Note the presence of membranous material (mm) between the parasite and the haemocytes (h). (Illustration prepared from Plate 3A of Rotheram & Crompton, 1972.)

(*c*) Electron micrograph through the surface of a young acanthella of *M. dubius* (a) in *Peri. americana* showing details of the membranous material (mm). (Illustration prepared from Plate 4A of Rotheram & Crompton, 1972.)

(*d*) Electron micrograph through the surface of an acanthor of *M. dubius* (a) cultured *in vitro* for 19 days. Note the membranous material (mm) associated with the surface of the parasite. (Illustration prepared from Plate 3A of Lackie & Rotheram, 1972.)

(*e*) Photomicrograph of dead acanthors of *M. dubius* in the haemocoele of *Nauphoeta cinerea*, infected orally and maintained at 28 °C. Note the evidence of haemocytic encapsulation and the formation of melanin. (Illustration supplied by J. M. Lackie.)

PLATE 3

(*a*) Photograph by S. Conway-Morris of *Ottoia prolifica* (US National Museum Catalogue No. 57619) from the mid-Cambrian Burgess Shale. (A photograph of this specimen was first published by Walcott, 1911, Plate 19, Fig. 1, and Walcott's photograph was later reproduced by Meyer, 1933, Fig. 380.)

(*b*) Photograph by S. Conway-Morris of *O. prolifica* (US National Museum Catalogue No. 57620) from the mid-Cambrian Burgess Shale. (A photograph of this specimen was first published by Walcott, 1911, Plate 19, Fig. 2, and Walcott's photograph was later reproduced by Meyer, 1933, Fig. 381.)

(*c*) Photograph by H. B. Whittington of the praesoma of *O. prolifica* (Geological Survey of Canada No. 40972) from the mid-Cambrian Burgess Shale.

PLATE I

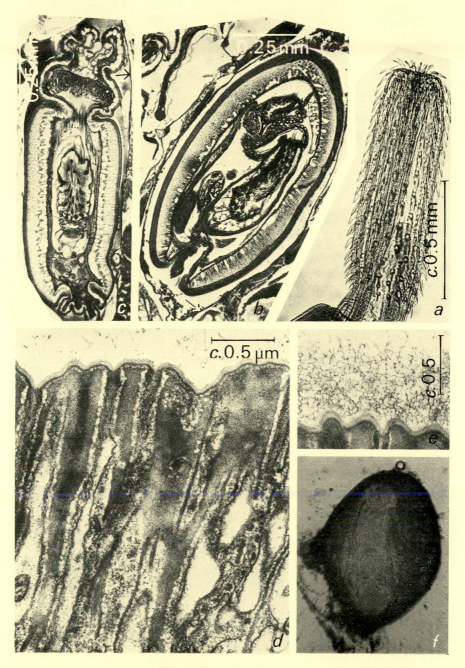

For explanation see p. 503

PLATE 2

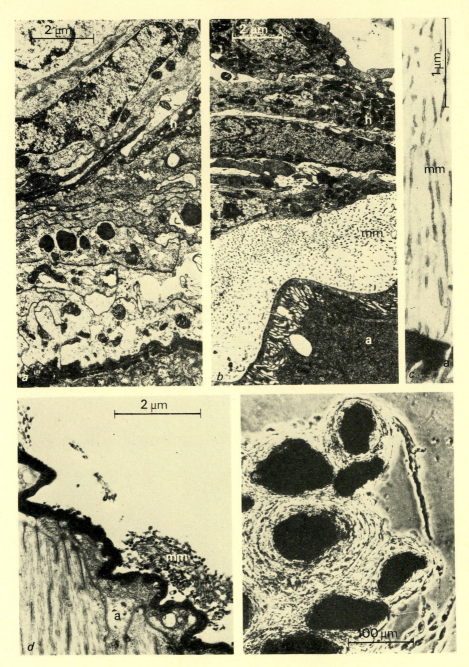

For explanation see p. 503

PLATE 3

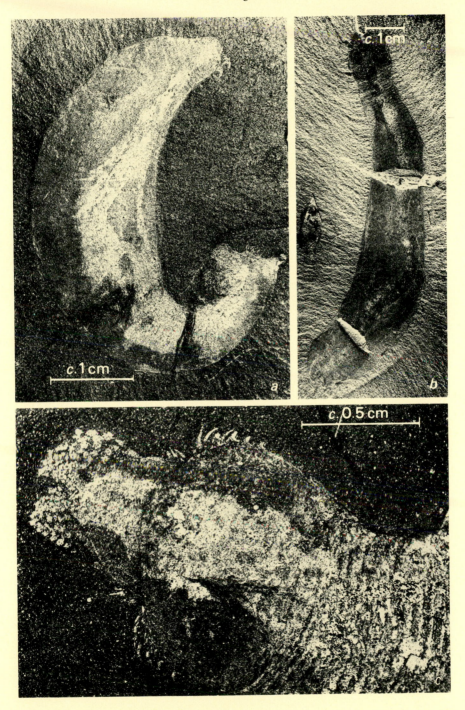

For explanation see p. 504

TAPEWORM–HOST INTERACTIONS

By C. ARME

Department of Zoology, Queen's University, Belfast, Northern Ireland

The study of a complex symbiotic association, such as endoparasitism, is fundamentally an ecological problem, complicated by the fact that the environment of one of the symbionts, the parasite, is a living organism, the host. A detailed analysis of this type of symbiosis ideally demands an examination of the environment provided by the host, in terms of its morphology, physiology, biochemistry, etc., together with a similar evaluation of the parasite. While the technical problems of working with the host may not be great, particularly if it is a warm-blooded vertebrate, studies on isolated parasites often present considerable difficulties. These in general relate to the necessity of obtaining large numbers of individuals drawn from a relatively homogeneous population, and their maintenance *in vitro* for a sufficient period of time to enable physiological observations to be made. It is hardly surprising that a symbiosis involving the laboratory rat and a tapeworm symbiont, *Hymenolepis diminuta*, which survives *in vitro* for a considerable time, has contributed much to our knowledge of tapeworm–host interactions. Standardised techniques for the in-vivo and in-vitro maintenance of this parasite has been described by Read, Rothman & Simmons (1963). Such is the degree of reproduceability within this system that experiments on nutrient uptake, conducted by undergraduate students in my laboratory, have yielded results almost identical to those presented by Read and co-workers, some 11 years ago.

There are however many other host–cestode symbioses which, although more difficult to study, and about which less is known, nevertheless pose problems of great biological interest. Such associations are to be found amongst the Pseudophyllidea, a tapeworm order, in which one of the larval stages, the plerocercoid, typically inhabits an extra-intestinal site in vertebrates.

HYMENOLEPIS DIMINUTA

Worm migration

The laboratory rat is commonly utilised as a host for adult *Hymenolepis diminuta*. A dynamic feature of the relationship between this parasite and its host is the active selection by the worm of suitable sites in the small intestine. This phenomenon of site selection has been recently reviewed by Holmes (1973) and Crompton (1973). It is evident that two separate factors are involved; an ontogenetic change in localisation and a more recently discovered, circadian migration, referred to by Holmes (1973) as a diel migration. Both these may themselves be influenced by other factors, for example, the intensity of infection, or the presence of other parasitic helminths (Holmes, 1961, 1962). Read & Kilejian (1969) studied circadian migration in rats harbouring single and multiple worm infections of *H. diminuta* of ages 10, 14 and 21 days. Food was given each day between 17.00 h and 8.00 h and the maximum number of scoleces and greatest worm biomass was located in the anterior 25 cm of the small intestine at 8.00 h. There was a posterior migration during a subsequent time period, the least worm biomass being found anteriorly at 16.00 h, and this migratory pattern was independent of worm age and intensity of infection. Migratory behaviour could be altered by varying host feeding times and the anterior migration was inhibited in starved animals; this latter point has been confirmed by Bailey (1971). Chappell, Arai, Dike & Read (1970) confirmed and extended the observations of Read & Kilejian described above. Using infections of 30 worms per rat they noted that migratory behaviour in young worms varied with parasite age. Food was withheld from the host between 8.00 h and 17.00 h, and worms aged 5–7 days (phase I) were found to migrate anteriorly within the first 20 cm of the small intestine during this period of host fasting. Eight day old worms (phase II) apparently failed to migrate and in older worms (phase III), a posterior migration occurred, involving the entire small intestine, as previously described by Read & Kilejian (1969). Chappel *et al.* (1970) related these age-dependent changes in migratory behaviour to the pattern of worm growth and a possible crowding effect (Fig. 1). Phase I is a period of exponential growth and phase II coincides with the inflexion of the growth curve and with the first changes in worm weight associated with a crowding effect. The latter is at a maximum from day 9 onwards and coincides with the asymptotic portions of the curves for worm glycogen and protein content and specific growth.

It is not known why *H. diminuta* undergoes diel migrations in the intestine of its host. Crompton (1973) commented that the migrations appeared to be of the exogenous type (i.e. in response to external stimuli) and suggested that such a stimulus might be gastrin, possibly affecting the muscular tone of host and parasite. Chappell *et al.* (1970) suggested that migration occurred in response to changes in the concentration of exogenous glucose and that the differences in migratory behaviour found in young and older worms could be explained on this basis. Young, and

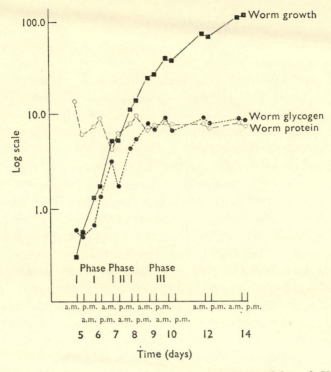

Fig. 1. Relationship between the phases of migratory activity of *Hymenolepis diminuta* in the rat intestine to the specific growth of worms, the glycogen content and the protein content of worms. The log scale of the ordinate is in units of (i) g wet wt for worm growth and (ii) glycogen and (iii) protein as per cent of wet wt of worm tissue. Each value represents the mean of at least four replicates (after Chappel *et al.*, 1970).

therefore smaller, worms experienced no competition for glucose and these parasites migrated to regions of optimum glucose concentration in the intestine. On the other hand, older and larger worms competed for space in regions of optimum glucose concentration, the less successful parasites migrating posteriorly. This resulted in a loss of worm weight when glucose concentrations became limiting (crowding effect). Mettrick (1971a) investigated the effects of host diet on migration of *H. diminuta*. He reported that the degree of anteriad migration of worms following feeding of the host with glucose, galactose or dextrin was virtually identical and that the most marked response occurred following feeding with olive oil. These data strongly suggest that worm migration is not a specific response to a single intestinal nutrient, as suggested by Chappell *et al.* (1970), but that several factors may be involved.

Nutrition and digestion

Absorption of nutrients

Almost all our knowledge of the nutrition of tapeworms has been derived from studies on adult *Hymenolepis diminuta*, largely by the efforts of the late C. P. Read and his associates at Rice University, Houston, Texas. This tapeworm is apparently impermeable to large molecules (Lumsden, Threadgold, Oaks & Arme, 1970) but a variety of low molecular weight organic nutrients are readily absorbed. Uptake of many of these has been shown to be mediated by transport mechanisms which often exhibit a high degree of substrate specificity. Read *et al.* (1963) studied the absorption of certain amino acids by *H. diminuta*, and their initial observations have been extended by Kilejian (1966), Harris & Read (1968), Laws & Read (1969) and Woodward & Read (1969). Their studies indicated that amino acids are absorbed by at least four distinguishable membrane transport mechanisms, viz a dicarboxylic amino acid system, a diamino amino acid system and two (possibly three) systems for the absorption of neutral amino acids. Neutral amino acids are absorbed via transport mechanisms which in many ways resemble the A and L systems described for Ehrlich ascites cells by Oxender & Christensen (1963). An interesting difference, however, between the parasite and tumour cell is that mono *N*-methylation of α-aminoisobutyric acid reduced its uptake by *H. diminuta*, whereas the converse was found in ascites cells (Laws & Read, 1969; Christensen, Oxender, Liang & Vatz, 1965). Extensive studies on the nutrition of *H. diminuta* have revealed the presence of the following additional uptake systems for other classes of nutrients: a glucose system which transports galactose but not fructose (Phifer, 1960; Read, 1961); a glycerol system (Pittman & Fisher, 1972); two systems for fatty acids, specific for short- and long-chain compounds respectively (Arme & Read, 1968; Chappell, Arme & Read, 1969); a riboflavin system which does not bind pyridoxine (Pappas & Read, 1972*b*); a thiamine system (Pappas & Read, 1972*a*), and systems for the transport of certain pyrimidines and purines (MacInnis, Fisher & Read, 1965; MacInnis & Ridley, 1969; Pappas, Uglem & Read, 1973*a*). In some of the above mentioned systems a diffusion component of uptake was also present. The extent of this and also the effects of metabolic poisons, bile salts, phlorizin, oubain, pH and cations on transport mechanisms in *H. diminuta* have been reviewed by Pittman & Fisher (1972) and additional information on anion and cation requirements for glucose and methionine absorption has been provided by Pappas, Uglem & Read (1974). The reports cited above have all involved the use of in-vitro techniques but there are also a number of experiments which have been carried out *in vivo*, the results of which further serve to illustrate the dynamic nature of the interaction between *H. diminuta* and the rat host. Both Hopkins & Callow (1965) and Arme & Read (1969) showed that worm amino acids rapidly equilibrated with those of the host. The latter authors demonstrated that when infected rats were injected intracardially with cycloleucine, an amino acid not metabolised by either worm or host, maximum concentrations were attained in both rat and tapeworm tissues after 15 min, the shortest time period that was examined. In worms pre-loaded with cycloleucine and then implanted into the small intestine of the host, there was a rapid efflux of the amino acid from the parasite into the gut lumen.

Hymenolepis diminuta has only a limited capacity for regulating the flow of free amino acids from the environment into its tissues. Any control that does occur is exerted at the transport locus level and is dependent upon the relative concentrations of amino acids in competition for a particular uptake system. This supply of amino acids is controlled by the host and in functional terms therefore the alimentary canal of the host can be regarded as the alimentary canal of the worm. Clearly the diet is an important source of luminal amino acids, but in 1950, Read put forward the concept of an exocrine–enteric circulation of amino acids, vitamins and perhaps other compounds in which certain organic components of the intestinal lumen were considered to be of endogenous origin. Much of the relevant data has been reviewed by Read (1971), in which it was argued that exogenous nitrogen in the diet was diluted in the lumen with endogenous protein derived from salivary, gastric, pancreatic and intestinal secretions, bile, and sloughed intestinal epithelium. Subsequent hydrolysis of endogenous protein would yield an amino acid mixture which would dilute the amino acids derived from the diet, and would result in a luminal pool of amino acids of relatively constant composition, in terms of their molar ratios, irrespective of the nutritional state of the animal. It has also been established that bidirectional fluxes of free amino acids occur between the mucosa and gut lumen. Read *et al.* (1963) stated that in a consideration of the functioning of transport processes, the molar ratios of amino acids were of more importance than their absolute concentrations. They suggested that the maintenance of a luminal amino acid pool in the host, of relatively constant composition, was possibly significant in the determination of host-specificity in parasites. Since the tapeworm was unable to modify its external environment appreciably, with respect to amino acid content, it was suggested that *H. diminuta* parasitised the homeostatic mechanism of the host. These views have been criticised by several workers (see Mettrick,

1971d, 1972). Mettrick suggested that the apparent constancy in molar ratios of amino acids, obtained by analysis of the contents of the entire small intestine, would not be found if different regions were examined separately. There is merit in this suggestion, since the use of pooled data ignores the fact of regional differences in the physiological properties of the small intestine. Further, if there is a homeostatic control of the relative molar ratios of luminal amino acids by the host, then it should not be possible to alter the free pool amino acids of worm or parasite growth rate by feeding an imbalanced diet to the rat. If such effects could be brought about, the homeostatic concept would be negated. Hopkins (1969) showed that imbalancing doses of dietary methionine affected the tapeworm amino acid pool for about 3 h in the anterior small intestine. Mettrick (1971a) demonstrated that, whereas egg albumen feeding was accompanied by changes in the composition of amino acids in the lumen of both the anterior and posterior small intestine, casein feeding only resulted in changes anteriorly. Chappell & Read (1973) demonstrated that a single dose of dietary proline resulted in an imbalance in the free amino acid pool of both host and parasite. A major point of interest arising from their observations, however, was that imbalance in the worm amino acid pool resulted in little change in the rates of lysine incorporation into protein *in vitro*. This supports the contention of Hopkins & Young (1967) that worm growth was unaffected by imbalance in host dietary amino acids, and conflicts with the evidence of Mettrick & Munro (1965) and Mettrick (1968), which suggested that an imbalanced host diet resulted in a retardation of parasite growth.

It is evident that there is little unanimity amongst workers in this field. It is accepted that the secretion of endogenous protein and amino acids into the intestinal lumen does occur, but the extent to which this is controlled by the host, and the significance of any such control to the parasite, is far from clear. If it is possible to override a homeostatically controlled amino acid balance in the rat intestine, the effects of imbalance on the growth of the tapeworm remains an open question. Chappell & Read (1973) have suggested that previous conflicting results might be explained on the basis of the different amino acids used in experiments. The possibility of amino acid toxicity, and complications which may arise from the peripatetic nature of the worm, must also be considered.

Although much of the above discussion concerns proteins and amino acids, it has been shown that the fatty acids of the rat intestine also appear to be largely derived from endogenous sources and that their composition differs in many respects from that of the diet (Kilejian, Ginger & Fairbairn, 1968).

Digestion

Arme & Read (1970) emphasised the fact that the tegument of *Hymenolepis diminuta* could be regarded as a digestive-absorptive surface, having many morphological and physiological features in common with the intestinal mucosa of vertebrates. In addition to the absorptive capability, described above, the tegument also possesses intrinsic digestive enzymes, which are membrane-bound and not released into the environment, and also contact digestive enzymes, which are extrinsic enzymes of host origin, adsorbed onto the tapeworm surface (Dike & Read, 1971b).

Intrinsic membrane-bound phosphohydrolases and disaccharidases have been demonstrated within the brush-border of mucosal cells (Granger & Baker, 1950; Miller & Crane, 1961b). There is no intrinsic disaccharidase activity in the tapeworm tegument, but intrinsic surface phosphohydrolase activity has been demonstrated (Arme & Read, 1970; Dike & Read, 1971a,b), and the results of Dike & Read (1971a), who used a variety of phosphory-lated substrates, indicated the presence of at least two of these enzymes. Miller & Crane (1961a) observed that the products resulting from intrinsic disaccharidase and phosphohydrolase activity in hamster intestine mucosal cells possessed a kinetic advantage in their subsequent absorption over identical compounds, free in the incubation medium. Dike & Read (1971b) provided evidence for a similar phenomenon in *H. diminuta*. They showed (a) that glucose-6-phosphate was hydrolysed at the tapeworm surface, (b) that the glucose transport and phosphohydrolase systems were separate, since one could be inhibited without affecting the other, and (c) that glucose liberated from glucose-6-phosphate was absorbed preferentially over free glucose in the medium. These data suggest that the transport and phospho-hydrolase systems, although separate, are in close proximity in the tegu-ment. Read (1971) has suggested that this arrangement of the tegument is of advantage to the worm in that it mitigates against the loss of free energy by diffusion.

The role of phosphatases in the biology of parasites has been the subject of some controversy. Although there is a correlation between the distribu-tion of phosphohydrolases and regions in which transport processes are known to occur, there is no experimental evidence to show that these enzymes participate directly in absorption. Phifer (1960) demonstrated continued glucose uptake in *H. diminuta* in the presence of molybdate ions which were found to inhibit phosphohydrolase activity completely. There are presumably a variety of phosphate esters in the lumen of the small intestine and, although these substances do not easily penetrate cell mem-branes, the products of their hydrolysis are readily absorbed, often by specific transport systems. It would therefore seem reasonable to assume that tegumentary phosphohydrolases function in a digestive capacity.

Cell surfaces in general, and the tegument of cestodes is no exception, appear to be covered by a mucopolysaccharide layer, external to the plasma membrane, termed a glycocalyx (Bennett, 1963; Ito, 1969; Oaks & Lumsden, 1971), although Curtis (1972) has suggested that this may be an artifact resulting from the processing of tissues for electron microscopy. Ugolev (1965) and Ugolev & De Laey (1973) claimed that pancreatic amylase could be adsorbed onto the surface of mammalian mucosal cells and that, following adsorption, the kinetic parameters of the enzyme were altered. The resulting observed enhancement of pancreatic amylase activity has been referred to as membrane or contact digestion. The usefulness of this concept is reduced, however, by Ugolev's failure to recognise the existence of an intrinsic membrane-bound mucosal amylase, and to distinguish between intrinsic and extrinsic amylolytic activity. Alpers & Solin (1970) carefully characterised the properties of an intrinsic membrane-bound amylase in rat mucosal cells and were able to distinguish it, on the basis of several characteristics, from adsorbed amylase of pancreatic origin. Their experiments provided no confirmatory evidence that the adsorption of pancreatic amylase by the intestinal mucosa enhanced its activity.

An enhancement of bacterial α-amylase activity, in the presence of three species of tapeworm, was reported by Taylor & Thomas (1968). This work was confirmed for *H. diminuta* by Read (1973), who used porcine pancreatic α-amylase. Read showed additionally that there was no intrinsic amylolytic activity associated with the tegument of *H. diminuta*, so that the possibility of ambiguities arising, similar to those described above for rat mucosa, was eliminated. An enhancement of α-amylase activity occurred in the presence of the worm and was apparently a function of the number of parasites present, suggesting indirectly that worm surface area was important. Adsorbed enzyme was readily removed from the worm surface by washing, and amylase adsorption was inhibited following pre-incubation of tapeworms with polycations, which themselves have been shown to adsorb onto the glycocalyx. Read concluded that enhancement of amylase activity was associated with a surface adsorption of the enzyme by the worm, accompanied by a change in molecular configuration of the protein which favoured its catalytic activity. It is not known whether contact digestion, involving extrinsic amylase, confers any advantage to the parasite, or whether the enhancement of enzyme activity is largely accidental, merely reflecting certain physico-chemical properties of the glycocalyx. It is possible that the overall effect of amylase adsorption by the worm is to increase the supply of disaccharides in the gut lumen of the host.

Although these cannot be further utilised by the parasite, since *H. diminuta* lacks intrinsic disaccharidases, their subsequent hydrolysis by membrane-bound mucosal enzymes will increase the amount of luminal glucose available to the worm. In this connection mention may be made of the work of Mead & Roberts (1972), who found that starch digestion following a liquid meal was greater in *H. diminuta*-infected rats than in controls; the presence of the tapeworm, however, had no effect on starch digestion following a solid meal.

The activity of other enzymes of pancreatic origin is known to be altered in the presence of cell surfaces. It has been shown that the intact mammalian mucosa contains both trypsin and chymotrypsin inhibitors, but that binding of the enzyme onto the cell surface does not occur. In contrast, suspensions of mucosal cells bind trypsin without causing any alteration in its kinetic properties (Goldberg, Campbell & Roy, 1968, 1969). This property of intestinal cells would of course be functionally important in the prevention of autodigestion. Intestinal cestodes also avoid the destructive effects of host digestive enzymes, and Reichenbach-Klinke & Reichenbach-Klinke (1970) demonstrated that intact *Proteocephalus longicollis*, an intestinal cestode of fishes, inactivated trypsin in gut contents and tissue extracts. These authors suggested that enzyme inactivation was brought about by the secretion of a trypsin inhibitor into the incubation medium, although no experimental data was offered to support this contention. Pappas & Read (1972c,d) reported that trypsin and α- and β-chymotrypsin were inactivated in the presence of *H. diminuta*. No evidence of intrinsic proteolytic activity, or the secretion of an enzyme inhibitor into the medium was found, and inactivation was irreversible and ceased upon removal of the worms from the incubation medium. Enzyme inactivation was progressive with time, related to worm weight rather than surface area, and was unaffected by preincubation of parasites with polyions. Although the mechanisms of inactivation of trypsin and α- and β-chymotrypsin were apparently similar, they differed with respect to pH. Optimum trypsin inactivation occurred at pH 7.2 whilst maximum effects on the chymotrypsins occurred over a broader pH range (6.3–7.5); these findings suggest that the two processes may be separate. It is clear that interactions between the parasite and host trypsin differ markedly from the interaction with amylase, and it seems unlikely that the results can be explained on the basis of simple adsorption. Pappas & Read (1972a) postulated that a labile inactivating substance was produced at a high rate at the medium–worm interface and that, following combination with trypsin, the enzyme–inhibitor complex detached from the worm, thus exposing fresh inactivator. A possible site for the production of inactivating substance was thought to be the glycocalyx, the renewal of which is thought to be rapid and continuous (Oaks & Lumsden, 1971). The advantages to the worm, of a mechanism whereby it can be protected from the potentially destructive action of host digestive enzymes, are obvious. The role of anti-enzymes in the protection of parasites has been discussed by von Brand (1966), and the results described above clearly add a new dimension to any consideration of this problem.

Interactions between *H. diminuta* and a third class of pancreatic enzyme, pig pancreatic lipase, have been described by Ruff & Read (1973). Their results differed from those obtained with trypsin or α-amylase because, although enzyme activity was reduced in the presence of the worm, the effect was reversible on removing the worms from the medium and was unaffected by pretreatment with polyions. It may thus be described as inhibition of enzyme activity rather than its inactivation. The mechanism whereby lipase inhibition is effected is not clear. The results suggest that a strong binding to the tapeworm surface is not involved, since inhibition is easily reversible and not affected by preincubation of parasites with polyions. Ruff & Read (1973) proposed that the enzyme underwent a loose, transitory attachment to the worm surface, by weak bonding involving van der Waal's forces. This would result in a temporary stabilisation of the enzyme in a configuration unfavourable for catalytic activity. No suggestions were offered of a possible role for lipase inhibition in the ecology of the parasite.

Pappas *et al.* (1973*b*) described ribonuclease activity associated with the tegument of intact *H. diminuta* but failed to demonstrate conclusively whether the enzyme was of extrinsic or intrinsic origin.

It is apparent from the above discussion that tapeworms undergo a number of complex interactions with their environment, and that the morphological basis for these interactions lies in the tegument and its associated structures. In many ways, both structurally and functionally, the tapeworm surface resembles the intestinal mucosa in that they both serve an absorptive, digestive and protective role. It is pertinent to note that most of the basic information on the biology of *H. diminuta* has been derived from in-vitro studies on the isolated adult parasite, and it is not known how relevant these data are in a consideration of the intact symbiosis. Recently discovered properties of the glycocalyx, viz its ability to adsorb protein molecules of host origin, may influence nutrient absorption *in vivo*. Results of preliminary experiments by undergraduate students in my laboratory, in which poly-L-lysine preincubation affected subsequent amino acid uptake, suggest that this is an aspect of tapeworm nutrition which merits further investigation. The peripatetic nature of the worm, and therefore its exposure to regions of the small intestine of differing physiological characteristics, must also be considered. For example, linear gradients of a variety of chemicals have been described in the small intestine (Mettrick, 1970; Read, 1971), and Mettrick (1971c) has shown that the pH of the anterior half of the intestine of infected rats varied between 5.17–6.95, with a mean of pH 6.29. This is considerably below the pH of 7.4 used in in-vitro investigations, and it has been shown that uptake systems of some classes of low molecular weight nutrients are pH-sensitive. Variations in pH were shown to affect the uptake of short-chain fatty acids (Arme & Read, 1968). It was shown that acetate uptake *in vitro* increased as a function of decreasing pH and that propionate inhibition of the mediated component of acetate uptake decreased with decreasing pH, over a pH range of 6.2–7.4. It was concluded that the undissociated form of acetate entered the worm primarily by diffusion, at low pH, and that mediated transport, involving the dissociated form of acetate, occurred at higher pH. If these circumstances pertain *in vivo* then at low pH the parasite will be unable to exert any significant control of acetate uptake at the membrane locus level.

Limitation of space permits only a brief mention of studies on the modification of the environment resulting from infection with *H. diminuta*. Mettrick (1971b,c, 1972) has shown that chemical gradients, hydrogen ion concentration and microbial flora in the intestine differ significantly in parasitised and non-parasitised rats. There have also been several investigations into immunological reactions associated with hymenolepid infections in rodents (see Hopkins, Grant & Stallard, 1973, for references). *Hymenolepis diminuta* in the rat has been considered non-immunogenic, but recently Harris & Turton (1973) have demonstrated parasite-specific antibody in the rat by homologous passive cutaneous anaphylaxis and an indirect fluorescent antibody method.

PSEUDOPHYLLIDEA

In contrast to the *Hymenolepis*–rat symbiosis, in which much information has been gained from studies on the isolated adult parasite, in the Pseudophyllidea most attention has been paid to the effects of larval stages on their vertebrate host. Adult pseudophyllids are found typically in the intestine of teleost fishes or fish-eating birds and mammals. Their life-cycle involves two intermediate hosts; the procercoid larva develops in the haemocoel of copepods and the plerocercoid often inhabits an extra-intestinal site in vertebrates. Three species in which tapeworm–host interactions have been particularly studied are *Ligula intestinalis*, *Spirometra mansonoides* and *Diphyllobothrium latum*.

Ligula intestinalis

The dominant phase in the life cycle of *Ligula intestinalis*, in terms of longevity, host specificity and host involvement, is the plerocercoid larva which parasitises the body cavity of fish. In the British Isles, infections are confined to the Cyprinidae (Orr, 1967), and common host species in which the effects of parasitism have been investigated are the roach (*Rutilus rutilus*), rudd (*Scardinius erythrophthalmus*), dace (*Leuciscus leuciscus*), bream (*Abramis brama*), bleak (*Alburnus alburnus*) and gudgeon (*Gobio gobio*). References to earlier literature have been cited by Arme & Owen (1968), Arme (1968) and Arme & Walkey (1970).

In multiple infections of *Ligula*, which are common, the weight of parasite tissue may exceed that of the host. This is very rare amongst helminth infections of vertebrates, being only found additionally in *Schistocephalus solidus* infections of *Gasterosteus aculeatus* (Arme & Owen, 1967). It is hardly surprising, therefore, that in infections of this magnitude many pathological effects in the host have been observed, although the most interesting, an effect on host reproduction, may be largely independent of worm burden.

In multiple infections of roach, the most studied fish host, a pronounced abdominal distension was observed, with internal organs often compressed and displaced (Plate 1*a*). The host responded to infection with a tissue reaction (Arme & Owen, 1968, 1970), which resulted in the parasites becoming enmeshed in sheets of connective tissue. In one instance, a heavily parasitised rudd, complete calcification of the body cavity was observed. The weights of liver and gonads, the packed-cell volume of erythrocytes and haemoglobin levels, were lower in parasitised than in non-parasitised roach (Arme & Owen, 1968). The most interesting consequences of infection, however, are the effects of the parasite on the gonads and pituitary gland of the host. These have been described by several authors and previous work has been confirmed and extended by Arme (1968). The effect on host gonads was so marked that a casual examination of the body cavity of an infected roach revealed little trace of gonadal tissue, the reduced reproductive organs being represented only by a thin strip of tissue which did not increase in relative size, even in older fish. Histological examination of the ovary of an infected female roach in the breeding season revealed only the presence of oogonia and oocytes in the first growth phase; later maturation stages were absent, and this condition was found irrespective of season, and apparently persisted for the life of the fish (Plate 1*b*). In contrast, the gonads of uninfected roach, in spring, were found to contain large yolky oocytes, in the last stages of maturation

(Plate 1c). In parasitised male roach, spermatogenesis never proceeded beyond the level of spermatogonia. A similar inhibition of gametogenesis has been observed in the limited number of rudd, dace and bream which have been examined. Observations on the pituitary gland of infected roach led to the identification of a basophil cell type in the mesoadenohypophysis thought to be responsible for gonadotrophin secretion. In uninfected fish these cells underwent cyclical changes in size and level of granulation which could be correlated with the breeding cycle. In infected fish, no seasonal variation in the appearance of these cells was observed, and there was a marked reduction in their size, level of granulation and apparent number, when compared to uninfected individuals (Plate 1d,e). The association of *Ligula*-infections with interference of the normal pituitary–gonadal axis of the host suggested several lines for experimental investigation. No infected fish survived the removal of worms from the body cavity, but when *Ligula*-plerocercoids were implanted surgically into the body cavity of previously uninfected fish, subsequent changes in the pituitary gland and gonads resembled those found in naturally occurring infections. Experiments involving the injection of *Ligula* homogenates into non-parasitised fish, or the administration of pituitary glands derived from uninfected fish to infected fish, yielded inconclusive results. In the latter case, however, histological examination of the gonads of two surviving parasitised female roach showed that vitellogenesis had commenced in some oocytes; this had never previously been noted in *Ligula*-infected roach from the wild. The parasite was found to survive for a limited period of time in the dorsal lymph sac of an unnatural host, the South African clawed toad, *Xenopus laevis*. Following *Ligula*-implantation changes occurred in certain cells of the pars distalis, designated type II basophils by Kerr (1965), and thought to be concerned with gonadotrophin secretion. These changes were very similar to those observed by Kerr (1965) following testosterone treatment. Gonads of implanted toads were apparently unaffected by the presence of the tapeworm.

There seems little doubt that changes in the pituitary gland and gonads of infected fish are directly attributable to the presence of *Ligula*. Further, the fact that the condition found in natural infections can be reproduced by the implantation of a single, relatively small, plerocercoid into a large fish seems to preclude the possibility that the changes are due to pressure effects on fish organs, or to a general debility of the host resulting from the metabolic demands of the parasite. It is possible, however, that an essential metabolite is preferentially absorbed or neutralised by the plerocercoid. The available evidence suggests that the lack of gonadal development in parasitised fish is related to a suppressive effect on certain pituitary gonadotrophins; this view is supported by the fact that gametogenesis in infected hosts is arrested at precisely those levels of development which are affected by hypophysectomy of the non-parasitised fish. It is suggested that *Ligula* produces a substance responsible for these effects, but the nature of such a compound is not known. In both natural and unnatural hosts, the results of experimental infections resemble those which follow testosterone treatment, but attempts to demonstrate steroid activity in *Ligula* have proved negative. It is possible that a non-steroidal compound with anti-gonadotrophic activity may be involved, and such a compound (α-methylallylthiocarbamyl -2-methylthiocarbamoylhydrazine) has been shown to block the action of pituitary gonadotrophins on the testes of three species of teleost fish (Hoare, Wiebe & Wai, 1967).

An effect of *Ligula* infection on the breeding behaviour of rudd has been described by Orr (1966). In 200 rudd removed from a shoal exhibiting courting behaviour only one was parasitised by *Ligula*, whereas in 75 rudd netted at random, 56 were infected. Orr concluded that infected fish were unable to participate in the normal breeding behaviour and suggested that parasitised rudd failed to produce the necessary sterohormones which are normally released into the environment to stimulate spawning.

It is of interest that the effects of *Ligula* on gudgeon differ in several respects from those described for other cyprinids. There is no apparent tissue reaction and gametogenesis proceeds further in infected gudgeon than in other species, although it has not been determined if infected gudgeon are capable of breeding. Whether these differences reflect differing host physiologies, or whether they may be due to the presence of a different physiological strain of *Ligula* which infects gudgeon alone is not known. However, Plate 2*a* shows the results of iso-electric focusing in a pH 3.5–10.0 gel of phenoxyethanol extracts of *Ligula* from gudgeon and roach, with subsequent staining with 1 % Coomassie Brilliant Blue R250 (Ferguson & Arme, unpublished). The general protein patterns differ in seven out of 43 fractions, which suggests that *Ligula* from these two hosts are genetically different and possibly separate species. *Ligula* infections of bleak (*Alburnus alburnus*) are also somewhat unusual in that only 44 % of fish examined in the breeding season showed a reduction in gonad size (Harris & Wheeler, 1974), and in one parasitised fish 'ripe ova' were observed. In parasitised roach all infected fish show a reduction in gonad size and mature ova are never found. Harris & Wheeler (1974) also failed to report a host tissue response in this host species.

A related cestode *Schistocephalus solidus* is found as a plerocercoid larva in the body cavity of the three-spined stickleback. No effects on the pituitary gland have been observed in this symbiosis although gross bodily distension, associated with heavy infections, interferes mechanically with normal nest building and courtship behaviour (Arme & Owen, 1967).

Spirometra mansonoides

A second pseudophyllidean cestode, the plerocercoid larval stage of which produces interesting and unusual effects in its host, is *Spirometra mansonoides*. Our understanding of the biology of this parasite results largely from the work of J. F. Mueller and his colleagues which has recently been reviewed (Mueller, 1974a).

Adult *S. mansonoides* is found in the small intestine of dogs, cats and racoons. Eggs, voided in the faeces, embryonate and hatch in water producing a motile coracidium larva which, when eaten by a copepod (*Cyclops* sp.) develops into a procercoid larva in the haemocoele. The plerocercoid larva (or sparganum) shows little host specificity and infects all classes of vertebrates with the exception of fishes; the Florida water snake, *Natrix* is a frequent natural host. If the plerocercoid is ingested by an unsuitable final host, it penetrates the intestine and re-establishes itself as a tissue parasite. Plerocercoids can be propagated by injecting scoleces subcutaneously into mice. The worms grow at a rate of 1 mm day^{-1} and on dissection they may be recovered loose under the mouse skin. These larvae can survive for long periods of time and, by transference of clipped scoleces from host to host, a longevity of 16 years has been achieved in the laboratory (Mueller, 1974b).

In 1963, Mueller observed that in mice used for the propagation of plerocercoids, there was an increase in size of infected individuals when compared to controls, and a similar effect was later observed in deer mice and hamsters (Mueller, 1965a). Parasite-induced weight gain in mice persisted for several months and the effect appeared to be additive (Fig. 2) (Mueller, 1963); the maximum growth response in a 20 g mouse was elicited by injection of 12 scoleces (Mueller, 1965a). That the increased growth in infected mice was not related to their increased food intake was demonstrated by Mueller (1965b); on a host-weight basis, food consumption, as well as urine and faecal output, was essentially the same in test and control animals. No growth stimulation was observed in parasitised intact male rats, possibly because the more rapid growth rate of these animals, already at or near its biological limit, was incapable of further stimulation. Growth in intact non-parasitised female rats normally plateaus after three to four months, at a weight of approximately 350 g. If female rats of this age were injected with *S. mansonoides* plerocercoids, growth resumed but only for a period of three to four weeks (Steelman, Glitzer, Ostlind & Mueller, 1971). Glitzer & Steelman (1971) investigated the effects of

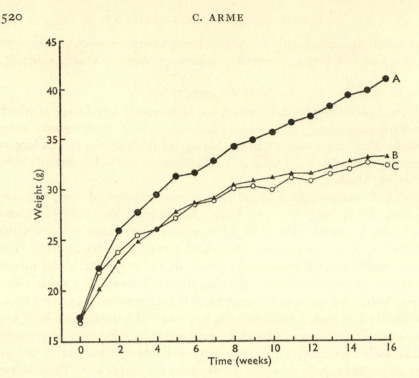

Fig. 2. Growth rates in mice infected with one (B) or six (A) plerocercoids of *Spirometra mansonoides*. The mice with six larvae show marked acceleration of growth over those with one or none. The control curve (C = o plerocercoids) and the single plerocercoid curve are confused at first but assume their expected position after the sixth week. Points are the mean of seven or eight mice (after Mueller, 1963).

infection on growing intact non-parasitised male and female rats in more detail. The most interesting feature of their results was the observation that pituitary gland weights in both sexes were depressed, possibly indicating an interference with pituitary growth and its ability to produce hormones. Measurement of plasma growth hormone indicated a reduction in levels in infected rats and these workers concluded that the production of a growth-promoting substance by the parasite suppressed the elaboration and/or secretion of endogenous growth hormone in the host.

A significant advance in our understanding of the biology of *Spirometra* occurred when the effects of the parasite on endocrine-deficient rats were studied. Mueller & Reed (1968) found that growth in control male rats, rendered hypothyroid by propylthiouracil (PTU) treatment, was slow and plateaued at 160–170 g in weight. Injection of scoleces resulted in an immediate resumption of growth which persisted for approximately five

PLATE I

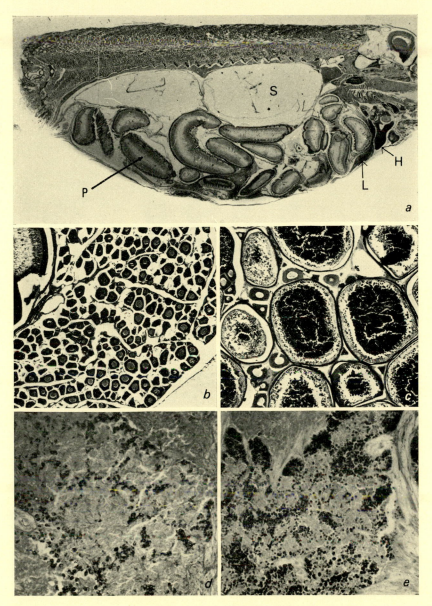

For explanation see p. 532

PLATE 2

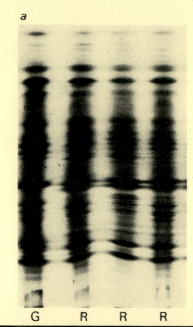

a

G   R   R   R

For explanation see p. 532

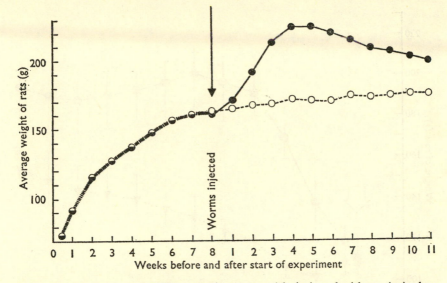

Fig. 3. Growth curves of PTU-treated rats ⊝, with (●) and without (○) plerocercoids of *Spirometra mansonoides* (after Mueller & Reed, 1968).

weeks, after which time the weight of most parasitised animals began to decline (Fig. 3). Similar results were obtained in weanling rats which had been surgically thyroidectomised and hypophysectomised (Mueller, 1968). Weight gains of up to 80 g over controls were recorded but, as in PTU-treated animals, the rate of growth declined four to six weeks after infection (Fig. 4, Plate 2*b*). The stimulation of growth in hypophysectomised rats is dose dependent. Mueller (1972*a*) investigated the effects of innocula of varying numbers of scoleces and found an increased response up to 40 worms per hypophysectomised rat; this intensity of infection produced a 65 % greater response than five scoleces. An interesting outcome of these experiments was that, whereas in most rats, growth rates declined some five weeks after infection (cf. growth curves in PTU-treated and thyroidectomised rats mentioned above), 25 % of animals continued to grow for longer periods, some for up to 50 weeks following injection.

It is clear from a consideration of the data from Mueller's publications, that the plerocercoid–hypophysectomised rat symbiosis possesses several advantages for the study of the growth-promoting properties of *S. mansonoides* over the plerocercoid–mouse system. In the latter, occasional animals fail to respond, due to worm encapsulation by host tissue, and weight differences between test and control animals only become clear-cut after a period of several weeks. With the altered rat, however, no failures

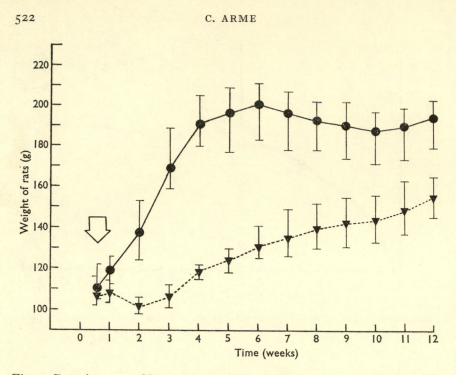

Fig. 4. Growth curves of hypophysectomised rats with and without plerocercoids of *Spirometra mansonoides*. Points are mean weights plus range (after Mueller, 1968). ●, 4 rats, 24 plerocercoids; ▼, 5 rats, ○ plerocercoids. Arrow indicates time of injection.

have been reported, and unequivocal evidence of weight gains are apparent within days following infection. There is an important difference between the growth curves of normal mice and both PTU-treated and surgically altered rats following infection. In mice, parasite-induced growth continues over long periods, whereas in a majority of rats growth stimulation persists for only a few weeks. Mueller & Reed (1968) suggested that this was due to an immune response by the host, neutralising the effect of a presumed growth-promoting substance produced by the parasite (plerocercoid growth factor: PGF). This hypothesis has received support from several sources. Mueller (1961) incubated plerocercoids *in vitro* with serum from infected mice and noted the formation of a precipitate along the entire length of the worm. Garland, Ruegamer & Daughaday (1971) found that when serum containing PGF was incubated with serum obtained from plateaued rats, the growth-promoting activity in the former was lost. In microbial infections of rats, precipitating antibodies are normally developed in less than half the time required for growth rates in plerocercoid-infected hypophysectomised rats to decline. It must be borne in mind, however, that PGF is possibly a soluble protein, and as such will be less efficient in inducing an immunological response in the host. On the other hand, although some rats did not plateau, their serum was found to contain precipitating antibodies which were identical to those produced by serum from plateaued rats using an Ouchterlony technique (Mueller, 1974a). It is also difficult to explain why the intact mouse and the majority of hypophysectomised rats should respond so differently to antigenic stimulation. Certain Asian strains of *Spirometra* have no growth-promoting effects on hypophysectomised rats but do induce weight gain in intact mice. Mueller (1972b) used these strains in an attempt to immunise hypophysectomised rats against *S. mansonoides*. These experiments were unsuccessful, indicating that either an immune response is not involved in the production of the plateau effect or that spargana of Asian strains and *S. mansonoides* produce substances which are immunologically distinct.

The available evidence suggests that plerocercoids of *S. mansonoides* produce a growth hormone-like substance which has been termed plerocercoid growth factor (PGF). A comparison of the effects of bovine growth hormone and PGF has been made by Steelman *et al.* (1970, 1971) and implantation of plerocercoids and injection of growth hormone both produced many similar effects in the hypophysectomised rat. These included an increase in body, thymus, and kidney weights, as well as a stimulation of bone size and tibial cartilage width. Their data confirmed that weight gain in experimental animals was not a result of oedema or obesity but reflected true skeletal growth. A detailed analysis of the metabolic effects of PGF revealed further similarities between it and growth hormone. Administration of both stimulated the incorporation of $^{35}SO_4$ into chondroitin sulphate of cartilage, increased the level of liver ornithine decarboxylase and accelerated the incorporation of $[^{14}C]$leucine into diaphragm muscle protein. An increase in liver glycogen also resulted from PGF and growth hormone treatment. However, previous reports that alkaline extracts of plerocercoids possessed insulin-like activity in rat epididymal fat pad (Harlow, Mertz & Mueller, 1967), were not confirmed. Infected mice were found to have high fasting blood-glucose levels (Sadun, Williams, Meroney & Mueller, 1965) but this effect was not observed in infected hypophysectomised rats. Garland *et al.* (1971) studied $[^3H]$thymidine uptake into costal cartilage *in vitro*. They concluded that PGF induced hypophysectomised rats to produce sulphation factor (somatomedin). It is unlikely that PGF contains sulphation factor itself, because there are several differences in the properties of these two substances (Steelman *et al.*, 1971).

Although in general properties, PGF and growth hormone appeared to have similar effects on the hypophysectomised rat, there were some important points of difference. Epididymal fat pads increased in weight in response to PGF, and plasma free fatty acids were slightly reduced. In contrast, growth hormone increased lipolysis, raising the levels of free fatty acids in the blood. An interesting observation by Steelman *et al.* (1970) was that PGF activity in the plasma of infected rats could be transferred to another uninfected rat, and that activity was retained, even after two such transfers. In contrast, plasma from growth hormone-treated, uninfected, controls failed to produce a significant response following transfer, indicating that growth hormone was less stable in plasma than PGF. In addition, no cross-reactivity between PGF and human or rat growth hormone was detected using immunochemical techniques.

Ruegamer & Mueller (1973) observed that growth was resumed in infected alloxan-diabetic rats even though blood glucose levels remained low. PGF was also found to possess lactogenic properties in so far as it resembled prolactin in its effects on pigeon crop sac mucosa (Phares & Ruegamer, 1973*b*). Both prolactin and growth hormone overlap in lactogenic activity and whereas the growth hormone-like properties of PGF have been extensively investigated, a study of its possible lactogenic properties is desirable.

The physicochemical properties of PGF have been little investigated. It is extremely potent, however, since 1 ml of plasma from a hypophysectomised rat implanted with 12–15 plerocercoids was equivalent in its effect to 100 μg of bovine growth hormone (Steelman *et al.*, 1970). Activity was lost when heated at 60–65 °C for 30 min and dialysis of plasma from parasitised rats had no effects on its growth-promoting properties. Steelman *et al.* (1971) concluded that PGF had a molecular weight of < 100 000 and that it was possibly protein. No growth-promoting effects were observed by injection of extracts of plerocercoids into hypophysectomised rats (Steelman *et al.*, 1971) but, when cultured in medium 199 enriched with serum (Chang, Raben, Mueller & Weinstein, 1973), PGF production continued for at least six months. An intraperitoneal injection of 0.3–1.0 ml of medium induced weight gain in hypophysectomised rats but no in-vitro growth of cell or tissue cultures was observed. Phares & Ruegamer (1973*a*) were subsequently able to concentrate PGF from simple saline solutions in which the worms had been incubated for 8 h, and this development should facilitate the isolation and identification of the growth factor.

It is evident that the interactions of *S. mansonoides* with its rodent hosts are complex and, despite significant advances over the past 10 years, much remains to be clarified. A key question is whether PGF itself brings about the observed effects, or whether its presence induces the host animal to produce factors which initiate the growth response. The fact that PGF is apparently able to substitute for thyroxine, pituitary growth hormone and to some extent insulin and prolactin, suggests that either several substances with highly specific properties may be involved, or alternatively, that PGF is a single compound which is able to affect different components of the host endocrine system. Further comment on this aspect of the biology of *Spirometra* must await the isolation and purification of the growth-promoting principle. Ultrastructural examination of the plerocercoid is limited to observations on the tegument (Lumsden, Oaks & Mueller, 1974), and further investigations to localise sites of PGF production and secretion are clearly desirable. Secretory material has been observed in the scolex of the plerocercoid of *Spirometra erinacei* (Kwa, 1972). Growth-promoting effects in *Spirometra mansonoides*, however, have been noted in mice implanted with headless fragments of plerocercoids (Mueller, 1963), so that the capacity for PGF production is not limited solely to the scolex.

Whether PGF is important in the physiology of the parasite, or whether it is a by-product of the worm whose specific effects on the host are largely accidental, is not known. However, it is established that some African and Far Eastern strains of *S. mansonoides* are equally successful, in terms of their growth and reproduction, but are without marked growth-promoting effects on hypophysectomised rats (Mueller 1972*b*; Opuni, Muller & Mueller, 1974).

Diphyllobothrium latum

I would like to mention briefly a third pseudophyllid, *Diphyllobothrium latum*, which is involved with the vitamin B_{12} metabolism of its host. In contrast to *Ligula* and *Spirometra*, however, these interactions have been most studied using the adult worm (Nyberg, 1963). Human infections of *Diphyllobothrium latum* are acquired by eating fish infected with the plerocercoid larval stage. A vitamin B_{12} deficiency results, which manifests itself as pernicious anaemia. Whereas in the non-parasitised subject, much of the dietary vitamin B_{12} is absorbed from the intestine, in parasitised hosts competition for the vitamin occurs, with the worm absorbing large amounts. So much is present that dried tapeworms can be used as a dietary substitute for vitamin B_{12}. The presence of the parasite has also been found to inhibit the absorption of the vitamin by the host mucosa. In the normal animal, prior to absorption, vitamin B_{12} attaches to Castle's intrinsic factor, and Nyberg (1963) demonstrated that extracts of *D. latum* were capable of splitting the B_{12}-intrinsic factor complex. Mueller (1968) has commented that both adult and larval *Spirometra mansonoides* contain large quantities of vitamin B_{12}, with the highest concentration in the plerocercoid being in the scolex. The extra-intestinal site of the plerocercoid may preclude the development of marked effects in the *Spirometra*-infected host.

CONCLUSION

It has been my intention, in the above review, to give some indication of the diversity of interactions that exist in tapeworm–host symbioses. Research into aspects of the four symbioses described has progressed along different lines, and generalised predictions for the future are not possible. For *Hymenolepis diminuta*, the extension of in-vitro studies on nutrition and digestion in the isolated adult worm need to be accompanied by in-vivo studies, in which the effects of the complex environment provided by the host intestine are examined. The phenomenon of enzyme sharing between host and parasite, and the desirability of isolating carrier molecules for transport studies, should perhaps attract the attention of molecular biologists. It must also be borne in mind that the adult worm in the rat represents only one phase of a complex life cycle. There is, in addition, an egg, which is essentially free living, and a cysticercoid larval stage, found in the haemocoele of insects. It would be of interest to compare aspects of the physiology of these different life-cycle stages and, if differences are found, to determine those environmental triggers which determine which physiological phenotype is expressed. Preliminary studies, in which aspects of cysticercoid and adult nutrition were compared, have revealed that there are remarkable similarities between transport systems for a variety of nutrients, in larval and adult worms (Arme & Coates, 1971, 1973; Arme, Middleton & Scott, 1973). Parasitism in the pseudophyllids described above, obviously involves a degree of physiological intimacy with the host not normally associated with parasitic infections. Yet is it reasonable to assume that complex interactions of this type are restricted to only two or three species of tapeworm? I would suggest not, and that future work will show that other cestodes are also capable of providing problems of comparable biological interest. A major difficulty facing investigators in this field is that any approach to the problem must be largely empirical at present. The growth-promoting effects of *Spirometra* were only observed some 30 years after Mueller's first observations on the parasite, and then only after the implantation of larvae into aberrant hosts.

ACKNOWLEDGMENT

I am indebted to Dr L. H. Chappell for comments on the manuscript.

Note

In the time between the completion of the manuscript and its submission to the printers, the following relevant articles have been noted.

PODESTA, R. B. & METTRICK, D. F. (1974). Components of glucose transport in the host parasite system, *Hymenolepis diminuta* (Cestoda) and the rat intestine. *Can. J. Physiol. Pharmacol.*, **52**, 183–197.

(1974). Pathophysiology of cestode infections: effect of *Hymenolepis diminuta* on oxygen tensions, pH and gastrointestinal function. *Int. J. Parasit.*, **4**, 277–292.

SOGANI, R. K., MATSUSHITA, S., MUELLER, J. F. & RABEN, M. S. (1972). Stimulation of ornithine decarboxylase activity in rat tissues by growth hormone and by serum growth factor from rats infested with sparganga of *Spirometra mansonoides*. *Biochim. Biophys. Acta*, **279**, 377–386.

TACHOVSKY, T. G., HARE, J. D., RITTERSON, A. L. & MUELLER, J. F. (1973). Enhanced growth of a polyoma transformed tumor cell in *Spirometra mansonoides*-infected hamsters. *Proc. Soc. exp. Biol. Med.*, **143**, 780–782.

REFERENCES

ALPERS, D. H. & SOLIN, M. (1970). The characterisation of rat intestinal amylase. *Gastroenterology*, **58**, 833–841.

ARME, C. (1968). Effects of the plerocercoid larva of a pseudophyllidean cestode, *Ligula intestinalis* on the pituitary gland and gonads of its host. *Biol. Bull. mar. biol. Lab. Woods Hole*, **134**, 15–25.

ARME, C. & COATES, A. (1971). Active transport of an amino acid by cysticercoid larvae of *Hymenolepis diminuta*. *J. Parasit.*, **57**, 1369–1370.

(1973). *Hymenolepis diminuta*: active transport of α-aminoisobutyric acid by cysticercoid larvae. *Int. J. Parasit.*, **3**, 553–560.

ARME, C., MIDDLETON, A. & SCOTT, J. P. (1973). Absorption of glucose and sodium acetate by cysticercoid larvae of *Hymenolepis diminuta*. *J. Parasit.*, **59**, 214.

ARME, C. & OWEN, R. W. (1967). Infections of the three-spined stickleback, *Gasterosteus aculeatus* L., with the plerocercoid larva of *Schistocephalus solidus* (Müller, 1776), with special reference to pathological effects. *Parasitology*, **57**, 301–314.

(1968). Occurrence and pathology of *Ligula intestinalis* infections in British fishes. *J. Parasit.*, **54**, 271–280.

(1970). Observations on a tissue reponse within the body cavity of fish infected with the plerocercoid larvae of *Ligula intestinalis* (L.) (Cestoda: Pseudophyllidea). *J. Fish Biol.*, **2**, 35–37.

ARME, C. & READ, C. P. (1968). Studies on membrane transport. II. The absorption of acetate and butyrate by *Hymenolepis diminuta* (Cestoda). *Biol. Bull. mar. biol. Lab. Woods Hole*, **135**, 80–91.

(1969). Fluxes of amino acids between the rat and a cestode symbiote. *Comp. Biochem. Physiol.*, **29**, 1135–1147.

(1970). A surface enzyme in *Hymenolepis diminuta* (Cestoda). *J. Parasit.*, **56**, 514–516.

ARME, C. & WALKEY, M. (1970). The physiology of fish parasites. In *Aspects of Fish Parasitology*. *Symp. Br. Soc. Parasit.*, **8**, 79–101 (ed. A. E. R. Taylor). Oxford and Edinburgh: Blackwell.

BAILEY, G. N. A. (1971). *Hymenolepis diminuta*: circadian rhythm in movement and body length in the rat. *Expl Parasit.*, **29**, 285–291.

BENNETT, H. S. (1963). Morphological aspects of extra-cellular polysaccharidases. *J. Histochem. Cytochem.*, **11**, 14–23.

BRAND, T. VON (1966). *Biochemistry of Parasites.* New York: Academic Press.

CHANG, T. W., RABEN, M. S., MUELLER, J. F. & WEINSTEIN, L. (1973). Cultivation of the sparganum of *Spirometra mansonoides in vitro* with prolonged production of sparganum growth factor. *Proc. soc. exp. Biol. Med.*, **143**, 457–459.

CHAPPELL, L. H., ARAI, H. P., DIKE, S. C. & READ, C. P. (1970). Circadian migration of *Hymenolepis* (Cestoda) in the intestine. I. Observations on *H. diminuta* in the rat. *Comp. Biochem. Physiol.*, **34**, 31–46.

CHAPPELL, L. H., ARME, C. & READ, C. P. (1969). Studies on membrane transport. V. Transport of long chain fatty acids in *Hymenolepis diminuta* (Cestoda). *Biol. Bull. mar. biol. Lab. Woods Hole*, **136**, 313–326.

CHAPPELL, L. H. & READ, C. P. (1973). Studies on the free pool amino acids of the cestode *Hymenolepis diminuta*. *Parasitology*, **67**, 289–305.

CHRISTENSEN, H. N., OXENDER, D. L., LIANG, M. & VATZ, K. A. (1965). The use of *N*-methylation to direct the route of mediated transport of amino acids. *J. biol. Chem.*, **240**, 3609–3616.

CROMPTON, D. W. T. (1973). The sites occupied by some parasitic helminths in the alimentary tract of vertebrates. *Biol. Rev.*, **48**, 27–83.

CURTIS, A. S. G. (1972). Adhesive interactions between organisms. In *Functional Aspects of Parasite Surfaces. Symp. Br. Soc. Parasit.*, **10**, 1–18 (eds. A. E. R. Taylor and R. Muller). Oxford and Edinburgh: Blackwell.

DIKE, S. C. & READ, C. P. (1971a). Tegumentary phosphohydrolases of *Hymenolepis diminuta*. *J. Parasit.*, **57**, 81–87.

(1971b). Relation of tegumentary phosphohydrolase and sugar transport in *Hymenolepis diminuta*. *J. Parasit.*, **57**, 1251–1255.

GARLAND, J. T., RUEGAMER, W. R. & DAUGHADAY, W. H. (1971). Induction of sulfation factor activity by infection of hypophysectomised rats with *Spirometra mansonoides*. *Endocrinology*, **88**, 924–927.

GLITZER, M. S. & STEELMAN, S. L. (1971). The effect of the implantation of *Spirometra mansonoides* spargana in the growing rat. *Proc. Soc. exp. Biol. Med.*, **138**, 610–613.

GOLDBERG, D. M., CAMPBELL, R. & ROY, R. D. (1968). Binding of trypsin and chymotrypsin by human intestinal mucosa. *Biochim. biophys. Acta*, **167**, 613–615.

(1969). Studies on the binding of trypsin and chymotrypsin by human intestinal mucosa. *Scand. J. Gastroent.*, **4**, 217–226.

GRANGER, B. & BAKER, R. F. (1950). Electron microscope investigation of the striated border of intestinal epithelium. *Anat. Rec.*, **107**, 423–442.

HARLOW, D. R., MERTZ, W. & MUELLER, J. F. (1967). Insulin-like activity from the sparganum of *Spirometra mansonoides*. *J. Parasit.*, **53**, 449–454.

HARRIS, B. G. & READ, C. P. (1968). Studies on membrane transport. III. Further characterisation of amino acid systems in *Hymenolepis diminuta* (Cestoda). *Comp. Biochem. Physiol.*, **26**, 545–552.

HARRIS, M. T. & WHEELER, A. (1974). *Ligula* infestation of bleak *Alburnus alburnus* (L.) in the tidal Thames. *J. Fish Biol.*, **6**, 181–188.

HARRIS, W. G. & TURTON, J. A. (1973). Antibody response to tapeworm (*Hymenolepis diminuta*) in the rat. *Nature, Lond.*, **246**, 521–522.

HOARE, W. S., WIEBE, J. & WAI, E. H. (1967). Inhibition of the pituitary gonadotropic activity of fishes by a dithiocarbamoylhydrazine derivative (I.C.I. 33828). *Gen. comp. Endocr.*, **8**, 101–109.

HOLMES, J. C. (1961). Effects of concurrent infections on *Hymenolepis diminuta* (Cestoda) and *Moniliformis dubius* (Acanthocephala). I. General effects and comparison with crowding. *J. Parasit.*, **47**, 209–216.

(1962). Effects of concurrent infections on *Hymenolepis diminuta* (Cestoda) and *Moniliformis dubius* (Acanthocephala). II. Growth. *J. Parasit.*, **48**, 87–96.

(1973). Site selection by parasitic helminths: interspecific interactions, site segregation and their importance in the development of helminth communities. *Can. J. Zool.*, **51**, 333–347.

HOPKINS, C. A. (1969). The influence of dietary methionine on the amino acid pool of *Hymenolepis diminuta* in the rat's intestine. *Parasitology*, **59**, 407–427.

HOPKINS, C. A. & CALLOW, L. L. (1965). Methionine flux between a tapeworm (*Hymenolepis diminuta*) and its environment. *Parasitology*, **55**, 653–666.

HOPKINS, C. A., GRANT, P. M. & STALLARD, H. (1973). The effect of oxyclozanide on *Hymenolepis microstoma* and *H. diminuta*. *Parasitology*, **66**, 355–365.

HOPKINS, C. A. & YOUNG, R. L. (1967). The effect of dietary amino acids on the growth of *Hymenolepis diminuta*. *Parasitology*, **57**, 705–717.

ITO, S. (1969). Structure and function of the glycocalyx. *Fedn Proc.*, **28**, 12–25.

KERR, T. (1965). Histology of the distal lobe of the pituitary of *Xenopus laevis* Daudin. *Gen. comp. Endocr.*, **5**, 232–240.

KILEJIAN, A. Z. (1966). Permeation of L-proline in the cestode, *Hymenolepis diminuta*. *J. Parasit.*, **52**, 1108–1115.

KILEJIAN, A. Z., GINGER, C. D. & FAIRBAIRN, D. (1968). Lipid metabolism in helminth parasites. IV. Origins of the intestinal lipids available for absorption by *Hymenolepis diminuta* (Cestoda). *J. Parasit.*, **54**, 63–68.

KWA, B. H. (1972). Studies on the sparganum of *Spirometra erinacei* – III. The fine structure of the tegument in the scolex. *Int. J. Parasit.*, **2**, 35–43.

LAWS, G. F. & READ, C. P. (1969). Effect of the amino carboxy group on amino acid transport in *Hymenolepis diminuta* (Cestoda). *Comp. Biochem. Physiol.*, **30**, 129–132.

LUMSDEN, R. D., OAKS, J. A. & MUELLER, J. F. (1974). Brush border development in the tegument of the tapeworm. *Spirometra mansonoides*. *J. Parasit.*, **60**, 209–226.

LUMSDEN, R. D., THREADGOLD, L. T., OAKS, J. A. & ARME, C. (1970). On the permeability of cestodes to colloids: an evaluation of the transmembranosis hypothesis. *Parasitology*, **60**, 185–193.

MacINNIS, A. J., FISHER, F. M. Jr & READ, C. P. (1965). Membrane transport of purines and pyrimidines in a cestode. *J. Parasit.*, **51**, 260–267.

MacINNIS, A. J. & RIDLEY, R. K. (1969). The molecular configuration of pyrimidines that causes allosteric activation of uracil transport in *Hymenolepis diminuta*. *J. Parasit.*, **55**, 1134–1140.

MEAD, R. W. & ROBERTS, L. S. (1972). Intestinal digestion and absorption of starch in the intact rat: Effects of cestode (*Hymenolepis diminuta*) infection. *Comp. Biochem. Physiol.*, **41A**, 749–760.

METTRICK, D. F. (1968). Studies on the protein metabolism of cestodes. 2. Effect of free dietary amino acid supplements on the growth of *Hymenolepis diminuta*. *Parasitology*, **58**, 37–45.

(1970). Protein nitrogen, amino acid and carbohydrate gradients in the rat intestine. *Comp. Biochem. Physiol.*, **37**, 517–541.

(1971a). Effect of host dietary constituents on intestinal pH and the migrational behaviour of the rat tapeworm *Hymenolepis diminuta*. *Can. J. Zool.*, **49**, 1513–1525.

(1971b). *Hymenolepis diminuta*: The microbial fauna, nutritional gradients, and physicochemical characteristics of the small intestine of uninfected and parasitised rats. *Can. J. Physiol. Pharmac.*, **49**, 972–984.

(1971c). *Hymenolepis diminuta*: pH changes in rat intestinal contents and worm migration. *Expl Parasit.*, **29**, 386–401.

(1971d). Discussion of 'Microcosm of intestinal helminths', see Read, C. P. (1971).

(1972). Changes in the distribution and chemical composition of *Hymenolepis diminuta* and the intestinal nutritional gradients of uninfected and parasitised rats, following a glucose meal. *J. Helminth.*, **46**, 407–429.

METTRICK, D. F. & MUNRO, H. N. (1965). Studies on the protein metabolism of cestodes. 1. Effect of host dietary constituents on the growth of *Hymenolepis diminuta*. *Parasitology*, **55**, 453–466.

MILLER, D. & CRANE, R. K. (1961a). The digestive function of the epithelium of the small intestine. I. An intracellular locus of disaccharide and sugar phosphate ester hydrolysis. *Biochim. biophys. Acta*, **52**, 281–293.

(1961b). The digestive function of the epithelium of the small intestine. II. Localisation of disaccharide hydrolysis in the isolated brush border portion of intestinal epithelial cells. *Biochim. biophys. Acta*, **52**, 293–298.

MUELLER, J. F. (1961). The laboratory propagation of *Spirometra mansonoides* as an experimental tool. V. Behaviour of the sparganum in and out of the mouse host and formation of immune precipitates. *J. Parasit.*, **47**, 879–881.

(1963). Parasite induced weight gain in mice. *Ann. N. Y. Acad. Sci.*, **113**, 217–233.

(1965a). Further studies on parasitic obesity in mice, deer mice and hamsters. *J. Parasit.*, **51**, 523–531.

(1965b). Food intake and weight gain in mice parasitised with *Spirometra mansonoides*. *J. Parasit.*, **51**, 537–540.

(1968). Growth stimulating effect of experimental sparganosis in thyroidectomised and hypophysectomised rats, and comparative activity of different species of *Spirometra*. *J. Parasit.*, **54**, 795–801.

(1972a). Further studies on sparganum growth factor in the hypophysectomised rat: response to large numbers of spargana. *J. Parasit.*, **58**, 438–443.

(1972b). Failure of oriental spargana to immunise the hypophysectomised rat against the sparganum growth factor of *Spirometra mansonoides*. *J. Parasit.*, **58**, 872–875.

(1974a). The biology of *Spirometra*. *J. Parasit.*, **60**, 3–14.

(1974b). Potential longevity of life history stages of *Spirometra* spp. *J. Parasit.*, **60**, 376–377.

MUELLER, J. F. & REED, P. (1968). Growth stimulation induced by infection with *Spirometra mansonoides* spargana in propylthiouracil treated rats. *J. Parasit.*, **54**, 51–54.

NYBERG, W. (1963). The effect of changes in nutrition on the host–parasite relationship. *Diphyllobothrium latum* and human nutrition with particular reference to vitamin B_{12} deficiency. *Proc. Nutr. Soc.*, **22**, 8–14.

OAKS, J. A. & LUMSDEN, R. D. (1971). Cytological studies on the absorptive surfaces of cestodes. V. Incorporation of carbohydrate containing macromolecules into tegument membranes. *J. Parasit.*, **57**, 1256–1268.

OPUNI, E. K., MUELLER, R. & MUELLER, J. F. (1974). Absence of sparganum growth factor in African *Spirometra* spp. *J. Parasit.*, **60**, 375–376.

ORR, T. S. C. (1966). Spawning behaviour of rudd, *Scardinius erythrophthalmus* infested with plerocercoids of *Ligula intestinalis*. *Nature, Lond.*, **212**, 736.

(1967). Distribution of the plerocercoid of *Ligula intestinalis*. *J. Zool., Lond.*, **153**, 91–97.

OXENDER, D. & CHRISTENSEN, H. N. (1963). Distinct mediating systems for the transport of neutral amino acids by the Ehrlich cell. *J. biol. Chem.*, **238**, 3686–3699.

PAPPAS, P. W. & READ, C. P. (1972a). Thiamine uptake by *Hymenolepis diminuta*. *J. Parasit.*, **58**, 235–239.

(1972b). The absorption of pyridoxine and riboflavin by *Hymenolepis diminuta*. *J. Parasit.*, **58**, 417–421.

(1972c). Trypsin inactivation by intact *Hymenolepis diminuta*. *J. Parasit.*, **58**, 864–871.

(1972d). Inactivation of α- and β-chymotrypsin by intact *Hymenolepis diminuta* (Cestoda). *Biol. Bull. mar. biol. Lab. Woods Hole*, **143**, 605–616.

PAPPAS, P. W., UGLEM, G. L. & READ, C. P. (1973a). The influx of purines and pyrimidines across the brush border of *Hymenolepis diminuta*. *Parasitology*, **66**, 525–538.

(1973b). Ribonuclease activity associated with intact *Hymenolepis diminuta*. *J. Parasit.*, **59**, 824–828.

(1974). Anion and cation requirements for glucose and methionine accumulation by *Hymenolepis diminuta* (Cestoda). *Biol. Bull. mar. biol. Lab. Woods Hole*, **146**, 56–66.

PHARES, C. K. & RUEGAMER, W. R. (1973a). *In vitro* preparation of a growth factor from plerocercoids of the tapeworm, *Spirometra mansonoides*. *Proc. Soc. exp. Biol. Med.*, **142**, 374–377.

(1973b). Pigeon crop sac stimulation due to a factor produced by plerocercoids of the tapeworm *Spirometra mansonoides*. *Proc. Soc. exp. Biol. Med.*, **143**, 147–151.

PHIFER, K. O. (1960). Permeation and membrane transport in animal parasites: On the mechanism of glucose uptake by *Hymenolepis diminuta*. *J. Parasit.*, **46**, 145–153.

PITTMAN, R. G. & FISHER, F. M. Jr (1972). The membrane transport of glycerol by *Hymenolepis diminuta*. *J. Parasit.*, **58**, 742–749.

READ, C. P. (1950). The vertebrate small intestine as a habitat for parasitic helminths. *Rice Inst. Pam.*, **37**, (2), 1–94.

(1961). Competitions between sugars in their absorption by tapeworms. *J. Parasit.*, **47**, 1015–1016.

(1971). The microcosm of intestinal helminths. In *Ecology and physiology of parasites*, 188–200 (ed. A. M. Fallis). Toronto and Buffalo: University of Toronto Press.

(1973). Contact digestion in tapeworms. *J. Parasit.*, **59**, 672–677.

READ, C. P. & KILEJIAN, A. Z. (1969). Circadian migratory behaviour of a cestode symbiote in the rat host. *J. Parasit.*, **55**, 574–578.

READ, C. P., ROTHMAN, A. H. & SIMMONS, J. E. JR (1963). Studies on membrane transport with special reference to host–parasite integration. *Ann. N. Y. Acad. Sci.*, **113**, 154–205.

REICHENBACH-KLINKE, H. H. & REICHENBACH-KLINKE, K. E. (1970). Enzymuntersuchungen an Fischen. II. Trypsin- und α-amylase-Inhibitoren. *Arch. Fischereiwiss.*, **21**, 67–72.

RUEGAMER, W. R. & MUELLER, J. F. (1973). Growth responses in thyroidectomised, hypophysectomised and alloxan diabetic rats infected with plerocercoids of the tapeworm *Spirometra mansonoides*. *J. Nutr.*, **103**, 1496–1501.

RUFF, M. D. & READ, C. P. (1973). Inhibition of pancreatic lipase by *Hymenolepis diminuta*. *J. Parasit.*, **59**, 105–111.

SADUN, E. H., WILLIAMS, J. S., MERONEY, F. C. & MUELLER, J. F. (1965). Biochemical changes in mice infected with spargana of the cestode *Spirometra mansonoides*. *J. Parasit.*, **51**, 532–536.

STEELMAN, S. L., MORGAN, E. R., CUCCARO, A. J. & GLITZER, M. S. (1970). Growth hormone-like activity in hypophysectomised rats implanted with *Spirometra mansonoides* spargana. *Proc. Soc. exp. Biol. Med.*, **133**, 269–273.

STEELMAN, S. L., GLITZER, M. S., OSTLIND, D. A. & MUELLER, J. F. (1971). Biological properties of the growth hormonelike factor from the plerocercoid of *Spirometra mansonoides*. *Recent Prog. Horm. Res.*, **27**, 97–120.

TAYLOR, E. W. & THOMAS, J. N. (1968). Membrane (contact) digestion in the three species of tapeworm, *Hymenolepis diminuta*, *Hymenolepis mirostoma* and *Moniezia expansa*. *Parasitology*, **58**, 535–546.

UGOLEV, A. M. (1965). Membrane (contact) digestion. *Physiol. Rev.*, **45**, 555–595.

UGOLEV, A. M. & DE LAEY, P. (1973). Membrane digestion. A concept of enzymic hydrolysis on cell membranes. *Biochim. Biophys. Acta*, **300**, 105–128.

WOODWARD, C. K. & READ, C. P. (1969). Studies on membrane transport. VII. Transport of histidine through two distinct systems in the tapeworm *Hymenolepis diminuta*. *Comp. Biochem. Physiol.*, **30**, 1161–1177.

PLATE 1

(*a*) Longitudinal section of *Rutilus rutilus* infected with plerocercoids of *Ligula intestinalis*. Note ventral distension of host. H, heart; L, liver; S, swim-bladder; P, plerocercoids. × 3.

(*b*) Transverse-section of ovary of *R. rutilus* infected with *Ligula*. × 30.

(*c*) Transverse-section of ovary non-parasitised *R. rutilus* at the same time of year. × 30.

(*d*) Longitudinal section of pituitary gland of *Ligula*-infected *R. rutilus* to show basophils of mesoadenohypophysis. × 100.

(*e*) Similar region of pituitary gland of non-parasitised fish (1*b*–1*e* after Arme, 1968). × 85.

PLATE 2

(*a*) Proteins of *Ligula* from *Rutilus rutilus* (R) and *Gobio gobio* (G).

(*b*) Hypophysectomised rats of the same initial batch operated at 90 g. Larger rats received 20 larval scoleces of *Spirometra mansonoides* approximately one month after the operation. Six months later it can be seen that parasitised animals outweight the controls by a factor of 3 or 4 (Mueller, 1974*a*).

THE ROLE OF BACTERIA IN THE METABOLISM OF RUMEN ENTODINIOMORPHID PROTOZOA

BY G. S. COLEMAN

Biochemistry Department, Agricultural Research Council,
Institute of Animal Physiology, Babraham, Cambridge CB2 4AT

INTRODUCTION

From the standpoint of a protozoon, the rumen is a strictly anaerobic and strictly temperature-controlled environment containing a thick soup of bacteria and relatively indigestible food particles such as cellulose and starch grains. The rumen ciliate protozoa, which must have been free-living and aerobic at one time, have adapted to their environment to such an extent that they are now found nowhere else. As the bacteria present rapidly metabolise most sugars, amino acids and other readily digestable materials, these compounds are normally present in low concentrations in the rumen. In addition, therefore, to requiring anaerobic conditions and a temperature of 35–42 °C, the rumen Entodiniomorphid protozoa usually metabolise low molecular weight compounds slowly but engulf rapidly readily available particulate matter including bacteria. The holotrich protozoa from the rumen behave similarly except that they also rapidly incorporate free glucose into an intracellular α-linked polysaccharide (Heald, Oxford & Sugden, 1952).

This paper examines the role of bacteria in the metabolism of rumen Entodiniomorphid protozoa in terms both of bacteria as a source of food and as symbiotic organisms living inside the protozoa.

GROWTH OF ENTODINIOMORPHID PROTOZOA
IN VITRO

Although there have been many reports of the growth of rumen Entodiniomorphid protozoa *in vitro* (e.g. Hungate, 1942, 1943; Coleman, 1960; Mah, 1964; Coleman, 1969a, 1971; Coleman, Davies & Cash, 1972), nobody has so far succeeded in growing them in the absence of bacteria. In established cultures of protozoa the principal bacteria are *Klebsiella aerogenes* and *Proteus mirabilis* (White, 1969). The protozoa are grown on a buffered salts medium of low redox potential to which is added, each day, intact starch grains (in the form of wholemeal flour or rice starch) and ground, dried grass. The amount of starch added is critical, for if it is too little, the protozoa die of starvation and, if it is too much, the protozoa cannot engulf it all and the excess is fermented by the bacteria. The protozoa then die presumably due to the accumulation of toxic or acidic compounds in the bottom of the culture tube where both protozoa and excess starch are found. Although some species, e.g. *Entodinium caudatum*, require the presence of 10 % prepared fresh rumen fluid (from which the protozoa have been removed) or autoclaved rumen fluid (Coleman, 1960) in the medium in order to grow to a high population density, the principal bacteria present in established cultures are the non-rumen organisms mentioned above, both of which have arrived adventitiously. No sterile precautions are ever taken in the maintenance of these cultures. As inferred above, the secret of successful cultivation of the protozoa is the maintenance of a balance between protozoal and bacterial growth. If a culture of *Entodinium caudatum* is growing well, the number of bacteria remains comparatively low (10^7–10^8 cm^{-3}; Coleman, 1962) because of (a) a lack of excess starch on which to grow and (b) engulfment by the protozoa (Coleman, 1964). In general the more turbid the culture, the fewer the number of protozoa present.

Any discussion on the requirement of the protozoa for living bacteria will be speculative at this time, but they must, apart from providing food, be important in the maintenance of a low redox potential in the medium. Experiments with washed suspensions of *Entodinium caudatum* have shown that although they will tolerate exposure to air at room temperature for at least one hour, they are killed by incubation in air-saturated salts medium for 15 min at 39 °C. Unless the redox potential of the medium is reduced to o mV or below inside 30 min, some of the protozoa die. Quinn, Burroughs & Christiansen (1962) have shown that the optimum redox potential for the growth of rumen ciliate protozoa is in the range – 200 to – 260 mV. Although by the use of reducing agents such as cysteine, oxygen-free gases and the appropriate precautions, it is possible to maintain redox potentials approaching these values without the use of live bacteria, it is easier with the presence in the medium of a vigorously growing bacterium such as *Klebsiella aerogenes*. This is especially true where many tubes have to be opened each day for the addition of food and then regassed.

In addition, in established cultures of *Entodinium caudatum*, 16–20 *Klebsiella aerogenes* and 8–10 *Proteus mirabilis* are found living in vesicles in the endoplasm of each protozoon (White, 1969), although neither is present in protozoa freshly isolated from the rumen. It is therefore considered likely that the successful cultivation of this protozoon *in vitro* has been due, initially, to the establishment of these bacteria in the medium and the protozoa, and secondly, to the ability of the bacteria to survive inside the protozoa without being killed. It is possible that these intracellular bacteria are important in removing any oxygen that might reach the protozoa. In a washed suspension of *Entodinium caudatum* set up in oxygen-free salts medium these bacteria are undoubtedly very important in rapidly reducing the redox potential to a level that allows survival of the protozoa.

On returning entodinia, that had been cultured *in vitro* for 10 years, to a sheep rumen that contained no protozoa, the entodinia rapidly became established (Coleman & White, 1970). The total number of bacteria inside each protozoon, six weeks after innoculation, was the same as that found *in vitro*, but the species present had changed completely to those typical of rumen contents. This shows that although intracellular bacteria are a constant feature, the nature of the bacteria appears to depend on and reflect those present in the environment.

When a culture of *Entodinium caudatum* is initiated from protozoa taken directly from the rumen, it is essential to add 10 % freshly prepared rumen fluid (from which the protozoa have been removed) to the medium. This cannot be replaced by autoclaved rumen fluid until several months have elapsed. Thereafter the culture will grow in the presence of autoclaved rumen fluid, apparently indefinitely. It is possible that the live bacteria in the fresh rumen fluid may be necessary to keep the redox potential down until the *Klebsiella aerogenes* and *Proteus mirabilis* have become established in the protozoa.

Epidinium ecaudatum caudatum, growing *in vitro*, differs from *Entodinium caudatum* in that, although it contains intracellular bacteria, it also has many bacteria attached to the outside of its pellicle (Plate 1). These are of two types, designated A and B, but neither has been isolated in pure culture. They are metabolically active and on incubation with tritiated materials, e.g. lysed ^{3}H-labelled *Bacillus megaterium*, it has been shown, by autoradiography in the electron microscope, that label was incorporated rapidly especially by the type B organism (Plate 2). It is not known if the bacteria are important to the protozoa but they are present on all species of epidinia examined and on an *Eremoplastron* sp.

THE METABOLISM OF INTRACELLULAR BACTERIA
BY *ENTODINIUM CAUDATUM*

The fate of the bacteria present inside *Entodinium caudatum* was followed initially by isolating pure cultures of the two bacteria and investigating their metabolism on addition to washed suspensions of the protozoa. The metabolism of these free bacteria was then investigated by incubating ¹⁴C-labelled bacteria (prepared by growth on a ¹⁴C-labelled substrate) with washed suspensions of protozoa, grown *in vitro*, and measuring the uptake of ¹⁴C by the protozoa (Coleman, 1964). Under these conditions all bacteria tested were engulfed by *Entodinium caudatum* and Fig. 1 shows a time

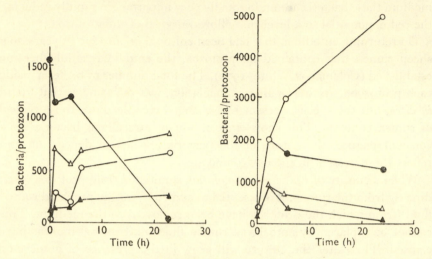

Fig. 1. Metabolism of 0.74×10^9 *Klebsiella aerogenes* cm^{-3} (labelled by growth with forced aeration on glucose-ammonia-salts medium containing [U-¹⁴C]glycine) by 3.4×10^5 *Entodinium caudatum*. △, bacteria in protozoa; ○, digested bacteria in supernatant fluid obtained by sonication and centrifugation of protozoa; ▲, digested bacteria in medium; ●, bacteria free in medium.

Fig. 2. Uptake of 2.0×10^9 *Klebsiella aerogenes* cm^{-3} (grown on [U-¹⁴C] leucine with forced aeration) by 6.5×10^5 *Entodinium caudatum*. ○, ●, bacteria taken up from ¹⁴C measurements; △, ▲, viable *Klebsiella aerogenes* present; ○, △, free bacteria present throughout experiment; ●, ▲, free bacteria removed at $2\frac{1}{4}$ h.

course for the uptake of *Klebsiella aerogenes* by this protozoon. The rate at which different bacteria, that had been engulfed, were killed and digested varied markedly between related organisms. For example, with Gram-negative bacteria being engulfed continuously, 35 % of *Klebsiella aerogenes*, 82 % of *Proteus mirabilis*, 60 % of *Serratia marcescens* and 0.7–12.5 % of *Escherichia coli* were still viable after $2\frac{1}{2}$ h (Coleman, 1967). Of the two bacteria found in the protozoa, *Proteus mirabilis* was obviously quite resistant to the killing action of the protozoa, whereas *Klebsiella aerogenes* was appreciably more sensitive and could have a half-life of less than one hour.

To investigate the metabolism of this latter bacterium in more detail, protozoa were incubated with (*a*) ¹⁴C-labelled bacteria continuously for 22 h and (*b*) with ¹⁴C-labelled bacteria for $2\frac{1}{4}$ h only and then for a further

20 h in their absence. Figure 2 shows that under condition (*a*) the amount of bacterial carbon in the protozoa increased continuously over 22 h and that, at $2\frac{1}{4}$, $5\frac{1}{4}$ and 22 h, 44 %, 22 % and 7 % respectively of these engulfed bacteria were still viable. If the free external bacteria were removed at $2\frac{1}{4}$ h [condition (*b*)], then over the next 20 h the amount of bacterial carbon present declined by 29 % and the number of viable bacteria by 93 % to a density that was only double that in protozoa that had been incubated throughout in the absence of bacteria in the medium. This latter result shows that over this period the bacteria have a half-life of approximately 5 h. However, in contrast, over half the bacteria engulfed in the first $2\frac{1}{4}$ h of the experiment were killed in this time. This suggests that *Klebsiella aerogenes* taken directly from the growth medium was very sensitive to the killing action of the protozoa, but that those bacteria, which survived and presumably metabolised inside the protozoa, were much more resistant. A possible explanation of this finding lies in the fact that this bacterium forms a capsule of a glucose-containing polysaccharide when allowed to metabolise glucose anaerobically. As the bacteria used for the experiment were grown with forced aeration, this capsule would not be present initially, although it could be formed by intracellular bacteria from glucose released from starch by the action of protozoal enzymes. If this capsule were relatively resistant to digestion by the protozoa, then this would explain why bacteria normally found inside the protozoa, and those that had been engulfed for several hours, were relatively resistant to killing.

To determine if the possession of a capsule made any difference to the rate at which bacteria were digested and if the capsule itself was digested easily, three different preparations of *Klebsiella aerogenes* were fed to the same protozoal suspension. The first preparation was made by incubating anaerobically a washed suspension of bacteria with 1 mmol l^{-1} [U-^{14}C]glucose for 30 min. Under these conditions ^{14}C was rapidly taken up by the bacteria and 54 % of the ^{14}C was incorporated into the polysaccharide compared with 21 % into the protein fractions. The second preparation was made by growing bacteria with forced aeration on glucose (0.2 %)–ammonia-salts medium containing [^{14}C]glycine. The resultant bacteria were labelled with ^{14}C in their protein and nucleic acid but had no capsule. The third preparation was the same as the second except that the bacteria were grown anaerobically and the glucose concentration was raised to 1 %. These bacteria were similarly labelled to the second batch but they were capsulated.

Figure 3*a* shows the percentage of the engulfed bacteria that was in the supernatant fluid after breakage of the protozoa and centrifugation of the

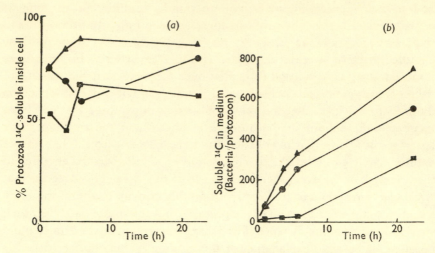

Fig. 3 (*a*) The digestion by *Entodinium caudatum* of *Klebsiella aerogenes* as shown by the proportion of the ¹⁴C in the engulfed bacteria that was present in the supernatant fluid after sonication and centrifugation of the protozoa. The bacteria were labelled with ¹⁴C by (i) incubation anaerobically of a washed suspension with [U-¹⁴C] glucose for 30 min (■); (ii) growth with forced aeration in the presence of [U-¹⁴C] glycine (▲); (iii) growth anaerobically in the presence of [U-¹⁴C] glycine (●). (*b*) The same as Fig. 3*a* except that digestion of the bacteria was followed by the appearance of soluble ¹⁴C in the medium.

homogenate at 7000 **g** for 20 min. Although the results were not conclusive, those bacteria which were labelled principally in the capsule were digested to a smaller extent than those without a capsule labelled in the remainder of the cell. ¹⁴C was also released by the protozoa into the medium and, by this criterion, the capsule was again digested more slowly than the remainder of the cell and bacteria with an unlabelled capsule were digested more slowly than those with no capsule (Fig. 3*b*).

If these indications are correct, then it suggests that *Klebsiella aerogenes* survives inside the protozoa because it possesses a capsule, the life of each bacterium depending on whether or not the bacterium can synthesise capsule faster than the protozoon digests it. As the only source of carbohydrate available to intracellular bacteria is probably the glucose derived from intracellular starch, starvation of the protozoa might be expected to decrease the number of viable bacteria inside each protozoon. This was investigated in an experiment in which *Entodinium caudatum* was starved for 18, 42 or 66 h before harvesting. As the number of viable bacteria present per protozoon was 37, 10 and 2 respectively, this result supports the above hypothesis.

Further support came from autoradiography, using the electron microscope, of sections of protozoa that had metabolised [6-³H]glucose for 5h. Under these conditions there were, in addition to intact labelled bacteria (Plate 3), some partially digested labelled bacteria in the protozoal endoplasm (Plate 4). As it is unlikely that dead organisms would incorporate glucose, this result suggests that, on harvesting the protozoa, live bacteria were present, that these incorporated glucose and some were then killed by the protozoa.

THE METABOLISM OF OTHER BACTERIA BY *ENTODINIUM CAUDATUM*

As mentioned above all bacteria and other particulate matter of a similar size are engulfed by *Entodinium caudatum* (Coleman & Hall, 1969). Quantitative studies on the uptake of bacteria have shown that the rate is similar for all small bacteria such as *Escherichia coli* or *Proteus vulgaris* and rates as high as 200 bacteria per protozoon min⁻¹ have been recorded (Coleman, 1964). A time course for the uptake of *Klebsiella aerogenes* is shown in Fig. 1. The maximum number of *Escherichia coli* (volume 2 μm³) taken up in a 3 h incubation was 10^4 per protozoon. Although many of the bacteria were digested during the 3 h incubation, studies in the electron microscope showed that the volume occupied by each bacterium remained approximately the same because the lipopolysaccharide part of the cell envelope remained undigested and spherical for several hours (Coleman & Hall, 1972). As the volume of a protozoon as determined from the number of protozoa in a packed pad was 4.7×10^4 μm³, it is obvious that each protozoon can engulf material until half its total volume is filled with bacteria. The number of a large bacterium, such as *Bacillus megaterium*, that was taken up was considerably less, although the total volume of material taken in was similar with all bacteria. Similarly, in experiments on competition between bacteria for engulfment by the protozoa, the effectiveness of any one organism in decreasing the uptake of another was proportional to its volume (Coleman, 1964). This means that bacteria were taken up in the proportion in which they were present in the medium and that there was no selective uptake of, for example, a species isolated from the rumen over one isolated from elsewhere.

As might be expected from the above, the number of bacteria, e.g. *Escherichia coli*, engulfed was not altered by boiling prior to carrying out the experiment, but the amount of bacterial carbon taken up was decreased markedly after disruption of the bacteria by sonication. In an experiment in which the bacteria were sonicated to produce an almost completely clear solution (for conditions see Coleman & Laurie, 1974a), the uptake of carbon from *Escherichia coli*, *Proteus mirabilis* and *Klebsiella aerogenes* was decreased by 67 %, 58 % and 61 % respectively by this treatment. This shows that *Entodinium caudatum* took up very small particles less readily than intact bacteria.

The uptake of bacteria by this protozoon is a very important process and experiments with individual bacterial species or mixed rumen bacteria, showed that sufficient bacteria were engulfed every 6–9 h to supply all the amino acids to enable the protozoa to divide every 24 h (Coleman & Laurie, 1974*a*).

The fate of the protein and nucleic acid of engulfed bacteria

As mentioned above the rate at which engulfed bacteria were killed by *Entodinium caudatum* varied markedly between species. As *Escherichia coli* was killed and digested rapidly after engulfment and could be easily labelled with specific amino acids etc., it was used as the test organism to determine the fate of various bacterial constituents inside the protozoa (Coleman, 1964).

In an experiment in which *Escherichia coli*, labelled specifically with [U-^{14}C]leucine, was fed to *Entodinium caudatum* for 11 h, 50 % of the ^{14}C was released into the medium as free leucine and 16 % was found in a soluble form inside the protozoon. Of this 70 % was present in protein as leucine (Coleman, 1967). Similar results were obtained with bacteria labelled with all the other amino acids tested, no evidence being obtained for any interconversion. Some amino acids were degraded to a small extent but no evidence was obtained for any biosynthesis from carbohydrate.

Bacterial nucleic acid was broken down to the nucleotide level inside the protozoa before being incorporated, without further degradation, into protozoal nucleic acid (Coleman, 1968).

THE METABOLISM OF BACTERIA BY *EPIDINIUM ECAUDATUM CAUDATUM*

Epidinium ecaudatum caudatum (see Coleman *et al.*, 1972, for description of different forms) also engulfed a wide range of bacteria, but there was considerable variation in the rates of engulfment between different species. Figure 4 which is a reciprocal plot of bacteria engulfed against bacterial population density, shows that *Proteus mirabilis* was engulfed much more rapidly than *Escherichia coli* at all population densities, although both bacteria were of similar size, shape and Gram reaction. A comparison of the rate of uptake of a number of bacterial species showed that although the

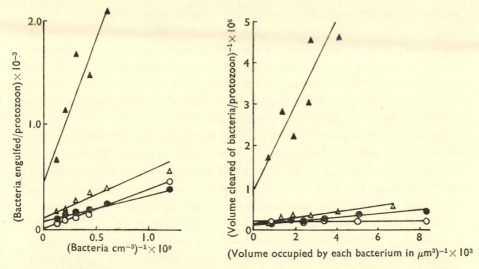

Fig. 4. Effect of sonication of bacteria on their uptake by *Epidinium ecaudatum caudatum*. ▲, Intact and △, sonicated *Escherichia coli*; ○, intact and ●, sonicated *Proteus mirabilis*.

Fig. 5. Effect of sonication of bacteria on their uptake by *Epidinium ecaudatum caudatum*. Symbols as for Fig. 4.

epidinia had five to six times the volume of *Entodinium caudatum*, they took up most species of bacteria at approximately the same rate. This means that the epidinia had less 'appetite' for bacteria and would obtain insufficient protein from this source to enable them to divide every 24 h. However, sonication or heating of the bacterial suspension prior to addition to the protozoa altered the rate of uptake of bacterial carbon, the actual effect depending on the bacterium and the population density.

Figure 4 shows that whereas sonication of *Proteus mirabilis* decreased the uptake of bacterial carbon at high population densities (above 1.6×10^9 bacteria cm^{-3}) and decreased it at low densities, sonication increased the uptake of *Escherichia coli*-carbon four to five times at all population densities. The result of these changes on sonication was that the differences between the rates of uptake of different bacteria were almost completely abolished. For example, at infinite bacterial population density intact *Proteus mirabilis* was taken up over fifty times as rapidly as *Escherichia coli*, whereas with sonicated organisms, the difference was only 1.4 times. Similarly the volume of medium cleared of 'bacteria' from a very dilute suspension of bacteria was widely different with intact, but almost the same with sonicated, organisms (Fig. 5). Although this phenomenon has not been investigated with a wide range of organisms, the uptakes of *Escherichia coli*, *Bacillus megaterium* and *Klebsiella aerogenes*, at least, were increased by sonication at all population densities. This suggests that, with these bacteria, epidinia find it easier to take up small particles rather than intact organisms. Why the converse is true with *Proteus mirabilis* is not known. Although epidinia have bacteria attached to the outside of their pellicles, no evidence has been obtained, by examination in the electron microscope of sections of the protozoa, for any attachment of *Proteus mirabilis* to the pellicle during incubation with the epidinia.

Entodinium caudatum feeds by passing, through ciliary action, a stream of medium in and out of its oesophagus and collecting particulate matter in vesicles at the bottom. It is uncertain if *Epidinium ecaudatum caudatum* feeds like this, but if it does, then the concept of protozoa removing organisms from a certain volume of medium is meaningful. The maximum volume cleared in any experiment by an epidinium was 16×10^6 μm^3 per protozoon h^{-1} which was obtained when the yeast *Saccharomyces fragilis* was present in a very dilute suspension (i.e. at around infinite dilution on a plot similar to that shown in Fig. 5). The volumes cleared at infinite dilution in the experiment shown in Fig. 5 were (in $\mu m^3 \times 10^{-6}$ h^{-1}): 0.35 with intact *Escherichia coli*, 2.1 with sonicated *E. coli*, 1.7 with intact *Proteus mirabilis* and 2.8 with sonicated *Pr. mirabilis*. The variation between these results means either, that the epidinia always pass the same volume of medium in and out of the oesophagus and remove particulate matter with varying degrees of efficiency, or that they remove all the matter from different volumes of medium. Of these two possibilities the former is considered to be the more likely, as it need involve no active change in the behaviour of the protozoa, only a change in the attraction between the bacteria or sub-bacterial particles and the cytoplasm at the site where the particulate matter enters vesicles. This should depend only on the surface properties of the particle involved and it could be postulated that the surface of *Escherichia coli* will not induce rapid vesicle formation, whereas that of the small particles will do so. However, it is possible that epidinia take up intact bacteria via the oesophagus and small particles by some other method. Some support for this idea comes from autoradiography, at the ultrastructural level, of the uptake of lysed, tritiated *Bacillus megaterium*. Tritium was taken up by the bacteria attached to the pellicle as well as by the remainder of the protozoal cytoplasm and occasionally tritiated material was present in the protozoal envelope (Plate 2). It is therefore possible that small particles could enter the cell through the pellicle, but much more work needs to be done before this can be considered as proved.

THE METABOLISM OF *BACILLUS MEGATERIUM* BY *EPIDINIUM ECAUDATUM CAUDATUM*

The experiments on the engulfment of bacteria by *Epidinium ecaudatum caudatum* showed that the bacteria tested could be divided into two groups, depending on whether or not the bacteria were lysed rapidly on incubation with washed suspensions of the protozoa. With most of the bacteria tested there was little evidence for the appearance of low molecular weight compounds in the medium, but with *Bacillus megaterium, B. subtilis* and *Micrococcus lysodeikticus* there was rapid lysis of the bacteria on incubation with the epidinia (Coleman & Laurie, 1974*b*). In one experiment of 2 h duration with [U-^{14}C]*Bacillus megaterium*, 90% of the added bacteria (equivalent to 5600 bacteria per protozoon) disappeared, whereas only eight bacteria were present inside each protozoon. As this was the maximum number of bacteria present in the epidinia during the incubation, it means that if the bacteria were engulfed and then digested, as was found with *Entodinium caudatum*, then each bacterium must have been present inside the protozoon for less than 10 s. As this was unbelievably short compared with periods of 30–60 min found with the entodinia (Coleman, 1964, 1972), it was obvious that the bacteria must be being lysed by some other mechanism, possibly the release of a lytic enzyme into the medium.

To investigate this possibility, a suspension of [U-^{14}C]*Bacillus megaterium* was incubated with three different preparations of epidinia and the release of low molecular weight material from the bacteria measured. The first was a normal protozoal suspension. The second was the same as the first except that the protozoa were removed by centrifugation after 30 min incubation with the bacteria. The third was the medium obtained after incubation of the epidinia for 30 min. Figure 6 shows that removal of the protozoa did not affect the rate of bacterial lysis and that, with the medium from above the protozoa, the rate was the same as with the protozoa. This shows that a lytic principle was released from the protozoa and that lysis of the bacteria did not involve their engulfment by the protozoa. The factor is considered to be an enzyme because it is heat sensitive and is only active against certain bacteria. Direct observation of the epidinia showed that they were apparently all intact, but it is possible that the enzyme was being liberated as the result of the disintegration of a small number of protozoa. It was therefore necessary to determine what proportion of the enzyme present in the protozoa was released and whether it was liberated rapidly, as might happen if the epidinia had suffered trauma, or slowly, as might occur if the release was physiological.

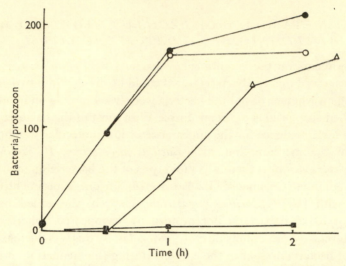

Fig. 6. Liberation of soluble ^{14}C into the medium from [U-^{14}C] *Bacillus megaterium* in the presence of *Epidinium ecaudatum caudatum*. ●, Epidinia present throughout; ○, epidinia removed at 30 min; △, epidinia incubated for 30 min in the absence of bacteria and then the supernatant fluid incubated with the bacteria; ■, ^{14}C incorporated by the protozoa in the presence of bacteria.

Time course for the release

Epidinia were incubated anaerobically in fresh salts medium and at intervals the enzyme activities in the supernatant fluid and sonicated protozoa were measured. After 0, 60, 200 and 300 min the amount of enzyme found in the supernatant fluid was 0.02 %, 0.5 %, 1.5 % and 3.5 % respectively of that present in the protozoa. In another type of experiment, the protozoa were incubated for 0, 60, 150 and 300 min, then harvested, washed under aerobic conditions, and incubated anaerobically for 30 min. The protozoa were then removed and measurements made of the lytic activity released during this second incubation. Taking the amount of enzyme released from protozoa harvested initially to be 1.0, then, with the protozoa harvested at 60, 150 or 300 min, the amount of enzyme released was 0.73, 0.38 and 0.30 respectively. These results show that the enzyme was released comparatively slowly into the medium either at a steady or declining rate. The finding of a declining rate of release in the second type of experiment, where the protozoa had to undergo the trauma of exposure to air and fresh medium between the first and second incubations, suggests that damage to the protozoa was not the cause of release.

Comparison of epidinia with other protozoa

This idea was supported by the finding that whereas all the epidinia examined released a lytic enzyme on incubation, *Entodinium caudatum* and

Polyplastron multivesiculatum did not. However, all these species of protozoa contained an enzyme, which was not released, that lysed *Bacillus megaterium* and all released protein into the medium on incubation in washed suspension (Table 1). The epidinia differ therefore from the other

Table 1. *The digestion of* Bacillus megaterium *by rumen ciliate protozoa*

	Entodinium caudatum	Epidinium ecaudatum caudatum	Polyplastron multi-vesiculatum
Bacteria digested per protozoon and released into medium*	31	1700	0
Bacteria inside each protozoon*	70	48	327
Lytic activity in each protozoon (bacteria min^{-1})	68	740	35 000
Lytic activity released per protozoon in 30 min (bacteria min^{-1})	0	46	0
Protein released in absence of bacteria (ng per protozoon)*	0.025	0.45	3.0
Protein in medium in presence of bacteria (ng per protozoon)*	0.31	1.17	3.2

* Duration of experiment: 2 h.

species in that they alone had the lytic enzyme present in the protein that was released. In the epidinia the release of enzyme could be a specific process not associated with the release of protein, the result of a slow generalised release of cellular protein or the result of the voiding of the contents of old food vacuoles, which might contain hydrolytic enzymes, through the anus. In order to distinguish between these three possibilities, the rate of release of the lytic enzyme, protein and an enzyme that is not involved in the digestion of food materials i.e. hexokinase, was measured. As the amount of hexokinase present was comparatively small, the incubations had to be carried out overnight in order to obtain sufficient in the medium to assay. Ampicillin was added to some tubes to control the bacterial growth that occurred during the prolonged incubation, but unfortunately only hexokinase could be measured in the medium in its presence. The results in Table 2 show that of the protein, lytic enzyme and hexokinase present in the protozoa initially 31 %, 37 % and 28 % respectively disappeared from the protozoa and 9 %, 11 % and 12 % appeared in the medium. In the presence of ampicillin less protein and hexokinase were lost from the protozoa, but the hexokinase was recovered quantitatively from the medium. In the absence of ampicillin, non-TCA-precipitable material, which gave a positive reaction in the Lowry (Lowry, Rosebrough, Farr & Randall, 1951) test for protein, equivalent to 13 % of the original soluble protein in the protozoa, also appeared in the medium.

Table 2. *The release of protein and enzymes by* Epidinium ecaudatum caudatum

	Protein	Non-protein nitrogenous material	Lytic enzyme	Hexo-kinase
			(arbitrary units)	
Present in protozoa				
Initially	14.6 ng	0	12 600	2.9
After 18 h	10.1 ng	0	8000	2.1
After 18 h + ampicillin	11.5 ng	0	–	2.6
Present in medium				
Initially	0.10 ng	0	0	0
After 18 h	1.34 ng	1.93 ng	1350	0.35
After 18 h + ampicillin	–	–	–	0.33
Loss from protozoa (%) in absence of ampicillin	31	–	37	28
Amount in medium (% of that in protozoa initially)	9	13 (of initial protein)	11	12

These results suggest that an appreciable amount of the protein released by the protozoa must have been degraded by the bacteria present during the incubation. As approximately the same proportion of the soluble protein and the enzymes present initially in the protozoa were lost, this is taken as evidence that there was a general release of cellular proteins. The mechanism is at present unknown but the evidence presented in the previous section suggests that the release could be physiological rather than the result of trauma.

Utilisation of the products of bacterial lysis

As the epidinia engulf comparatively few bacteria and are apparently unique among the Entodiniomorphid protozoa in releasing this lytic enzyme, it was important to discover if the protozoa could utilise the products of bacterial lysis. However, it must be remembered that none of the sensitive bacteria are found either in the rumen or in the culture media. Experiments were carried out with [U-14C]*Bacillus megaterium* lysed by both the action of the soluble enzyme inside the cell and the enzymes released into the medium. With enzymes from both sources 77 % of the 14C in the bacteria was solubilized and on chromatography appeared to be principally high molecular weight material, there being little evidence for the presence of free amino acids. Both the soluble and insoluble materials were then incubated with a fresh suspension of epidinia and with other protozoa to determine how much could be taken up. In a 5 h incubation with the soluble material (derived from 60×10^6 bacteria cm^{-3}) obtained by the action of the enzyme in the medium, the following values for uptake (in bacterial equivalents per protozoon) were obtained: 570 with *Epidinium ecaudatum caudatum*, 270 with *Epidinium ecaudatum tricaudatum*, 3260 with *Polyplastron multivesiculatum*, 1580 with an *Ophryoscolex* sp. and 3 with *Entodinium caudatum*. The rate of 114 bacterial equivalents/protozoon h^{-1} with *Epidinium ecaudatum caudatum* compares favourably with a rate of 50 bacteria h^{-1} from a suspension of 75×10^6 intact *Bacillus megaterium* cm^{-3}. Increase in substrate concentration increased the rate of uptake of all fractions and the results shown in Table 3 were obtained from double

Table 3. *The uptake by* Epidinium ecaudatum caudatum *of the products of the action of the lytic enzyme on* Bacillus megaterium

	Source of enzyme			
	Extracellular		Intracellular	
Product of enzyme action	Soluble	Particulate	Soluble	Particulate
Uptake (in bacteria from which fraction derived) per protozoon				
h^{-1} at infinite concentration	135	218	89	465
at 10^9 cm^{-3}	126	212	87	454
ng protein taken up per protozoon				
h^{-1} at 1 mg protein cm^{-3}	0.087	0.093	0.106	0.177
ng protein in soluble fraction of protozoa*	0.044	0.015	n.d.	n.d.
Division time of protozoa	286 h	830 h	n.d.	n.d.

* Supernatant fluid obtained by sonication of the protozoa followed by centrifugation at 7000 g. n.d., Not determined.

reciprocal plots of uptake against substrate concentration and are expressed both in terms of bacteria from which the fraction was derived and of protein. On the latter basis similar uptakes were obtained from the supernatant and pellet fractions (produced by the intracellular or extracellular enzymes) when these were added at 1 mg cm^{-3}.

Of the material taken up from the supernatant fraction produced by the action of the extracellular enzyme, 66 % was found in the soluble fraction inside the protozoon and of this, 74 % was present as protein. Acid hydrolysis of both the starting supernatant fraction and the protozoal soluble protein showed the presence of an apparently complete set of amino acids in both materials. This suggested that the lytic products could be a valuable source of amino acids for the protozoon. However, as shown in Table 3, only 0.044 ng protein per protozoon h^{-1} were incorporated into the protozoal soluble protein and this would only allow for division every 286 h compared with a measured time of 48 h (Coleman *et al.*, 1972).

Examination of possible sources of amino acids for protozoal growth showed that of the 600 pg h^{-1} required for each epidinium to divide every 24 h, 150 pg could be derived from intact bacteria, 50 pg from free amino acids, 1 pg by biosynthesis from starch and ammonia, and 50 pg from lysed *Bacillus megaterium* or other sensitive bacterium, if present (Coleman & Laurie, 1974a). There must therefore be another source of amino acids which could be the protein in the wholemeal flour added each day. The epidinia therefore differ from the entodinia in not relying solely on the engulfment of bacteria to provide amino acids for growth.

THE ROLE OF INTRACELLULAR BACTERIA IN THE METABOLISM OF SOLUBLE COMPOUNDS BY *ENTODINIUM CAUDATUM*

The metabolic activities of the living *Klebsiella aerogenes* and *Proteus mirabilis* present in vesicles in the protozoal endoplasm are obviously important in any consideration of the incorporation of ^{14}C from a ^{14}C-labelled compound. These bacteria can take up material from the vesicle in which they are present. Unfortunately it is not known if these vesicles are in equilibrium with the main protozoal pool and it has only been possible to determine what soluble compounds are present in the protozoal TCA-soluble pool, which contains material from both sources, and the kinetics of uptake of these, and the original substrate, by the bacteria. Most of the work has been carried out with glucose and maltose (Coleman, 1969b) although there have been some investigations with amino acids (Coleman, 1963, 1967).

Incorporation of ^{14}C from [^{14}C]sugars by Klebsiella aerogenes and Proteus mirabilis

These two bacterial species were isolated from the entodinia, grown up in conventional laboratory media with forced aeration (Coleman, 1969b) and then washed suspensions prepared. When these suspensions were incubated anaerobically with [U-^{14}C]glucose at a population density of 10^9 bacteria cm^{-3}, the rate of uptake of glucose with *Klebsiella aerogenes* was approximately five times that with *Proteus mirabilis*, but with both bacteria there was a sharp break in the curves at 10–20 mmol l^{-1} on a reciprocal plot of uptake of glucose against glucose concentration (Fig. 7). *Klebsiella aerogenes* incorporated maltose-carbon at up to twice the rate of glucose-carbon when the two sugars were present separately at equimolar concentrations, whereas *Proteus mirabilis* did not metabolise maltose. On incubation of *Klebsiella aerogenes* with either sugar at concentrations in excess of 1 mmol l^{-1}, a fluffy layer of a glucose polymer was found associated with the bacteria on centrifugation. Over 90 % of the glucose incorporated was present as this material. As there were approximately twice as many

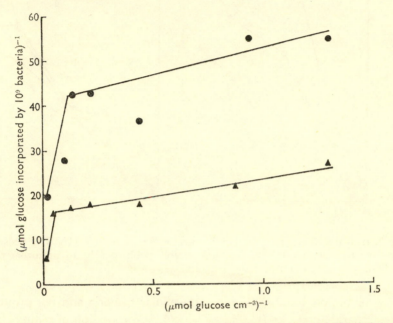

Fig. 7. Effect of substrate concentration on the incorporation of [U-^{14}C]glucose for 20 min by washed suspensions of *Klebsiella aerogenes* (▲) and *Proteus mirabilis* (●).

Klebsiella aerogenes as *Proteus mirabilis* present inside each *Entodinium caudatum* organism, these results suggest that this latter bacterium was comparatively unimportant as far as protozoal carbohydrate metabolism was concerned.

Attempts to correlate the behaviour of the bacterial suspensions with that of the bacteria liberated from the protozoa on sonication were partially successful in that the bacteria in the sonicate incorporated carbon from both sugars but only took up maltose at 145 % of the rate with glucose. This difference could be due to the different conditions under which the bacteria were grown.

The metabolism of sugars by intact Entodinium caudatum

Figure 8 shows that on incubation of *Entodinium caudatum* with 0.1 mmol l^{-1} or 67 mmol l^{-1} [U-^{14}C]glucose, ^{14}C was incorporated initially into the

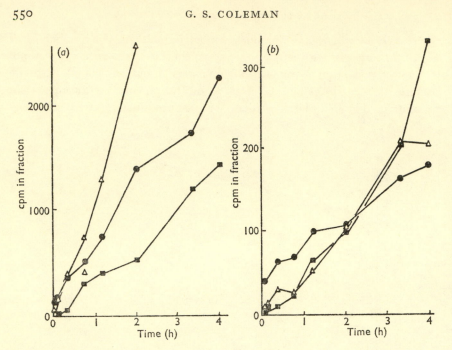

Fig. 8. Incorporation of 0.1 mmol l⁻¹ (a) or 67 mmol l⁻¹ [U-¹⁴C]glucose by *Entodinium caudatum*. ●, ¹⁴C in TCA-soluble pool; ■, ¹⁴C in protozoal poly-saccharide; △, ¹⁴C in intracellular bacteria.

TCA-soluble pool and then into intracellular bacteria and the protozoal polysaccharide granules (Coleman, 1969b). Incorporation of tritium from [6-³H]glucose into an intracellular bacterium is shown in Plate 3. The relative incorporation into these three fractions depends both on the time of incubation and on the glucose concentration. Figure 9 shows that, on a reciprocal plot of incorporation against substrate concentration, there was a discontinuity in the curves for incorporation into the TCA-soluble pool and the bacteria but not for incorporation into protozoal polysaccharide. The break in the curve for uptake into the bacteria occurred at 20 mmol l⁻¹, which was at the same concentration as was found with pure cultures of the two predominant bacterial species, and may therefore not be associated with any protozoal process. The discontinuity in the curve for incorporation into the pool (at 2.5 mmol l⁻¹) and the fact that the left-hand part passed through the origin suggested that there may be an 'active' (predominant at low glucose concentrations) and a passive (predominant at high glucose concentrations) uptake process as was found for amino acids (Coleman, 1967). On the other hand, if what is called the passive uptake represents external medium that was trapped on removing the protozoa from the incubation medium, and which was not removed by washing,

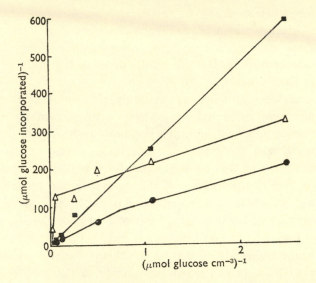

Fig. 9. Effect of substrate concentration on the incorporation of [U-^{14}C]glucose for 45 min by *Entodinium caudatum*. ●, ^{14}C in TCA-soluble pool; ■, ^{14}C in protozoal polysaccharide; △, ^{14}C in intracellular bacteria.

then it must have been present in 1–2 % of the protozoal volume. When protozoal metabolism was inhibited by iodoacetate, this figure increased to 11 % and it has been suggested (Coleman, 1969*b*) that this uptake could be into the contractile vacuole and that, in the presence of inhibitor, the vacuole did not contract and remained full of medium.

There is some further support for this from the results of the experiment shown in Fig. 8*a,b*. At low glucose concentration (Fig. 8*a*) some ^{14}C was incorporated into the pool in 1 min and this incorporation increased thereafter with time. Incorporation into the bacterial fraction was lower at 1 min but then increased rapidly, suggesting that uptake into the pool may have to precede incorporation by the bacteria. Incorporation into the protozoal polysaccharide was approximately half of that into the pool. In contrast, at high external glucose concentration (67 mmol l^{-1}) (Fig. 8*b*) the initial uptake (in the first minute) of ^{14}C into the pool was large (principally as glucose), but increased comparatively slowly thereafter. Uptake into the bacteria was relatively much slower than at low glucose concentrations suggesting that the additional glucose in the pool was not available for incorporation by the bacteria. As the discontinuity in the curve for the uptake of glucose by *Klebsiella aerogenes* was at 20 mmol l^{-1} (Fig. 7), the uptake should have been rapid if the glucose reaching the bacteria was above this concentration. These results are therefore consistent with the presence of glucose, taken up passively into a structure similar to a contractile vacuole and thus not available to the bacteria. There is, of course, no firm evidence for this, but the results do show that there is more than one pool in the protozoon.

Maltose was taken up more rapidly by the protozoa than glucose. When the protozoa were incubated for 45 min with each of the [^{14}C]sugars separately at 0.4 mmol l^{-1}, the incorporation into the pool, the protozoal polysaccharide, and the bacteria, from maltose was 138 %, 65 % and 196 % respectively of that from glucose. This higher incorporation into the bacteria than the remainder of the protozoon suggests that maltose, rather than glucose, was being metabolised by the bacteria. Investigation of the incorporation of maltose into the protozoal pool showed that it was rapidly taken up, the uptake reaching a maximum at 5 min after which time it remained constant. The time course for the appearance of [^{14}C]-glucose from [^{14}C]maltose was similar except that the final level was only 60–70 % of that of the maltose. If it is assumed that the medium does not have direct access to the bacteria in the vesicles, then the above findings could be taken as evidence that the fluid in the vesicles may be in equilibrium with the main protozoal pool (i.e. the pool to and from which the protozoon returns or removes low molecular weight compounds in the degradation or biosynthesis of macromolecules).

The effect of inhibitors on the uptake of sugars

A number of sugars and sugar analogues inhibit the uptake of glucose and maltose by *Entodinium caudatum* (Coleman, 1969b). The most useful proved to be 2-deoxyglucose which, in one experiment over 4 h at 67 mmol l^{-1}, decreased the incorporation of ^{14}C from [U-^{14}C]glucose into the protozoal pool, protozoal polysaccharide, and bacteria, to 7.7, 5.6 and 7.8 % respectively of that in the absence of inhibitor. In the same experiment, incorporation of ^{14}C by the intracellular bacteria in broken protozoal preparations was decreased to 14 % in the presence of 67 mmol l^{-1} 2-deoxyglucose when the method of breakage was gentle sonication and to 46 % when a Potter homogeniser (Potter & Elvehjem, 1936) was used. This suggests that sensitivity of these bacteria to the inhibitor depended on the extent to which they were damaged by the breakage procedure, the less drastic treatment in the homogeniser producing the less effect.

Similar experiments were therefore carried out with washed suspensions of the *Klebsiella aerogenes* and *Proteus mirabilis* isolated from the protozoa and with neither organism was incorporation inhibited by 2-deoxyglucose. As this difference may have been the result of the exposure of the intra-

PLATE I

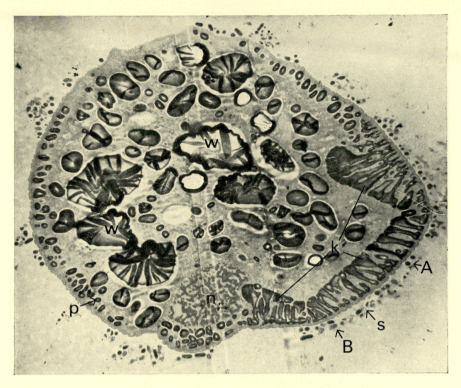

For explanation see p. 557

PLATE 2

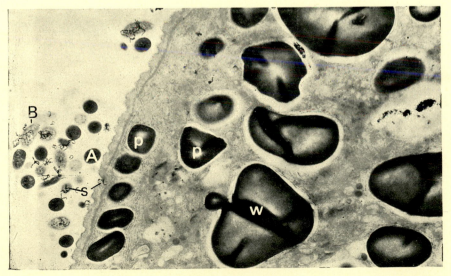

For explanation see p. 557   (Facing p. 552)

PLATE 3

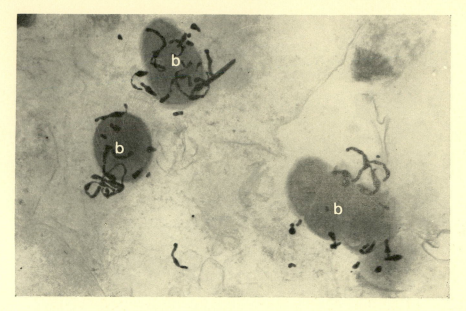

For explanation see p. 557

PLATE 4

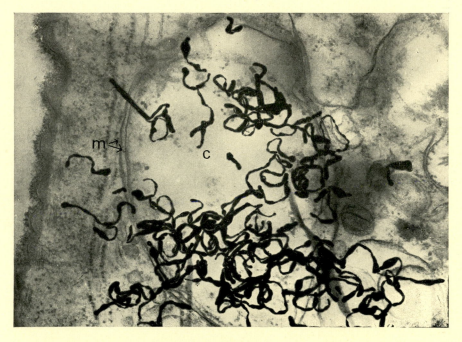

For explanation see p. 557

PLATE 5

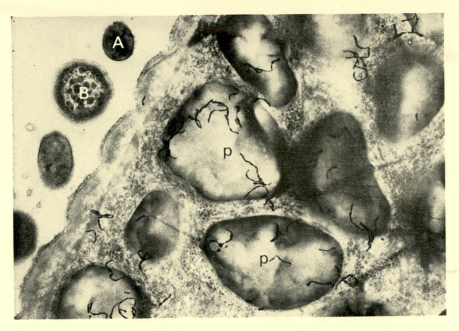

For explanation see p. 558

PLATE 6

For explanation see p. 558

cellular bacteria, but not the washed suspensions, to ultrasonic treatment, the effect of such treatment of washed suspensions on their ability to incorporate [^{14}C]glucose was examined. On gentle ultrasonic treatment (Coleman, 1969b) of suspensions of *Klebsiella aerogenes* for 30 s, the ability of the bacteria to incorporate [^{14}C]glucose in a 30 min incubation was decreased by 20 %, and 67 mmol l^{-1} 2-deoxyglucose inhibited uptake by these sonicated bacteria by 30 %. After breakage in the Potter homogeniser, 2-deoxyglucose inhibited uptake by 10 %.

If it is assumed that 2-deoxyglucose does not penetrate into the vesicles in the endoplasm, then the finding that it inhibits incorporation into all fractions, including the bacteria, equally, suggests that the inhibitor acts by preventing the uptake of glucose by the protozoa and that the bacteria take up glucose from the pool. If, however, 2-deoxyglucose can penetrate into the vesicles then the differential sensitivity of the protozoa and bacteria is important. The problem is complicated by the finding that bacteria damaged by the methods used to break protozoa have increased sensitivity to the inhibitor and means that results obtained with broken protozoa may be meaningless. If the bacteria in the vesicles behave like washed suspensions of undamaged bacteria and are insensitive to 2-deoxyglucose, the results mean that the glucose in the medium cannot reach the bacteria directly, because incorporation into the bacteria in intact protozoa is sensitive to the inhibitor. Therefore the bacteria probably take up glucose directly or indirectly from the protozoal pool.

The effect of antibiotics on the uptake of sugars

Incubation of *Entodinium caudatum* with 1400 units penicillin and 570 μg neomycin sulphate cm^{-3} for four days killed most of the bacteria present in the endoplasm (Coleman, 1962) but produced inactive protozoa that did not metabolise soluble compounds. Subsequently it was found that incubation with these antibiotics for only 16 h reduced the rate of glucose incorporation to 16 %, while leaving an appreciable number of viable bacteria present. The effect of antibiotics had therefore to be examined over shorter periods and it was found that, during a 60 min incubation, the presence of penicillin and neomycin inhibited incorporation into the pool by 26 %. In a 4 h experiment (Table 4), where the total uptake into the protozoa was reduced by 76 % in the presence of antibiotics, incorporation into the pool was decreased by 74 %, into the protozoal polysaccharide by 53 %, and into the bacteria by 83 %. Incorporation by the bacteria in the sonicated protozoal preparation was reduced by 92 %. These results show that incorporation into the intracellular bacteria was more sensitive than uptake into the pool and that bacteria freed from the protozoa were more

Table 4. *The effect of antibiotics and of sonication on the incorporation of* [^{14}C]*glucose or* ^{14}C-*labelled amino acids by rumen ciliate protozoa*

	Radioactivity (cpm)					
	Intact protozoa				Broken protozoa	
	− Antibiotics		+ Antibiotics		− Antibiotics	+ Antibiotics
Broken cell fraction	Soluble	Particulate*	Soluble	Particulate*	Particulate*	Particulate*
Entodinium caudatum						
Glycine	2 900	1 440	2 920	840	12 000	190
Glucose	2 280	6 040 (1 430)	596	1 440 (665)	6 360	470
Epidinium ecaudatum caudatum						
Serine	700	1 240	930	1 020	4 710	260
Glucose	23 700	40 700 (3 200)	27 000	27 500 (3 200)	67 100	7 400

The figures in brackets are the incorporation into protozoal polysaccharide.

* Fraction contains any bacteria associated with the protozoa.

sensitive to antibiotics than those present inside intact protozoa. They are also consistent with the postulate that the external medium penetrated only slowly into the vesicles and that the antibiotics inhibited the mechanism for the uptake of glucose. Similar results were obtained for the uptake of [^{14}C]glycine except that the effects with intact protozoa were less marked.

THE METABOLISM OF GLUCOSE BY
EPIDINIUM ECAUDATUM CAUDATUM

Since many of the bacteria associated with the epidinia were attached to the outside of the pellicle (Plates 1, 2 and 5), it was of interest to investigate the metabolism of glucose by suspensions of *Epidinium ecaudatum caudatum*. Table 4 shows that breakage of the protozoa increased the incorporation of ^{14}C from [U-^{14}C]glucose into the associated bacteria by 80 %, which indicates that many of the bacteria that incorporated ^{14}C from [^{14}C]glucose were not freely in contact with the medium. This was supported by autoradiography, at the ultrastructural level, of sections of protozoa that had incorporated [^{3}H]glucose (Plates 5 and 6), where it is apparent that protozoal polysaccharide in skeletal plates near the cuticle and intracellular bacteria were more heavily labelled than the bacteria attached to the outside of the pellicle (Coleman & Hall, 1974). That the bacteria taking up [^{14}C]glucose were not freely in contact with antibiotics in the medium was also shown by the observation, that the uptake of [^{14}C]glucose by bacteria associated with intact protozoa and into the same bacteria liberated by sonication of the protozoa, was decreased to 65 % and 11 % respectively by the antibiotic, ampicillin. The finding, that incorporation into the soluble material in the protozoa or the protozoal polysaccharide was only slightly or not affected by ampicillin, suggests that it had little effect on the mechanism for glucose uptake by the protozoa and acted only on the bacteria.

CONCLUSIONS

Although the rumen ciliate protozoa are completely dependent on the host animal for the provision of the habitat in which they live, the ruminant can be reared successfully in the complete absence of protozoa in its rumen. In contrast, the bacteria and protozoa present in in-vitro cultures, and probably also in the rumen, are mutually beneficial to each other. The *Entodinium* spp. digest starch and proteinaceous material making sugars and amino acids available to the bacteria in the medium and the protozoal endoplasm. The bacteria grow on these materials and maintain the low redox potential essential for protozoal growth. The protozoa then engulf the free bacteria and digest both these and the intracellular bacteria, utilizing the bacterial protein as a source of amino acids for the synthesis of protein. Some of the bacterial amino acids are also liberated into the medium where they are fermented by the remaining bacteria. The net result of the presence of these protozoa in the bacterial culture or the rumen is, therefore, to increase the rate at which the bacteria grow and are killed i.e. to increase the rate of turnover of bacterial carbon and nitrogen. In the rumen, under appropriate conditions, 1% of the bacteria can be killed each minute.

ACKNOWLEDGEMENT

The author thanks Mr F. J. Hall for taking and providing the electron micrographs.

REFERENCES

COLEMAN, G. S. (1960). The cultivation of sheep rumen oligotrich protozoa in vitro. *J. gen. Microbiol.*, **22**, 555–563.

(1962). The preparation and survival of almost bacteria-free suspensions of *Entodinium caudatum*. *J. gen. Microbiol.*, **28**, 271–281.

(1963). The growth and metabolism of rumen ciliate protozoa. *Symp. Soc. gen. Microbiol.*, **13**, 298–325.

(1964). The metabolism of *Escherichia coli* and other bacteria by the rumen ciliate *Entodinium caudatum*. *J. gen. Microbiol.*, **37**, 209–223.

(1967). The metabolism of the amino acids of *Escherichia coli* and other bacteria by the rumen ciliate *Entodinium caudatum*. *J. gen. Microbiol.*, **47**, 449–464.

(1968). The metabolism of bacterial nucleic acid and of free components of nucleic acid by the rumen ciliate *Entodinium caudatum*. *J. gen. Microbiol.*, **54**, 83–96.

(1969a). The cultivation of the rumen ciliate *Entodinium simplex*. *J. gen. Microbiol.*, **57**, 81–90.

(1969b). The metabolism of starch, maltose, glucose and some other sugars by the rumen ciliate *Entodinium caudatum*. *J. gen. Microbiol.*, **57**, 303–332.

(1971). The cultivation of rumen Entodiniomorphid protozoa. In *Isolation of Anaerobes*, 159–176 (eds., D. A. Shapton and R. G. Board). London and New York: Academic Press.

(1972). The metabolism of starch, glucose, amino acids, purines, pyrimidines and bacteria by the rumen ciliate *Entodinium simplex*. *J. gen. Microbiol.*, **71**, 117–131.

COLEMAN, G. S., DAVIES, J. I., & CASH M. A. (1972). The cultivation of the rumen ciliates *Epidinium ecaudatum caudatum* and *Polyplastron multivesiculatum in vitro*. *J. gen. Microbiol.*, **73**, 509–521.

COLEMAN, G. S. & HALL, F. J. (1969). Electron microscopy of the rumen ciliate *Entodinium caudatum*, with special reference to the engulfment of bacteria and other particulate matter. *Tissue & Cell*, **1**, 607–618.

(1972). Fine structural studies on digestion of bacterial species in the rumen ciliate *Entodinium caudatum*. *Tissue & Cell*, **4**, 37–48.

(1974). The metabolism of *Epidinium ecaudatum caudatum* and *Entodinium caudatum* as shown by autoradiography in the electron microscope. *J. gen. Microbiol* **85**, 265–273.

COLEMAN, G. S. & LAURIE, J. I. (1974*a*). The metabolism of starch, glucose, amino acids, purines, pyrimidines and bacteria by three *Epidinium* spp. isolated from the rumen. *J. gen. Microbiol.*, **85**, 244–256.

(1974*b*). The metabolism of *Bacillus megaterium* and the release of a lytic enzyme by three *Epidinium* spp. isolated from the rumen. *J. gen. Microbiol.*, **85**, 257–264.

COLEMAN, G. S. & WHITE, R. W. (1970). Re-establishment of *Entodinium caudatum*, cultured *in vitro*, in the rumen of a defaunated sheep. *J. gen. Microbiol.*, **62**, 265–266.

HEALD, P. J., OXFORD, A. E. & SUGDEN, B. (1952). A convenient method for preparing massive suspensions of virtually bacteria-free ciliate protozoa of the genera *Isotricha* and *Dasytricha* for manometric studies. *Nature, Lond.*, **169**, 1055–1056.

HUNGATE, R. E. (1942). The culture of *Eudiplodinium neglectum*, with experiments on the digestion of cellulose. *Biol. Bull. mar. biol. Lab. Woods Hole*, **83**, 303–319.

(1943). Further experiments on cellulose digestion by the protozoa in the rumen of cattle. *Biol. Bull. mar. biol. Lab. Woods Hole*, **84**, 157–163.

LOWRY, O. M., ROSEBROUGH, N. J., FARR, A. L. & RANDALL, R. J. (1951). Protein measurement with the Folin phenol reagent. *J. biol. Chem.*, **193**, 265–275.

MAH, R. A. (1964). Factors influencing the *in vitro* culture of the rumen ciliate *Ophryoscolex purkynei* Stein. *J. Protozool.*, **11**, 546–552.

POTTER, V. R. & ELVEHJEM, C. A. (1936). A modified method for the study of tissue oxidations. *J. biol. Chem.*, **114**, 495–504.

QUINN, R. W., BURROUGHS, W. & CHRISTIANSEN, W. C. (1962). Continuous culture of ruminal micro-organisms in chemically defined medium. 2. Culture medium studies. *Appl. Microbiol.*, **10**, 583–592.

WHITE, R. W. (1969). Viable bacteria inside the rumen ciliate *Entodinium caudatum*. *J. gen. Microbiol.*, **56**, 403–408.

EXPLANATION OF PLATES

PLATES 1 AND 2

Electron micrographs of autoradiograms of a section through *Epidinium ecaudatum caudatum* that had been incubated with [^{3}H]*Bacillus megaterium* for 2 h. Note the presence of bacteria of two types (A and B) attached to the pellicle all the way round the protozoon. Radioactivity in the section is shown by the presence of silver grains (s) which are associated principally with the B-type bacteria. p = Protozoal polysaccharide; k = polysaccharide in the skeletal plate; n = nucleus; w = wholemeal starch grains on which the protozoon had fed. Plate 1 × 2890; Plate 2 × 7300.

PLATES 3 AND 4

Electron micrographs of autoradiograms of a section through *Entodinium caudatum* that had been incubated with [6-^{3}H]glucose for 4 h. Note the presence of label associated with intact bacteria (b) and a partially digested bacterium (c). m = Bacterial membranes. Plates 3 × 31 000; Plate 4 × 32 000.

PLATES 5 AND 6

Electron micrographs of autoradiograms of a section through *Epidinium ecaudatum caudatum* that had been incubated with [6-³H]glucose for 4 h. Note the presence of label associated with the protozoal polysaccharide granules (p) and the intracellular bacterium (b) rather than the bacteria (A and B) attached to the pellicle. Plate 5 × 24 000; Plate 6 × 28 000.

SYMBIOTIC RELATIONSHIPS BETWEEN TERMITES AND THEIR INTESTINAL MICROBIOTA

By J. A. BREZNAK

Department of Microbiology and Public Health,
Michigan State University, East Lansing, Michigan 48824, USA

A variety of micro-organisms have long been known to exist symbiotically with an equally diverse array of insects (Brooks, 1963; Buchner, 1965; Koch, 1967). In fact, experimental analysis of symbioses had its beginnings in studies of insects and their microbial symbionts. Symbiotic micro-organisms may reside within the insect body. As such they may be extracellular and inhabit the alimentary tract, Malpighian vessels, lymph, and fat body, or they may be intracellular and reside within specific insect tissue cells called mycetocytes. On the other hand, some symbionts (especially fungi) proliferate primarily outside of the insect body, being only temporarily stored in special body organs for the purpose of dissemination. Fungi cultivated by ambrosia beetles, and on which the beetles browse, are a good example (Baker, 1963; Francke-Grosmann, 1967). The occurrence of insect–microbe symbioses, and the origin and development of studies in this area, have been treated in detail by Baker (1963), Brooks (1963), Buchner (1965), Francke-Grosmann (1967), and Koch (1967).

A marked decrease in the vigor and developmental performance of insects often results from removal of their microbial symbionts. In some cases the biochemical basis for the response is known, and is associated with the omission of a growth factor normally provided by the micro-organisms. For example *Rhodnius prolixus*, a blood-sucking bug, is dependent on bacterial symbionts for its supply of B vitamins (Baines, 1956). Similarly, the retarded development of aposymbiotic cockroaches (*Blattella germanica*) can be ameliorated by supplying certain polypeptides in the diet (Brooks & Kringen, 1972). Presumably the polypeptides are normally provided by intracellular bacterial symbionts. Pupation in ambrosia beetles (*Xyleborus ferrugineus*) depends in part on ergosterol which is synthesized by symbiotic fungi (Chu, Norris & Kok, 1970; Norris, 1972). However, the role of symbiotic microbes in supplying growth factors may be contrasted with one in which they function as catabolic agents, supplying energy sources derived from the breakdown of food ingested, but not digestable, by the insects. Such a relationship exists between xylophagous termites and their mutualistic, cellulolytic, intestinal protozoa.

The purpose of this paper is to discuss symbiotic relationships between termites and their intestinal microbiota. The paper will include a resumé of the digestive system of termites and the role of mutualistic protozoa in digestion, then it will focus on recent investigations in this writer's laboratory concerning bacterial symbionts of termites and two phenomena which appear to be mediated by bacteria – nitrogen fixation and methane formation. Associations between termites and fungi will not be included here; this topic was recently reviewed by Sands (1969).

TERMITES AS ANIMALS

Termites are insects belonging to the order Isoptera meaning equal (*iso*) winged (*ptera*), and reflecting the fact that the fore and hind wings of sexually mature alates are similar in size and veination (Weesner, 1965). They are fascinating animals which live in highly integrated social units, or colonies, exhibiting a division of labor among morphologically distinguishable castes. Although often referred to as 'white ants', because of their size, greyish white color, and social habits, they are not ants at all. The latter belong to a completely different order of insects, Hymenoptera.

Colonies of termites live in nests, consisting of a complex architectural system of galleries and runways fashioned from soil, plant material, detritus, and salivary cementing material. Nests may be subterranean or arboreal and inconspicuous, or may protrude above ground in the form of spectacular mounds up to several meters in height. The latter are particularly common in the tropics. The size, shape, and location of nests is often highly distinctive, and characteristic of a particular termite species (Lee & Wood, 1971).

About 1900 living and fossil species of termites have been described (Krishna, 1969). They are frequently thought of only with respect to the damage they cause, and this is not surprising. Annual world-wide termite damage and control costs have been estimated at $1.2 billion (Hickin, 1971), yet less than 10 % of the named species have actually been recorded as pests. Unfortunately, their ecological significance is often overlooked. Termites, along with other soil macro- and micro-organisms, constitute important scavengers of cellulose-based plant litter, degrading such material to simpler carbon compounds. In this capacity they play a significant role in the carbon cycle of nature. The biology of termites, and their role as soil animals, have recently been the subjects of excellent comprehensive reviews (Krishna & Weesner, 1969, 1970; Lee & Wood, 1971).

THE DIGESTIVE SYSTEM OF TERMITES

The digestive system of termites, its comparative anatomy, cytology, and physiology, have been discussed in detail by Noirot & Noirot-Timothée (1969), however a digression on the major features of digestion in termites seems appropriate here.

The gut of a typical worker termite is composed of three major divisions: the foregut, midgut, and hindgut. The foregut consists of the esophagus, crop, and muscular gizzard, and terminates with an esophageal valve. The midgut follows the foregut. It is a tube of uniform diameter, except in certain species possessing anterior ceca. Its histological structure is almost invariant, and consists of epithelial cells which rest on a distinct basement membrane. This in turn is surrounded by bundles of circular and longi-tudinal muscle fibers. The apical poles of epithelial cells possess a brush border. At each molt the epithelium is sloughed off as a unit, and is reformed from regenerative cells which remain in place close to the base-ment membrane. Intestinal absorption at the level of the midgut is probably limited to easily hydrolyzable compounds, or soluble compounds derived from the ingesta. The midgut terminates at the proctodeal valve, the point of connection of Malpighian tubules, and is followed by a well-developed hindgut of which five segments are discernable. The bulbous 'paunch' segment of the hindgut is particularly important for the present discussion since it is in this dilated region that the horde of bacteria and cellulolytic protozoa (in those species which possess protozoa) reside. Comparative anatomy of the gut indicates that some species of evolutionarily 'higher' termites possess a mixed segment which is an elongated extension of the midgut extending along, but separated from, the first segment of the hindgut. When a mixed segment is present, it is usually filled with bacteria which are so similar in morphology as to suggest a pure culture.

The nutritive regimen of termites is varied, but dominated by cellulosic materials. Wood, either sound or in various stages of decomposition, is favored by evolutionarily 'lower' termites. 'Higher' termites prefer wood which has been partially decomposed by fungi, and in fact, the Macro-termitinae cultivate fungi for this purpose. 'Higher' termites also feed on leaves, grasses, and humus, and some extract nutrients from soil par-ticles. Crude nutrients are obtained by the worker caste. Distribution of food to other castes (e.g. young larvae, soldiers, reproductives) takes place by the transfer of stomodeal food (regurgitated raw food mixed with saliva) or proctodeal food (paunch contents) from the worker larvae. Since newly hatched larvae and post-molt larvae are essentially devoid of cellulolytic protozoa, this same mechanism (particularly proctodeal transfer) serves to 'inoculate' recipients with the normal complement of microbes.

Passage of food through the alimentary tract takes about 24 h. Although some digestion may take place in the midgut, the hindgut is the primary site of cellulose digestion. In 'lower' termites, cellulose digestion is performed primarily, if not solely, by mutualistic protozoa, and will be discussed below. In 'higher' termites bacteria presumably perform this function, at least in part: the evidence for synthesis of cellulase by termites is inconclusive at the present time. Cellulose undergoes substantial degradation in termites, as do hemicelluloses. The problem of lignin decomposition is still unsettled. Whereas lignin appears to be degraded, determination of the extent of degradation has been confounded by the coprophageal habit of termites which subjects the compound to repeated digestion.

Absorption of the major products of wood digestion occurs through the hindgut, which is lined with a relatively thin epithelium. Whether absorption is by passive diffusion or active uptake has not been satisfactorily shown. The enteric valve prevents reflux of nutrients from the paunch to the midgut for absorption. In most insects the midgut represents the primary site of nutrient absorption, whereas the hindgut is secondary. Termites represent an exception to this general rule. The overall mechanism of wood digestion in termites can be described by the 'cecal model' (Hungate, 1972). The model is one in which a *partial* digestion and absorption of food is accomplished by the host (termite), but is followed by a prominent microbial fermentation in an enlarged cecum (in this case the paunch). Fermentation products, which may be waste products as far as the gut microbiota are concerned, are absorbed by the host and oxidized for energy. Microbial protoplasm may become available to the animal, as a nutrient, through coprophagy.

BIOCHEMICAL BASIS FOR MUTUALISM BETWEEN XYLOPHAGOUS TERMITES AND THEIR HINDGUT PROTOZOA

Xylophagous termites of the four lower families (Rhinotermitidae, Kalotermitidae, Hodotermitidae, and Mastotermitidae) harbor an abundant flora of symbiotic bacteria and anaerobic flagellate protozoa in their paunch. So great is the biomass of protozoa that during termite intermolt periods they may account for one seventh to one third of the body weight of larvae (Hungate, 1955). An impression of the profusion of protozoa in the paunch can be obtained by scanning electron microscopy (Plate 1).

Most studies of termite gut microbes have focused on the protozoa, and it is easy to see why this is so. Besides their abundance, they represent unique genera and species of oxymonad, trichomonad, and hypermastigote flagellates found only in termites and *Cryptocercus* (a wood-eating cockroach closely related to termites; Cleveland, Hall, Sanders & Collier, 1934). Moreover, they are spectacular in both morphology and motility, and some have been shown to be cellulolytic.

Termite-associated protozoa, and their role in cellulose digestion, have been extensively reviewed (Hungate, 1946a, 1955; Honigberg, 1967, 1970). It was primarily the work of Cleveland (reviewed by Hungate and Honigberg) which established the relationship between termites and their cellulose-digesting protozoa to be one of mutualism: protozoa digesting cellulose in wood to a form utilizable by the insects, and the insects providing an anaerobic chamber (the paunch) replete with food (wood) for the protozoa. Termites could be defaunated (i.e. made to lose their protozoa) by a variety of techniques including starvation, or incubation at elevated temperatures or under hyperbaric oxygen. When this was done they would perish after several weeks on a diet of sound wood or cellulose. Their survival was prolonged, however, if fed glucose or wood extensively degraded by fungi. If refaunated, they could again thrive on sound wood or cellulose. Since bacteria were observed in the guts of defaunated termites it was felt that they were not crucial insofar as cellulose digestion was concerned. Likewise, as pointed out by Honigberg (1970), not all protozoa in xylophagous termites are cellulolytic. Therefore the term mutualistic should be reserved, for the time being, for cellulolytic species.

While mutualism between termites and their cellulolytic protozoa had been fairly well established by the late 1920s, the biochemical basis was not clear. Of course it was known that cellulose decomposition by protozoa was central to the mutualistic phenomenon. Indeed, in the years to follow, Trager (1932, 1934) succeeded in cultivating *Trichomonas* (*Trichomitopsis*) *termopsidis* (a xylophagous flagellate from the gut of *Zootermopsis angusticollis*) in a cellulose-containing medium, and demonstrated cellulase activity in cell extracts. But what *products* of cellulose decomposition served as substrates for termites? Cleveland (1924) postulated that protozoa hydrolyzed cellulose to sugars, e.g. glucose, which were then absorbed by termites and completely oxidized to carbon dioxide and water for energy. This hypothesis was consistent with the ability of glucose to prolong the survival of defaunated termites (Cleveland, 1924). However, mere hydrolysis of cellulose to glucose would provide little, if any, available energy for the protozoa. Moreover, glucose could not be demonstrated in gut contents of termites or in culture fluids of protozoa (Hungate, 1943).

It remained for other workers, principally Hungate, to elucidate some of the biochemical details of mutualism. In a series of elegant experiments Hungate demonstrated that indeed protozoa, and not bacteria, molds, or termite enzymes, were the most important agents of cellulose decomposition (Hungate, 1936, 1938). He then prepared suspensions of cellulolytic protozoa (primarily *Trichonympha*) by squeezing out the hindgut contents of *Zootermopsis*, and showed that they were anaerobic, and capable of fermenting cellulose to carbon dioxide, hydrogen, and acids (primarily acetic acid) in amounts accounting for 70–75 % of the cellulose decomposed (Hungate, 1939, 1943). Hydrogen did not appear to be used by termites. Therefore, knowledge of the acetate: hydrogen ratios resulting from cellulose fermentation by suspensions of protozoa *in vitro* permitted him to estimate acetate production *in vivo* (i.e., in intact termites) by measuring hydrogen emission. He also measured oxygen uptake by termites. When this was done it was found that the magnitude of oxygen uptake by termites corresponded closely with the theoretical amount required for complete oxidation of acetate to carbon dioxide and water (Hungate, 1943). Since acetic acid could only be demonstrated in hindgut contents of *Zootermopsis*, and not in the food (wood) or fecal pellets, and since the hindgut was permeable to acetic acid (Hungate, 1943), the following scheme for symbiotic cellulose utilization in termites emerged: wood particles are engulfed by protozoa and the cellulose fermented intracellularly to carbon dioxide, hydrogen, and acetic acid as major products. The last compound is absorbed through the hindgut wall and oxidized by termites for energy. This scheme is conceptually appealing because it allows for acquisition of energy (ATP) by *both* protozoa (via anaerobic fermentation of cellulose) *and* termites (via aerobic oxidation of acetate to carbon dioxide and water). According to this scheme defaunated termites would be expected to survive if fed acetate, yet this has not been accomplished (Cook, 1943; Hungate, 1946*a*). However, rather than cast doubt on the validity of Hungate's scheme, it suggests that the biochemical interrelationships of mutualism are more intricate than heretofore supposed.

BACTERIAL SYMBIONTS OF TERMITES

Bacterial inhabitants of termite guts have been generally neglected, and consequently are ill-understood. This is unfortunate because they are abundant, and would appear to constitute a biochemically important component of the gut microbiota. For example, 10^6 to 10^7 bacteria per termite gut have been reported (Krasil'nikov & Satdykov, 1969; Breznak, Brill, Mertins & Coppel, 1973). Since the volume of gut fluid is on the order of microliters this amounts to a bacterial concentration of 10^8 to 10^{10} cells cm^{-3} – a sizeable population indeed. Furthermore, these are probably minimum estimates because, as will be discussed, dense populations of bacteria adhere to the gut wall and could be completely overlooked in examination of gut *fluid*. Evidence will be cited to suggest that bacteria are important for the growth and vitality of termites. However, pending definitive data the term symbiont, as applied here to bacteria, must be regarded in a broad sense as employed by de Bary (1879) i.e. one of a number of dissimilar organisms living in close proximity on a permanent or semipermanent basis.

Spirochetes

Spirochetes are a group of bacteria which are defined morphologically. They possess a coiled or undulate protoplasmic cylinder which is bound by the cell wall, and which houses the cytoplasm and nucleus. Inserted at each end of the protoplasmic cylinder are one or more axial fibrils which wind around or together with the protoplasmic cylinder and overlap in the central region of the cell. The protoplasmic cylinder–axial fibril complex is in turn enveloped in a membranous 'sheath'. Spirochetes exhibit lashing, spinning, and serpentine-like movements, and can effect net translocation presumably via the activity of axial fibrils. Some species cause diseases such as syphilis, whereas others are innocuous, free-living forms which inhabit marine and fresh waters and soil. Still others exist in a nonpathogenic relationship within a variety of animals (Breznak, 1973).

Spirochetes have been reported in a number of insects, but they are particularly abundant in the gut of termites. The literature on the spirochetes of termites has recently been reviewed (Breznak, 1973), however little is known about the organisms since they have never been isolated in pure culture. Although present in the gut as individual entities, spirochetes also adhere to the surface of protozoa, and have been observed to exhibit active motility within the cytoplasm of termite flagellates, particularly when the insects were fed on filter paper (Kirby, 1941). Actually a variety of organisms, including algae and fungi, are known to form some type of ecto- or endosymbiosis with protozoa, and it is not unusual for an assortment of symbionts to be associated with a protozoan at the same time (Kirby, 1941; Ball, 1969). Grassé (1938) suggested that phoretic associations resulted from nutrient secretion by protozoa. Consistent with this notion is the striking degree of localization of adherent organisms on specific areas of the protozoan surface.

In one case a role for adherent spirochetes has been demonstrated. During a study of *Mixotricha paradoxa* (a large, wood-eating polymastigote flagellate which inhabits the gut of the termite *Mastotermes darwiniensis*) Cleveland & Grimstone (1964) found that its surface was not covered with cilia, as Sutherland (1933) originally thought, but was adorned with a vestment of spirochetes and rod-shaped bacteria. The prokaryotes were attached to bracket-like structures on the protozoan surface formed, in part, by the protozoan plasma membrane. Rod-shaped bacteria were also demonstrated in the cytoplasma of *M. paradoxa*. The most fascinating aspect of the association was the demonstration that the co-ordinated undulations of spirochetes actually propelled the protozoan through the gut fluid. Although *M. paradoxa* possessed four flagella, it used them merely for steering! Obviously the protozoa benefit from the association – they expend less energy for locomotion. But it also seems likely that spirochetes benefit, perhaps by receiving nutrients from the protozoa, thereby elevating the association to one of mutualism. A role for the rod-shaped bacteria was not found. Bloodgood, Miller, Fitzharris & McIntosh (1974) recently described a nose-like extension of spirochetes which may aid their attachment to *Pyrsonympha*. Adherence of spirochetes to *M. paradoxa*, and their locomotive role, have prompted Margulis (1970) to suggest the evolution of eukaryotic flagella via spirochete ectosymbionts.

Other bacteria

Pierantoni (1936) believed that bacteria residing within termite gut protozoa were the actual agents of cellulose digestion, but there is no good evidence for this. Most studies of termite gut bacteria have, in fact, investigated their involvement in cellulose digestion (Beckwith & Rose, 1929; Dickman, 1931; Tetrault & Weis, 1937; Hungate, 1936, 1946a; Pochon, de Barjac & Roche, 1959; Mannesmann & Weldon, 1971). These have been legitimate pursuits because some of the termites examined were 'higher' termites which either lacked, or had a drastically reduced, protozoan fauna. However, these studies frequently involved enrichment cultures which told little or nothing about the quantitative (and therefore biochemical) significance of cellulolytic bacteria in the gut. The overall weight of evidence is either against or not supportive of a significant role for bacteria in cellulose decomposition in 'lower' termites (which possess cellulolytic protozoa). Bacteria are probably more significant in this respect in wood-eating 'higher' termites (Hungate, 1946a). Hungate (1946b) isolated an anaerobic, cellulolytic actinomycete from the gut of *Amitermes minimus* and named it *Micromonospora propionici*. The organism fermented either cellulose or glucose to carbon dioxide and acetic and propionic acids as major products. Hungate recognized the need for quantification and determined that 500 colony-forming units of *M.*

propionici were present per larva. However, because of their apparent low density, and slow growth rate in laboratory culture, he concluded that *M. propionici* was probably of limited significance in symbiotic cellulose digestion.

Enumeration of bacteria in the guts of transcaspian and Turkestan termites has recently been done, but little was said of their possible role (Krasil'nikov & Satdykov, 1969).

This author and colleagues have recently discovered that bacteria in the gut of several species of termites, and a wood-eating cockroach (*Cryptocercus punctulatus*) fix atmospheric nitrogen (Breznak *et al.*, 1973) and evolve methane (Breznak, Mertins & Coppel, 1974). Benemann (1973) has also demonstrated nitrogen fixation in termites. These activities were demonstrable in intact, living insects and, to our knowledge, constitute the first clear evidence of such phenomena. The following account will deal with nitrogen fixation and methane formation in termites and *Cryptocercus* and will include some of our recent, though preliminary, data concerning the isolation of bacterial symbionts, and in-situ electron microscopy of gut microbes.

Nitrogen fixation in termites and a wood-eating cockroach

The low nitrogen content of wood (the diet of many termites), plus the ability of termites to thrive on a diet of cellulose, prompted Cleveland (1925*a*) to postulate nitrogen (N_2) fixation in termites. However, he failed to correlate gas (air) uptake by living termites with N_2 fixation. Pierantoni (1937) suggested that intracytoplasmic bodies of termite gut protozoa were N_2-fixing bacteria, but he obtained no proof for his suggestion. Hungate (1941, 1944) evaluated the nitrogen economy of termite colonies. He, too, obtained no evidence for net N_2 fixation, but noticed that the addition of nitrogen sources to wood increased the growth of termites (1941). On the other hand, Tóth (1948) observed a net increase in combined nitrogen when macerated termites were incubated in a nitrogen-free organic broth: an obvious enrichment technique which could have selected for N_2-fixing bacteria in the medium. Greene & Breazeale (1937) briefly reported the isolation of N_2-fixing bacteria from termites, but their methods were not stated in detail.

We decided to reopen the question of N_2 fixation in termites by exploiting the sensitivity of the acetylene reduction assay, a reliable indicator of N_2 fixation (Postgate, 1971). Our results (Breznak *et al.*, 1973) indicated that bacteria in the gut of termites fixed N_2. All termites tested were 'lower' termites which had an abundant protozoan fauna in their hindgut. When termites were fed antibacterial antibiotics, N_2-fixing activity and hindgut bacteria (but not protozoa) disappeared precipitously. Protozoa eventually disappeared after several weeks, however, indicating that the presence of bacteria was somehow important for their viability. N_2-fixing activity varied inversely with the amount of nitrogen contained in the termites' diet, and the response of the activity to a change in dietary nitrogen occurred quickly (within 5 h). These data indicated that the N_2-fixing system in termites was efficient, and potentially capable of quickly supplying the termites with combined nitrogen should the dietary level suddenly drop.

A variety of other insects tested showed no evidence of N_2 fixation (Breznak *et al.*, 1973). The wood-eating *Lyctus planicollis* and *Xyletinus peltatus* also gave negative results (unpublished). However, *Cryptocercus punctulatus* reduced acetylene at rates comparable to those of termites on a body weight basis (Breznak *et al.*, 1974). This was rather expected inasmuch as the biology of *C. punctulatus* resembles that of termites, and the two groups of insects are believed to be closely related phylogenetically (Cleveland *et al.*, 1934). As with termites, antibacterial drugs quickly abolished N_2-fixing activity and bacteria (but not protozoa) from the gut of the cockroach. N_2 fixation in termites and *C. punctulatus* compares favorably with other natural samples (Table 1).

Table 1. N_2 fixation in termites and other natural samples[a]

Insects	μg N fixed day^{-1} g^{-1} wet wt
Coptotermes formosanus worker larvae	0.16–49.39
Reticulitermes flavipes worker larvae	0.04
Zootermopsis sp. worker and brachypterous larvae	0.06
Cryptotermes brevis brachypterous larvae	0.38
Kalotermes minor worker larvae	0.80–18.87
Cryptocercus punctulatus adults and larvae	0.01–0.12
Other samples	
Human feces	0.17
Sheep rumen contents	0.33
Goat rumen contents	0.13
Florida (USA) estuary sediment	0.04
Lake Erie (USA) sediment	0.28
Rice paddy soil[b]	10^{-5}–2×10^{-4}
Glycine max (legume)	269–9542
Alnus species (non-legume)	296–1747
Peltigera canina (lichen)[b]	6.7×10^4
Blue-green algae[b]	1.5×10^3–5.0×10^4

[a] Data for insects from Breznak *et al.* (1973, 1974 and unpublished) except for *K. minor* (Benemann, 1973). Data for other samples from Hardy, Burns & Holsten (1973).

[b] Activity expressed per gram *dry* weight.

PLATE I

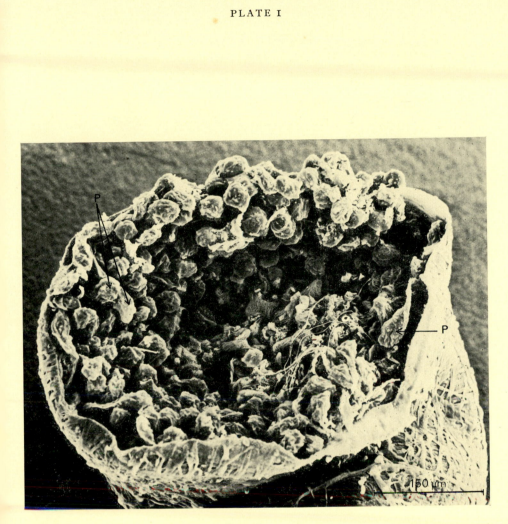

For explanation see p. 579

(Facing p. 568)

PLATE 2

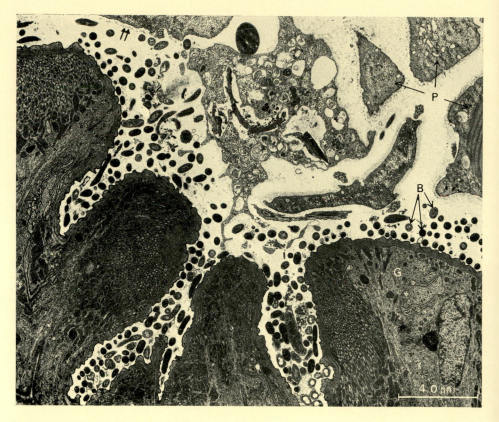

For explanation see p. 579

PLATE 3

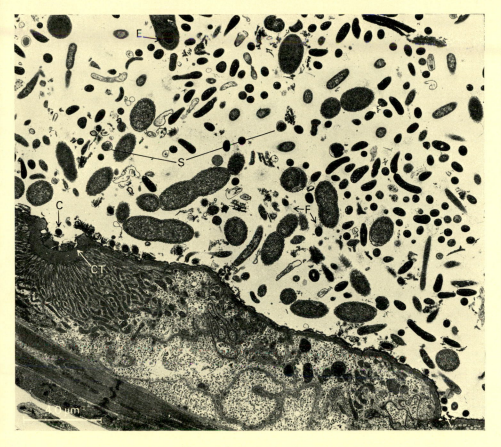

For explanation see p. 579

PLATE 4

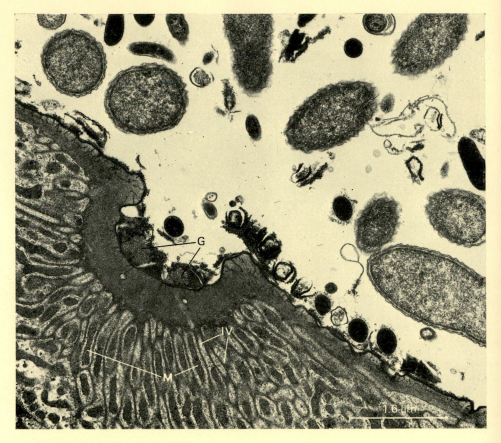

For explanation see p. 579

PLATE 5

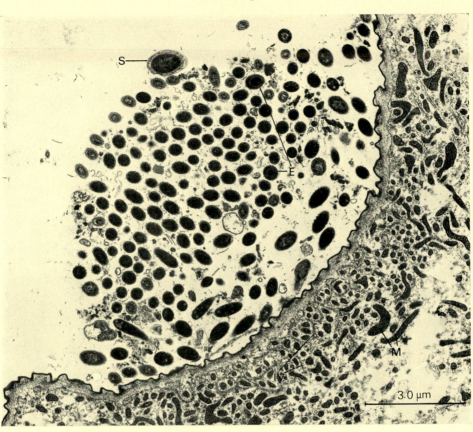

For explanation see p. 580

PLATE 6

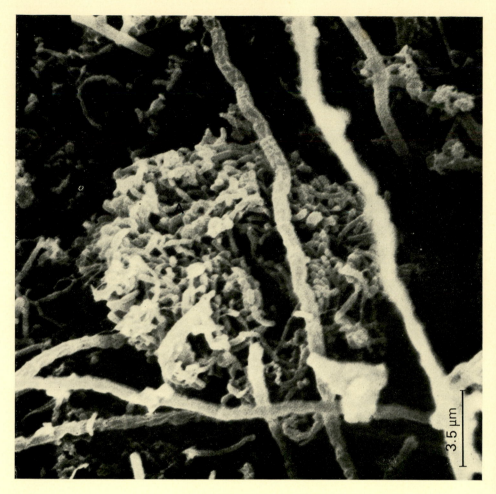

3.5 μm

For explanation see p. 580

Our findings suggested that N_2-fixing bacteria, or their metabolic products, might serve directly as a nitrogen source for the termites, or for their cellulose-digesting protozoa, or both. N_2-fixing bacteria might also (or alternatively) provide vitamins or other growth factors for the termites. Unpublished data from this laboratory are consistent with the former possibility. A recently collected colony of *Coptotermes formosanus* contained many young larvae (average fresh weight, 1.9 mg per larva) which had acetylene-reducing activities of 220.5 nmol C_2H_4 formed $h^{-1} g^{-1}$ fresh wt. This rate was equivalent to 18.03 mg N fixed $year^{-1} g^{-1}$ fresh wt, and was *317-fold greater* than that of older larvae (average fresh weight, 4.0 mg per larva) used previously (Breznak *et al.*, 1973)! The reason for this remarkable increase in acetylene-reducing activity is not known, but a tenable hypothesis is that *young, growing* larvae must exhibit a positive nitrogen balance; N_2 fixation by gut bacteria may markedly increase in young larvae and provide a significant amount of combined nitrogen to the insects. Assuming that the nitrogen content of *C. formosanus*, like *Zootermopsis*, is 21.3 mg g^{-1} fresh wt (Hungate, 1941), the amount of N_2 fixed by young larvae could, over the period of a year (see estimate above), allow them to grow to the extent of almost *doubling* their nitrogen content if the N_2 fixation rate remains constant. Benemann (1973) also observed considerable variation in acetylene reduction rates between different batches of *Kalotermes minor* and considered that the age of the larvae might be a contributing factor.

The actual *importance* of N_2 fixation to termites, and determination of the pathway of nitrogen flow from atmospheric N_2 to combined nitrogen in termite protoplasm, are two questions being pursued in this writer's laboratory. Certainly the potential exists for intestinal N_2 fixation to provide significant amounts of combined nitrogen to some termites. However, the degree to which this potential is exploited in natural environments is unknown, and undoubtedly depends on the dietary nitrogen availability to termites, which in turn, could be influenced by N_2 fixation occuring in wood itself (e.g. see Seidler, Aho, Raju & Evans, 1972; Cornaby & Waide, 1973). Cannibalism is probably another influential factor (Moore, 1969).

Methane formation in termites and a wood-eating cockroach

During our studies of N_2 fixation, we discovered that termites and specimens of *C. punctulatus* evolved methane (Breznak *et al.*, 1974). On a body weight basis similar amounts of methane were produced by *Reticulitermes flavipes* and *Cryptocercus punctulatus*, whereas *Coptotermes formosanus* routinely emitted considerably less (Table 2). Methane emission by *Cryptotermes brevis*, while comparable to that of *R. flavipes* and *C.*

Table 2. *Methane emission by termites and* Cryptocercus punctulatus[a]

Assay	Insect[b]	Pre-assay diet[c]	nmol CH$_4$ emitted h^{-1} g^{-1} fresh wt
1	*Reticulitermes flavipes* larvae	Nest wood	27.0–73.1
2	*R. flavipes* larvae	FP : 4 days	149.0
3	*R. flavipes* larvae	FP : 8 days	1340.0
4	*R. flavipes* larvae	FP + AB : 8 days	0.2
5	*Coptotermes formosanus* larvae	Nest wood	0.3
6	*Cryptotermes brevis* brachypterous larvae	Paper towel : 1 day	11.1
7	*Cryptocercus punctulatus*	Nest wood	13.9–24.3
8	*C. punctulatus*	FP + AB : 8 days	12.6
9	*C. punctulatus*	FP + 10 × AB : 8 days	16.0

[a] Insects were contained in rubber-stoppered 10 ml serum vials. Samples of gas phase were periodically removed and assayed for methane by using gas chromotography. All insects remained alive during assay. Data from Breznak, Mertins & Coppel (unpublished).

[b] The following animals emitted no detectable methane: *Culex pipiens* (mosquito); *Oncopeltus fasciatus* (milkweed bug); *Tribolium confusum* (flour beetle); *Aecheta domestica* (cricket); *Supella supellectilium* (brown-banded cockroach); *Malacosoma americanum* (tent caterpillar); *Argia* sp. (damselfly); *Camponotus* sp. (carpenter ant); *Tracheoniscus rathkei* (wood louse); *Callipus lactarius* (millipede).

[c] Symbols: FP, Whatman no. 2 cellulose filter paper impregnated with a trace salts solution; AB, a mixture of penicillin G, streptomycin sulfate, chloramphenicol, and tetracycline at concentrations used previously (Breznak *et al.*, 1973).

punctulatus, may have been influenced by a paper-towel diet. Methane was apparently a constituent of the insects' flatus because it was formed by excised guts, but not by de-gutted bodies. On a molar basis, about 9.5 % as much methane as carbon dioxide was produced by intact *R. flavipes* larvae.

In termites the origin of methane appears to be gut bacteria, since anti-bacterial drugs abolished methane formation and detectable bacteria (but not protozoa) from guts (Table 2; assay 4). Antibacterial drugs, while ridding the gut of detectable bacteria (but not protozoa), had no effect on methanogenesis by *C. punctulatus* (Table 2; assay 8 and 9). The origin of methane is less clear for this insect. Although one might suspect that protozoa are methanogenic agents in cockroaches, the only organisms shown to produce methane under natural conditions are bacteria (Wolfe, 1971). A variety of other animals tested gave no indication of methane formation (Table 2).

The amount of methane produced by the xylophagous insects reported herein is relatively large. For comparison, intestinal methane production by a 68 kg man ranges from 2.1 to 16.5 nmol CH$_4$ h^{-1} g^{-1} body wt depending on his diet (Steggerda & Dimmick, 1966). Ruminants may form 200 liters of methane per day (Wolfe, 1971) which is equivalent to 749.8 nmol CH_4 h^{-1} g^{-1} body wt (assuming a 453.6 kg ruminant). Table 2 reveals that methane emission by *R. flavipes* and *C. punctulatus* is generally greater than that by humans, and under some conditions (Table 2; assay 3) may even surpass that of ruminants. Increased methanogenesis in *R. flavipes* following a diet of filter paper may reflect a selective increase of methane-forming bacteria in the gut, or may indicate that filter paper is degraded to substrates for methane formation (e.g. carbon dioxide and hydrogen) faster than wood.

The anaerobic nature of termite hindguts could be deduced from the oxygen-sensitivity of resident protozoa (Cleveland, 1925b; Hungate, 1939). However, based on our understanding of methane bacteria (Wolfe, 1971) methanogenesis in termites and *C. punctulatus* suggests that the hindgut is not only anaerobic, but possesses a relatively low (i.e. electronegative) redox potential. It is noteworthy that Ritter (1961) has implicated gluta-thione as the natural reducing agent in the gut of *C. punctulatus*.

To this reviewer's knowledge ours is the first demonstration of methane formation in insects. Cook came closest to demonstrating methanogenesis in termites in 1932. He stated that termites (*Zootermopsis nevadensis*) evolved an 'undetermined gas' which was possibly 'hydrogen or methane, or a mixture of both'. Production of the gas depended on the integrity of the intestinal protozoan fauna, and was not evolved by defaunated termites.

The biochemistry of symbiotic cellulose utilization in termites is super-ficially similar to that in ruminants, a fact recognized by Hungate years ago (1946a). Plant cellulose ingested by the host is fermented by microbes in the rumen (cattle) or hindgut (termites) to yield products, particularly organic acids, which are in turn oxidized by the host for energy. Carbon dioxide, hydrogen, and formate produced in the rumen serve as growth substrates for methanogenic organisms. The horde of organisms which proliferates during rumen cellulose fermentation is then passed on to the stomach and intestine where it is digested and serves as a valuable source of proteins and vitamins for the animal (Hungate, 1966). The role, if any, that methanogenic organisms play in the vitality of termites (and *C. punctulatus*) remains to be established.

In-situ electron microscopy of microbial symbionts

In an effort to obtain a better understanding of the microecology of bacteria and protozoa in termite guts, the writer is employing an approach using electron microscopy. This aspect is being done in collaboration with H. S. Pankratz. For obvious reasons we are focusing our attention on the paunches of worker larvae. Our results are preliminary, and have not yet been submitted for publication, but they are intriguing.

For transmission electron microscopy guts were removed and quickly fixed in 2 % glutaraldehyde in 0.1 M phosphate buffer, pH 7.2. They were then cut with a razor at the level of the paunch to facilitate penetration of fixative into the lumen of the paunch. Subsequent steps were as described (Beaman, Pankratz & Gerhardt, 1972) except Epon 812 was used as the embedding resin. For scanning electron microscopy, fixation was done as described above. This was followed by dehydration in ethanol and isoamyl alcohol, critical point drying with carbon dioxide, and gold casting prior to examination.

A morphologically heterogeneous population of bacteria was present in the paunch of both *R. flavipes* and *C. formosanus*. Bacteria were present in the lumen, but also in dense populations adhering to the cuticle of the gut wall. Relatively flat regions of the gut wall were colonized as well as crevices (Plate 2). Of the protozoa, there was a tendency for *Pyrsonympha* to be situated closest to the gut wall. This was readily apparent by phase contrast microscopy of Epon-embedded thin sections, and was also observed by scanning electron microscopy (Plate 1). However, no clear evidence for an attachment organelle on *Pyrsonympha* was obtained. 'Tree'-like projections on the surface of *Pyrsonympha* (Smith & Arnott, 1973; Bloodgood, 1974; Bloodgood *et al.*, 1974) were also observed, and can be discerned on a portion of a *Pyrsonympha* cell in Plate 2.

Dense, hemispherical aggregates of bacteria were conspicuously associated with cup-like depressions present in the apical ends of paunch epithelial cells (Plates 3, 4, 5). Bacteria in the aggregates ranged in diameter from about 0.3 μm to 1.2 μm. Endospore-forming bacteria were present, as were curved rods, and spirochetes. Fibrous appendages were present at the poles of many rod-shaped organisms (Plate 3). 'Cups' in *R. flavipes* were about 1.7 μm in diameter, and frequently contained granular material of an undetermined nature (Plate 4). The epithelial cell membrane in the region of cups displayed numerous invaginations which in turn were interposed with elongate mitochondria (Plate 4). Such an arrangement is strikingly similar to that in kidney tubules (Sjöstrand & Rhodin, 1953; Fawcett, 1966), and suggests that cups are sites of active transport of ions. Could these be sites of transport of nutrients (e.g. acetate) from the lumen? Depressions were also observed in the gut of *C. formosanus* (Plate 5), but their diameter (5 to 15 μm) was frequently larger than that of cups in *R. flavipes*. Because of this, bacterial aggregates tended to be situated within, rather than over the depressions. Scanning electron microscopy also revealed the presence of hemispherical aggregates of bacteria (Plate 6) which were generally more compact than those observed by transmission electron microscopy. Preparative procedures for transmission electron microscopy may loosen the aggregate. Baccetti (1963) observed similar depressions in the hindgut of *R. lucifugus*, and noted the association of bacteria with them, but did not study the microbes in detail. He observed pinocytotic vesicles in the region of cups and concluded that cups were sites of nutrient absorption. Whatever the function of the depressions, they clearly represent sites of dense colonization by bacteria.

Adherence of bacteria to the gut wall of termites is not a unique phenomenon; it has been observed in other animals (e.g. see Savage, 1972). Such adhesion should retard the washout of bacteria and help them compete successfully in the gut environment (particularly if their growth rates are slow). Moreover, sites of physical intimacy probably also mean biochemical intimacy between bacterial species, and between bacteria and the termite. Biochemical intimacy could be manifested as cross-feeding of growth factors or energy sources, detoxification of waste products, or creation of a favorable redox potential (Bryant, 1972), and could be a crucial factor in maintaining the compositional integrity of the gut microbiota and the vitality of the insects.

Isolation of bacterial symbionts of termites

Attempts are being made in this laboratory to isolate, identify, and quantify bacterial symbionts of termites. An anaerobic glove box (Aranki, Syed, Kenney & Freter, 1969) is being used for removal of termite guts, accompanied by strict anaerobic techniques (Hungate, 1950) for cultivation of bacteria. Preliminary results indicate that 0.3 to 1.3 × 10^6 colony forming units/gut can be isolated from *R. flavipes* worker larvae by using readily-prepared laboratory media, e.g. a medium containing peptone, yeast extract and glucose (PYG of Holdeman & Moore, 1972). About 80 % of the isolates obtained with this medium have been tentatively identified as lactic acid bacteria of the genus *Streptococcus*; the remainder are mostly anaerobic, Gram-negative rods. The presence of a significant population of streptococci raises questions concerning the energy source(s) which supports their growth. Known lactic acid bacteria are primarily saccharolytic, although some species will ferment glycerol, citrate and malate (Deibel, Lake & Niven, 1963; Whittenbury, 1965), and arginine (Deibel, 1964). Carbon dioxide, hydrogen, and acetate (presumably formed by protozoa of *R. flavipes*) should not support significant growth of streptococci. Nutritional characterization of the isolates, coupled with biochemical assays of gut fluid, should suggest fermentable substrates which are available *in situ*. *Streptococcus faecalis* was always found in the guts of

Anacanthotermes species (Krasil'nikov & Satdykov, 1970), and comprised about 30 % of all isolates. They were considered to be part of the normal flora of termites. Martin & Mundt (1972), however, found enteric streptococci associated with 20 of 36 different insects obtained from undisturbed, natural habitats, but recovered no streptococci from *R. flavipes*. The origin or diet of termites may markedly influence the numbers of streptococci in their guts.

We have obtained nitrogen-fixing bacteria from the guts of *C. formosanus* larvae, but are uncertain of their numerical significance. The organisms are Gram-negative, rod-shaped bacteria which reduce acetylene when growing anaerobically with N_2 gas as sole nitrogen source. Because they were enriched from relatively low dilutions of gut contents (10^{-2}, 10^{-3}) they may not be the primary N_2-fixing agents in *C. formosanus*. On the other hand, if the organisms *in situ* adhere tenaciously to the gut wall, the extent to which growth would be observed in dilution tubes would be dependent on the size and number of gut particles used as inoculum, which in turn would depend on the efficacy of the grinding procedure used to disrupt guts. Conceivably, our estimates could be low by at least two to three orders of magnitude. Further studies are in progress to distinguish between these alternatives.

CONCLUSION

The study of symbiosis between termites (and *Cryptocercus*) and their intestinal microbes is at an interesting stage. Technological advances over the last fifty years now permit the sensitive measurement of processes such as nitrogen fixation and methane formation in living insects, and the observation of fine details of the 'architecture' of the gut microbiota by high resolution electron microscopy. These are far cries from the facilities available when Cleveland began his studies. One fact that has emerged from recent investigations is the probable importance of the *bacterial* component of the gut microbiota. Consequently, one purpose of this paper has been to call attention to this oft-neglected group of micro-organisms. The intention has not been to undermine the importance of protozoa, nor to imply that every bacterium in the gut is of equal significance. However, it seems logical that in order to accurately assess the biochemical interrelationships of the intestinal microbes with each other, and with the insects, the bacteria cannot be ignored.

Recent efforts (Smythe, 1972; Smythe & Mauldin, 1972; Mauldin, Smythe & Baxter, 1972; Mauldin & Smythe, 1973) have focused on the effect of an abnormal protozoan fauna on the biochemical (cellulose catabolism; protein and lipid synthesis) and developmental capacity of termites. While it is clear from these studies that gut microbes are influential, the biochemical details of this influence, and the roles played by bacteria are unclear. This is particularly true because treatments used to remove protozoa also eliminated bacteria (Mauldin & Smythe, 1973). Speck, Becker & Lenz (1971), on the other hand, have employed antibiotics to eliminate groups of symbionts selectively. Such an approach may be more suitable. As more becomes known about the microbial flora of termites, we may one day be able, selectively, to remove specific organisms from the gut, and thereby assess their significance as symbionts with high precision.

ACKNOWLEDGEMENTS

I am grateful to W. J. Brill, H. C. Coppel, J. C. Ensign, and J. W. Mertins for encouragement. Journal article no. 6918 of the Michigan Agricultural Experiment Station.

REFERENCES

ARANKI, A., SYED, S. A., KENNEY, E. B. & FRETER, R. (1969). Isolation of anaerobic bacteria from human gingiva and mouse cecum by means of a simplified glove box procedure. *Appl. Microbiol.*, **17**, 568–576.

BACCETTI, B. (1963). Ricerche sull'ultrastruttura dell'intestino degli insetti. III. Il mesentero ed il colon nell'operaio de *Reticulitermes lucifugus* Rossi. *Symp. Genet. Biol. ital.*, **11**, 230–255.

BAINES, S. (1956). The role of the symbiotic bacteria in the nutrition of *Rhodnius prolixus* (Hemiptera). *J. exp. Biol.*, **33**, 533–541.

BAKER, J. M. (1963). Ambrosia beetles and their fungi, with particular reference to *Platypus cylindrus* Fab. *Symp. Soc. gen. Microbiol.*, **13**, 232–265.

BALL, G. H. (1969). Organisms living on and in protozoa. In *Research in Protozoology*, vol. 3, 565–718 (ed. T.-T. Chen). New York: Pergamon Press.

DE BARY, A. (1879). *Die Erscheinung der Symbiose.* Strassburg: Trubner.

BEAMAN, T. C., PANKRATZ, H. S. & GERHARDT, P. (1972). Ultrastructure of the exosporium and underlying inclusions in spores of *Bacillus megaterium* strains. *J. Bact.*, **109**, 1198–1209.

BECKWITH, T. D. & ROSE, E. J. (1929). Cellulose digestion by organisms from the termite gut. *Proc. Soc. exp. Biol. Med.*, **27**, 4–6.

BENEMANN, J. R. (1973). Nitrogen fixation in termites. *Science, Wash.*, **181**, 164–165.

BLOODGOOD, R. A. (1974). Unique cell surface structures on *Pyrsonympha. Arch. Microbiol.*, **95**, 275–278.

BLOODGOOD, R. A., MILLER, K. R., FITZHARRIS, T. P. & McINTOSH, J. R. (1974). The ultrastructure of *Pyrsonympha* and its associated micro-organisms. *J. Morph.*, **143**, 77–106.

BREZNAK, J. A. (1973). Biology of nonpathogenic, host-associated spirochetes. *CRC crit. Rev. Microbiol.*, **2**, 457–489.

BREZNAK, J. A., BRILL, W. J., MERTINS, J. W. & COPPEL, H. C. (1973). Nitrogen fixation in termites. *Nature, Lond.*, **244**, 577–580.

BREZNAK, J. A., MERTINS, J. W. & COPPEL, H. C. (1974). Nitrogen fixation and methane production in a wood-eating cockroach, *Cryptocercus punctulatus* Scudder (Orthoptera: Blattidae). *University of Wisconsin Forestry Research Notes*, no. **184**.

BROOKS, M. A. (1963). Symbiosis and aposymbiosis in arthropods. *Symp. Soc. gen. Microbiol.*, **13**, 200–231.

BROOKS, M. A. & KRINGEN, W. B. (1972). Polypeptides and proteins as growth factors for aposymbiotic *Blattella germanica* (L.). In *Insect and Mite Nutrition*, 353–364 (ed. J. G. Rodriguez). Amsterdam: North-Holland.

BRYANT, M. P. (1972). Interactions among intestinal micro-organisms. *Am. J. Clin. Nutr.*, **25**, 1485–1487.

BUCHNER, P. (1965). *Endosymbiosis of Animals with Plant Micro-organisms*. New York: Wiley-Interscience.

CHU, H.-M., NORRIS, D. M. & KOK, L. T. (1970). Pupation requirement of the beetle, *Xyleborus ferrugineus*: sterols other than cholesterol. *J. Insect. Physiol.*, **16**, 1379–1387.

CLEVELAND, L. R. (1924). The physiological and symbiotic relationships between the intestinal protozoa of termites and their host, with special reference to *Reticulitermes flavipes* Kollar. *Biol. Bull. mar. biol. Lab. Woods Hole*, **46**, 178–227.

(1925*a*). The ability of termites to live perhaps indefinitely on a diet of pure cellulose. *Biol. Bull. mar. biol. Lab. Woods Hole*, **48**, 289–293.

(1925*b*). The effects of oxygenation and starvation on the symbiosis between the termite, *Termopsis*, and its intestinal flagellates. *Biol. Bull. mar. biol. Lab. Woods Hole*, **48**, 309–327.

CLEVELAND, L. R. & GRIMSTONE, A. V. (1964). The fine structure of the flagellate *Mixotricha paradoxa* and its associated micro-organisms. *Proc. R. Soc. B*, **159**, 668–686.

CLEVELAND, L. R., HALL, S. R., SANDERS, E. P. & COLLIER, J. (1934). The wood-feeding roach *Cryptocercus*, its protozoa, and the symbiosis between protozoa and roach. *Mem. Am. Acad. Arts Sci.*, **17**, 184–342.

COOK, S. F. (1932). The respiratory gas exchange in *Termopsis nevadensis*. *J. cell. comp. Physiol.*, **4**, 95–110.

(1943). Nonsymbiotic utilization of carbohydrates by the termite, *Zootermopsis angusticollis*. *Physiol. Zool.*, **16**, 123–128.

CORNABY, B. W. & WAIDE, J. B. (1973). Nitrogen fixation in decaying chestnut logs. *Pl. Soil*, **39**, 445–448.

DEIBEL, R. H. (1964). Utilization of arginine as an energy source for the growth of *Streptococcus faecalis*. *J. Bact.*, **87**, 988–992.

DEIBEL, R. H., LAKE, D. E. & NIVEN, C. F. JR (1963). Physiology of the enterococci as related to their taxonomy. *J. Bact.*, **86**, 1275–1282.

DICKMAN, A. (1931). Studies on the intestinal flora of termites with reference to their ability to digest cellulose. *Biol. Bull. mar. biol. Lab. Woods Hole*, **61**, 85–92.

FAWCETT, D. W. (1966). *An Atlas of Fine Structure, The Cell, Its Organelles and Inclusions*. Philadelphia: Saunders.

FRANCKE-GROSMANN, H. (1967). Ectosymbiosis in wood-inhabiting insects. In *Symbiosis*, vol. **2**, 141–205 (ed. S. M. Henry). New York: Academic Press.

GRASSÉ, P. P. (1938). La Vêture schizophytique des flagellés termiticoles: *Parajoenia, Caduceia* et *Pseudodevescovina*. *Bull. Soc. zool. Fr.*, **63**, 110–122.

GREENE, R. A. & BREAZEALE, E. L. (1937). Bacteria and the nitrogen metabolism of termites. *J. Bact.*, **33**, 95–96.

HARDY, R. W. F., BURNS, R. C. & HOLSTEN, R. D. (1973). Applications of the acetylene–ethylene assay for measurement of nitrogen fixation. *Soil Biol. Biochem.*, **5**, 47–81.

HICKIN, N. E. (1971). *Termites, a World Problem.* London: Hutchinson.

HOLDEMAN, L. V. & MOORE, W. E. C. (1972). *Anaerobe Laboratory Manual.* Blacksburg: Virginia Polytechnic Institute and State University.

HONIGBERG, B. M. (1967). Chemistry and parasitism among some protozoa. *Chem. Zool.*, **1**, 695–814.

(1970). Protozoa associated with termites and their role in digestion. In *Biology of Termites*, vol. **2**, 1–36 (eds. K. Krishna and F. M. Weesner). New York: Academic Press.

HUNGATE, R. E. (1936). Studies on the nutrition of *Zootermopsis*. I. The rôle of bacteria and molds in cellulose decomposition. *Zent. Bakt. ParasitKde Abt. II*, **94**, 240–249.

(1938). Studies on the nutrition of *Zootermopsis*. II. The relative importance of the termite and the protozoa in wood digestion. *Ecology*, **19**, 1–25.

(1939). Experiments on the nutrition of *Zootermopsis*. III. The anaerobic carbohydrate dissimilation by the intestinal protozoa. *Ecology*, **20**, 230–245.

(1941). Experiments on the nitrogen economy of termites. *Ann. ent. Soc. Am.*, **34**, 467–489.

(1943). Quantitative analysis on the cellulose fermentation by termite protozoa. *Ann. ent. Soc. Am.*, **36**, 730–739.

(1944). Termite growth and nitrogen utilization in laboratory cultures. *Texas Acad. Sci. Proc. Trans.*, **27**, 1–7.

(1946a). The symbiotic utilization of cellulose. *J. Elisha Mitchell Sci. Soc.*, **62**, 9–24.

(1946b). Studies on cellulose fermentation. II. An anerobic cellulose-decomposing actinomycete, *Micromonospora propionici*, n. sp. *J. Bact.*, **51**, 51–56.

(1950). The anaerobic mesophilic cellulolytic bacteria. *Bact. Rev.*, **14**, 1–49.

(1955). Mutualistic intestinal protozoa. In *Biochemistry and Physiology of Protozoa*, vol. **2**, 159–199 (eds. S. H. Hutner and A. Lwoff). New York: Academic Press.

(1966). *The Rumen and its Microbes.* New York: Academic Press.

(1972). Relationships between protozoa and bacteria of the alimentary tract. *Am. J. clin. Nutr.*, **25**, 1480–1484.

KIRBY, H. Jr (1941). Organisms living on and in protozoa. In *Protozoa in Biological Research*, 1009–1113 (eds. G. N. Calkins and F. M. Summers). New York: Columbia University Press.

KOCH, A. (1967). Insects and their endosymbionts. In *Symbiosis*, vol. **2**, 1–106 (ed S. M. Henry). New York: Academic Press.

KRASIL'NIKOV, N. A. & SATDYKOV, S. I. (1969). Estimation of the total bacteria in the intestines of termites. *Microbiology*, **38**, 289–292.

(1970). Bacteria of termites' intestines. *Microbiology*, **39**, 562–564.

KRISHNA, K. (1969). Introduction. In *Biology of Termites*, vol. **1**, 1–17 (eds. K. Krishna and F. M. Weesner). New York: Academic Press.

KRISHNA, K. & WEESNER, F. M. (eds.) (1969). *Biology of Termites*, vol. **1**. New York: Academic Press.

(eds.) (1970). *Biology of Termites*, vol. **2**, New York: Academic Press.

LEE, K. E. & WOOD, T. G. (1971). *Termites and Soils.* New York: Academic Press.

MANNESMANN, R. & WELDON, D. (1971). Biological studies of subterranean termites at the Texas Forest Products Laboratory. In *Wood Products Insects*, Minutes of the 16th Southern Forest Insect Work Conference. Mississippi: The Mississippi Forestry Commission.

MARGULIS, L. (1970). *Origin of Eukaryotic Cells.* New Haven: Yale University Press.

MARTIN, J. D. & MUNDT, J. O. (1972). Enterococci in insects. *Appl. Microbiol.*, **24**, 575–580.

MAULDIN, J. K. & SMYTHE, R. V. (1973). Protein-bound amino acid content of normally and abnormally faunated Formosan termites, *Coptotermes formosanus*. *J. Insect Physiol.*, **19**, 1955–1960.

MAULDIN, J. K., SMYTHE, R. V. & BAXTER, C. C. (1972). Cellulose catabolism and lipid synthesis by the subterranean termite, *Coptotermes formosanus*. *Insect Biochem.*, **2**, 209–217.

MOORE, B. P. (1969). Biochemical studies in termites. In *Biology of Termites*, vol. **1**, 407–432 (eds. K. Krishna and F. M. Weesner). New York: Academic Press.

NOIROT, C. & NOIROT-TIMOTHÉE, C. (1969). The digestive system. In *Biology of Termites*, vol. **1**, 49–88 (eds. K. Krishna and F. M. Weesner). New York: Academic Press.

NORRIS, D. M. (1972). Dependency of fertility and progeny development of *Xyleborus ferrugineus* upon chemicals from its symbiotes. In *Insect and Mite Nutrition*, 299–310 (ed. J. G. Rodriguez). Amsterdam: North-Holland.

PIERANTONI, U. (1936). La simbiosi fisiologica nei Termitidi xilofagi e nei loro flagellati intestinali. *Arch. Zool. Ital.*, **22**, 135–173.

(1937). Osservazioni sulla simbiosi nei Termitidi xilofagi e nei loro flagellati intestinali. II. Defaunazione per digiuno. *Arch. Zool. Ital.*, **24**, 193–207.

POCHON, J., DE BARJAC, H. & ROCHE, A. (1959). Recherches sur la digestion de la cellulose chez le termite *Sphaerotermes sphaerothorax*. *Annls Inst. Pasteur, Paris*, **96**, 352–355.

POSTGATE, J. (1971). The acetylene test for nitrogenase. In *The Chemistry and Biochemistry of Nitrogen Fixation*, 311–315 (ed. J. R. Postgate). New York: Plenum.

RITTER, H. JR (1961). Glutathione-controlled anaerobiosis in *Cryptocercus*, and its detection by polarography. *Biol. Bull. mar. biol. Lab. Woods Hole*, **121**, 330–346.

SANDS, W. A. (1969). The association of termites and fungi. In *Biology of Termites*, vol. **1**, 495–524 (eds. K. Krishna and F. M. Weesner). New York: Academic Press.

SAVAGE, D. C. (1972). Associations and physiological interactions of indigenous micro-organisms and gastro-intestinal epithelia. *Am. J. clin. Nutr.*, **25**, 1372–1379.

SEIDLER, R. J., AHO, P. E., RAJU, P. N. & EVANS, H. J. (1972). Nitrogen fixation by bacterial isolates from decay in living white fir trees [*Abies concolor* (Gord. and Glend.) Lindl.]. *J. gen. Microbiol.*, **73**, 413–416.

SJÖSTRAND, F. S. & RHODIN, J. (1953). The ultrastructure of the proximal convoluted tubules of the mouse kidney as revealed by high resolution electron microscopy. *Expl Cell Res.*, **4**, 426–456.

SMITH, H. S. & ARNOTT, H. J. (1973). Scales associated with the external surface of *Pyrsonympha vertens*. *Trans. Am. microsc. Soc.*, **92**, 670–676.

SMYTHE, R. V. (1972). Feeding and survival at constant temperatures by normally and abnormally faunated *Reticulitermes virginicus* (Isoptera: Rhinotermitidae). *Ann. ent. Soc. Am.*, **65**, 756–757.

SMYTHE, R. V. & MAULDIN, J. K. (1972). Soldier differentiation, survival, and wood consumption by normally and abnormally faunated workers of the Formosan termite, *Coptotermes formosanus*. *Ann. ent. Soc. Am.*, **65**, 1001–1004.

SPECK, U., BECKER, G. & LENZ, M. (1971). Ernährungsphysiologische Untersuchungen an Termiten nach selektiver medikamentöser Ausschaltung der Darmsymbionten. *Z. Angew. Zool.*, **58**, 475–491.

STEGGERDA, F. R. & DIMMICK, J. F. (1966). Effect of bean diets on concentration of carbon dioxide in flatus. *Am. J. clin. Nutr.*, **19**, 120–124.

SUTHERLAND, J. L. (1933). Protozoa from Australian termites. *Q. Jl microsc. Sci.*, **76**, 145–173.

TETRAULT, P. A. & WEIS, W. L. (1937). Cellulose decomposition by a bacterial culture from the intestinal tract of termites. *J. Bact.*, **33**, 95.

TÓTH, L. (1948). Nitrogen-binding by *Kalotermes flavicollis* (Isoptera) and its symbionts. *Hung. Acta Biol.*, **1**, 22–29.

TRAGER, W. (1932). A cellulase from the symbiotic intestinal flagellates of termites and of the roach, *Cryptocercus punctulatus*. *Biochem. J.*, **26**, 1762–1771.

(1934). The cultivation of a cellulose-digesting flagellate, *Trichomonas termopsidis*, and of certain other termite protozoa. *Biol. Bull. mar. biol. Lab. Woods Hole*, **66**, 182–190.

WEESNER, F. M. (1965). *The Termites of the United States, a Handbook.* New Jersey: The National Pest Control Association.

WHITTENBURY, R. (1965). The differentiation of *Streptococcus faecalis* and *S. faecium*. *J. gen. Microbiol.*, **38**, 279–287.

WOLFE, R. S. (1971). Microbial formation of methane. In *Advances in Microbial Physiology*, vol. **6**, 107–146 (eds. A. H. Rose and J. F. Wilkinson). New York: Academic Press.

EXPLANATION OF PLATES

PLATE 1

Scanning electron micrograph of a section through the paunch of *Reticulitermes flavipes*. The profusion of protozoa within the paunch give the latter a 'cornucopia' appearance. Protozoa of the genus *Pyrsonympha* (P) are situated close to the gut wall.

PLATE 2

Electron micrograph of a thin section of a portion of the paunch of *R. flavipes*. Gut epithelia (G), bacteria (B), and protozoa (P) can be seen. 'Tree'-like projections (double arrows) can be discerned on a portion of a *Pyrsonympha* cell in the upper left area of the micrograph.

PLATE 3

Electron micrograph of a thin section of the paunch of *R. flavipes* showing a cup-like depression (C) in the epithelium, and a portion of a hemispherical aggregate of bacteria situated above the cup. The cup is formed in part by the cuticle (CT) of the epithilium. Among the bacterial aggregate can be discerned endospore-forming bacteria (E), and spirochetes (S). Fibrous appendages (F) are present on some bacteria.

PLATE 4

A higher magnification micrograph of the same cup-like depression shown in Plate 3. Granular material (G) is situated within the cup. Numerous invaginations of the cell membrane (IV), interposed with elongate mitochondria (M), are abundant in the vicinity of the cup.

PLATE 5

Electron micrograph of a thin section of a depression in the paunch wall of *C. formosanus*. Elongate mitochondria (M) are abundant in the area below the depression. Invaginations of the epithelial cell membrane, though present (IV), are not as sharply defined as those in *R. flavipes*. A hemispherical bacterial aggregate is situated within the depression. The aggregate contains endospore-forming bacteria (E) and a spirochete (S) among other bacteria.

PLATE 6

Scanning electron micrograph of a hemispherical bacterial aggregate (A) situated on the paunch wall of *R. flavipes*.

THE ROLE OF BEHAVIOUR IN MARINE SYMBIOTIC ANIMALS

By H. W. FRICKE

Max Planck Institut für Verhaltensphysiologie,
8131 Seewiesen und Erling-Andechs, W. Germany

INTRODUCTION

Symbiotic animal relationships are common in marine environments, especially in tropical coral reefs. The proximity and high density of various species give rise to a whole range of interspecific interactions having selective pressure, so that stable symbiotic relationships evolve.

The behavioural study of marine symbiotic animals gives rise to a number of extensive research projects, each of which can be assigned to an appropriate field in the behavioural sciences (Reese, 1964); for instance

(1) The study of adaptedness in behaviour.
(2) Its ecological significance and survival value for the species.
(3) Physiological mechanisms underlying the behaviour.
(4) Communication and other kinds of stimulus–response relations.
(5) Ontogeny and phylogeny of behaviour.

As marine behavioural sciences are still in their infancy it is not possible to give a comprehensive review of all these fields. But to demonstrate adaptation in social behaviour, ecological significance, communication, attraction signals and host recognition by behavioural analysis, five cases of symbiosis are described. These are

(a) The clownfish–anemone mutualism;
(b) cleaning symbiosis in fish and shrimp;
(c) mutualism of shrimps with gobiid fish;
(d) feeding commensalism of goatfish and other fish;
(e) symbiosis between fish, shrimp and sea urchin.

TYPE OF SYMBIOSIS AND ITS ECOLOGICAL SIGNIFICANCE

Clownfish–anemone mutualism

Mutualism between clownfish, genera *Amphiprion* and *Premnas*, and sea anemones, mostly genera *Physobrachia*, *Radianthus* and *Stoichactis*, was

[581]

observed a hundred years ago (Collingwood, 1868; Crespigny, 1869). Mucus transferred to the fish from the surface of the sea anemone tentacles protects it from the anemone's nematocysts (Schlichter, 1970; Mariscal, 1971). During the night the fish hides between the tentacles or in the anemone's body cavity; by day he feeds on plankton in the water column above the host, returning to it frequently. It is accepted that the fish gets shelter from the anemone (Mariscal, 1970) but any benefit accruing to the anemone has long been the subject of speculation. Allen (1972), Fishelson (1965) and others observed fish (mostly Chaetodontids) feeding on anemone tentacles. My own field experiments in the Red Sea showed that *Amphiprion bicinctus* chases away predatory fish (*Chaetodon fasciatus*), thus protecting its anemone host (Fricke, 1974). Another possible benefit for the anemone must be considered. Most species are nocturnal, perhaps to avoid diurnal predators, but the symbiotic species are diurnally active. Whether this shift is a behavioural adaptation of the anemone, evolved through the predator-chasing of *Amphiprion*, has yet to be investigated.

Cleaning symbiosis

Cleaning symbiosis is common in both freshwater and marine environments all over the world. Numerous marine organisms exist permanently or temporarily by cleaning other marine organisms which benefit thereby (Limbaugh, 1961; Hobson, 1969 and others). Ectoparasites, bacteria and necrotic tissue are removed. Feder (1966) reported 45 species of cleaner-fish and six of cleaner-shrimp. Both cleaners and customers display in various ways to attract each other. Cleaner territoriality and customer migration to cleaning stations are well-known behaviour patterns (Eibl-Eibesfeldt, 1955 and others). Little experimental work has been done on the ecological importance of the symbiosis. In the Bahamas, Limbaugh (1961) removed all cleaners from two isolated patches of reef, and within two weeks the fish population decreased rapidly, territorial fish developing skin diseases. Although Youngbluth (1968) repeated the experiments in Hawaii with negative results, it is generally agreed that cleaners play an important part in the maintenance of the coral reef community.

Mutualism of shrimps with gobiid fish

The mutualism of *Alpheus* shrimps with different gobiid fish has been observed in the Pacific Ocean, Indian Ocean, Red Sea and the West Indies (references in Karplus, Tsurnamal & Szlep, 1972). On grainy and sandy bottoms the shrimps, mostly living in pairs, dig burrows which are used by the fish as temporary shelter by day and resting place by night (Magnus, 1967*a*; Karplus *et al.*, 1972). The fish acts as a warning system for the purblind shrimp, which uses antennac contact to detect movement, thus being warned in case of any disturbance. Magnus (1967a) discussed the ecological value of the mutualism. The digging activity of the shrimp affords the gobiids a new ecological niche on flat and otherwise shelterless bottoms, in an environment almost free of food competitors. The shrimp for its part benefits in acquiring an optical warning system, which allows him to leave his subterranean habitat for short periods to feed on sediment on the bottom. The mutualism enlarges the shrimp's feeding habitat. The evolution of this mutualism led to the development of a differentiated communication system (Magnus, 1967a and Karplus et al., 1972).

Commensalism of goatfish and other fish

In tropical and subtropical waters all over the world goatfish (mostly *Mulloidichthys*, *Mullus* and *Pseudupeneus*) dig in sandy bottoms for food (Fricke, 1970a). The resulting cloud of sand attracts numerous other fish to search in it for particles (Plate 1). In the Gulf of Akaba 15 fish species of nine different families were found to feed in the sand cloud. Even when there is no attractive sand cloud they follow the goatfish for long distances. The goatfish *Pseudupeneus cyclostoma*, however, swim after various labrids and sarids, maintaining body contact. This species is specialized for feeding on encrusted rocky or coral bottoms, on which the labrids and scarids also browse, dropping particles which are snapped up by the goatfish. The hosts, sand-digging goatfish in the first case and labrids and scarids in the second, have no benefit from the association, but their 'customers' enlarge their feeding habitat.

Symbiosis of shrimp and fish with sea urchins

Creeping ctenophores, various crustaceans, snails, squid and many fish inhabit the spine forest of sea urchins as a suitable habitat (Magnus, 1964, 1967b and others). Adult cardinal fish of the genus *Siphamia* are constantly found in association with tropical *Diadema* or *Astropyga* sea urchins (Eibl-Eibesfeldt, 1961; Magnus, 1967b). The dark colouring of the fish is an excellent camouflage. They live mainly singly or in small groups. Under conditions of over-population, however, *Siphamia argentea* form dense groups around the sea urchin *Astropyga radiata*, the whole then resembling an enormous sea urchin (Fricke, 1970b). This collective mimetic behaviour, therefore, protects *Siphamia* against predators. *Siphamia permutata* of the Red Sea inhabit *Diadema setosum*. Magnus (1967b) observed that the fish remain motionless in the upper region of the spine forest, head pointing down towards the body of the host, to which the fish can then swim in case of danger. My own dummy experiments indicate that this kind of behaviour serves as optimal camouflage. *Siphamia* never voluntarily leaves its host, the fish feed on planktonic particles drifting among the sea urchin spines. Magnus (1967*b*) used dummies to demonstrate that the fish uses the sea urchin as a hard-bottom structure. The sea urchin has no benefit from the association, but it enables the fish to invade sandy and otherwise shelterless open areas where there is decreased competition for planktonic food.

Many shrimp species also live in these spine forests. The Indopacific shrimp *Tuleariocaris zanzibarica* swims erect with head down (Fricke & Hentschel, 1971), matching the host spines perfectly in coloration and motion. In coral reefs of Madagascar I released several shrimps, without sea urchins, on open sandy bottoms close to a coral reef; control experiments provided sea urchins. The first shrimps were snapped up by predators, the second were unmolested. This indicates that the shrimp is protected by its host, but there is no evidence of a possible benefit (for instance cleaning) for the sea urchin.

ADAPTATION OF BEHAVIOUR: INFLUENCE OF THE SYMBIONT ON SOCIAL BEHAVIOUR

The degree of adaptation of behaviour to environmental factors can be investigated by comparative studies of ecology on the one hand and the behaviour of related species on the other (for example Wickler, 1971). The symbiont is part of the environment and should therefore influence the evolution of behavioural adaptations in the host. In a comparison of symbiotic and closely related non-symbiotic animals, behavioural divergences partly influenced by the symbiont should be apparent. This method, long used by ethologists, can be a useful tool to study the influence of the symbiont in the evolution of host behavioural, anatomical and physiological features.

At present I am investigating the ecology and behaviour of closely related damselfish of the Red Sea (Fricke, 1973*a*). A comparative analysis revealed that the symbiotic clownfish *Amphiprion bicinctus*, belonging to this family of fish, evolved at least 47 behavioural and physiological characteristics not occurring in any other pomacentrid (Fricke, 1974). An example is the social structure. Investigation of group structure in 21 pomacentrid species showed that only *Amphiprion* lives in permanent monogamy. Has this a selective advantage over other possible social structures, e.g. polygamous groups, harems or even each sex living a solitary way of life? The evolution of social structure depends largely on the ecology of the species, therefore some knowledge of *Amphiprion's*

PLATE 1

For explanation see p. 594

PLATE 2

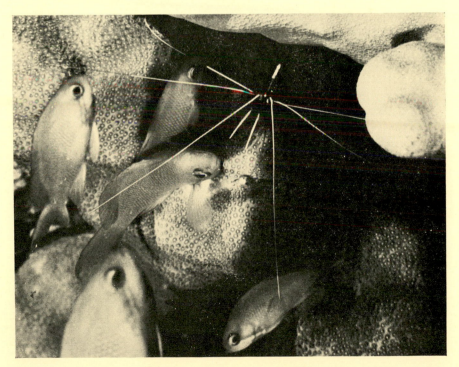

For explanation see p. 594 (*Facing p.* 584)

PLATE 3

For explanation see p. 594

ecology is needed before we can understand the monogamy, and the influence of the symbiotic partner on the behaviour development. Field investigations on a fringe reef in the Gulf of Akaba (Eilat, Israel) revealed that *Amphiprion* occupied almost all available anemones. Adults occurred singly or in pairs, with never more than two adults in a group. Table 1 summarizes some data. Obviously, space limitation allows no more than two adults per anemone; scarcity of anemones leads to strong competition in this particular reef, and tenanted anemones are immediately defended against conspecifics.

Table 1. *Distribution of the anemones* Gyrostoma helianthus *and* G. quadricolor, *and the symbiotic clownfish* Amphiprion bicinctus, *in different zones of a fringing reef in the Red Sea (Eilat, Israel)*

Depth (m)	Lagoon 0-1	Reef edge 1-5	Fore reef 10-15	Deep reef 20-25
Total number of anemones	9	109	53	44
Occupied anemones	9	109	52	43
Total number of juvenile *Amphiprion*	3	34	28	34
Total number of adult *Amphiprion*	12	145	59	27
Single *Amphiprion* (%)	33	24	22	48
Pairs of *Amphiprion* (%)	67	76	78	52

In these ecological conditions it would be calamitous if a sexually mature animal had to leave its host during reproduction to find a mate, because

(*a*) it might not find another anemone in the vicinity;

(*b*) a conspecific could occupy the vacant anemone in its absence;

(*c*) it might not find a mate in the same stage of reproductive readiness;

(*d*) it is exposed to predators while searching for a mate.

In this ecological setting monogamy has quite obviously a selective advantage over all other kinds of social structure. Monogamous mates can stay to defend their anemone against competitors, the danger of predation is reduced, the reproductive cycle can be synchronized. I must emphasize, however, that correlations between ecology and behaviour are not proofs of causal relations, but offer *possible* explanations about forces of selection. The monogamy of *Amphiprion* could be an adaptation for life on a very limited habitat – the symbiotic anemone. The process of adaptation is subject to many selective forces (Tinbergen, 1970) and alters not only the social structure but the entire behavioural repertoire of a species.

COMMUNICATION IN SYMBIOSIS

Losey (1971) regards it as an indication of information transfer or communication when a change in the probability of a behavioural response occurs in an animal and can be correlated with a stimulus generated by the behaviour of another animal. Behavioural analyses are therefore important for the investigation of communication. What messages are transmitted? Which signal channels are taken for their transmission? Signals in symbiotic marine animals are mainly transferred on tactile, chemical and optical cues. Here are some examples.

Karplus *et al.* (1972) and Karplus (personal communication) investigated the chain-signal system of the shrimp–gobiid mutualism. The shrimps are chemically attracted to the fish; they leave their subterranean burrows only when the fish is at the mouth. A typical behaviour sequence of the two is recognizable. The shrimp approaches the entrance, from the inside, head first; when it reaches the motionless fish one antenna contacts the caudal region of the fish, and only then the shrimp emerges from the burrow, maintaining constant antennal contact (Fig. 1). Retreat responses are easily released by alarm stimuli; according to the strength of these the fish enters the burrow head or tail first, whereas the shrimp always backs in. Observations of Magnus (1967a) and Karplus *et al.* (1972) reveal that

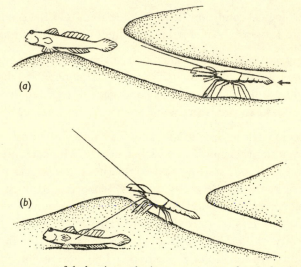

Fig. 1. The sequence of behaviour during emergence from the burrow in the mutualism of the shrimp *Alpheus djiboutensis* and the fish *Cryptocentrus cryptocentrus*. (*a*) The shrimp moves head-first toward the entrance of the burrow; antennae are somewhat raised at an oblique forward angle with the tips 2–3 cm apart. (*b*) Outside of the burrow the shrimp keeps one antenna in contact with the fish (after Karplus *et al.*, 1972).

warning signals by tail-flicking of the fish are transferred from fish to shrimp through the antenna. In the absence of antennal contact no information is transferred to the shrimp from the retreating fish.

In this signal system, alarm signals are transferred by tactile stimuli. Similar transfer seems to occur in the cleaning symbiosis between cleaner-shrimp and its fish customers. The shrimp *Hippolysmata grabhami* first touches the customer with the antenna (Plate 2). The customer adopts a motionless body position of differing orientation. The antenna contact allows transfer of information, by tactile signals, concerning the readiness, for cleaning, of the customer.

Optical signals play an important role in the cleaning symbiosis of fishes. Eibl-Eibesfeldt (1955), Limbaugh (1961), Limbaugh, Pedersen & Chace (1961), Wickler (1963) and others indicate that colouration, and swimming movements (the 'dance') of the cleaner, are optical signals releasing the cleaning-pose of the customer. The 'dance' and the cleaning inspection trigger the behaviour of the customer, who then adopts various poses to invite the cleaner. Eibl-Eibesfeldt (1955) observed cleaner-fishes entering the mouths of groupers to clean parasites and food remains. When cleaning is terminated, the grouper closes its mouth repeatedly, forcing the cleaner to leave the mouth cavity.

I performed experiments with wooden dummies to investigate the function of cleaner and customer signals (Fricke, 1966). To test the relative releasing value of the cleaner-dance, a painted, motionless cleaner-dummy was presented to various customers, eliciting cleaning postures. The dance was seen to be superfluous for this end. In another experiment wooden 'customer' dummies presented to the cleaner without the typical pose nevertheless elicited the inspecting behaviour of the cleaner. These simple field experiments indicate that there is a wide variety of signal responsiveness in cleaning symbiosis. Losey (1971) made a detailed experimental analysis of communicative stimuli in cleaning symbiosis. He pointed out that environmental and internal stimuli affect the variability of behavioural responses of the signal receivers (cleaner and customer). Shape, colouration and swimming motions of the cleaner influence the probability of occurrence of the customer pose, just as shape, colouration and behaviour pattern of the customer affect the probability of cleaner-inspecting. External stimuli, e.g. the presence of a cleaning station, temporary deprivation of cleaning or increased parasitism increase the probability that the customer will pose. Losey discussed the similarity of cleaner's and of customer's behaviour in different areas of the world, comparing information transfer in social behaviour and in cleaning symbiosis. He postulated that communication in the latter should not require specialized signals, as observed in the social behaviour of vertebrates, because 'the necessary inclusion of many species into the evolutionary scheme may hinder or even reduce the adaptive value of signal specialization'. Such a system probably favours 'behaviour with less phylogenetic programming and a greater dependency on learning mechanism'. This agrees with Abel (1971) who emphasized that the behaviour of cleaner and customer is probably influenced by learning modified releasing mechanisms ('EAM' Schleidt, 1962). Losey compared message transfer in the cleaning symbiosis and in other kinds of symbiosis. In the latter, signals are mainly confined to attraction and recognition, and are given for purposes other than inter-specific communication. Messages in the communication system of the shrimp–gobiid mutualism and in the cleaning symbiosis were 'coded' by the transmitter and addressed. The next section gives examples of signal systems in which the signals are not 'coded' by the transmitter, nor are they explicitly addressed to a receptor.

ATTRACTION SIGNALS AND RECOGNITION OF THE HOST

To study the susceptibility to stimuli, ethologists present animals with stimulus situations differing in some parameters. The intensity of the response indicates which parameters are significant for it. I shall give two examples from my own work.

Attraction signals in commensalism of goatfish

The digging goatfish produces a sand cloud in which the customer feeds. The feeding customer is confronted with various stimuli: the sand cloud plus the optical features of the host (shape, behaviour, colouration). Which part or parts of this complex stimulus situation are critical dis-criminative and have signal value for the customer? Four different stimulus parameters were tested with the goatfish *Pseudupeneus macronema* by means of dummies (Fig. 2).

The reactions of the customers, *Cheilio*, *Cheilinus*, *Lepidaplois* and *Hemibalistes* were recorded during dummy presentation. When a diver with a weighted line produced a moving sand cloud all customers except the trigger fish *Hemibalistes* approached the cloud but hesitated to come close. The customers did not feed in the sand cloud. When a yellow dummy of goatfish shape was mounted on the line I observed the same reaction, with one important behavioural difference: customers were attracted over a greater distance. A sand cloud with a natural-type dummy elicited approach and feeding behaviour, only the trigger fish hesitating to come

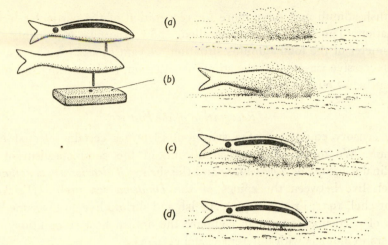

Fig. 2. Investigations of attraction and recognition signals in feeding commensalism of *Pseudupeneus macronema*. The dummies are mounted on a piece of lead. The lead sinks in the sand producing a sand cloud when moved by a diver. (*a*) Sand cloud without goatfish; (*b*) sand cloud with a yellow dummy of goatfish shape; (*c*) sand cloud with a natural-type goatfish dummy; (*d*) natural-type goatfish dummy without sand cloud.

Table 2. *Reaction of feeding customers to different stimuli (explanation in text and Fig. 2)*

	Reaction of the customer				
Stimulus situation	*Cheilio*	*Cheilinus*	*Coris*	*Lepidaplois*	*Hemibalistes*
Sand cloud without host	2	2	2	2	2
Sand cloud + body shape of the host	3	3	3	3	3
Sand cloud + natural coloured host	4	4	4	4	3
Natural coloured host without sand cloud	I	I	I	I	I

Reaction of customer: (1) no approach to test situation; (2) approach to test situation but no feeding in sand cloud; (3) approach to test situation over greater distance (3–5 m) but no feeding in sand cloud; (4) approach and feeding in sand cloud.

very close. The customer reactions are summarized in Table 2. Negative results were obtained with a natural-type goatfish dummy without sand cloud. Probably the customers recognize a complex pattern of goatfish stimuli including a behaviour pattern which could not be imitated with the dummies. However, the field observations showed that the customers follow the goatfish even in the absence of the sand cloud, but that the latter is a strong attractive stimulus. The differences in behaviour responses hint that the sand cloud functions as an attraction signal, while the specific goatfish stimuli may serve more for recognition of the particular feeding situation. The goatfish commensalism shows that we can identify the parameter of the total stimulus situation relevant to a particular response by successive variation of the stimulus situation, and assessing the subsequent behavioural response.

Recognition of the host

The sensory equipment of a symbiont identifies certain selected host-stimuli. We investigated optical stimuli in the commensalism of the cardinal fish *Siphamia permutata* and the shrimp *Tuleariocaris zanzibarica*, which live between the spines of the *Diadema* sea urchin (Fricke & Hentschel, 1971; Fricke, 1973*b*). As fish and shrimp live on the same host, it is of particular interest to compare the optical host recognition parameters in animals with different visual systems.

Fish and shrimp have a strong scototactic response to large, dark objects. This might favour detection of distant sea urchins contrasting strongly with their backgrounds. At closer range, form and pattern-specific stimuli of the sea urchin release the host-selection response in fish and shrimp. In a binary choice situation, two stimuli were presented simultaneously (Plate 3). Stimuli with a greater releasing value should command a stronger orientation response. Both test species recognized the sea urchin by its spine forest, but not by body form. The complex shape of the whole spine forest can vary in different visual parameters. Some data are summarized in Fig. 3. Both shrimp and fish distinguish colour, number, diameter and orbital position of the spines; the fish is also capable of evaluating distances between spines (Fricke, 1973*b*). The most effective stimulus configuration consists of many dark, parallel, verticle spines, the more numerous and darker the spines and the stronger their contrast with the background, the more marked is the response.

Comparison of the various stimulus parameters indicates that the fish and shrimp detect, with their anatomically different visual systems, nearly the same host-stimuli, although slight variations occur in the choice of individual parameters. The shrimp response is more sensitive to the orbital position of the spines; the fish response is stronger for increasing number. This might correlate with the respective behaviour of the animals, the shrimp habitually swimming head down and parallel to the spines, for camouflage, so that the optical system could be expected to be sensitive to orbital position, which is less important for the fish, finding protection as it does among many spines.

These are good examples of the adaptation of the visual system to the particular need of the species. Discrimination of host-sign stimuli deter-

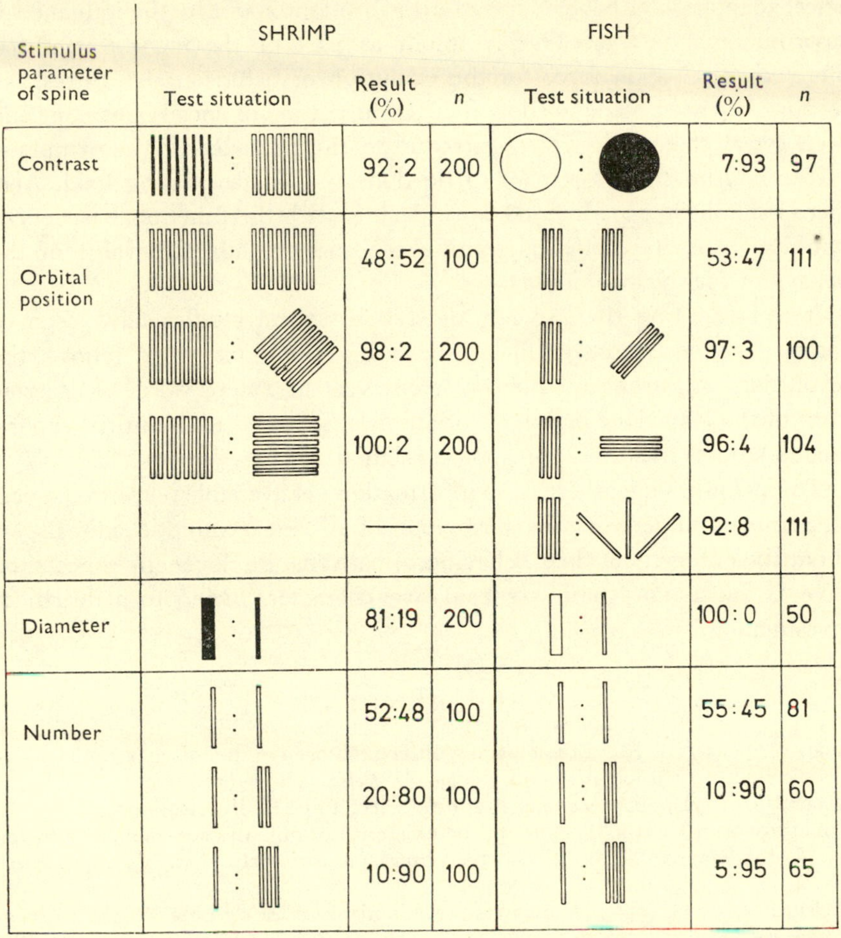

Stimulus parameter of spine	SHRIMP			FISH		
	Test situation	Result (%)	n	Test situation	Result (%)	n
Contrast		92:2	200		7:93	97
Orbital position		48:52	100		53:47	111
		98:2	200		97:3	100
		100:2	200		96:4	104
	—	—	—		92:8	111
Diameter		81:19	200		100:0	50
Number		52:48	100		55:45	81
		20:80	100		10:90	60
		10:90	100		5:95	65

Fig. 3. Comparative study of optical parameters in recognition of a sea urchin by the shrimp *Tuleariocaris zanzibarica* and the fish *Siphamia permutata*. Solid black indicates a black spine; white indicates a white spine.

mines host selection. Large dark objects trigger an approach response at greater distances, at close range special form and pattern stimuli become material for host recognition.

CONCLUSION

In this paper some aspects of recent behavioural studies of marine animals have been reviewed, demonstrating the importance of behaviour for symbiotic species. As in the case of anatomical or physiological characteristics, adaptation of behaviour pattern is brought about by the influence of environmental factors. The symbiont as part of the environment has therefore a selection force for the evolution of behavioural features. An attempt has been made to show that the formation of social behaviour and even social structures are in part due to the presence of the symbiotic partner. More studies remain to be done in this fascinating field. Also there is too little experimental work dealing with the functional aspects of symbiosis, i.e. its ecological significance and the adaptive value of the behaviour for the symbionts.

It is interesting that, so far, most behavioural studies have been of communication, sign stimuli and releasing situations. This follows the 'evolution' of European ethology, whose very first steps were the descriptions of the behaviour repertoire of animals, followed up by initial experimental studies in stimulus–response relations.

To my knowledge, there is no information yet available on physiological mechanisms underlying the behaviour of marine symbiotic animals, or about the ontogeny of their behavioural patterns. Phylogenetic aspects too have as yet a more or less speculative character, owing to a dearth of investigation.

REFERENCES

ABEL, E. F. (1971). Zur Ethologie von Putzsymbiosen einheimischer Süsswasserfische im natürlichen Biotop. *Oecologia, Berl.*, **6**, 133–151.

ALLEN, G. (1972). *Anemonefishes*. Neptune City: T. F. H. Publications.

COLLINGWOOD, C. (1868). Note on the existence of gigantic sea-anemones in the China Sea, containing within them quasi-parasitic fish. *Ann. Mag. nat. Hist.*, **4**, 31–33.

CRESPIGNY, C. C. (1869). Notes on the friendship existing between the malacopterygian fish *Premnas biaculeatus* and the *Actinia crassicornis*. *Proc. zool. Soc. Lond.*, 248–249.

EIBL-EIBESFELDT, I. VON (1955). Über Symbiosen, Parasitismus und andere zwischenartliche Beziehungen bei tropischen Meeresfischen. *Z. Tierpsychol.*, **12**, 203–219.

(1961). Eine Symbiose von Fischen (*Siphamia versicolor*) und Seeigeln. *Z. Tierpsychol.*, **18**, 56–59.

FEDER, H. M. (1966). Cleaning symbiosis in the marine environment. In *Symbiosis*, vol. **1**, 327–380 (ed. S. M. Henry). New York and London: Academic Press.

FISHELSON, L. (1965). Observations and experiments on the Red Sea anemones and their symbiotic fish *Amphiprion bicinctus*. *Bull. Sea Fish Stn Israel*, **39**, 1–14.

FRICKE, H. W. (1966). Zum Verhalten des Putzerfisches *Labroides dimidiatus*. *Z. Tierpsychol.*, **23**, 1–3.

(1970a). Zwischenartliche Beziehungen der tropischen Meerbarben *Pseudupeneus barberinus* und *Pseudupeneus macronema* mit einigen anderen marinen Fischen. *Nat. u. Museum*, **100**, 71–80.

(1970b). Ein mimetisches Kollektiv – Beobachtungen an Fischschwärmen, die Seeigel nachahmen. *Mar. Biol.*, **5**, 307–314.

FRICKE, H. W. & HENTSCHEL, M. (1971). Die Garnelen-Seeigel-Partnerschaft. Untersuchung der optischen Orientierung der Garnele. *Z. Tierpsychol.*, **28**, 453–462.

(1973*a*). Behaviour as part of the ecological adaptation. *Helgoländer wiss. Meeresunters.*, **24**, 120–144.

(1973*b*). Die Fisch-Seeigel-Partnerschaft. Untersuchung optischer Reizparameter beim Formenerkennen. *Mar. Biol.*, **19**, 290–297.

(1974). Öko-Ethologie des monogamen Anemonenfisches *Amphiprion bicinctus*. *Z. Tierpsychol.*, **36**, 429–513.

HOBSON, E. S. (1969). Comments on certain recent generalizations regarding cleaning symbiosis in fishes. *Pacific Sci.*, **13**, 35–39.

KARPLUS, I., TSURNAMAL, M. & SZLEP, M. (1972). Associative behavior of the fish *Cryptocentrus cryptocentrus* (Gobiidae) and the pistol shrimp *Alpheus djiboutensis* (Alpheidae) in artificial burrows. *Mar. Biol.*, **15**, 95–104.

LIMBAUGH, C. (1961). Cleaning symbiosis. *Scient. Am.*, **205**, 42–49.

LIMBAUGH, C., PEDERSEN, M. & CHACE, F. A. (1961). Shrimps that clean fishes. *Bull. Mar. Sci. Gulf Carib.*, **11**, 237–257.

LOSEY, G. S. (1971). Communication between fishes in cleaning symbiosis. In *Aspects of the Biology of Symbiosis*, 45–76 (ed. T. C. Cheng). Baltimore: University Park Press.

MAGNUS, D. B. E. (1964). Zum Problem der Partnerschaften mit *Diadem* Seeigel. *Verh. dt. zool. Ges. München 1963, Zool. Anz.*, **27**, 404–417.

(1967*a*). Zur Ökologie sedimentbewohnender *Alpheus* Garnelen (Decapoda, Natantia) des Roten Meeres. *Helgoländer wiss. Meeresunters.*, **15**, 506–522.

(1967*b*). Ecological and ethological studies and experiments on the echinoderms of the Red Sea. *Stud. trop. Oceanog.*, **5**, 635–664.

MARISCAL, R. N. (1970). The nature of the symbiosis between Indo-Pacific anemone fishes and sea anemones. *Mar. Biol.*, **6**, 58–65.

(1971). Experimental studies on the protection of the anemone fishes from the sea anemones. In *Aspects of the Biology of Symbiosis*, 283–314 (ed. T. C. Cheng). Baltimore: University Park Press.

REESE, E. S. (1964). Ethology and marine zoology. *Oceanogr. & mar. Biol.*, **2**, 455–488.

SCHLEIDT, W. (1962). Die historische Entwicklung der Begriffe 'Angeborenes auslösendes Schema' and 'Angeborener Auslösemechanismus'. *Z. Tierpsychol.*, **19**, 697–722.

SCHLICHTER, D. (1970). Chemischer Nachweis der Übernahme anemoneneigener Schutzstoffe durch Anemonenfische. *Naturwissenschaften*, **57**, 312–313.

TINBERGEN, N. (1970). Umweltbezogene Verhaltensanalyse Tier und Mensch. *Experientia*, **26**, 447–456.

WICKLER, W. (1963). Zum Problem der Signalbildung, am Beispiel der Verhaltensmimikry zwischen *Aspidontus* und *Labroides*. *Z. Tierpsychol.*, **20**, 657–679.

(1971). Soziales Verhalten als ökologische Anpassung. *Verh. dt. zool. Ges.*, **64**, 291–304.

YOUNGBLUTH, M. J. (1968). Aspects of the ethology and ecology of the cleaning fish, *Labroides phthirophagus* Randall. *Z. Tierpsychol.*, **25**, 915–932.

EXPLANATION OF PLATES

PLATE I

Commensalism between *Pseudupeneus macronema* and the feeding customers *Hemibalistes chrysopterus* and *Cheilinus triolbatus* in coral reefs of the Red Sea.

PLATE 2

Cleaning of the fish *Anthias squamipinnis* by the shrimp *Hippolysmata grabbami* in the Red Sea. The cleaning shrimp makes contact by its antennae.

PLATE 3

Binary choice experiment with the fish *Siphamia permutata* and sea urchin dummies with one and three spines. At the beginning of an experiment the start dummy (sea urchin in foreground) is lifted. The test fish – formerly sitting between the spines of the start dummy – react immediately by swimming toward the choice dummies. The Plate shows the end of a choice when the test fish swim back to the start dummy which is presented immediately after a successful approach of a dummy.

AUTHOR INDEX

Figures in bold type indicate pages on which references are listed.

SUBJECT INDEX

Acanthocephala, 468–70
  evolution of relation of host and, 494–5
  relations of, with intermediate hosts
    (arthropods): effects on hosts, 473–7;
    host resistance, 477–81; process of
    infection, 470–3
  relations of, with definitive hosts (ver-
    tebrates): distribution in alimentary
    tract, 481–5, 486; nutrition, 491–4;
    process of infection, 485, 487–491
Acanthocephalus dirus, 474
Acanthocephalus jacksoni, 471, 481, 482
Acanthocephalus lucii, 482
Acanthocephalus ranae, 483, 487, 492, 494
Acanthotermes, Streptococcus faecalis in gut
  of, 573–4
Acarospora sinopica, 390
acetate
  produced from cellulose by symbionts
    of termites, 564, 566
  utilizes by germinating fungal spores,
    298, 299, and by symbiotic dino-
    flagellates, 273–4
acetylene reduction assay, for nitrogen
  fixation, 567
acid phosphatase: associated with non-
  living particles, but not with
  symbionts, in digestive cells of
  Tridachia, 236, and in vacuoles of
  Paramecium bursaria, 190
Acinetobacter, varied associations of
  Bdellovibrio with, 104
actinomycin D, inhibitor of RNA synthesis, 58, 301
adenine nucleotide carrier, absent from all
  bacteria, present in all mitochondria, 40
adrenal glands, iron-containing protein in
  (adrenodoxin), 54
Aerobacter aerogenes, food of *Paramecium
  aurelia*, 133, 142
Aiptasia, growth of dinoflagellates in parabiosis with, 271–2
alanine
  release of, by *Gymnodinium*, 270, 271
  transfer of, between *Platymonas* and
    Convoluta, 218–19
algae
  free-living, in lichen habitats, 385
  green, in lichens, transfer polyols to
    fungi, 374
  number of genera of, in lichens, 373
  see also blue-green algae, *and individual
    species*

alpha symbionts of *Paramecium aurelia*
  (*Cytophaga caryophila*), non-killer,
  127, 128, 169
Alpheus shrimps, mutualism between
  gobiid fish and, 582–3
Alpheus djiboutensis, communication
  between fish *Cryptocentrus* and, 586–
  587
α-amanatin, sensitivity of RNA polymerases to, 58, 59, 60, 61
amides, transferred from alga to *Convoluta*, 221, 222
amino acids
  absorption of: by Acanthocephala, 492;
    by *Convoluta*, 215; by *Hymenolepis*,
    508; by lichens, 389
  made available to bacteria by rumen
    protozoa, 555
  in media for rust fungi, 303, 305
  release of, by *Gymnodinium* in culture,
    270
  sources of: for *Entodinium*, 540; for
    Epidinium, 548
  transfer of: between alga and *Convoluta*,
    218–19, 221, 222; between host and
    Hymenolepis, 509, 510; not transferred
    from alga to fungus in lichens, 383
p-aminobenzoic acid: diets deficient in,
  for controlling infections with intraerythrocytic protozoa, 435
α-aminoisobutyric acid: effect of methylation of, on uptake by *Hymenolepis*
  and by ascites cells, 508
aminopeptidase, of Acanthocephala, 492–3
Amitermes minimus, actinomycete in gut of,
  566
ammonia
  not taken up by *Convoluta*, 213, 214,
    215; transfer of, between *Convoluta*
    and alga, 221
  uptake of, by *Peltigera*, 383, 384
amoeboids (mycoplasm-like fermenters),
  postulated as original host cells of
  symbiotic cell types becoming
  organelles, 22
Amphidinium chattonii, A. klebsii (symbiotic dinoflagellates), 216
  growth of, in axenic culture, 268, 273
  growth rates of, in host and *in vitro*,
    269, 270
  in parabiosis with hosts, 271–2
amphids, receptors on head of nematodes,
  354